民國建築工程期刊匯編

MINGUO JIANZHU GONGCHENG QIKAN HUIBIAN

50

《民國建築工程期刊匯編》編寫組 編

廣西師範大學出版社
GUANGXI NORMAL UNIVERSITY PRESS
·桂林·

第五十册目録

南大工程

The Journal Of

The Lingnan Engineering Association

Vol. 1, No. 1 Jan. 1933

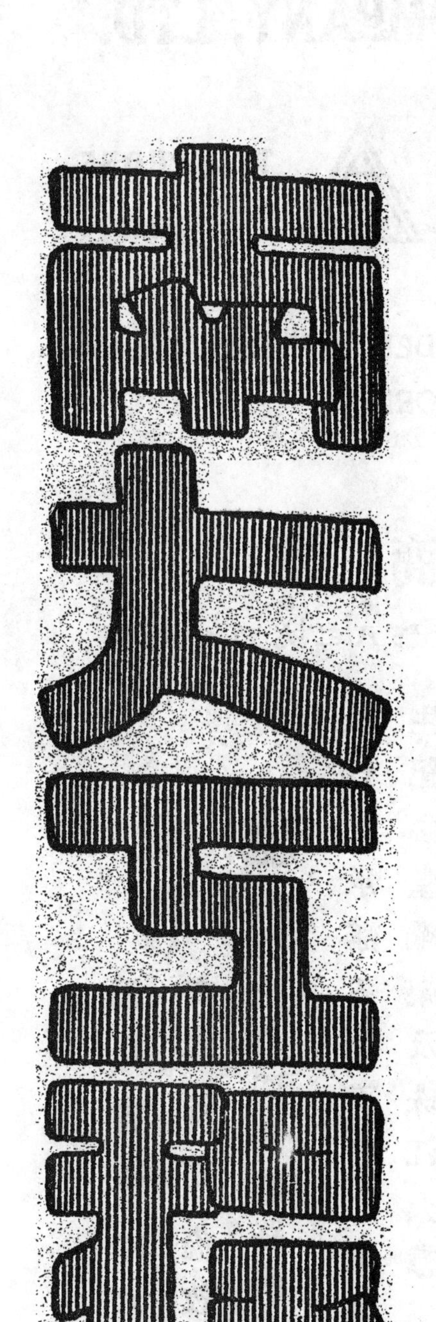

第壹卷 第壹期

嶺南大學工程學會發行

24996

南大工程

嶺南大學工程學會會刊

助編輯　黎樹仁　鄒漢新　梁卓芹

總編輯　龍寶鎏

廣告主任　余伯長
校對主任　梁寶琨
發行主任　廖健文

第一卷第一期目錄

嶺南大學工程學會發行

本會會址......................廣州嶺南大學校內
自動電話......................30057
中文電報及無線電碼......................0190
英文電報CABLE...................... "LINGNANUNI" CANTON

投稿簡章

本刊歡迎投稿，且鑒於外國文稿件已爲國內僅有之二三工程刊物所摒棄，在作者方面，每苦於專門名詞未有標準之遂譯，恆習慣直接用外文撰稿，本刊在中央編譯館未將各標準名詞公佈以前，外文稿件亦樂爲登載，爰錄簡章數則於後：

(一) 投寄之稿，自撰或翻譯，均以關於工程者爲限，文體文言白話不拘。

(二) 外國文稿件，暫以英德法三文爲限。

(三) 來稿請繕寫清楚，如有附圖附表，尤須書寫清潔，外文稿件并打字寄來尤佳。

(四) 如係譯稿，請書明原著書名篇目出版地點日期。

(五) 稿末請注明姓名住址，以便通訊，至揭載時如何署名，聽投稿者自便，作者能將詳細履曆敘明尤佳。

(六) 投寄之稿在二千字以上者，如不揭載，編輯部於每期出版後，按址郵寄回。

(七) 來稿經揭載後，酌酬本刊是期多本。

(八) 來稿編輯部或酌量增删之，但投稿人不願他人增删者，請豫先聲明。

(九) 來稿請逕寄廣州嶺南大學工程學會出版部收。

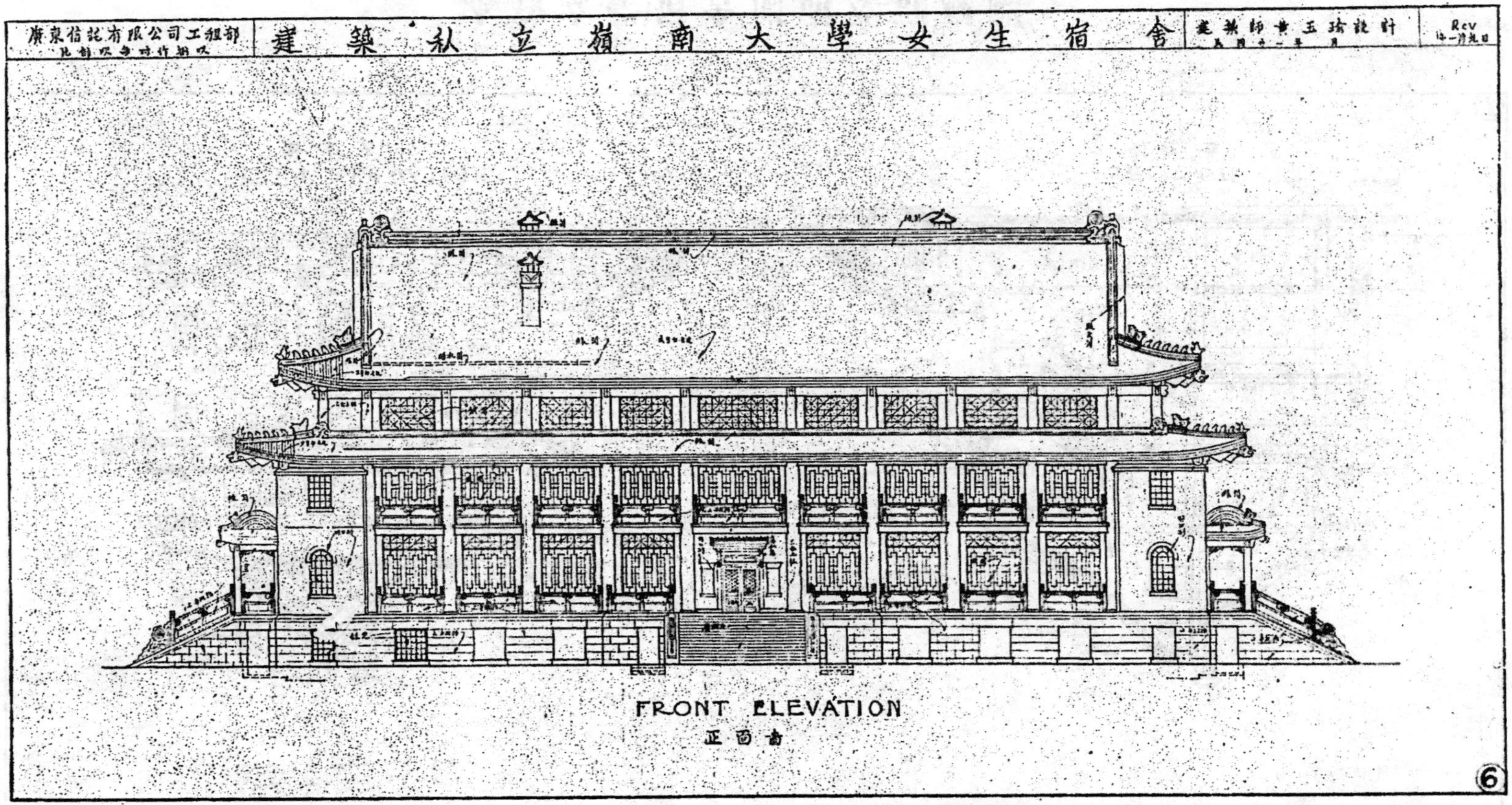

廣東信記有限公司工程部

建築私立嶺南大學女生宿舍

建築師黃玉瑜設計

Rev

FRONT ELEVATION

正面圖

⑥

建築中之本校女學宿舍平面圖　（參看本期附錄）

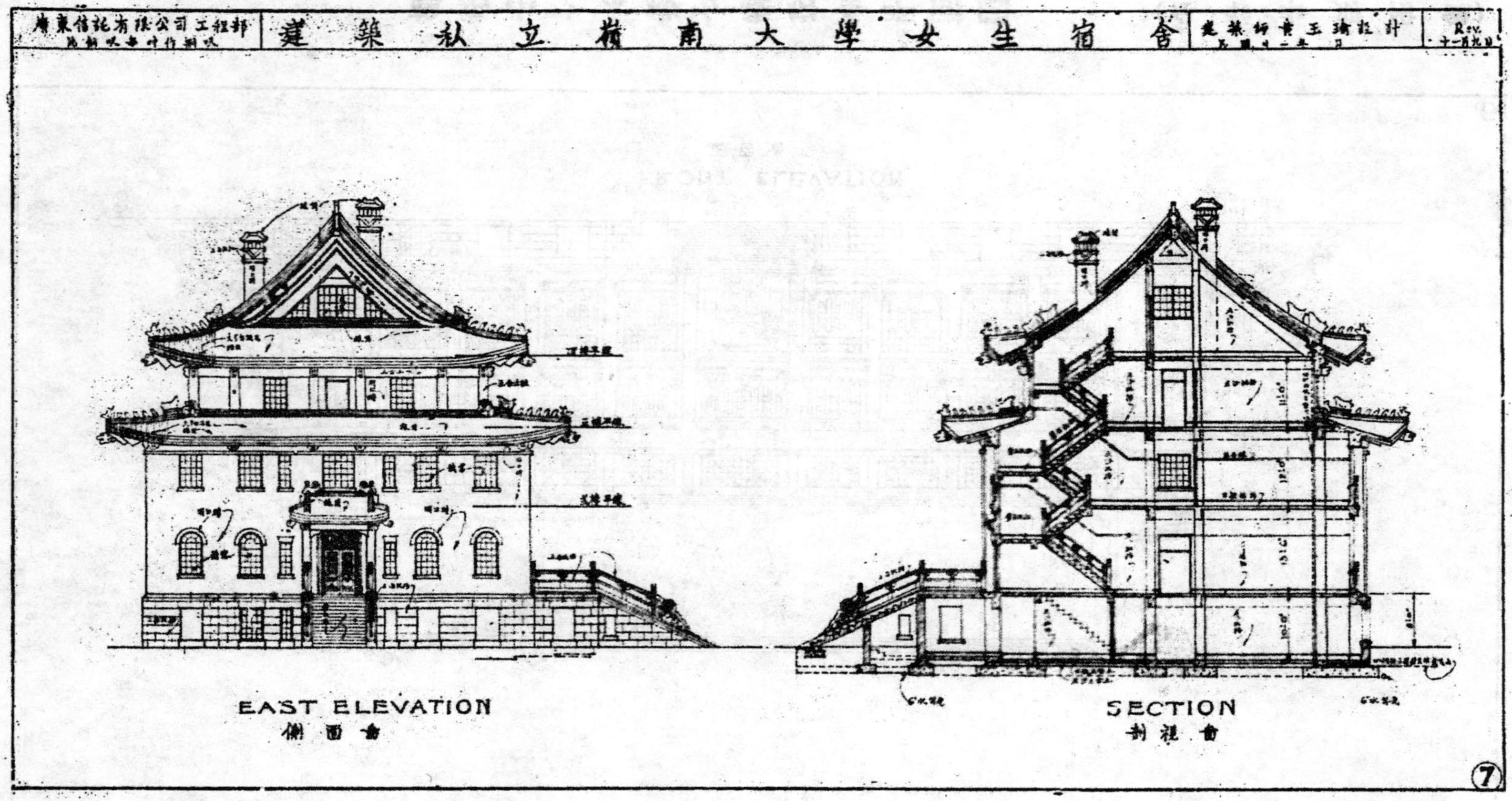

廣東信記有限公司工程部

建築私立嶺南大學女生宿舍

建築師黄玉瑜設計

EAST ELEVATION
側面圖

SECTION
剖視圖

⑦

本校女學宿舍側面及剖視圖

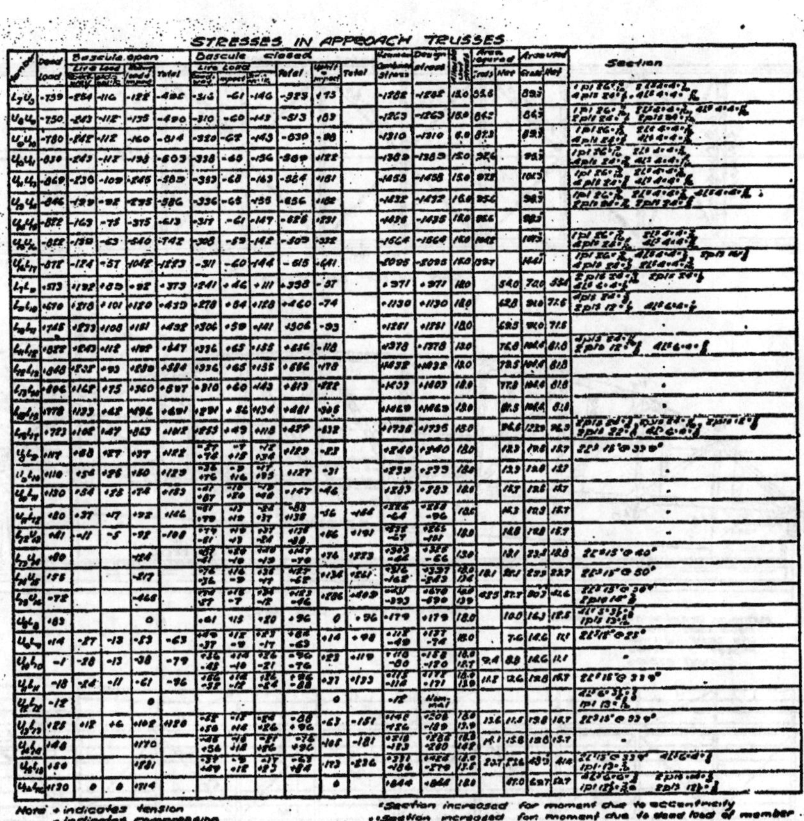

STRESSES IN APPROACH TRUSSES

Member	Dead load	Bascule open: Live load, track way	Bascule open: Live load, side walk	Bascule open: Dead load & impact	Bascule open: Total	Bascule closed: Live load, road way	Bascule closed: Impact	Bascule closed: Side walk	Bascule closed: Total	Uplift + impact	Total	Maximum combined stress	Design stress	Allowable unit stress	Area required: Gross	Area required: Net	Area used: Gross	Area used: Net	Section
U_7U_8	-139	-254	-116	-122	-495	-316	-61	-146	-523	+173		-1282	-1282	12.0	95.6		89.3		1 pl 26"×½, 4 pls 24"×½, 4 L 6×4×⅝
U_8U_9	-750	-243	-112	-135	-490	-310	-60	-143	-513	+103		-1263	-1263	12.0	86.2		86.3		[illegible]
U_9U_{10}	-780	-242	-112	-160	-514	-320	-62	-143	-530	+98		-1310	-1310	12.0	87.3		89.3		[illegible]
$U_{10}U_{11}$	-830	-243	-112	-138	-503	-338	-65	-156	-559	+122		-1389	-1389	15.0	92.6		93.3		[illegible]
$U_{11}U_{12}$	-869	-236	-109	-245	-589	-353	-68	-163	-584	+181		-1453	-1453	15.0	97.2		100.3		[illegible]
$U_{12}U_{13}$	-846	-139	-92	-295	-586	-336	-65	-155	-556	+182		-1432	-1432	15.0	95.6		98.3		[illegible]
$U_{13}U_{14}$	-822	-163	-75	-375	-613	-317	-61	-147	-525	+231		-1435	-1435	15.0	95.6		98.3		
$U_{14}U_{15}$	-822	-150	-63	-540	-742	-303	-59	-142	-509	+332		-1564	-1564	15.0	104.2		107.3		[illegible]
$U_{15}U_{16}$	-878	-124	-57	-1048	-1223	-311	-60	-144	-515	+441		-2095	-2095	15.0	139.7		146.1		[illegible]
L_7L_8	+373	+132	+85	+92	+373	+241	+46	+111	+398	-57		+771	+771	18.0		54.0	72.0	55.2	[illegible]
L_8L_9	+670	+218	+101	+120	+439	+278	+54	+128	+460	-74		+1130	+1130	18.0		62.8	81.0	71.6	[illegible]
L_9L_{10}	+745	+233	+108	+181	+492	+306	+59	+141	+506	-93		+1251	+1251	18.0		69.5	91.0	71.5	
$L_{10}L_{11}$	+822	+243	+112	+192	+547	+336	+65	+155	+556	-118		+1378	+1378	18.0		76.6	100.6	81.0	[illegible]
$L_{11}L_{12}$	+848	+235	+93	+250	+584	+336	+65	+155	+556	-178		+1432	+1432	18.0		79.5	100.6	81.0	
$L_{12}L_{13}$	+896	+162	+75	+360	+597	+310	+60	+143	+513	-222		+1493	+1493	18.0		77.8	100.6	81.0	
$L_{13}L_{14}$	+978	+133	+62	+496	+691	+291	+56	+134	+481	-303		+1669	+1669	18.0		91.3	100.6	81.0	
$L_{14}L_{15}$	+723	+102	+47	+863	+1012	+255	+49	+118	+422	-532		+1735	+1735	18.0		96.6	120.0	96.3	[illegible]
U_7L_8	+117	+88	+27	+37	+152	+27 +78	+7 +15	+12 +36	+129	-23		+240	+240	18.0		12.3	17.6	13.7	2 L 15"@ 33.9"
U_8L_9	+110	+56	+26	+50	+129	+26 +76	+5 +16	+12 +35	+127	-31		+239	+233	18.0		12.3	17.6	13.7	
U_9L_{10}	+130	+54	+25	+74	+153	+31 +87	+7 +20	+16 +40	+147	-46		+283	+283	18.0		14.5	17.6	13.7	
$U_{11}L_{12}$	+80	+37	+17	+92	+146	+41 +79	+13 +18	+25 +37	+138	-56	-166	+186 -64	+258 -96	18.0		14.3	17.6	13.7	
$L_{12}U_{13}$	+41	-11	-5	-92	-108	[illegible]	[illegible]	[illegible]	[illegible]	+86	+191	[illegible]	[illegible]	18.0		14.8	17.6	13.7	
$L_{13}U_{14}$	+80			-124		[illegible]	[illegible]	[illegible]	[illegible]	+76	+253	[illegible]	[illegible]	13.0		18.1	23.5	18.0	2 L 15"@ 40"
$L_{14}U_{15}$	+75			-317		[illegible]	[illegible]	[illegible]	[illegible]	+136	+261	[illegible]	[illegible]	[illegible]	18.1	20.1	29.5	22.7	2 L 15"@ 50"
$L_{15}U_{16}$	-72			-468		[illegible]	[illegible]	[illegible]	[illegible]	+186	+403	[illegible]	[illegible]	[illegible]	42.3	37.7	80.7	41.6	[illegible]
U_8L_8	+83			0		+61	+15	+20	+96	0	+96	+179	+179	18.0		10.0	16.3	12.5	[illegible]
U_9L_9	+14	-27	-13	-53	-63	[illegible]	[illegible]	[illegible]	[illegible]	+14	+98	[illegible]	[illegible]	18.0		7.6	12.6	11.1	2 L 15"@ 25"
$U_{10}L_{10}$	-1	-28	-13	-38	-79	[illegible]	[illegible]	[illegible]	[illegible]	+23	+119	[illegible]	[illegible]	[illegible]	9.4	8.9	12.6	11.1	
$U_{11}L_{11}$	-18	-24	-11	-61	-96	[illegible]	[illegible]	[illegible]	[illegible]	+37	+133	[illegible]	[illegible]	[illegible]	11.8	12.6	17.6	13.7	2 L 15"@ 33.9"
$U_{12}L_{12}$	-12			0						0		-12	Nominal						[illegible]
$U_{13}L_{13}$	+25	+12	+6	+102	+120	[illegible]	[illegible]	[illegible]	[illegible]	-63	-151	[illegible]	[illegible]	[illegible]	13.6	11.5	17.6	13.7	2 L 15"@ 33.9"
$U_{14}L_{14}$	+48			+170		[illegible]	[illegible]	[illegible]	[illegible]	-105	-181	[illegible]	[illegible]	[illegible]	14.1	15.6	17.6	13.7	
$U_{15}L_{15}$	+50			+251		[illegible]	[illegible]	[illegible]	[illegible]	-173	-236	[illegible]	[illegible]	[illegible]	23.7	22.6	43.9	41.6	[illegible]
$U_{16}L_{16}$	+130	0	0	+114						0		+844	+844	18.0		47.0	62.7	52.7	[illegible]

Note + indicates tension
- indicates compression
Stresses given in 1000#
Areas given in sq in

*Section increased for moment due to eccentricity
• Section increased for moment due to dead load of member
‡ Section increased for lateral forces
∘ Pl not used for section

STRESSES IN FLOOR BEAMS & STRINGERS

Member	Shears in 1000 lbs: Dead load	Shears: Live load	Shears: Impact	Shears: Total	Moments in 1000 ft lbs: Dead load	Moments: Live load	Moments: Impact	Moments: Total	Section
Stringer S3	8.4	14.0	3.5	25.9	318	620	164	1140	7-ʃ 15"@ 54.9"
Stringer S4	8.0	14.0	3.5	25.5	274	618	153	1039	7-ʃ 15"@ 54.9"
Stringer S5	2.1	3.3	0	5.4	15.1	16.3	0	31.4	1-ʃ 12"@ 30.5"
Stringer S6	1.9	2.7	0	4.6	10.4	14.9	0	25.3	1-ʃ 12"@ 30.5"
Floor beam F7	34.5	25.6	12.8	72.9	361	133	70	570	[illegible]
Floor beams F8 to F15	63.6	62.8	15.4	140.4	750	807	202	1759	[illegible] 1 web pl 48"×⅝
Floor beam F16	63.0	595.0*	13.0	803.0	873	1483*	298	2665	[illegible] 1 web pl 78"×⅝, 4 pls 8"×[illegible]
Floor beam F17	36.5	62.8	31.8	130.1	371	634	317	1322	[illegible] 1 web pl 46"×⅝

*Stress from Rolling Load

CAMBER DIAGRAM

Camber Note:-
Truss members are to be lengthened or shortened as shown + indicates lengthening, - indicates shortening

Note
For general notes
see sheet #2

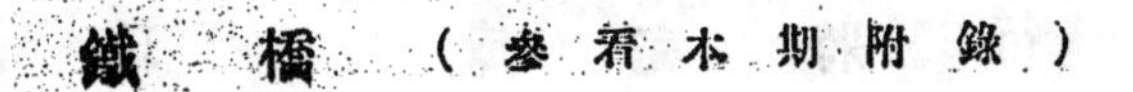

鐵　橋　（參看本期附錄）

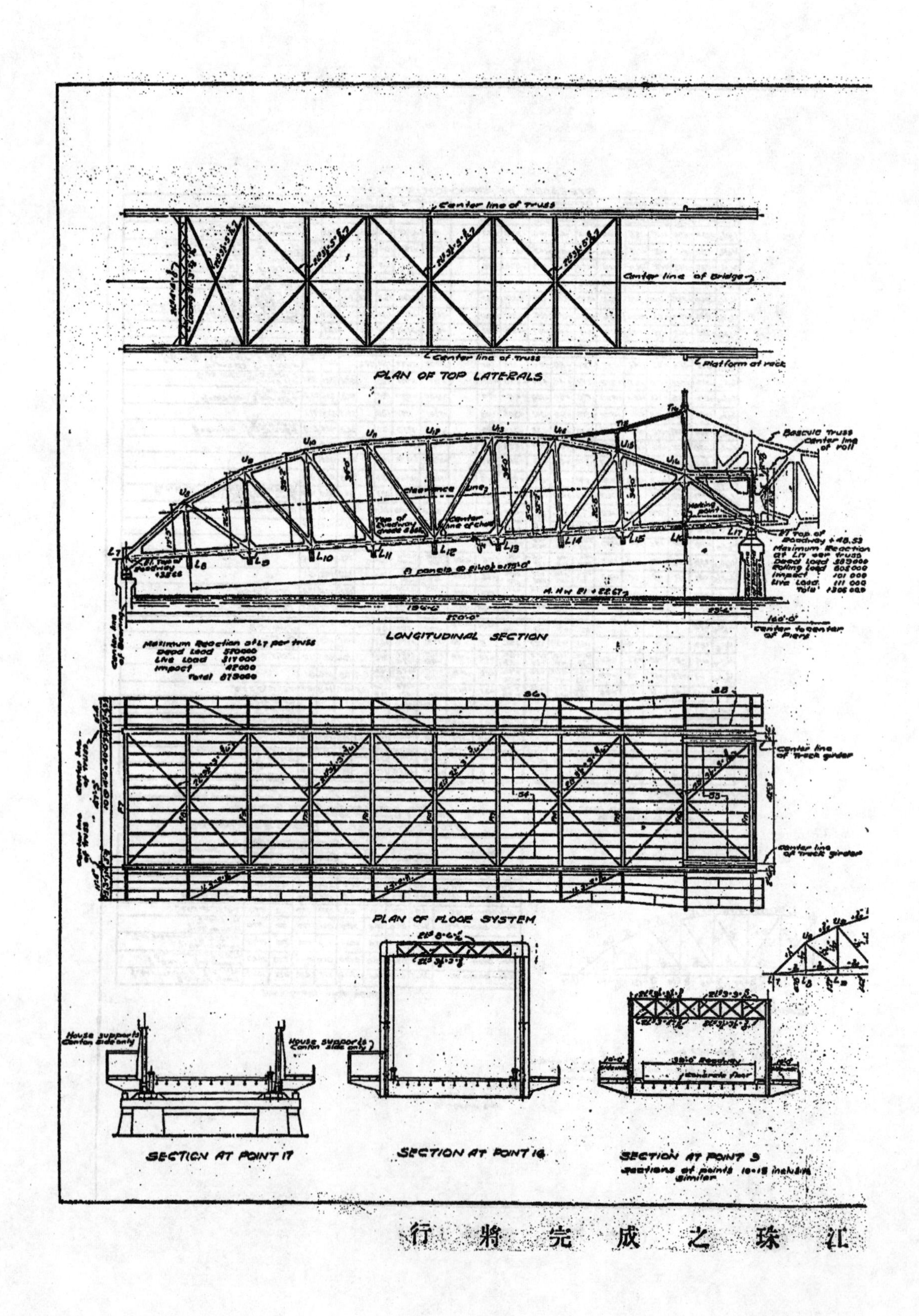
Center line of truss
Center line of Bridge
Center line of truss
Platform at rack
PLAN OF TOP LATERALS
Bascule truss
Center line of roll
clearance line
LONGITUDINAL SECTION
Maximum Reaction at L7 per truss
Dead Load
Live Load
Impact
Total
center to center of Piers
Center line of track girder
Center line of track girder
PLAN OF FLOOR SYSTEM
House supports Canton side only
House supports Canton side only
Roadway
Concrete floor
SECTION AT POINT 17
SECTION AT POINT 16
SECTION AT POINT 3
Sections at points 10-15 inclusive similar

行將完成之珠江

THE STUDY OF CIVIL ENGINEERING

by Dr. A. R. Knipp

The Need in China for the Civil Engineer.

It is generally agreed that one of the most pressing needs of the new China is the introduction of the applied sciences. The Chinese who travels in Western countries can not fail to be impressed by the higher standard of living of the West and by the fact that this higher standard of living has been made possible chiefly through the widespread applications of science to industry and transportation. It is the rightful aim of many Chinese to modernize their own country by introducing into China the scientific methods of the West.

In these first efforts at modernization the civil engineer will inevitably play an extremely important part. Already there is an effective demand for trained civil engineers to supervise the construction of roads and bridges, to survey the districts where modern highways are needed and to design and install modern systems of sanitation.

China is feeling the need as never before for political unity. But in a country which has the size and the population of that of China, the attaining of political unity is slow and difficult unless there are ready means of communication and transportation from one part of the country to another. Until travel in China becomes much easier and less expensive than it is at present, the people of the north and those of the south will lack the feeling of community of interest which underlies national unity. Ease of travel and modern facilities for transporting the proudcts of one province to another far distant will help greatly to overcome regional jealousies and prejudices.

Already throughout the southern part of China many cities and towns have widened their streets. Hundreds of miles of well built roads are making possible travel by motor-car and motor-bus. But this means a considerable start has already been made toward provincial unity in Kwongtung and Kwongsi. There is every reason to expect that this program of road building will be continued at an accelerated pace.

Unfortunately as yet only a small beginning has been made in China toward providing even the larger cities and towns with pure water and with modern systems of sanitation. The next large-scale development in modern engineering will probably be along this line, for it is of such major importance in conserving the health of the younger generation.

These two features of modern sanitation and the building of highways are of course only two of the numerous fields of endeavor in which the civil engineer will find scope and demand for his services.

The Training of the Civil Engineer.

Like the other branches of engineering, civil engineering is dependent upon the use of the scientific method for the solution of its problems. On this account the student in the civil engineering college should realize the great importance of becoming thoroughly familiar with the fundamental principles of mathematics, physics and chemistry. Without an understanding of the fundamental scientific principles which underlie all the branches of engineering, he will be unable to profit from the study of his professional courses.

It should be realized by the young student that in starting upon his civil engineering course he has taken it upon himself to undergo a rigorous training in a line of work which requires long and persistent application. The mistakes which an engineer may make if he is careless or if his knowledge of fundamental principles is inadequate, are likely to be disastrous to himself and to the enterprise on which he is working.

The best engineering colleges in Europe and America deservedly have the reputation of graduating only the most thoughtful and diligent students. Their courses have the reputation of being very difficult, because of the large number of students who after enrollment find the course of study too exacting and are obliged to withdraw.

It is not enough for the student of civil engineering to be satisfied with his ability to graduate from college. Perhaps the fact of his graduation and the degree which he is awarded may make it possible for him to get employment with an engineering firm or in a government enterprise. But in order for him to become an engineer in fact as well as in name, he will have to continue his study after graduation just as enthusiastically and diligently as he did during his college course. He will be able to learn a great deal from the more experienced engineers under whom he is working. But especially he will need to keep informed about recent engineering developments by reading the current engineering journals.

No engineering college of itself can produce experienced civil engineers. It is the habits of study and of scientific thinking, directed along professional lines, which are formed in the civil engineering college and which are continued throughout the engineer's career, which make the engineering course of study have such great importance in his professional training.

ANALYSIS OF RESTRAINT ARCHES

LO SHIH-LIN. (羅石麟)

1. General Discussion. In the following treatment we concern ourselves with ribbed, hingeless symmetrical and unsymmetrical arches whose springings are supported on elastic abutments. The degree of restraint of the springings due to the elasticity of the constituent elements is in every case determined by the amount of yield of the supports and the foundation. This movement of the springing may be considered as consisting of (1) rotation of the section and (2) horizontal and vertical displacements. For the present we will consider only the former and leave the latter to a further discussion. We shall chose the angle of rotation τ_A of the tangent to the arch axis at A (springing) due to an unit moment applied at A and similarly τ_B for the springing B as our measure of the degree of restraint of the supports. It is assumed that these values τ_A and τ_B are known. It is also assumed that the cross section is small as compared with the radius of curvature and that all loads are applied in the plane of the arch so that torsion is completely excluded. (Figure 1.)

2. Introduction of two hinged arch as basic system. A hingeless arch is statically indeterminate to the third degree, there being three unknown reactions on each support making a total of six. Ordinary statics furnishes us three equations of equilibrium for a non-concurrent coplanar force system. The three other equations necessary for a complete determination of the reactions may be obtained by finding the relation between th elastic deformation of the arch and the internal and external forces. Through the assumptions which we have just made i.e. both vertical and horizontal displacements of the springings are equal to zero, we will have to find one more relationship for an additional equation. If we can only establish a relationship between the rotation of the end tangents of the arch axis and the moments therein the whole problem of finding the reactions of a restraint arch is solved.

In order to obtain this relationship in the simplest manner we will introduce hinges at the springings of the arch and use the resulting two hinged arch as our statically indeterminate basic system. On this system we will apply the external loads as well as the springing moments of the hingeless arch as external forces which will in turn, produce additional stresses and reactions on our two hinged basic system. Before proceeding any further we are faced with the problem of evaluating these springing moments of a restraint arch, which, however, may be expressed as functions of the angles of rotation of the end tangents. This relationship in addition to that of unyielding supports are all that are necessary for the solution of our problem.

3. The two hinged arch. (Fig. 2) In a two hinged arch under external loads the reactions developed at the supports may be resolved into two components, one of which is the vertical component like that in simple beams and the other a thrust H_o acting along the line joining the two hinges. To evaluate this statically indeterminate value recourse may be had of the fact that the supports cannot be displaced. So that, if M_x is the moment and N_x the normal thrust at any section

$$\Delta L = \int \frac{M_x}{EI} y\,ds - \int \frac{N_x}{EA} \cos \eta_x\,ds = 0$$

In this ΔL is the change length of the span

E is the modulus of elasticity of concrete

I is the moment of inertia of any section

A is the area of any section

y is the perpendicular distance of any point on the arch axis above the line joining the two hinges

ds is the length of the arch elment measured along the axis.

η_x is the angle between the tangent to the arch axis of the element and the line joining the two hinges.

The effect of shear is neglected and the term involving it is omitted.

Putting $M_x = M_o - H_o y$ where M_o is the simple beam moment

$N_x \cos \gamma_x = H_o$ while $N_x \sin \gamma_x = 0$, we have

$$\int \frac{M_o}{EI} y\,ds - H_o \int \frac{y^2}{EI} ds - H_o \int \frac{ds}{AE} = 0$$

giving
$$H_o = \frac{\int \frac{M_o}{EI} y\,ds}{\int \frac{y^2}{EI} ds + \int \frac{ds}{AE}} = \frac{\int M_o y\,dw}{\int y^2 dw + \int dv} \qquad (1)$$

Where $\frac{ds}{I} = dw$ and $\frac{ds}{A} = dv$.

The simplification $N_x \sin. \gamma_x = 0$ is attributed to Moersch. E. Smulski gives the following discussion in Taylor and Thompson—Concrete, Plain and Reinforced Volume 11 p. 593. The term $N_x \gamma_x \sin$ is actually zero at the crown of the arch where $\sin \gamma_x = 0$ increasing towards the springings "but even at its maximum assumes only a small value."

Were all arch forms and the variation of cross sections expressible by simple equations Equation 1 may well stand as it is but for practical uses we may best express it as follows.

$$H_o = \frac{\Sigma M_o y w}{\Sigma y^2 w + \Sigma v} \qquad (1a)$$

With H_o known the stress in the two hinged arch can best be done by means of the three static equations of equilibrium.

Having dealt with the two hinged arch we will now proceed to a treatment of the restraint arch and derive the necessary elastic equations.

4. The Elastic Equations. (Fig. 3). The derivation of the elastic equations to determine the statically unknown values of the moments at the springings of a restraint arch may be based on the relationship that the sum of the rotations of the end tangents of our two hinged basic system due to external as well as internal loads must be equal to the rotation of the end tangents of

the restraint arch due to the springing moments actually developed there. The values of the springing moments depend upon the angle of rotations and the given degree of fixed endedness.

Let M_A = moment at springing A

M_B = moment at springing B

φ = angle of rotation of the end tangent at A of the two-hinged arch due to external loads.

ψ = angle of rotation of the end tangent at B of the two hinged arch due to external Loads.

αaa = angle of rotation of the end tangent at A of the two hinged arch due to M_A =1 applied at A.

βab = angle of rotation of the end tangent at B of the two hinged arch due to application of M_A = 1 at A.

βbb = angle of rotation of the end tangent at B of the two-hinged arch due to M_B = 1 applied at B.

αba = angle of rotation of the end tangent at A of the two hinged arch due to M_B = 1 applied at B.

Then

$$M_A \alpha_{aa} + M_B \alpha_{ba} + \phi = -M_A T_A$$
$$M_A \beta_{ab} + M_B \beta_{bb} + \psi = -M_B T_B \qquad (2)$$

The convention of signs adopted is

1. Moments are positive which produce tension at the bottom of the arch rib.
2. Angles of rotation which are in the same sense as the moment are positive. The sign on the right hand side of the equation is negative because the angle of rotation T of the tangent at the springing is opposite in sense to the moment at the support.

Using Maxwell's theorem of reciprocal displacement we may put $\alpha ba = \beta ab$ and denoting them as $\alpha aa = \alpha$ and $\beta bb = B$

and $\alpha ba = \beta ab = \gamma$ the simplified equations take the form

$$M_A \alpha + M_B \gamma + \phi = -M_A T_A$$

$$M_A \gamma + M_B \beta + \psi = -M_B T_B$$

and solving for M_A and M_B we have

$$M_A = -\frac{\phi(\beta + T_B) - \psi\gamma}{(\alpha + T_A)(\beta + T_B) - \gamma^2} \qquad (3a)$$

$$M_B = -\frac{\psi(\alpha + T_A) - \phi\psi}{(\alpha + T_A)(\beta + T_B) - \gamma} \qquad (3b)$$

or

$$M_A = C_3\psi - C_1\phi \qquad (4a)$$

$$M_B = C_3\phi - C_2\psi$$

The moment constants C_1, C_2 and C_3 in the above expressions are independent of the external loads and are functions of the arch form only. Their values are

$$C_1 = \frac{\beta + T_B}{(\alpha + T_A)(\beta + T_B) - \gamma^2} \qquad (5a)$$

$$C_2 = \frac{\alpha + T_A}{(\alpha + T_A)(\beta + T_B) - \gamma^2} \qquad (5b)$$

$$C_3 = \frac{\gamma}{(\alpha + T_A)(\beta + T_B) - \gamma^2} \qquad (5c)$$

To solve these expressions we must know the values of each individual angle of rotation and the following will lead us to determine all of them.

For the case of a fixed ended arch the values of T_A and T_B are each equal to zero. If in addition the arch is symmetrical $\alpha = \beta$. For a symmetrical fixed-ended arch the expressions take the simplified form

$$C_1 = C_2 = \frac{\alpha}{\alpha^2 - \gamma^2} \qquad (5d)$$

$$C_3 = \frac{\gamma}{\alpha^2 - \gamma^2} \qquad (5e)$$

原刊缺第一至二頁

交通河濱風景，均有極大之增益。又在此道路系統中。其橫出河道兩岸之路線多成爲相對之勢，如是則河道兩岸之交通，或架橋樑，或用輪渡，均可收直接聯絡之效。

（七）市區內各部之交通聯絡

大抵每一都市繁榮發達以後，各區之人物運輸必異常繁重，尤以車站碼頭商業區等處爲尤甚。若不有多數幹綫以疏導之，則將來在一二道綫之上，其交通之壅塞勢將不可避免。循至影響及市內各部之發達，事後始圖補救，能無恨晚。故本系統規畫之始，即以聯絡貫通全市各部之主要交通道線爲目標，并主張多設主要及次要幹線以爲疏導。此項幹線，可分爲兩種：一爲直達幹線，一爲環形幹線。直達幹線之設計，以徼捷直達爲主。查本市河北部份之幹綫，東西相距者長約十七公里。河南部份之幹線，東西相距者長約十六公里。此項路綫均須徼捷直達，以利長距離之交通。若一路之交通已臻擁滯之程度，則有其他之平行幹線以調劑而補助之。

環形幹線之設計，蓋用以聯絡市區各縱橫幹線而設。故本系統於環繞舊城之附近，設一主要之環形幹線，以聯絡河北河南及芳村大坦沙一帶。而此環形幹線與各路線之聯絡，約形成一多足蟹形。將來交通之運動行狀，況即可就此形勢而想像及之。其在此環形幹線橫過河道之處，則應架設橋樑，以收直達之效。

除此主要環形幹線之外，其他規劃之路線，倘有作環形或曲形者，是則或因地勢之高下使然，或因須與天然之堤岸線平行，以增進其地勢之價值，或因須參設曲線以增加市道之美觀，庶免過量拘泥於直線道路。

（八）幹道交叉點之避免與減少

查往昔計劃市政道路類多趨重於多設道路交义點，將數路或十數路之路線使之交义於壹點。此項計畫，固亦具特殊優點，如該交义點在特繁榮區域則城市之美觀點綴，亦可增進。

惟晚近之城市設計者，因近世之車輛交通愈恒發達，交义點地位以內交通情形過於衝繁擠擁，極難管理及疏導，以致車輛傷人及意外事項次數增多，實屬不妥。本系統設計之初，即注意避免此種弊病，但仍於相當範圍之內，酌設交义點，并擴大之使成廣塲，以爲城市道路之點綴。

（九）道路系統範圍內之支線及房屋地點之分配

道路系統規劃之大意旣如上所述，惟支路之計劃及房屋地段之分配須俟道路系統確定後，始易於進行。查本道路系統因地勢及附近情形之不同，故道路系統內之道路線縱橫距離頗難預定一固定不易之標準，因而各地段內之支線畫分及房屋地段分配，亦惟有因地制宜，妥爲分配。又晚近之市政設計原則亦不主張固定一種之標準方式，以免過於拘泥，有一成不變之勢也。查本市市民建屋習慣，每注重南北向者，故將來內街之分配，宜多設南北向之地段。又於交通繁重之區，內街與幹線之橫貫交通交叉點，亦宜酌量設法避免或減少之，以減輕市內車輛之不幸事件。

道路系統內之地段，其大者約爲三千呎乘二千三百呎，其小者約爲二千呎乘一千五百呎。將來該項地段之分配，（除闢爲公園或其他公共建築者不計外，）即將其劃分爲若干支衢若干橫街，以與道路系統聯貫。至該範圍房屋地段大小之規畫，亦應視該分區之需要及狀況而定。房屋之深度有由五六十呎而至百餘呎以上者，濶度有由十餘呎而達三十餘呎者，均以能適合各該區需要爲度。

（十）總理實業計劃之依循

總理實業計劃之內關於改良廣州市河南附近之水道及發展河南芳村一帶之計劃已樹立楷模。故本市道路系統之規畫即循此規範以期逐漸謀此計畫之實現其壹切設施以不至與實業計劃發生抵觸爲原則。

（十壹）市郊公路交通之聯

考城市之發展與四郊之公路交通有莫大之關係。良以市區一地實爲四郊附近市鎮之中心市場舉凡貨物之運銷食糧之輸送均應有良好之交通以資聯絡庶免日後各處一經發達而市與郊之交通樞要地點有壅塞不堪之虞。本道路系統亦注意及此已成郊外公路及將來其他各部公路建設完成均當以新市區爲集中之點本市道路系統確定後屆時市區道路與四郊公路之聯接在交通上當可收重大之果也。

（十二）道路之寬度

凡闢馬路應有精密之交通調查以作根據然後確定其寬度。設任意規定必至狹不敷用或大而無當則於工程經濟及交通原則大相違背。是以計劃一路首應考查其目的何在性質何屬。如屬住宅區則路兩旁之建築物最高應爲若干尺其需要之光線空氣應有若干尺寬之馬路始足以供給。其他若車行路之寬度是否足供必經之車輛通過而無危險人行路之寬度是否足供行人來往而不擠擁。又通地下之渠道水喉電話線等之位置分配等均應先事妥爲預算者也。人行過之寬度視乎步行人數之多寡而定。雖則此項預算尙無方式惟現在之交通情形與及將來預料之發展多足供參考。如冷靜之橫街行人稀少最多兩人並肩或對面行往則人行路之寬度一五公尺實已敷用普通馬路交通不甚繁盛可留四個人或三公尺之人行路商業繁盛之馬路應留四五至九公尺之人行路，但仍視其繁盛程度如何而定也。照普通規定，每邊人行路應占全馬路寬度之四分一至六分一。

車行路寬度，應適合於該路最密車輛之時間各種車輛能通行而不致擠擁停滯爲准。計算寬度，可以各種車輛寬度而推算之。普通汽車於行駛時，來往共佔五五至六公尺，如住宅區之冷靜橫馬路，只供該路居民之汽車出入，其寬度五五公尺卽已足用。商業區馬路，爲一一至二二公尺，因兩旁應預備各有汽車一架停放，而中間至少須有兩架汽車行駛也。至於需要植樹或種草時則普通植樹須濶一　五至二公尺，草地須濶壹至壹五公。

上述各點僅就交通而言，當未論及光線及空氣在濶大之馬路，此點當無問題。惟在交通不繁之區，若祗據交通，而路之寬度則恐馬路太窄不足以供給兩旁建築物充足之光線及空氣是以美國波士頓等地，有限制建築物不得高過馬路寬度之一倍半，紐約對於住宅之規定，是不得高過馬路之寬度。綜上各点，擬權定下列各種馬路寬度：

商業區馬路

（界）堤岸馬路　　三十至四十公尺

（乙）幹道　　二五至三十公尺

（丙）普通馬路　　十二至二十公尺

工業區馬路

（甲）堤岸馬路　　二十五至三十公尺

（乙）幹道　　二十五至三十公尺

（丙）普通馬路　　十二至十五公尺

園林住宅區馬路

（甲）堤岸馬路　　二十五至三十公尺

（乙）林蔭大道　　二十五至三十公尺

（丙）住宅馬路　　十二至二十公尺

普通住宅馬路

（甲）幹道　　二十公尺

（乙）住宅馬路　　八五至十五公尺

行政區馬路

（甲）大道　　三十公尺

（乙）幹道　　二十至二十五公尺

（丙）普通馬路　　十二至十五公尺

（十三）總結

根據上述之原則規劃，現在擬就之道路系統，其在河北則有經德宜路天官里直達東圍城之東西幹綫，其在河南亦有横貫東西而直達黄埔之幹綫，聯絡全市各部則有環形幹綫，此本系統之大綱，而爲其他各道路之主腦，亦卽本系統之重要基點也。以此道路系統爲根據，從而斟酌情形實行分區，幷設置醫院市場屠場公厠，與夫學校公園運動場游泳場跑馬場圖書館博物院市立劇院以供市民之享用，又從而計畫無軌電車及公用汽車與整理水上交通，以便民行，則皆有待於本系統之確定矣。

廣州小北水災情形及今後救濟方法

去秋廣州小北一帶，洪水爲災，禍區廣濶，死傷甚夥，災情奇重，爲本市數年來所僅見。工務局對此次水災起因已調查清確，對於今後之救濟亦有具體辦法。茲將該局呈市政府文錄後：上略本市東北一帶，因山洪暴雨，一時齊來，低窪之地，頓成澤國，廬宇坍塌，居民流離悽慘之狀，殊堪憫惻。當時局長開耗，督同屬員於是日晨親到該區巡査審察，覺洪水橫流，實所罕見。其時越秀北路以東，即東濠一帶之水平線，與越秀北路以西，即小北一比較，實有超過數尺之多。質言之，即小北一帶之水，其時不特不能流出東濠，而東濠之水反有灌入小北之狀態。因此認定東濠上游之宣洩，必有絕大之障礙。隨即派員趕赴下游調査，果見船艇竹排木排等物，爲之梗塞，尤其是錢路頭附近之柴店，被水漂流之柴把障礙更甚。以致上游之水，宣洩延滯。覩此情形，立即督率工人百餘名，將船艇竹排木排柴把等物，悉數搬清各種障礙物。旣經除去，水流始行通暢，至下午東濠之水勢悉退，小北一帶，即銅關以內之水，乃能向關外流出，迨至午後十一時許退清，此爲當日水災狀況，及辦理經過之情形。事後派員前往災區，將水勢測量，計錢路頭一帶，水高三十三尺七一，（依照本市規定水準標誌計算，下仿此。）小北一帶，水高三十尺九一。兩相比較，即銅關以外之水，高過銅關以內之水三尺八寸。水低處不能向水高處傾流，此理至爲顯然。因此可以証明此次小北一帶之水災，實緣東濠下游被船艇竹排木排柴把等物梗塞，加以白雲山之山洪來勢太猛，且在深夜中，一時防止不及所致。而市民間有以爲六脈渠銅關等處，渠道去水之寬度不足者，不無悞會。以現在法政路渠道之寬度計算，其於宣洩小北一帶之水，實足應付有餘，斷無宣洩不足之理。假使當時水勢，東濠低過銅關以內，（在水準式十五六尺之間，）則宣洩通暢，小北一帶自無淹浸之患。或謂銅關以外水勢退後數小時，銅關以內之水，尚未退盡，以此認爲法政路渠道太細，不敷宣洩之用。不知當東濠水漲時，其水由法政路渠灌入小北一帶，渠徑愈大，灌水愈多，灌入之水，輒比原有雨水多至數倍，故東濠水退後，法政路渠除排洩原有雨量外，尚須排洩灌入之水。時間延長。此爲勢所必然之事。故以時間而論。關外之水由是日早弍時起灌入關內直至正午。關外之水方退至與關內齊平。其灌入時間歷十小時。關內之水。由下午起流出直至午後十一時許。方能退淸。其排洩時間共需十一小時。灌入與排洩之時間約略相等。此更可証明其渠口之非細。故救濟辦法不在銅關渠口之改濶。而在浚深東濠。淸除淤積。並嚴重取締船艇竹排木排等物之停泊梗塞。同時改善小北一帶內街道。并將地勢畧爲提高。或於銅關設置活動水閘。使東濠之水。不能灌入。至白雲山流下之山洪則擬另闢其他涌溝導入珠江。以分東濠之水勢。夫如是。小北一帶自可免除水患。而市民亦可安居矣。

（下畧）

建築珠江鐵橋之緣起及經過情形

世稱模範都市之廣州市。中隔珠江壹水。界分南北。交通往來既難厲揭而行。徒恃舟楫之濟。今海珠堤岸不日建成矣。洲頭咀內港堤岸亦將告竣矣。地面形勢當然爲之一新。商務交通當然爲之一變。矧商務與交通實有連帶關係。因勢利導。捨溝通南北聯成整個的都市。以求交通無阻。則商務地位。祇可以言保守。未足以言進步及繁榮。攷廣州與國外早已通商。商業雖興。水路交通雖便。論者且謂廣州占國中首要地位。而形格勢禁。非改造無以利民生。市政當局。因而有建築珠江鐵橋之成議。務將交通窒礙點打破。使廣州市壹躍而爲世界商港模範都市之名益彰。玆將經過工作分段臚列。俾知梗概焉。

（一）北岸第壹橋墩鉅工之經過情形

維新路口橋墩工程。自十八年十二月壹日興工。先將橋墩位置用鋼板長樁圍擋，致免水淹，然後施行工作。圍樁而後抽水出外，施以挖泥。面層浮泥掘去之時，發現亂石無數，隨即施工炸碎，纔以起重機器抽石出圍。續又現出石碎士敏三合土地基，復以人力鑿碎起去。最下一層尚有舊時橋址木樁滿佈，全由人工拔起，纔續掘泥。同時並用水泵抽泥出浮泥，加以水力機器衝動，乃易施工。迨浮泥出淨，驗明實土，遂用十二吋方形松木樁條，施以汽鎚壓下，汽鎚汽壓力為五噸重，壓至不能再下時然後停止。樁既完全打妥，修正樁頭，始落石碎士敏三合土。該橋墩先後經過兩載，現在已告完成。又開始設計安裝鐵橋工作矣。

（二）北岸第二部西便橋躉工之經過情形

河中西便橋座，自前年興工建築，其始妥定位置，先將橋座基礎外圍裝妥。該外圍係用鋼板構造，深入河床之下，約有二十餘呎，連上層高度總共五十二呎，以為阻攔河水淹入，便於工作而設。裝妥後興工挖泥，并用水力機器衝動浮泥，同時亦用水泵連同浮動泥沙一併抽出。抽至實土時施行打樁工作。所打之樁木，為英呎十二吋方，施用汽鎚壓下，汽壓力計有三噸重，壓至該樁不能再下時為止。壓妥後遂落石碎士敏三合土基礎，約深九呎，時當尚未賡續造上層工作，意欲俟該基礎堅結後，再行施工。不料外面水力過強，致將鋼板圍擋浮起。緣該鋼板圍擋內部空虛，壓力輕薄，不敵外面重大水力，兩者互相比較，內外相差，約有一百餘噸。更加汽輪往來，水力衝動，又受風力搖蕩，地基樁木，連帶搖動，圍擋因而浮起。工程經過數月，一旦化為烏有。此乃去年一月秒之事。隨將所造之石碎士敏三合士基礎鑿碎，並將所壓下之樁木一併起出清除，纔用水泵水力機抽出餘泥，抽至紅色硬土，本應行打樁工作，惟何底紅色硬土性質堅實，樁嘴雖經打斷，未能打入毫釐，不得已而擬將打樁計劃取銷。第是地腳土質所受壓力若干，未能明瞭。故用一鋼樁試驗，試驗時鋼樁切近硬土處，面積共有二英尺，經於樁頂施以一十四噸之重量壓下。歷時五日，視其變動結果，僅始初之二三日時間，壓下約英寸七分。迨後則絕無變動，試驗若此，計劃自應變更。遂即飭工下水，將該地脚四圍審察，絕無沙石浮土，始行落石碎土敏三合土，約深一十六尺，作爲地脚頭層之用。旋復賡續接連造上層工作。西東兩橋座，亦如法建築，今已完全告竣。近者設計安架橋樑鐵件，現已裝安完妥。其經過時日亦已兩年，所經建築情形有如上說。

（三）南岸第一部橋躉鉅工與第四第五部之橋座工程，早已工竣。其建築工程情形，畧如北岸。今已設計安架橋樑，連月各工程均積極進行，該橋完全工竣時想在今春三月之後。

廣州市珠江鐵橋之概要

查建設珠江鉄橋，原巳利河南北之交通，惟是橫跨珠江，如遇河水高漲，對於輪渡之往來，不無窒碍，遂有橋頭用開合式之設計。

（一）橋身之長寬度及橋底至水面高度

該橋身之長度，擬定六百英呎，南岸橋頭直臨南華馬路，左右貫通中西各路。北岸接駁維新南路，直達中央公園，均連東西各大馬路，橋之橫面中段係車輛往來之路，寬度四十呎，兩傍行人路，寬度十呎，祗准行人，全座寬度總共六十呎。橋底至水面距離高率平均共二十七呎四吋。

（二）橋身之載重

橋身載重之限制，亟宜審愼。茲定橋之中段路面，係任載車輛往

來祇準二十噸重之貨車。兩傍行人路每平方呎任重量八十磅

(三)橋身之構造

該橋之構造係用鋼料造成，南北兩頭堤岸，橋躉至河中橋座，兩相距離，由中線起係二百二十英呎。該段橋面用鋪石碎士敏三合土，再用腊青沙蓋面。其高度接駁堤岸斜坡，而斜坡斜度，北岸每百尺叁尺，南岸每百尺四尺，車輛往來直通維新路而東西兩頭堤岸，欲由橋上過河行人可由步級上升橋面。

(四)斜坡至橋與堤岸交通

斜坡底至堤岸路面高度有十三英尺，車輛均可由斜坡下通行，如要上橋，可由維新南路斜坡口直達橋上而通河南廠前街馬路，河中橋身中段開合處，兩橋座中線距離一百六十英呎。該開合處橋身亦係鋼料造成。而橋面鑲木，取橋身重量減少，體質堅結。其開合機關係用電力機得以盪動快捷約兩分鐘時即能開起，輪渡往來，交通無阻。此建築珠江鐵橋之概要也。

嶺南女大全校設計畧說

本校女學全部建築，爲本學院教員黃玉瑜先生所設計，校址在黃岡與沙岡全校有大會堂一座宿舍七座青年會一座家庭經濟實驗所二座游泳池一所。全校建築費約二百萬元。大道由南至北直通廣東公路，路旁遍植花菓，以增美觀。其他貫通東西道路其濶度比大道者略小。最宏偉之大會堂，建築于沙岡之上，內容室內運動場一間，劇場或演講台一座，下層爲會食堂，第二層爲會客室與辦公室，第三層則有客房數間。現在興工建築中之宿舍，如圖中所示者，即土名黃岡之地，預備爲大學第十年

級生所用者，建築費約十七萬元，約于本年中完成，內容學生一百名，但預計將來第二學期有學生一百二十名時，各學生不免感覺地方狹窄，然不得不暫時容忍，以俟將來增建第二宿舍也。宿舍共五層，地下將暫借爲會食堂實驗室等用途，第一層爲教職員餐室客室醫藥室養病室及睡房數間，第二三兩層則全爲學生寢室，頂層爲鋼琴練習室四所及大儲物室一所。宿舍之外觀結構，全仿中國皇宮式樣，但其內部之設備，則採用歐美最新之方法。女學已向有紫禁城之稱，將來工竣，稱爲南中國首屈一指之學生宿舍，亦無不可也。

私立嶺南大學建築材料力量試驗章程

本校工學院所辦土木工程一科其材料試驗室設備尚週置有建築各種材料力量試驗機士敏土力量試驗機等均屬最新式者以之試驗力量極爲快捷準確授聘各教授俱留學歐美領有高級學位學識豐富且曾在國內外任工程要職多年對於試驗建築材料極富經驗玆爲應社會需求及提倡使用國產建築材料起見特訂定辦法公開代工商各界試驗各種建築材料凡屬國產之士敏土磚瓦石階磚木材等物依照下定辦法請代試驗者尤所觀迎

試驗建築材料力量辦法

(一) 凡公司團體或人委託本學院試驗建築材料無論該項材料係國產抑屬外貨經本院認爲可代辦者即與接受

(二) 委託試驗之材料無論磚石鋼鐵灰木或士敏土概由委託人自行供給註明出產地或製造廠及物名分量須足敷三次試驗之用如係士敏土或灰必須封固免受濕氣發生變

化所有材料試驗後留院陳列概不發還

(三) 試驗三合土或灰沙等製造品爲求準確起見委託人須將各原料交本學院配合若委託人自行配合後始交來試驗者必須將各項詳細紀錄連同交來

(四) 試驗費暫酌收如下

士敏土牽力試驗　　廿五元

士敏土全部試驗　　一百元

士敏三合土壓力試驗　　廿五元

磚或石磨擦試驗　　廿五元

木材或鋼鐵或鋼筋三合土之牽壓剪扭屈各種力試驗每種　　五十元

(五) 其他材料之試驗費視乎手續之繁簡及消耗電力或其他物料之多寡而定均臨時面討

(六) 本學院接受委託人材料及試驗費後即發回收據收據係兩聯式一給委託人一留院存查

(七) 本學院接受委託後即發交主任教授試驗工作完畢即將試驗結果詳細說明送交委託人妥收如未得委託人同意本學院永遠代守秘密不向外發表

(八) 委託人如居住本市以外可將材料及試驗費由郵局掛號寄交本學院惟委託人姓名住址必須詳細註明以便寄發報告所有一切郵寄費用均由委託人自理

(九) 凡土木工程師會會員委託試驗其試驗費照七折計算

會　　務

工程學會之成立經過及最近會務

梁卓芹

本會之成立及經過· 廣州工科太落後吧!四年前任誰都是這樣說.在一九二九年秋本校始創工科——也是南中國第一個工科學院的成立,當時因爲初辦而又在風雨飄搖的時期,成績當然說不上「大有可觀」;同時學生人數也並不多。迨一九三十年秋學生的人數也跟着工學院的年齡而增加。在這個時候我們感覺到工科同學太散渙,不特對於智識的交換,感情的聯絡欠缺,同時也許有些見面不相識。當時乃由三數同學倡組織工程學會之議結果得全體同學讚同,於是舉出籌備委員以便起草會章,呈請學校立案,及進行一切。是年冬得學校批準立案,工程學會於是正式成立,並舉行成立典禮,是日到會者有鍾校長,工學院長,學生自治總會代表,大學學生自治會代表及本校各團體代表,濟濟一堂極一時之盛。

一九三一年春曾舉行大規模之參觀!當時由溫其濬先生帶領,除三數同學因事不克參加外,幾得全工科同學參加,參觀地點在韶關樂昌一帶之株韶段鐵路建築,韶平公路建築,及富國煤礦等地.是次參觀誠不可多得,且對於我們智識上都有很大的補助。

一九三一年秋至一九三二年夏,全年除曾往參觀珠江鐵橋西村自來水廠河南士敏土廠及磚窰等,無別項工作,也許因本學院初改廿四學分制,各同學功課極忙,致未暇顧及。

最近之會務　自本屆職員就任以來,蒙各師長先進之指導,更得各同學之讚助,會務始稍有進展。九月卅日曾舉行一次全體新舊會員大會,是日到會員生約五十餘人,會中由主席報告工作經過,及本學期未來計劃;並得工學院院長李權亨先生訓話,及議決案多件。隨着是新舊會員聯歡會的舉行,先由各人自行介紹,及遊戲譜談等,秩序也算豐富,最後乃飽啖鷄粥而散。關於本期工作方面,大約擇要畧述於後:(1)組織體育委員會以便振興本會體育,曾與本校各院同學作籃球友誼比賽,結果皆屬本會勝利。(2)關於參觀方面面,曾往西村士敏土廠,及自來水廠參觀。二三年級同學曾往韶關樂昌等地作數日之長期旅行,曾參觀株韶鐵路建築,韶平公路建築及富國煤礦等,日間並擬往珠江鐵橋作實地參觀,及作地質學之實地調查。(3)關於學術演講,曾請粵漢鐵路株韶段路局局長凌鴻勛先生來校演講。並擬極力聘請國內工程學者來校多作演講(4)刋行工程學報。我們因爲功課比較別院同學實在忙得多,所以除實際有益之工作外,其他如交際會等祗得少些舉行,這就本期工作的大概茲將本屆職員並本會會章附列於後。

主席　梁卓芹

文書　陳銘珊

總務　關祖舜

司庫　陳景洪

交際　鄒漢榮

出版　龍寶鎏

研究　林植萱

英文書記　梅國超

體育委員會　黃樹邦,鄧樹翟,許寶照

廣州私立嶺南大學工程學會會章

（一）定名　本會定名爲「廣州私立嶺南大學工程學會

（二）宗旨　本會以研究學術交換智識及聯絡感情爲宗旨

（三）會員　1. 凡屬私立嶺南大學工學院之同學皆得爲本會會員

2. 凡屬私立嶺南大學之同學有志趣於工程學而願遵守本會會章及得本會會員二人之介紹經常務委員會通過者得爲本會會員

（四）權利　本會會員皆有選舉被選提議表決與享受本會一切設施之權利

（五）義務　本會會員皆有遵守本會會章及服務本會之義務

（六）組織　1 本會設委員七人組織常務委員會內分主席一人副主席兼財政主任一人總務主任一人文牘主任一人研究主任一人交際主任一人出版主任一人以一學期爲一任如再被選得繼任本職一次

2. 本會各部職員均先由上屆常務委員會介紹候選人二名由大會用雙記名法投票選舉之

3. 本會常務委員於學期終結前二星期由大

會依法選舉之執行職務至下學期新職員選出交代清楚後始得卸責

（七）職務 (1)主席委員對外為本會代表對內為本會會議之主席及執行本會一切之議决案

(2)總務主任主理本會一切總務事宜

(3)研究主任主理本會一切研究宜事

(4)文牘主任主理本會一切文牘及記錄

(5)交際主任主理本會一切交際事宜

(6)財政主任主理本會一切進支事宜

(7)出版主任主理本會一切出版事宜

（八）顧問 本會得敦請校內教授或校外有名之工程學者為顧問

（九）會費 凡本會會員須於每學期繳納常費一元(毛銀)於開學時徵收之如遇特別事宜須徵收特別費時可由常務委員會通過佈告徵收之但其數不得超過廣東毛銀二角如超過二角以上者須由全體會員通過始得徵收

（十）會議 (1)全體大會——全體會員大會為本會最高之機關於每學期開課後二星期及學期終結時各舉行一次如遇必要時得由委員三人以上或三份以上之會員連署得隨時召集之

(2)會務會議——會務會議由本會主席及各委員組織之於每月舉行二次由主席委員召集之於必要時得由主席委員隨時召集之

(3)法定人數——全體大會以會員三份二以上出席為法定人數會務會議以委員四人

以上出席爲法定人數

（十一）附則　本會會章如有未盡事宜經會員五人之提議三份二以上之會員決議得修改之

世界最大之特聶泊水電廠上月竣工開幕

蘇俄五年計劃之中心事業，舉世屬目之特聶泊大電廠已於十一月十日正式開幕，特聶泊堤堰及其所屬全部工程，始於一九二七年十一月，迄今恰爲五年計用工人二萬五千以上，需款八萬萬盧布，該廠每年可供給八十一萬匹馬力，三十萬萬啓羅華德小時之電力，新實業區域內之全部工業及交通上用電，以及南烏克蘭之冶金業所用電力：均仰給於是.按美國尼加拉瀑布發電廠僅能供給四十三萬匹馬力，北美最大之台尼西河水電廠，亦僅供給六十二萬匹馬力，而此特聶泊廠則可供給八十一萬匹馬力之電力，故爲世界上最大之水力電廠云.

編後話

鎣

本期稿件，篇數似略嫌其少，惟其中數篇，頗可珍貴，編者亦以此稍自告慰。Analysis of Restraint Arches一文，作者詳論環拱之受力，參以例證，與我們一有系統之供獻。討論Diesel Engine之文章，在各國工程雜誌極不多覯，本期Diesel Engine in Practice中，解釋該機發明之理論及進展，施工之實際及困難爲一不可多得之作。Modern Architecture一文，敘述現代建築設計之趨向及其變遷之緣因，指出Des Pradelle派之功用主義(Functionalism),Frank Lloyd Wright之長平體，及德國之立方體等派別之歧異，足資設計者之借鏡。

本校工學院剏國民政府鐵路部出資委辦，南中國之工程大學亦以本校設立爲最早，本會對於華南工程進展，應負相當之促進責任，本期爲南中國工科大學學刊之嚆矢，故中文稿件，畧側重西南之工程建設，以求普遍。「廣州地質載重能率」一文，作者供獻其歷年之經驗，本市附近之坭土之負重能率，亦可由此推測得之。本校非文氏氣象觀測所，歷史至爲深遠，所藏廿餘年氣象記錄，爲南中國所僅有，本期特將近十年來記錄平均列表刊出。市內各種重要建築物料市價，本期亦特約專員調查登載。之上三文，對於市內工程師，尤其初來之外國或外省工程師，或成爲重要參考資料。至於南方之工程建設，如建築中或計劃中之鐵路或大建築，本市道路溝渠系統等，均盡量採訪刊登。

本期付印倉卒，致特約撰述之文稿數篇，如陳銘珊梁卓芹合作之工程材料力量試驗結果，韋金信之劇場座位與視線之研究，李文泰之求centroid一法，及編者之工廠之空氣清潔問題，均未及脫稿，弗克於本期登載，謹向著者及閱者道歉。付印時適值學期試驗，功課忙廹，文稿之修改及校對均缺乏時日，加以編者知識簡陋，力量棉薄，本刊又出版伊始，錯誤之處，在所不免，大雅宏達，幸有以進之。

澳僑置業有限公司

AUSTRALIAN OVERSEA REAL ESTATE CO LTD

77 TAI PING ROAD SCUTH CANTON CHINA

陶朱致富。治產積居。計然遺規。勝算獨握。同人法師前哲。步步爲營。生意固鋭意經營。同心一德。公司則自置產業。樓高五層。況復地處太平南路。交通最爲利便。存欵異常穩當。按揭利息相宜。至代理買賣產業。代理收租及信托等事業。更注意顧全利益。以副雅愛。如蒙惠顧。無任歡迎

監理余榮

司理許承瑞

司庫劉五明

地址：廣州市太平南路七十七號

自動電話：壹肆〇式壹

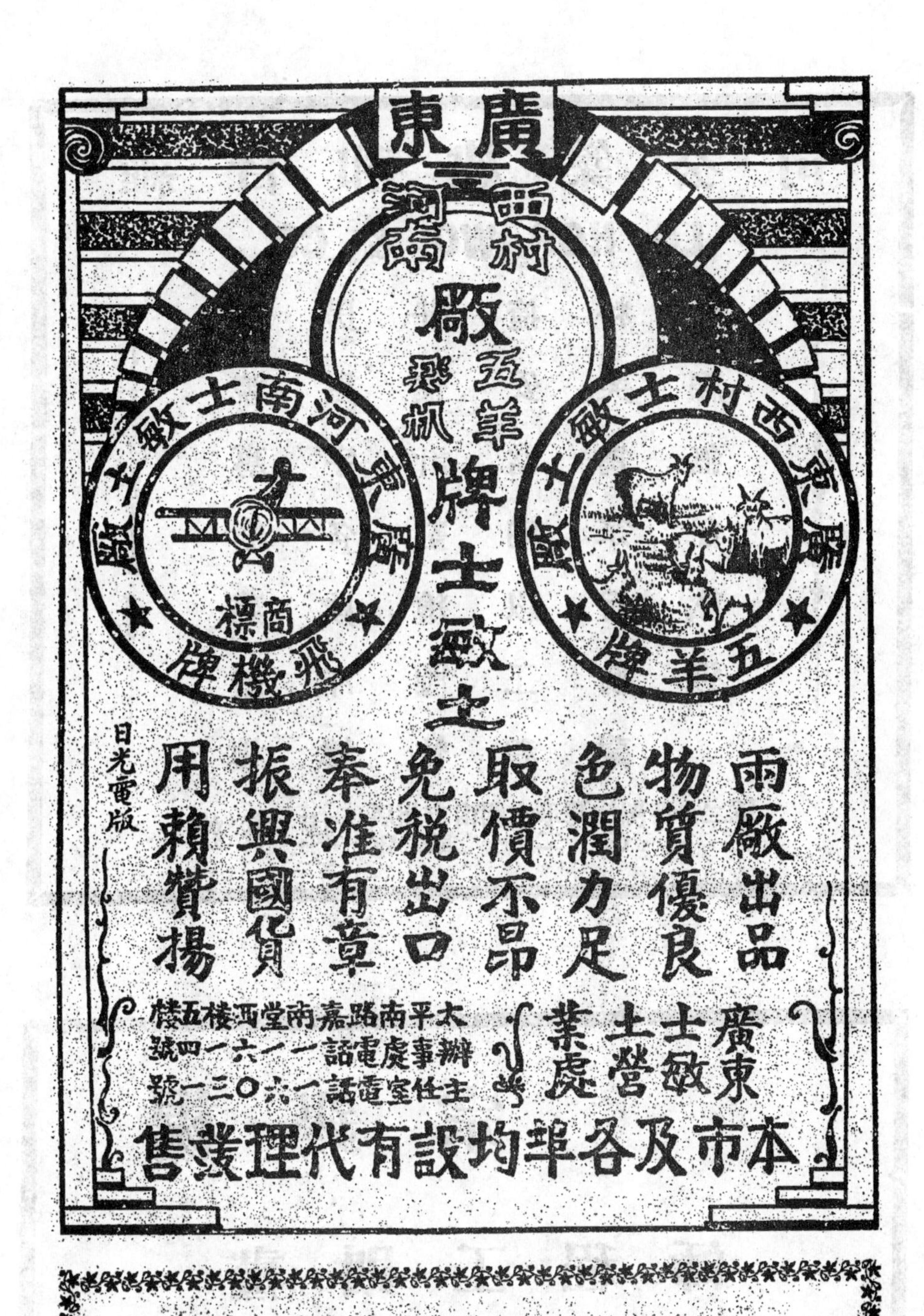

三民興洋服店

新款兼便宜

鋪在財政廳前

THE JOURNAL
OF
THE LINGNAN ENGINEERING ASSOCIATION

VOL. I NO. 1

Editor-in-chief LUNG PO YUK

Associate Editors: LAI SU YAN, CHOU HON SAN, LEUNG CHUK KAN

Advertising Manager YUE PAAK CHEUNG

Proof-reader LEUNG PO KWAN

Circulating Manager LIU KIN MAN

中華民國二十二年一月十日出版

南大工程

第一卷 第一期

（每冊三角） 郵費••本市二分••國內五分••國外三角

編輯者……………………龍寶淯

出版者……………………南大工程學會

發行者……………………南大工程學會

總發售處……………………南大書局

分售處……………………各大書局

25039

LAM CONSTRUCTION CO.,

HONG NAME (WAH YICK)

16 LUEN FAT STREET, HONG KONG. TEL. No. 26125.

3 CHING PING BRIDGE, CANTON. TEL. No. 13957.

香港

華益建築公司

啟者本公司承接建造樓房屋宇各欵洋樓地盆碼頭堤磡橋樑所有海陸一切工程連工包料并代計劃繪圖快捷妥當如蒙 光顧祈為留意

香港聯發街十八號

電話弍六壹弍五號

廣州市沙基清平橋三號

電話壹三九五七號

25040

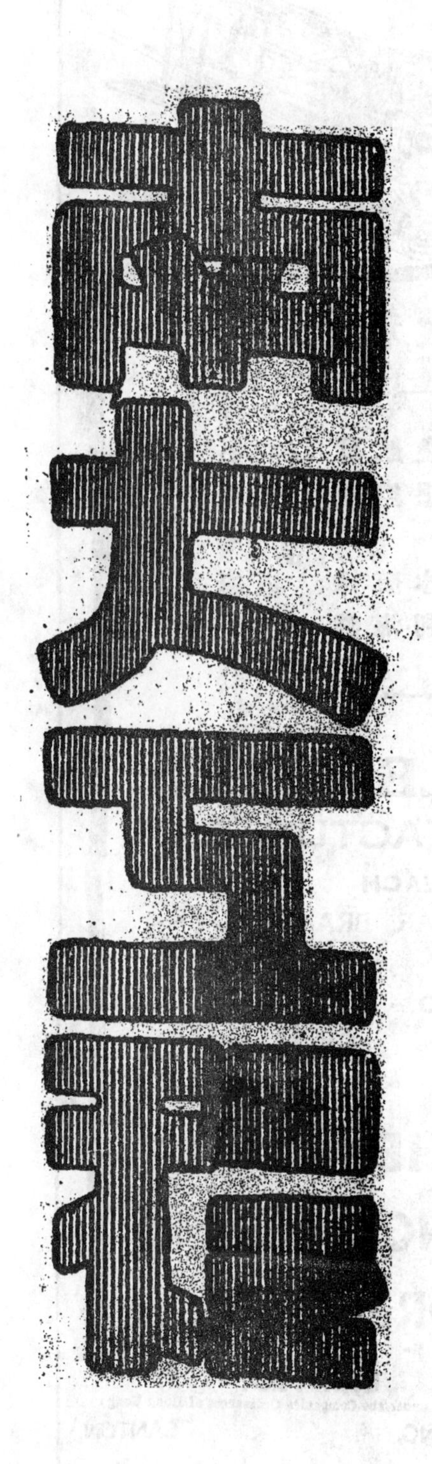

The Journal Of

The Lingnan Engineering Association

Vol. 2, No. 1 Feb. 1934

第二卷 第壹期

嶺南大學工程學會發行

英商怡和機器有限公司(即渣甸洋行)

經理各國名廠專家出品

發電採礦航空航海農墾水利紡織救火鐵路印刷各種機械及製冰機冷氣機升降機潔浄具軍用品等諸君惠顧無任歡迎

香港必打街十四號

廣州沙面英租界

SIXTY-FIVE OF THE **LEADING**
ENGINEERING MANUFACTURERS
OF THE WORLD EACH
SPECIALISING IN A PARTICULAR BRANCH
OF ENGINEERING
ARE REPRESENTED
IN CHINA BY

THE JARDINE ENGINEERING CORPORATION

LIMITED.

(Incorporated under the Companies Ordinances of Hong Kong)

HONG KONG, 14, Pedder St.

CANTON, Shameen.

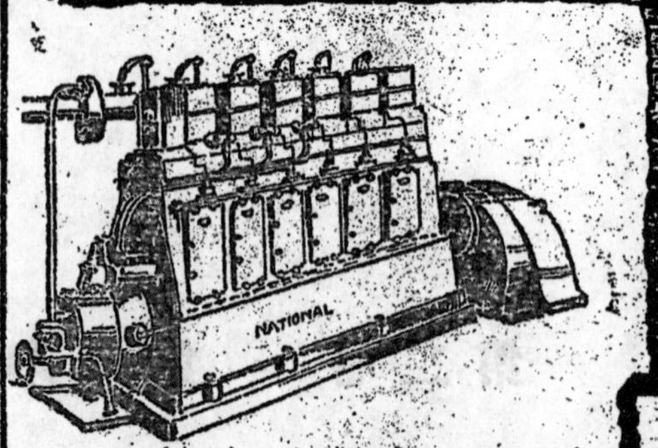

THE NATIONAL GAS ENGINE Co. LTD
ASHTON-UNDER LYNE.

25042

中
國
旅
行
社
本社經售
廣九廣三粵漢各鐵路局頭二三等火車票
太古怡和招商局各輪船公司唐餐樓官艙房艙船票
香港至上海南洋歐美各外國郵船公司頭二三等船票
廣州至汕頭廈門福州温州上海飛機航空客票
兼辦各輪船公司航空公司及鐵路局定價不另取手續
費並兌售上海商業儲蓄銀行國內旅行支票分社遍設
國內各大埠車站馬頭派有着制服招待及小工搬運行
李並理報關保險如蒙 委辦旅行一切事項無任歡迎
地址：太平南路 四十七號
電話：一零八八零號

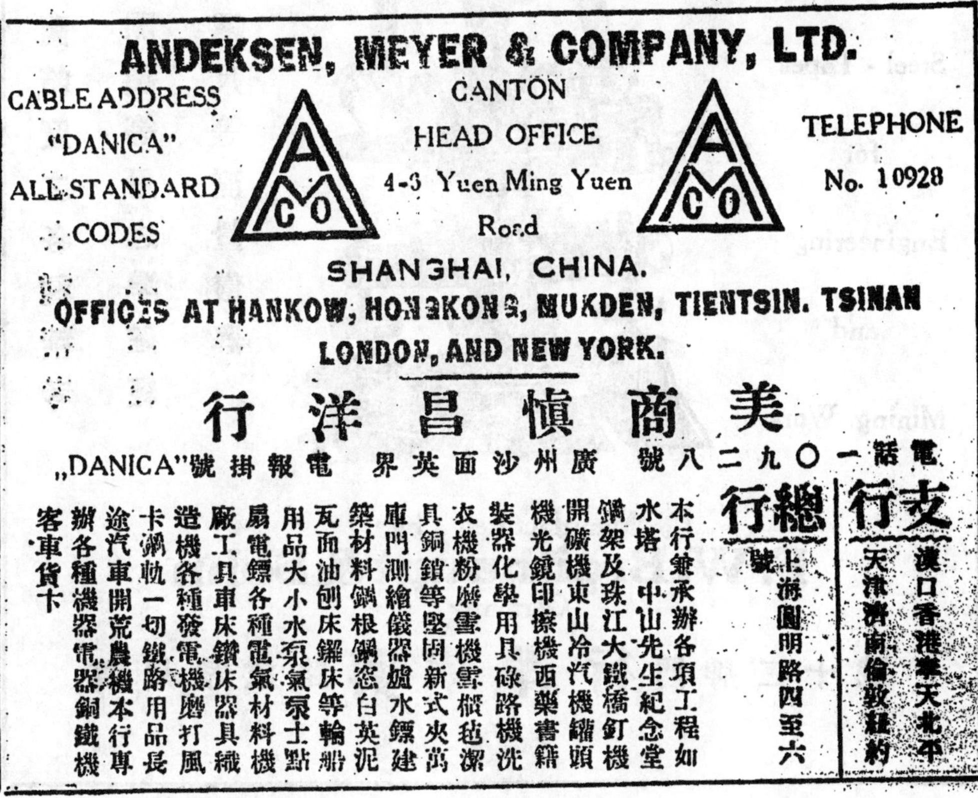
ANDEKSEN, MEYER & COMPANY, LTD.
CABLE ADDRESS
"DANICA"
ALL STANDARD
CODES
CANTON
HEAD OFFICE
4-6 Yuen Ming Yuen
Road
SHANGHAI, CHINA.
TELEPHONE
No. 10928
A M CO
OFFICES AT HANKOW, HONGKONG, MUKDEN, TIENTSIN. TSINAN
LONDON, AND NEW YORK.
美商慎昌洋行
電話一〇九二八號 廣州沙面英界 電報掛號"DANICA"
支行 漢口香港奉天北平天津濟南倫敦紐約
總行 上海圓明路四至六號
本行兼承辦各項工程如水塔 中山先生紀念堂鋼架及珠江大鐵橋釘機開礦機東山冷汽機罐頭機光鏡印擦機西藥書籍裝器化學用具碌路機洗衣機粉磨雪機雪櫃毡潔具銅鎖等堅固新式夾萬庫門測繪儀器爐水鏢建築材料鋼根鋼窗白英泥瓦面油刨床鑼床等輪船用品大小水泵氣泵士點扇電鏢各種電氣材料機廠工具車床鑽床器具織造機各種發電機磨打風卡鋼軌一切鐵路用品長途汽車開荒農機本行專辦各種機器電器銅鐵機客車貨卡

南大工程

嶺南大學工程學會會刊

助編輯 陳景洪 梁卓芹 李文泰

總編輯 劉載和

廣告主任 劉文漢
校對主任 李卓傑
發行主任 盧健文

第二卷第一期目錄

投稿簡章

本刊歡迎投稿且鑒於外國文稿件,已為國內僅有之二三工程刊物所摒棄,在作者方面,每苦於專門名詞未有標準之迻譯,恆習慣直接用外文撰稿,本刊在中央編譯館未將各標準名詞公佈以前,外文稿件亦樂為登載,爰錄簡章數則於後:

(一) 投寄之稿,自撰或翻譯,均以關於工程者為限,文體文言白話不拘。

(二) 外國文稿件,暫以英德法三文為限。

(三) 來稿請繕寫清楚,如有附圖附表,尤須書寫清潔外文稿件打字寄來尤佳。

(四) 如係譯稿,請書明原著書名篇目出版地點日期。

(五) 稿末請注明姓名住址,以便通訊,至揭載時如何署名,聽投稿者自便,作者能將詳細履歷敘明尤佳。

(六) 投寄之稿在二千字以上者,如不揭載,編輯部於每期出版後,按址郵寄回。

(七) 來稿經揭載後,酌酬本刊是期多本。

(八) 來稿編輯部或酌量增删之,但投稿人不願他人增删者,請豫先聲明。

(九) 來稿請逕寄廣州嶺南大學工程學會出版部收。

本校工學院

Public Water Supplies

By

Dr. Wen-Wei Huang

I.—Historical Sketch

The earliest method of artificially obtaining a water-supply was doubtless by the digging of wells. In the vicinity of the pyramids there still exists wells which were in use when those great works were constructed. Joseph's well at Cairo is perhaps the most famous of all ancient wells. Remains of such works are numerous in ancient Greece, Assyria, Persia, and India. Perhaps the deepest wells were dug by the Chinese[1], depths of 1500 ft. or more being reached by methods almost identical with those now in use.

Besides the digging of wells, the ancients executed many works for the storage and conveyance of water. In Jerusalem underground cisterns were built for the storage of rain water; and other reservoirs were constructed near the city to store the water which was brought thither in masonry conducts. The remains of this kind of work are still seen around the Western-Hills in Peiping. Aqueducts were also built in the ancient Greece. Works for irrigation in Egypt., Assyria India and China were established on an immense scale. Dai Yü administered water works, thrice passed his door without entering it. That was the first thing of this nature, dated back in about 2000B.C., mentioned in the Chinese history.

Among ancient systems of water-supply the works of no other nations equaled those of the Romans, either in point of size or number. The water was usually brought by the Romans from long distance through conduits or aqueducts first passed into large cisterns, and from these was distributed through lead pipes to other cisterns, and to the fountains, baths, and various public buildings and to private consumers.

The ancients have some clear notions concerning the quality of water-supplies. In his time, Hippocrates[2] knew something of the danger of drinking water which had passed through lead pipes, and even recommended the boiling and filtering of polluted water. The fall of Rome brought with it the destruction of the aqueducts and the general neglect of the entire subject of water-supply. The terrible ravages of pestilence during the Middle ages, it was claimed, were to a great extent due to the use of badly polluted water. General improvement was not made in sanitary matters until about the end of the sixteenth Century. During the seventeenth, eighteenth and the first half of the nineteenth Centuries progress, on the whole, was very slow, even in the cities of Paris and London[3]. For the last thirty years of the nineteenth Century, however, the development in all countries has been very great, and the rate of growth has constantly increased.

For many years the larger pipes were usually of wood, made by boring out logs to a diameter of 6 or 7 inches. Cast iron pipes came into general uses about 1800 in England,

When water first began to be supplied to each house in Europe the water was turned on for only a few hours a day. For sanitary reasons, and as a matter of covenience, the constant-supply system came into general use in London in 1873[4].

Filtration on a large scale was first inaugurated by the Chelsea Company in London[5], which in 1829 started the first large sand filte similar to those now in use. In the last 25 or 30 years the use of such filters has rapidly extended until now it is seldom to find an European city without any modern means of water-supply.

In the United States the first work for the supply of water to towns were those of Boston, built in 1652.

The first instance where machinery was used was at Bethelhem Pa.[6], the works of which were put into operation June 20, 1754. Wooden pipes were also first used, but it is stated by Charute[7] that cast iron pipes were used in Philadelphia as early as 1804. The growth in number of water-works in the U.S. since 1850, given in the "Manual of American Water-Works" was 83 in 1850 and 3196 in 1896.

In China the problem of water-supply never had been one of mportance until the last 30 or 40 years when people began to learn the, significance of a pure supply to the community. The thing that the quality of a water-supply never bothered us much is that we always boiled the water before we use it for drinking purposes. We alss learned the simple method of clarifying water by putting a few pieces of alum into the water storage tank in case the water is too turbid to use.

II.—Value of A Public Water-Supply

The most important use of a public water-supply is that of furnishing a suitable water for domestic purposes. The absolute necessity of a supply of some sort for such purposes in a large city is well appreciated, but the value of purity is, by many, not rated as high as should be. The transmission of certain diseases such a cholera and typhoid fever by polluted water is now universally recognized, and the value to a city of a pure supply when compared to one constantly polluted by sewage can scarecely; be overestimated.

A public supply of pure water is of great values not only in large cities, but in the smaller towns and villages. However, people in small places seldom pay any attention other than to get a goofd supply of water for fire protection and expect to depend on walls for drinking purpose as before.

Another highly important function of a water-supply is that o- furnishing the necessary flushing-water for a sanitary system of drainl age and fire protection and adequate soft water in place of hard welf water, a part of considerable importance to both domestic and industria users.

The commercial value of a good water-supply is appreciated only when one considers various industrial projects, which demand for their operation, large quantities of suitable water.

III.—Sources of Supply

Classification. The sources of water-supply may be classified according to the source and the methods of collecting:

1. Surface waters..
   (a) Water from rivers
   (b) Rain-water collected from roofs, etc.
   (c) Water from natural lakes.
   (d) Water collected in impounding reservoirs.
2. Ground waters...
   (e) Water from springs.
   (f) Water from shallow wells.
   (g) Water from deep and artesian wells.
   (h) Water from horizontal galleries.

The kind of water which a region can furnish depends on its climatic, geologic topographic features. Of all the surface waters, those drawn from lakes and rivers occupy an important position in water supply and will continue to furnish a large and increasing number of cities. However, they will not be as a rule very satisfactory, unless some methods of purification is applied. Ground water are as a whole of better quality from a sanitary point of view than surface waters, but in many cases they contain a great deal of iron and hardness, as impurities.

IV. —Quality of Water Supply

In securing a water supply for public or private use, the question of quality is of supreme importance. Pure water from the standpoint of the chemist is not to be found in nature; neither is it desirable that such should be furnished for general purposes, for the presence of certain salts in water makes it more palatable and better for use than distilled water.

The origin of all water-supplies is primarily to be traced to the rainfall. As the water vapor condenses in the atmosphere and falls to the earth surface, it begins to absorb impurities; and its whole history from the time it is precipitated until finally finds its way back into the air through evaporation is marked by the absorption of substances which pass into solution or are held in suspension, as well as the precipitation or elimination of the same or other ingredients. Some of these changes are harmless, others are of much consequence, depending upon the requirements to which the water-supply is subjected.

The keynote of sanitary science is to be noted in the relation that exists between communicable or transmissible diseases and public water-supplies. That disease producing germs may find their way into the human body through water and so possibly cause outbreaks of different maladies has been known from time immenserable. Of all the transmissible diseases, only a limited number are likely to be distributed through the medium of water. There are known as water-borne diseases in contradistinction to those that are disseminated through the air or find an entrance by means of wounds. The most important water-borne diseases to consider in this connection are typhoid fever and cholera. These are distinctively water-borne diseases; and while others, such as anthrax, gastro-intestinal catarrhs, dysentery, and malaria may be traceable to a similar origin, but these troubles are often so imperfectly defined that they are not with certainty associated with any definite specific organism.

V.—Need of Water Purification

It is evident that water contains, both soluble and insoluble, living and dead some of them benefical but most of them undesirable. However, it is still hard to impress people, the extent to which any of these effects our daily life. If this can be brought to the attention of the people by way of illustrations and figures, perhaps the need of water purification will receive a better attention than it used to. The first of these to be discussed here is the soap waste due to hard water.

Soap waste.. The precipitation of soap by calcium and magnesium salts is probably one of the most commonly recognized disadvantages of hard water. Suppose one gallon per capita per day is taken as the amount of water used for washing. And on this basis it was reported by the water commissioner of St. Louis, 1922[8] that the removal of 72 p.p.m. of hardness from the water of the city of St. Louis resulted in a saving of $920 worth of soap per day. He has estimated that in a town of 40,000 inhabitants using a water of 300 p.p.m. hardness a ton of soap is wasted per day. Buswell's[9] figure of 0.1 lb. of soap wasted per p.p.m. hardness per 1000 gallons of water is said to be very close to laundry experience. It is general conceded that the saving in soap alone is greater than the cost of municipal water softening. The other benefits of soft water are clear gain. The waste of soap is not the only objection to hard water, the lime soap tend to stick to the fabric and thus cause spots and stains on the laundered articles.

The second is boiler scale caused by hard water. An equally serious effect of hard water is the formation of scale by hard water when heated. Parr[10] stated that a conservative estimate would place the loss in fuel efficiency at 10% per each 1/16 in. thickness of scale. The report of the Water Service Committee of American By Engineers' Association for 1925 gives a saving per locomotive per year due to softened water as averaging $1000 for each 10 grains of hardness or $5.88 for each p.p.m. of hardness. Savings in localities where raw water is very bad have run as high as $8000 per locomotive per pound of scale forming material entering the boiler.

The third is hard water in special industries..

Textile Finishing: One of the chief disadvantages of hard water in textile industry is the formation of insoluble compounds of lime with the soap which must be used in many of the processes. The lime soaps coat the fibers and render the cloth stiff and hard and sometimes spotted.

Tanning: Lime and magnesium carbonates combine with

certain chemicals used, more especially the tannins, causing considerable waste. Iron is harmful since it causes stains and discolorations."

In addition to textile and tanning, for canning, fermentation industries, artificial ice, paper mills, glue factories, sugar refineries, starch works, concrete, various chemical works, etc., the quality of the water used is an important problem.

The fourth and the last is the high death rate due to polluted water; C.P. Hoover[11], superintendent of Columbus, Ohio, Water Works, in his paper on "The treatment of The Water Supply Of The City Of Columbus, Ohio", gives figures on typhoid fever death rate per 10,000 population, to be 70.3 during a period from 1904-1908 before filtration, and 3.2 during 1924-1926 after filtration. In 1908, the typhoid fever death rate reached 138 per 100,000 population, During the summer of that year there were 1500 cases of typhoid. There were five deaths from typhoid during 1926, three of which were resident and two non-resident. Assuming the population to be 285,000, would give the city a total death rate from typhoid of 1.7 per 100,000 a resident death rate of 1. per 100,000, and a non-resident death rate of 0.7 per 100,000. The death rate for New York City[12] was 13.5 during a puriod from 1906-1910 before using disinfectant and 2.9 during 1916-22 after using, a reduction of 79%. Cincinnati, Ohio, was 30.9 during 1906-1910 before using and 3.5 during 1916-22 after using, a reduction of 88%.

So you may see for yourself that the need of water purification is not one of those things that can be put aside for a whole, but is a thing of absolute necessity to the community; and I can produce a good many of instances from actual plant operations to show that such is the case.

VI.—Processes of Purification

Having discussed the need of water purification, the processes of purification will be the next thing to be brought forth for consideration. Different waters require a different methods of handling; and the qualities of the water depend much on the process selected. By water purification is meant totally or partially removing impurities from it or else so changing them that their presence is less objectionable. An illustration of the former is the removal of suspended clay by sedimentation or filtering; an illustration of the latter is the bleaching of coloring matter or the destruction of bacteria by sterilizing agents.

Plain Sedimentation:

Sedimentation is the cheapest and generally the best method of removing the coarser suspended matters. The finer the matter the longer it takes to settle. The time for complete sedimentation is very different for different waters, and to determine this period recourse must be had to actual experiments. For plain sedimentation a period of 24 hours' subsidence is about the minium limit adopted, but this will seldom give a clear water. A considerably longer time is often necessary to give acceptable results. At St. Louis, 24 hours is the standard, but often the results are not good. At Cincinati it is planned to allow at least three days, but here the treatment is intended as a preparation for filtration. At Louisville, Fuller[13] found that the economical limit of plain subsidence is about 24 hours, during which time 75% of the suspended matter is removed. There is always a reduction of Bacterial count in water after a period of sedimentation. Fuller found in his experiments that about 75% of the bacteria removed by three days' subsidence in Cincinati. The monthly results obtained by the Chelsea Water Co. of London in 1896 for 12 days storage were 97.85% average reduction by subsidence and 99.86% average reduction by subsidence and filtration. Results obtained by other London works ranged from 49-85% average reduction for periods of subsidence varying from 3.3 to 15 days.

Sedimentation with coagulation:

The use of alum for coagulation was developed long before any other chemical treatments of water. It was found in record that the Chinese from earliest times have treated water to be used in paper making with alum in order to remove the turbidity and undersirable coloring matter. Its application in connection with public water-supply, up to 1900, was almost confined entirely to their employment in connection with mechanical filters, and it was quite recently in a few places where sedimentation is the only treatment. If a water can be satisfactorily purified the greater part of the year by plain sedimentation, the use of a coagulant at other times as an aid in the process is well worth consideration. The rate of sedimentation depends greatly upon the amount of coagulant employed. It takes place much more quickly than where no coagulant is used, so that a large part of the action will occur in a few hours. Where the water contains large amounts of sediments, it will often be more economical to allow the coarser particles to settle before applying the coagulant. This will reduce the cost of chemicals and give a more satisfactory result. The efficiency is a function of the time, amount of coagulant and character of the sediment. The bacterial efficiency follows in a general way the efficiency with respect to the susnded matter.

Sand Filtration:

The first filter of which we have any record was established by Mr. James Simpson[14] in 1829 for the Chelsea Water Company of London. The chief object of this filter was to remove turbidity, and in this it was a success.

When efficient chemical methods of water analysis were derived about 1870[15] and applied to the subject of filtration, it was found that but little purification, chemically was effected by the process. The result was disappointing. After the establishment of the germ theory of disease and the application of modern bacteriological methods to water filtration by Prof. P.F. Frankland in 1885, the subject was put upon an entirely new and substantial basis; for it was found, fortunately, that the sand filter, although showing imperfect results from a chemical standpoint, ~~was~~ excellent medium for removing bacteria.

In the slow sand filter, as employed almost universally in Europe, the sand bed is constructed in large water-tight reservoirs, either open or closed, each having usually an area of from 1/2 to 1 1/2 acres. On the bottom of the reservoir is first laid a system of drains, then above this are placed successive layers of broken stone and gravel of decreasing size, and finally the bed of from 2 to 5 ft. of sand which forms the true filter. The water flows by gravity, or is pumped, upon the filter, passes through the underdrains to a collecting-well, and hence to a consumer. Water containing much sedment is usually first passed through settling-basins, where a large part of the sedment is removed. The rate of filtration is, maintained, inspite of the loss of head caused by friction, nearly uniform by suitable regulating devices, which vary the head according to the resistance. When the working head has reached a certain fixed limit of a few feet, the water is shut off, the filter drained, and the surface cleaned by removing a thin layer of clogged sand. The operation is resumed. The period of service is the time that elapses two scrapings of the filter. In practise it varies from a few day if the conditions are specially bad, to 5 or 6 weeks where optimum conditions prevail. The amount of water filtered between cleanings ordinarily ranges from 40-80 million gallons per acre of filter area. Clean sand is added from time to time to the filter in order to keep the thickness of sand almost constant. The rate of filtration does usually not exceed 2 or 3 million gallons per acre per day. Bacterial efficiency per cc. ranges from 98 to 99% or above.

Mechanical Filter:

This type of filter is commonly called the American filter or the rapid filter. This was first used for municipal supplies in 1885, and was first established on a sound scientific basis by the investigation of Geo. W. Fuller at Louisville, in 1895-1898, and by others during the next few years.

The water after has been treated with the coagulant and a great part of the suspended matter has settled out, the water now passes through a sand bed in order to free itself from any of such matters.

As the water passes through this at the rate of 100 to 150 million gallons per acre per day, or about 12 to 19 vertical ft. per hour, silt and coagulant collect at the top of the filter in increasing amount, until finally water passes through it too slowly for economical working, when the sand is washed by reversing the flow of water, the upward flow carrying out most of the silt and coagulant jelly, the dirty water being drawn off to a sewer. The standard plan is to arrange the filter beds on two sides of a central gallery, in which gallery are pipes for bringing the coagulated water to the beds, removing the filtered water, supplying filtered water and air for washing, and drains for removing the dirty wash water. They are all controlled by gate valves. The filtered water passes by gravity through the strainers at the bottom of the Sand bed into the filtered water main and in this flows to a clear water reservoir.

The rapid filters above described are known as the gravity type. A number of municipal and more commercial plants use filters. These act on the same principal as gravity filters, but the sand bed and strainers, being under pressure, are enclosed in a steel cylinder with a limited capacity about a million a day.

Aeration:

This is employed for adding oxygen to the water or for releasing gases. It is effected by passing water in thin sheets over weirs or inclined planes, by spraying in the air from fountains, by allowing to flow through perforated trays, to trickle through coke or other coarse grained filters, to flow down a channel filled with stones which violently agitate it, etc. Removing gases requires more thorough agitation than introduction of oxygen. The gases most commonly removed by aeration are CO_2, H_2S and certain hydrogen gases that cause odors in water. When iron is present in water as carbonate or in other easily oxidizable form, it can be removed by applying any methods of aeration, which renders the iron insoluble, when it may be removed by filtration. In certain case organic matter tends to hold the iron in suspension, even when oxidized; in which cases it may be possible to remove it by use of alum, sedimentation and filtration. Iron in the form of $FeSO_4$ has been removed by the use of lime, aeration, and filtration. Iron and manganese found difficult to remove by other methods have been removed by oxidizing in a coke filter used as a constant bed, then allowing it to settle and filtering.

Softening :

1. Soda-Lime process....Softening is effected by adding to the hard water lime in the form of calcium hydroxide and soda ash in the form of sodium carbonate. After these were added and thoroughly mixed with water, the latter flows to the settling basins. The mud and precipitated chemicals settle to the bottom of these basins. Just before the clear softened water leaves the settling basins, CO_2 gas in some plants, for instance, in Columbus, Ohio, U.S.A., is applied in order to neutralize any excess lime. After carbonation, the water discharges from the settling basins into the filter. As it discharges from the filters into the filtered water reservoir a small amount of chlorine gas is added to insure sterilization.

2. Zeolite process....In practice the Zeolite or Permutite (artificial zeolite) which is insoluble, granular solid, is put into a vertical tank of suitable size for the volume of water to be treated. This tank is equipped with the necessary pipes and valves for controlling the flow of water and for introducing and washing out the regeneration salt solution, sodium chloride.

3. Combining Lime softening with the permutite process[16].... This process has been patented by the Permutite Company in the U.S. The idea is to precipitate the bicarbonate of calcium with lime and leave in the water the magnesium and the sulphates, chlorides, etc. of

non-carbonate origin to be removed by zeolite. Zero hardness is attained with but much less sodium salt to the water than if the whole softening had been accomplished with permutite.

Disinfection:

The use of chlorine and chlorine compounds was first developed to disinfect sewage[17]. One of the electrolytic plants errected near London in 1899 was for this purpose. The first plant in U.S. for sewage treatment was at Brewster, N.Y., where in 1893, a small plant was errected to protect the water of Croton Lake, which supplies New York City with water. Houston and McGowan of the English Royal Commissioner on Sewage Disposal in 1905 were perhaps the first to use hypochlorite in water for Lincoln London. Disinfectant has taken a very important place in water treatment since 1910, when Major Darnall of Medical Corps. U.S. Army, demonstrated the practical use of chlorinc gas. The whole process of disinfection consists in subjecting the water to some sort of treatment that will kill a large part of the bacteria. It produces no clarifying effect or removal of suspended matter.

Three rather different methods of disinfection have been employed: (1) Chemical disinfectants, (2) actinic rays (ultra violet) and (3) heat. Chemical disinfectants are more generally used at present than the other two. A large number have been suggested from time to time but those which have found practical application are stated in the following: Chlorine, hypochlorites, chloramine, permangante, ozone, and "excess lime". In addition, iodine, bromine, chloramine-T (toluio-sulfochloramide) silver flouride, iodine tincture, and various other substances have been used in water disinfection for treating small quantities, for trips in the field, expeditions, etc., and of course boiling is always an effective though an impractical method when much water is to be treated.

The application of disinfectants is a simple matter. Some of them may be introduced into the water in the form of solution and others may be put in dircetly.

Lingnan Water Works, a combination of different processes:

The filtration plant of Lingnan University is of the gravity, rapid-sand filtration type with a capacity of 1,000,000 U.S. gallons daily. Two of the three filters are only equipped for present use. The raw water supply is secured from the Pearl River and is taken in through a trash rack and sluiceway to a settling basin or reservoir. By means of a tide gate, the reservoir will be filled automatically twice a day at high tide. With a capacity of over 1,000,000 gallons, the retention provided in the reservoir will allow the heavier grit and sediment to settle out as the first step of the treatment.

From the reservoir the water flows by gravity through an eighteen inch concréte pipe line terminating in a screen chamber and suction pit for the low lift pump. A coarse bar screen and a fine brass screen are provided. Two motor driven centrifugal units (0.5 million gallons daily) pump the water through a pipe terminating in a spray system, into a basin, aerator, with a capacity of 200,000 gallons. Aeration was considered necessary on account of occacionally high organic content and necessary removal of hydrogen sulfide, free Ammonia, free Carbon dioxide and iron in the water.

From the aerator basin the water flows by gravity to a mixing basin located in the filter house. Here it is treated on entering with principally alum and frequently alum and lime or soda ash, dependent upon conditions require for satisfactory treatment. Two dry feed chemical machines, capable of varying rates of feed supply the required amount of chemicals to the aerated water. A thorough mixing of the water and chemicals is secured by a motor driven prepella agitator. After a detention of 15 min. under maxium flow, the mixed water passes to the coagulating basin.

The coagulating basin is constructed in two units and provides a detention period of about 4 hours under maxium conditions. The two units can be operated in series or in parallel in either direction. The addition of chemicals to the water results in the formation of flocculent precipitates, which tend to settle out together with suspended particles and a great deal of bacteria. After a period of settling, the treated water passes to the filters. The filters located in the filter house, of the open type, constructed of concrete and consisting of a system of cast-iron underdrains, and above 18 inches of graded gravel and 30 inches of filter sand. After passing through the beds the filtered water passes through a pipe line to a filtered water basin located below.

The cleaning of the filters is accomplished by washing the beds with a reverse flow of water, A supply of wash water under pressure from the elevated tank is introduced into the underdrains and passing up through the bed washes and agitates the sand so that the waste material is carried over the wash trough and away through the drain.

Final sterilization of the filtered water is secured by treating with liquid chlorine as it passes to the filtered water basin. From the filtered water basin the filtered water is pumped by two motordriven centrifugal-high lift pumps (0.5 M. gal. daily) into an elevated tank of 30,000 gal. capacity, one hundred-twenty-five feet high, located above the filter house; or to the distribution system as the necessity requires.

Complete laboratory equipment has been provided for making all chemical and bacteriological tests and a close check will be kept upon the purity of the treated water.

Chemistry of Water Purificatin.

Coagulation with alum.

Alum for water treatment should be that known as aluminum sulfate or basic sulfate of alumina, containing no free acid. As it is added to water, it will react with the natural alkalinity of the water and give aluminum hydroxide as follows:

$$Al_2(SO_4)_3 + 3\,CaCO_3 \cdot H_2\,CO_3 \longrightarrow Al_2\,(OH)_6 + 3\,CaSO_4 + 6\,CO_2.$$

If it is used in conjunction with lime or soda ash the reactions will be

$$Al_2(SO_4)_3 + 3\ Ca(OH)_2 \rightarrow Al_2(OH)_6 + 3\ CaSO_4$$

$$Al_2(SO_4)_3 + 3\ Na_2CO_3 + 3\ H_2O \rightarrow Al_2(OH)_6 + 3Na_2SO_4 + 3\ CO_2$$

If sufficient alkalinity, natural or artificial, is not present to react with the aluminum sulfate, basic sulfate will form. Some of these are soluble, so that no coagulation may appear. The reactions using natural alkalinity, are:

$$Al_2(SO_4)_3 + CaCO_3.H_2CO_3 \rightarrow Al_2(SO_4)_2.(OH)_2 + CaSO_4 + 2\ CO_2$$

$$2Al_2(SO_4)_3 + 3\ CaSO_3.H_2CO_3 \rightarrow Al_2(SO_4)_3.Al_2(OH)_6 + 3\ CaSO_4 + 6\ CO_2$$

$$Al_2(SO_4)_3 + 2\ CaCO_3.H_2CO_3 \rightarrow Al_2SO_4(OH)_4 + 2\ CaSO_4 + 4\ CO_2$$

Any one of these reactions may take place, depending on the conditions. This accounts for the difficulty of obtaining coagulation sometimes especially in winter, when the reactions are slow and the coagulum formed will combine with sulfate of alumina still in solution. Under cold weather condition, this will pass through the filters in basic solution form. The reaction will complete in the clear water basin, causing the formation of minute specks of coagulum in the filtrate.

Hydrate of alumina, which does the work of clarifying water, is amphoteric in character, that is, it acts both as a base and as an acid. It may be ionized in two ways

$$\underline{Al(OH)_3} \rightleftharpoons Al(OH)_2^- + H^+$$

$$(precipitate) \rightleftharpoons Al(OH)_2^+ + OH^-$$

giving off one hydrogen ion as monobasic acid or one OH as a monoacid base. A second and third hydrogen are also given off to a slight extent and likewise a second and third OH ion. If enough H ions are added, the second and third OH ions will be removed from the aluminum hydroxide according to the law of Mass Action and all the aluminum will be dissolved in the form of aluminum ions (Al^{+++}). On other hand if the OH ion concentration is made large enough all the aluminum will go into solution as aluminate ion $[Al(OH)_2^+O^-]$. In other words, aluminum will dissolve as Al ion in solution with low pH and as Aluminate in solution of high pH. But somewhere in between it will be precipitated more or less completely. The point at which all Al $(OH)_3$ is most insoluble, is known as iso-electric point of aluminum hydroxide. In water work practice, different waters give a different pH value at which Al floc will be found most insoluble. As it has been found out both by laboratory experiments and plant practice the existence of a zone between pH 4.5-7.0, out of which a good floc is seldom obtained. This, however, is not of general application. In the majority of the Ohio waters,[18] for example, the optimum conditions of flocculation are found in the region of pH 7.4-8.4. On the other hand, the optimum range for coagulation of the Pearl River water was found between pH 6.5-7.0. In other words, every water has its optimum range of pH for coagulation and it is the job of the chemist in charge to find out just what that should be.

Acration (removal of iron)

Most soils contain iron, in the form of hematite, Fe_2O_3. As surface waters pass through the soil, containing organic matter, they are rapidly deprived, by the oxidation of the organic matter, of the free oxygen which they contain. In this condition, the organic matter will attack the insoluble ferric oxide, leaving it as ferrous oxide, FeO. This combines with CO_2 in the water to form soluble ferrous bicarbonate $(Fe(HCO_3)_2)$, which is carried off in solution It is removed along with calcium and magnesium in water softening. When water is not to be softened, the usual process is aeration and filtration. Aeration introduces oxygen into the water causing the oxidation of ferrous ion to ferric ion. At the same time CO_2 is released raising the pH and favoring the precipitation of the basic bicarbonate or oxide. The reactions as explained by Foulk[19] as follows:

$$Fe(HCO_3)_2 \xrightarrow{-CO_2} Fe(OH)_2 + 2CO_2$$

$$Fe(OH)_2 \xrightarrow{O_2} Fe(OH)_3 \downarrow$$

Softening

Soda-Lime process:

Dr. Thomas Clark[20], an English Chemist, by the use of lime developed the first rational system of water softening treatment for removing calcium and magnesium carbonates. Later another chemist, Dr. J. H. Porter, developed the soda ash treatment of water to reduce the permanent hardness, subsequently, the two systems were combined, and when necessary the two treatments are applied simultaneously. The Porter-Clark process, developed and perfected into what is known today, the Soda-lime process.

This process depends principally upon the relative insolubility of $CaCO_3$ and $Mg(OH)_2$. By the addition of the proper amount of lime and soda ash (Na_2CO_3) these insoluble compounds are formed and the "hardness" thereby removed. The chemical reactions are as follows

(a) With Lime.

1. $H_2CO_3 + Ca(OH)_2 \longrightarrow \mathbf{CaCO_3} + 2H_2O$
2. $Ca(HCO_3)_2 + Ca(OH)_2 \longrightarrow \mathbf{2CaCO_3} + 2H_2O$

3*a*. $Mg(HCO_3)_2 + Ca(OH)_2 \longrightarrow \mathbf{MgCO_3} + CaCO_3 + 2H_2O$

b. $MgCO_3 + Ca(OH)_2 \longrightarrow \mathbf{CaCO_3} + \mathbf{Mg(OH)_2}$

4. $MgSO_4 + Ca(OH)_2 \longrightarrow \mathbf{Mg(OH)_2} + CaSO_4$

(b) With Soda ash.

1. $CaSO_4 + Na_2CO_3 \longrightarrow$ **$CaCO_3$** $+ Na_2SO_4$.
2. $Ca(NO_3)_2 + Na_2CO_3 \longrightarrow$ **$CaCO_3$** $+ 2NaNO_3$.
3. $CaCl_2 + Na_2CO_3 \longrightarrow$ **$CaCO_3$** $+ 2NaCl$, etc.

Zeolite process:

Softening with zeolite consists in passing the hard water through an artificially prepared exchange silicate which has the property of taking out all the Calcium and Magnesium and giving up an equivalent of sodium salt. After some time this action ceases and silicate is then generated by treating it with a solution of salt, NaCl. An exchange is in the reverse direction. Calcium and Magnesium salts go into solution and sodium compounds enter into the solid zeolite. The process can be repeated almost indefinitely without loss of the original material. The water is softened at the expense of common salt.

With hardness of the water.

$$2SiO_2 \cdot Al_2O_3 \cdot Na_2O + Ca(HCO_3)_2 \longrightarrow 2SiO_2 \cdot Al_2O_3 \cdot CaO + 2NaHCO_3$$

$$2SiO_2 \cdot Al_2O_3 \cdot Na_2O + CaSO_4 \longrightarrow 2SiO_2 \cdot Al_2O_3 \cdot CaO + Na_2SO_4$$

$$2SiO_2 \cdot Al_2O_3 \cdot Na_2O + Mg(HCO_3)_2 \longrightarrow 2SiO_2 \cdot Al_2O_3 \cdot MgO + 2NaHCO_3$$

$$2SiO_2 \cdot Al_2O_3 \cdot Na_2O + MgSo_4 \longrightarrow 2SiO_2 \cdot Al_2O_3 \cdot MgO + Na_2SO_4$$

Regeneration of zeolite.

$$2SiO_2 \cdot Al_2O_3 \cdot \begin{matrix} CaO \\ MgO \end{matrix} + 2NaCl \longrightarrow 2SiO_2Al_2O_3 \cdot Na_2O + \begin{matrix} CaCl_2 \\ MgCl_2 \end{matrix}$$

Disinfection:

Liquid Chlorine. Of the various known disinfectants those which have been proposed for use in water treatment are with one exception, "excess lime", all strong oxidizing agents.

Chlorine gas liquified by pressure is now very largely used for sterizing water. Its germicidal effect results from the liberation of

25070

nascent oxygen is well known. The reactions will be represented as follows:

$Cl_2 + H_2O \longrightarrow HOCl + HCl$

$HOCl \longrightarrow HCl + O$

Besides oxidation there are three other probable reactions between chlorine and organic matter. It may substitute for hydrogen on a carbon atom thus[21]

$RCH_3 + Cl_2 \longrightarrow RCH_2Cl + HCl;$

It may add directly to an unsaturated compound thus

$$R_2C{=}CR_2 + Cl_2 \longrightarrow \underset{\displaystyle Cl}{R_2C}{-}\underset{\displaystyle Cl}{CR_2};$$

or it may substitute for hydrogen on nitrogen thus:

$RNH_2 + Cl_2 \longrightarrow RNHCl + HCl$. Any of these reactions might well be as effective as oxidation in disintegrating the bacterial cell. As a matter of fact gelatin and peptone are rendered insoluble by the action of chlorine, or bromine, and an analytical method has been based on this reaction[22].

Hypochlorites[23].

Hypochlorires are either in the form of Calcium hypochlorite, bleaching powder or sodium hypochlorite. The latter is more soluble and more stable. In reacting with water and with acids it forms hypochlorous acid from there on the reaction with organic matter appear to be the same as those of chlorine:

$$2Ca\begin{matrix}\diagup Cl\\ \diagdown OCl\end{matrix} \longrightarrow Ca(OCl)_2 + CaCl_2 \text{ (bleach to calcium hypochlorite)}$$

$Ca(OCl)_2 + CO_2 + H_2O \longrightarrow CaCO_3 + 2H_2O$; then

$HClO \longrightarrow HCl + O$ and

$CaCO_3 + 2HCl \longrightarrow CaCl_2 + CO_2 + H_2O$. Or

$2Ca(OCl)_2 + 4H_2O \rightleftarrows 2Ca(OH)_2 + HOCl + HCl$; then

$Ca(OH)_2 + 2HCl \rightleftarrows CaCl_2 + 2H_2O$.

$2NaClO \longrightarrow NaCl + NaOH + HClO$ (sodium hypochlorite)

Chloramine.

A number of investigators have obtained that the presence of Ammonia and nitrogenous organic compounds increased the bactericidal action of chlorine. This has been due to the formation of chloramine (NH_2Cl) according to the reaction

$Cl_2 + NH_3 \longrightarrow NH_2Cl + HCl$.

The chemical reactions between Cl and NH_3 have been extensively studied by Noyes and co-workers, and by Bray and Dowell[24]. The Cl in chloramine is in a similar state to that in hypochlorous acid, that is, it is an oxidizing and chlorinating agent. Its greater effect on bacteria has been accounted for on the supposition that it is more soluble than Cl and penetrates the bacterial cell more easily. It does not decompose spontaneously as does hypochlorous acid and therefore its entire strength is available for disinfection.

Ozone.

The production of ozone ($3O_2 \rightleftarrows 2O_3$) by a high tension discharge has long been known. Its strong oxidizing power is responsibile for its bactericidal property.

Ultra-Violet Rays.

The harmful effect of short wave radiations ($400\mu\mu$ and less) on living cells is well known, as is the mercury lamp or arc for producing these radiations. There seem to be two explanations of this action of ultra-violet rays. One is based on the fact that these rays produce detectable tracts of ozone and hydrogen peroxide in aerated water and the action is due to the oxidizing effect of these compounds on bacteria. The amount of compounds seem to be so trifly small, that it would seem more probable that there is a direct photochemical effect upon the dacterial cell material analogous perhaps to the action of actinic rays on silver salts or to the more familiar but unpleasant effect on the skin, sunburn.

Copper sulphate..

Copper sulphate has been used to prevent the growth of algea or other vegetations, that impart disagreeable odors and tastes to water storing in large reservoir for domestic uses. It was pointed out by Drs. Moore[25] and Kellerman of the Bureau of Plant Chemistry, U.S. Department of Agriculture that one part of copper sulphate to ten million parts of water is sufficient to kill some types of algea, and one part in one million is sufficient to kill most of the common type upon which experiments were made. No harmful effects will result from a proper treatment with copper sulphate. The entire amount added if not in excess of 1 p.p.m. would not compare unfavorably with the copper content in some of our most common foods: Almond, 36.8 p.p.m., egg white 7.2 p.m., figs 15.1 p.p.m., potato 2.8, p.p.m. of copper.

Iodine tincture..

This is not used on a large scale for treating water, but it is a good thing for campers and travellers to know that if one drop of ordinary tincture of iodine (7% strength)[26] is mixed wish one quart of contaminated water, the water so treated with usually be safe for drinking purpose after thirty minutes.

Importance of labortory control

After a plant for the purification of water has been installed, the problem of maintenance comes into play. The most important thing is not to keep every part of the machinery well oiled and valves in pertect condition. Of course, these are important, but not so much as compare to the maintenance of a water of the best quality. If there is no way of telling the quality of the water so treated, no matter how good is everything, the one idea of securing a safe pure and soft water is not yet realized. Therefore, in order to keep a plant at its highest efficiency and able to locate any discrepancy in any step of the purification process and to keep a close check on the purity of the water, a strict laboratory control is indispensable. It is a very known fact that the mineral contents bacteriological quality of water may vary even as much as every hour, this is especial true with surfare water, depending upon the weather conditions, chances of pollution, etc. The thing which can be depended on, to detect such a fluctuation on the character of the water, is nothing but laboratory control' which may be chemical alone, or chemical and bacteriological in a better equipped plant. Of all, this is the only means by which the purity of the water can be maintained. Unless this is done, the quality of the treated water is far from certainty.

Biblipgaphy

A. F. E. Turneeaure and H. L. Russel, Public water-Supplies, 1st. Ed., John Wiley and Sons, N. Y. (1907), 746 pp. References; Nos. 1, 2, 3, 4, 5, 6.

B. Chanute, "Annual Addrsss" Srans. Am. Soc. C. E., IX, p. 220 (1880); Turneaure, ans Russell, "Public Supplies 1st. ed. Ref. 7

C. Arthur M. Buswell, "The Chemistry of Water and Sewage Treatment", Am. Chem. Soc. Monograph Series (1928) 362 pp., The Chemical Catalog Co., N.Y. Ref. 8, 9.

D. Parr. "Fuel, Gas, Water and Lubricants, p. 116; A. M. Buswell, "The Chemistry of Water and Sewage Treatment." Ref. 10.

E. Charles P. Hoover, "The Treatment of The Water Supply of the City of Columbus, Ohio," Reprinted from Proceedings of Am. Soc. C.E., (Feb. 1928). Ref. 11.

F. Harrson E. Howe, "Chemistry In Industry," II, The Chemical Foundation Co., N. Y. (1927) 392 pp. Ref. 12, 20, 21, 22, 25, 26.

G. M. F. Stein, "Water Purification Plants And Their Operation," 3rd. ed., revised and enlarged, John and Sons, N. Y. (1926), 316 pp. Ref. 13, 14.

H. C. W. Foulk, "Geological Survey of Ohio, Industrial Water Supplies of Ohio, The Kelly-Springfield Printing Co., Springfield, Ohio, (1925), 406 pp. Ref. 15, 16, 18, 19.

I. Joseph Race, "Chlorination of Water," 1st. ed., John Wiley and Sons, N. Y. (1928), 158 pp. Ref. 17, 21, 22, 23, 24.

CHINESE ARCHITECTURE IN THE PAST AND PRESENT

Dr. ERNST BOERSCHMAN

This is an Illustrated Lecture Delivered before the Arts and Science Club of Lingnan University by Dr. Ernst Boerschman who is Professor of the History of Oriental Art at the Technical University in Berlin. Dr. Boerschman, who has specialized in the study of Chinese auchitecture, is the author of a number of large and magnificently illustrated works in this field, notable among which are Chinese Architecture (in two volumes), Chinese Pagodas, and Architecture and Landscapes in China. Pictures of unusual excellence from the last mentioned volume, shown during the lecture are not printed here, however the *discription* is so vivid that you can see the picture *between the words—editor*.

Ladies and Gentlemen:

I consider it a great honor for me to be permitted to speak here to-night as guest of this outstanding institution of learning in New China. May I express my sincere thanks for this privilege to you and especially to your chairman, Dr. Baldwin Lee.

The invitation to speak about Chinese architecture I accept with great pleasure, especially right here in Canton. This leading city of South China is situated at the focal point not only of the political currents of the day, but also of progress in all fields of culture and industry. Through the decisive change in the architectural appearance of the city, with all the impetus of last few years and the present day, Canton because actually a model for the cultural rejuvenation from Yesterday to Today.

It is important to remember the fact that architecture is most closely related to all activities of life. The art of building, including both architecture and engineering construction, is not something external. The art of building is a reflection of a people, of an epoch. It is just as possible to recognize a people, by its architecture, its residential buildings, its state monuments, its religious shrines, as to recognize a man by the external forms in his life, by his home, his clothing, his surroundings. The forms whether good or bad, have their counter-effect upon sentiments and acts, and are therefore themselves full of vitality. For that reason also the architectural form are an important part of the living spirit of a people and should be an expression of it. This expression, however can be found only by such artists as know their field thoroughly and at the same time have the knowledge and fine culture which are the possession of their whole people.

Like an individual in his own life, so also a people creates for itself thru the centuries of its development a characteristic personality, a culture. This issues in part from its inherent capacities but also on the other hand from a conscions and faithful expression of self. There is, however, always a clear-cut connection with tradition. Even if for the time being, the development appears to be as stormy as that now in Canton, the connection nevertheless with the traditions and the heritage continues to be of primary importance. A people should put every effort into going forward in the right paths while at the same time it remains firmly anchored in its own nature. For only straight lines of development, a firm charater, and self-assertion will lead a people like the Chinese, who feel as a nation, to the next high level. It is the duty of every people to express itself and to perfect itself in accord with its native traits. Each must create the best that is in him. This in turn leads to mutual respect among men and nations.

China has the good fortune to be able to look back upon a great tradition. The foundation principles and forms of expression of its high culture have been known to other people for long and been admired by them. The experts on China and her friends have brought these spiritual values home to Europe and America. You will permit me to mention the fact that Germany has held a leading place in this process. Even at times when in the conflicts of current Chinese history and culture there has been much which we have only with difficulty been able to understand, we have nevertheless found it possible to recognize the strong spiritual foundation which the great Chinese sages have given from earliest times to their rulers and statesmen and to themselves. I am confident that because of this firm cultural foundation China, even though it be again engaged to day in battles and struggles with others and with itself, will enter upon a new and lofty period of cultural bloom.

In order to understand this, we must attempt to grasp the living forces which are eternally effective. And if we take as our example "Chinese architecture," in which the culture must be faithfully reflected, then we must become conscious, especially in the confusion of the present day, of those great values which has been transmitted to the Chinese people during its great past. We Germans have been forced in most recent times to rediscover ourselves and to become conscious again of the strong roots of our own utility. Exactly in in the same way it is my wish that the China to which I am devoted thru study and by affection may experience a similar realization of and return to itself.

In the mirror of its architecture we can discover many characteristics which we can also discover from our general knowledge of your fatherland. Among these are the unity and breadth of Chinese culture, great perseverance, as well as creative verssatility, deligance, and the striving after harmony and rhythms. All these characteristies can also be found in the architectural forms: the Chinese hall, the house, especially in North China, the roof, although variously shaped and curved, the substructure under hall and house, wooden columns, the shape of the beams, of the ornamentation, of the altar, and of furniture, all are in their nature similar throughout China. Only in degree are there differences. But those characteristics held in common are so numerous that I see therein a strong conformity with the social homogeneousness which from earliest times has been the outstanding quality of the Chinese people and which by the late revolution of Sun Yat Sen was given even greater development. From the communal nature of families in farms and villages, and in small and large towns, sprang the feeling for justice, patience, and simplicity. In spite of all necessary differences between rich and poor, between the ordinary people and the leaders, and in spite of many negative examples, these qualities have always been a sign for the recognition of the Chinese community. It is possible to derive from the general picture of Chinese architecture exactly these traits of character, namely justice, patience' and simplicity, just as the individual nations in Europe in the same way reveal their innermost souls by the nature of their architectural forms.

In the case, however, of the architectural forms of the Chinese there seems to be an additional characteristic which is peculiar only to them. That is the close relation between architecture and nature. Theough the centuries this relation has been felt more and more intensely and found expression more and more clearly in the concepts and symbol of the sages and the popular religion. Finally there developed an adaptation to nature and an obedience to it and its laws so intimate that they are found not only in the spirit of the culture and its people, and in the social community and the nature of the government, but also directly in the forms of its architecture. This may he termed the "symbolical" in Chinese architecture, and it is a distinct quality which in this form is found in no other known people of the world, but which in china has been a permanent source of great strength. Innumerable symbols are at all times and in all places correctly used, in ornamentation as well as in the construction and placing of buildings in the landscape. We find those foundation principles of male and female and the endless dualism, so often pictured, of natural natural and spiritual forces. We find also the four points of the compass, the seasons, and the creative forces, frequently represented in pictures of godheads but actually with this was the system of number symbolism, by which nearly all resulting from natural and accurate deliberation. Closely bound up numbers were explained and which were used in buildings and decorations. There is probably not a single one of the smaller numbers which does not have its definite place in the spiritual explanation of buildings and works of art. Essentially, however, the favorable position of a house or grave, of a temple or of a whole city, yes, even of a landscape, stimulated the adjustment of experiences and symbols to organic union.

of the building with the landscape. This has been designated "Feng shui", a term which has frequently been ridiculed abroad, but which at various times has been investigated scientifically and as a result has been admired as a very beautiful and artistically a most fruitful method.

This close tie with nature, and also with ethical, political, and social forces, as revealed in all structures right down even to the simplest hut, where we find an accurate orientation in direction and in whose axis is hidden its family altar, has brought into Chinese architecturet as a whole and in its details, a very strong inner life and therefore also beautiful forms. In this respect China can allow herself to be compare,d with all other countries without fear and can in fact be said to march right at the front of all epochs usually known as classical, for instance Egyptian, Grecian, Roman, or the Gothic, Renaissance, or more modern styles. The great Chinese architecture of the old style which reaches right down to our own times, should be termed in every way classical. We still have today in all parts of the great expanse of China an inestimably large number of buildings and monuments of buildings and monuments of all kinds and of purest beauty reflecting the great spirit of ancient China, a spirit which even today is by no means dead. For you and all of us are still standing in the middle of the stream of traditions tied up in a thousand ways with a treasured past. There remains only the question: How should these possessions of a fully-developed culture be used and made fruitful at the time of the transition into the new age with its new problems, requiring of course new solutions? That the old is not yet dead is shown by the energy of the whole people, and its willingness to try new paths. But the pioneers of this new age, including you in this hall, ladies and gentlemen, have all been born with he old powers. You have obtained the best part of your education from your past. Now the present brings with it the duty to create something new on that foundation, without destroying the old.

Let us return to architecture. This then is our task: To find what was good and fruitful in the earlier forms, and what can and must be carried over into the new age. I shall return to that later. First you will permit me to point out thru a number of pictures a few main characteristics of Chinese construction, so as to arouse a general feeling for the greatness and beauty of Chinese architecture. The pictures follow the course of the trips which I made during five years thru fourteen provinces of China. This was between 1902 and 1909, when shortly before the end of the last dynasty, only a few foreign building had been erected in the interior of China. At that time the development of modern metropolitan cities could be observed only in the large port cities. Buildings according to the old Chinese style were still everywhere the normal thing.

I should like to make this further preliminary observation that the monumental buildings themselves very seldom possess great age. It is a strange fact that while Chinese civilization, notwithstanding the development which it has undergone, has nevertheless come directly down to us intact in its foundations these thousands of years, yet monuments seldom date can be traced beyond the Sung dynasty. Dating from the Tang period we have several examples, from the Han period a very few, and from earlier times actually only grave mounds exist as examples of the building art. These may at some future date give archaeologists perhaps important clues, but today are only honored memorials and very interesting curiosities. Beyond that one must rely upon the old literature and its descriptions of palaces and cities. But a correct picture of the true architectural forms cannot be obtained from them. At any rate the descriptions to point out the fact that at least in the early Han Dynasty a number of characteristics has already been worked out in the style found in later structures. Here also the main line of dovelopment has been broken as little as in the general culture.

When we therefore today look at these examples of Chinese architecture, we should remember that they are of a more recent date but are in direct connection with the classic period. Essential enriching of the building forms and a greater feeling for the monumental appear with the introduction of Buddhism shortly after the birth of Christ in the later Han Dynasty. But here also there resulted a far-resching amalgamation with the old Chinese style, so that the style which we now designate as old Chinese took definite shape only in the Tang and Sung Dynasties since when it has been is vogue up to our own day.

In the pictures I shall touch only briefly upon Pekin, which is sufficiently familiar to all of you, and which by the way is inexhaustible as a world city containing architecture of the highest order. Furthermore I can only touch upon the wonderful connections of religious architecture with landscapes and homes as attractive, a task as a more complete treatment would be. We shall confine ourselves to the monumental structures exemplifying religious architecture, because here we can most clearly observe the striving for style and the most most highly developed forms.

The pictures were all taken by me and are being reproduced tonight from my ilustrated work: "China, a Journey through Fourteen

Provinces", which is a volume of the comprehensive collection "Orbis Terrarum."

1. The great wall near Peking, north of Nankou Pass. The greatest piece of construction in the world, extending from the sea to the western border of Kansu Province. For this reason it is known as the 10,000 li wall. The first parts of this wall were begun at about the time of Confucius in Shansi Province. Having been frequently allowed to fall to ruin and then reconstructed again as a protection against the invading tribes from Inner Mongolia the present structure is in its main parts derived from the Ming Dynasty (1368-1644).

2. The Temple of Heaven in Peking, the most distinguished part of which is the great round hall for the petition to heaven at the time of the annual harvest celebration. It was constructed at the beginning of the Ming Dynasty around the year 1420. Its characteristic features are the round terrace on three levels made of white marble, the threefold roof of blue glazed tiles, the golden knob at the top, the bottom floor of twelve divisions colored red, and the vari-colored connections between the beams. The most impressive and unforgettable monument in the world, in its clear lines and colors.

3. A corner tower of the city wall of Peking of the later Ming period. Notice the strong lines of the four rows of firing holes, the two roofs, and the massive foundation, behind which are visible projecting bastions.

4. Somewhat west of Peking is standing the proud and beautiful Pagoda near the Westgate of about the year 1580 in the Ta Ming Wan Li era. It is the sister pagoda of a somewhat similar but much older one standing south of it and dating from arond 1100 in the Sung Dynasty.

The pagoda which you see, belongs to a special group of pagodas, which are distributed all over North China, but in the South are represented by only a few small examples. They may be recognized by the following characteristics: three main parts, a large foundation structure the main part of the structure in which relics are often reserved, and a superstructure of two or four, or six, eight, ten, or twelve quite narrow circular stories with cornices. The top of the Pagoda is brought to completion by a knob, often quite large, with the so-called fire pearl. In the case of this pagoda, 50 meters or 160 feet in height, there are twelve circular stories. Much terracotta, partially glazed, and stucco are used. This is a very excellent example of pagoda architecture, a subject which I am at present wording up in a special treatise, of which volume I recently appeared.

5. The great Buddhist Temple of the Buddhist vow terraces, in the Western Hills near Peking is one of the many temples and monasteries of which there are several hundred smaller ones and a number of rather large ones decorating the otherwise bare Western Hills. From here one looks down upon distant Peking. This is one of she most famous monasteries, with numerous terraces, pagodas and grave yards of priests dating back 700 years. It is supposed to have been founded in the Tang Dynasty.

6. This is the five-towered marble pagoda of the monastery of the jade clouds in the Western Hills of Peking. It is of the Ch'ien Lung period, 1750. It is the last division of an extensive temple. From the highest terrace one looks down right over the temple straight to Peking. You are undoubtedly acquainted with the fact that the body of your great Sun Yat Sen after his death in Peking, was ceremoniously

kept is state for two years in the upper chapel of this pagoda, until it it was finally transferred to Nanking.

7. This is the approach to the Ming Tombs, north of Peking. The two celestial columns and the Sacred Way, an avenue lined with the figures of animals. The impression of this grand entrance to the tomb temples protected by the northern hills, is tremendous.

8. Here we have the approach to a tomb temple in the western imperial graves of the late Ts'ing Dynasty. From a great bridge the so-called "holy way" passes through the dragon and phoenix gate with its three sections to the grave mound and the tower in front of the distant mounteins. It is a faithful carrying out of ancient Chinese motives.

9. One of the nine magnificent Buddhist Lama temples, which in a large circle surround the park at Jehol. This one is Potala, called after its approximate proto-type in Lhassa, Tibet, the seat of the Dalai Lama. Also the buildings in this park hold to the Tibetan type, a massive cube with several stories. Nevertheless the fundamental lay-out is quite Chinese, and the pavilions on the great top block of the wall almost determine the whole group of buildings as Chinese. It was built about 1760 under Ch'ien Lung.

10. The famous monastery of the living mounatin top, in Shantung Province, north of the holy mountain T'ei shan, south of Tsinanfu. Already started before the Tang Dynasty, this very beautiful pagoda dates from the Sung Dynasty, around 1000 A.D. Placed in a deep valley surrounded by high mountains this monastery is one of the most beautiful spots in China and guards rich memories. The pagoda shows distinctly outlined steps and low stories, and it may be climbed. In its upward striving line it corresponds to the character of the landscape.

11. The entrance to the extensive main temple of the holy mountain oi T'ai shan, in the city of T'aianfu. The threefold entrance gate thru the wall surrounding the temple and the richp'ailou in front of it with its outstanding relief work have come down from the Ming Dynasty. The temple and the city lie at the foot of the mountain in the valley.

12. The summit of T'ai shan with several of the temples on the topmost peak. The altitude is about 1000m. or 4500 feet. The temple at the left has been recently rebuilt. It is an apt example of the clear and simple conception in which such constructions are built. On the highest peak is the temple for the Jade Emperor.

13. The temple of Confucius, in Ch'ue fu, Shantung Province, the home of Confucius. At the corner of the main hall may be recognized the limestone columns with the dragon reliefs. Two dragons on each column are playing with the pearl. The year of erection was probably around 15?0 during the middle of the Ming era.

14. Shantung Province deserves special admiration because of its beautiful architecture in freestone. Stone p'ailous like this one in the city of Yenchoufu are exemplary in their construction and decorated with the best kind of sculpturing. It is of the Ming Dynasty.

15. May I go back to Jehol and the park there? This three-sectional bridge shows in its construction the the beautiful rhythm and the gracious vitality produced thru gentle oscillations, which always appear to us Europeans as the special characteristic of the Chinese garden style and as such unusually attractive.

16. The great Buddhist Holy Mountain Wu t'ai shan in Shansi Province. Here a total of seventy-two Buddhish temples, nearly all

of them Lama Temples, are concentrated upon a broad plateau at an altitude of about 2200 meters or 6600 feet. This upper plateau receives special emphasis thru the striking white Lama pagoda in the middle. It is sixty meters or 180 feet high and dates in its present form from approximately the year 1410 during the reign of Yung Lo in the Ming Dynasty.

17. One of the large monasteries of the Wu t'ai shan contains at the end a terrace with two massive buildings in Indo-Chinese style, Between them at the end point of the main axis a bronze pavilion, and in front of it five very beautiful bronze pagodas, partially gilded, each about six meters or eighteen feet high, of the Ta Ming Wan Li era, about 1580 A.D. In the picture only three pagodas are visible, of peculiar shapes, with very rich decorative work in ornament and figures. Together with the other two pagodas, these five bronze monuments represent the five peaks of the Holy Mountain itself, which surround this valley.

18. Shansi Province is known for its rich use of bronze and iron in very artistic forms. These two bronze lions stand in a temple of T'ai Yuen fu, the capital of Shansi. The eastern one with its opened jaws is considered a male, the western one with the closed jaws a female. In their bearing and perfection of technical execution to the very smallest detail, they are masterpieces of pictorial art in bronze; Late Ming Dynasty.

19. In Shansi Province, in the Mienshan Mountains, is an unusual monastery with its many closely crowded buildings placed in the cave right under an over hanging rock ledge. In spite of the cramped space there are delightful building and details in great number

which lend to the mighty mountain range a religious tone of great attraction.

20. A city gate of Sianfu, ancient Ch'ang An, the capital of Shensi Province, after having been newly reconstructed. This shows the great and sober lines of defensive construction in Chinese cities; but it shows also a fine artistic expression thru the rounded roofs of the tower halls.

21. This is one of the numerous thousand Buddha rocks in China. It is situated in Szech'uan Province. It was built in accordance with an Indian prototype, but these numerous groups, chapels, and figures were composed in Chinese severity and with added harmony and beauty. These rocks were shaped always in the most distinguished spots of the landscape.

22. This temple hall in Ch'engtufu, the capital of Szech'uan Province, displays great calm and nobility, elegance and beauty and withal utmost simplicity in the main body. The clear lines of the verticals and horizontals, the smooth surfaces of the roof, and the latticed windows cannot be emphasized with greater architectural effect.

23. One of the many suspension bridges made of bamboo ropes over pile piers customarily found in western China, and truly made to live by the swinging railing posts placed at regular intervals. But the nobility of the total structure is derived from the pleasing bridge pile in the middle and from similar bridge heads at the extremities, where religious chapels have been placed.

24. Similarly in western Szech'uan the traveller passes many such attractive village entrances and places where a lively but simple panelling and smooth roofs, though at times overly-ornamented, provide

an effective relief to the spacious, graceful tower which is an incense pagoda wherein the smoke rises and at the top escapes thru the mouth of a three-legged turtle into the atmosphere.

25. The capricious and wilful art of the Szech'uanese is very noticeable in contract to that of other provinces and may be seen even in the memorial p'ailous, whose red sandstone allows of the fantastic The architectonic severity which is so characteristic of all elaboration of the points of the roof and the centre pieces.

26. A similar striving appears in the upward curving corners of the roofs which are constructed with great difficulty and may be found for instance in theatrical stages,

27. These theatrical stages also have wonderful long curved lines and very rich ridges on the roofs which correspond somewhat to the South Chinese ones, especially when, as in this example, their joyousness is heightened by the use of colored porcelain pieces.

28. The monumental great hall of the main temple near the southern Holy Mountain, Hengshan, in Hunan Province, proves that in spite of the upcurving corners of the roofs in the South Chinsee style the monumental form of expression of this type of hall is not confined to the north. Right here in Canton we find further proofs of this in several of the old halls for instance of the Hoit'ung ji in Honan.

29. In this connection it is instructive to show at least one picture of the old and yet ever new Canton, the Ng Chang Lau, or five story pagoda as it appeared in all its glory in 1909, with its horizontals and verticals and the only slightly curved corners. The modernisation of this tower-like hall was found necessary in most recent times, but it

has most happily kept its old character intact although a new, modern note has been introduced.

30. The architectonic severity which is so characteristic of all of Chinese architecture in spite of much grace and imagination, is shown even in the pagoda of the drogon blossoms. Shanghai, one of the most perfect examples of gallery pagodas with their exterior galleries and parapets. It has a substructure and six upper stories, which progressively become lower and narrower. These stories are so ightly fitted out with overlaid roofs curving upwards that the lofty point with its division into rings and the three-fold fire pearl appear as a natural rounding out of the whole structure. The building has been repeatedly reconstructed, but in its first form is supposed to date back to the earliest times, long before the Tang Dynasty.

31, Restraint and thoroughly modern elegance are shown in the inner room of the well-known library at Sai Wu, the West Lake at Hangcho. The feather-like decorations on the cases and the engulfed network of the parapets introduce that quality of life which one expects in true book learning.

32. In conclusion there is a bridge of the island of P'u t'o shan, the holy island of the goddess of mercy, near Ningpo, to which I devoted my first large published work. It is of the Tao Kuang period early 12 century, and exhibits a noble clarity in the surfaces and parapets with their restrained decorations. On the insides of the parapets. however, very lively reliefs have been worked out.

33. In the same temple on the terrace of the main shall, the tablets of the parapets bear the oft-repeated twenty-four representations of filial piety, which in this case are of exquisite attraction. The boy

is embracing the tombstone of his mother, who had always feared thunder storms. The composition in its artistic and heartfelt conception and execution cannot be matched. The clearness and depth which we feel in this simple and heart-touching pictorial composition are indicative of all of the old architecture of China; they belong to the Chinese character, and it is for this reason that we end this series of examples of monuments of architecture with a masterpiece from the sister art, sculpture.

Ladies and Gentlemen! These few examples from the limited field of monumental and essentially religious architecture of the highest order, which I have been able to show you may be easily duplicated by you from your own knowledge, especially if you include the large field of dwelling houses and official and other public buildings. Everywhere you will find that your ancient art created very practical, beautiful, and comfortable rooms and courts, that, looked at from a purely architectonic point of view, the architectural forms, whether rich or simple, North Chinese or Southe Chinese, are complete and must in their noblest forms be given the designation of classic.

Now in the last decades the picture has changed. At about the turn of the century more than thirty years ago, there was released throughout the world an impetuous urge for a reconstruction, a renewal of everything. Political and economic developments parallel the striving after new paths in the cultural field. This is true of all nations of the world. However, China has taken an especially intensive part in the search for a new era. In fact, by its revolution of twenty-two years ago, China inaugurated this new era with a definite event. In Europe it was necessary to have a world war and its decisive results first, before the philosophy of life and the general culture were definitely influenced. As the greatest motive force, the national and nationalistic element arose in the peoples and it strengthened the will for self-assertion and for the development of their own natures. This tendency even received in many cases a clearly religious note and thus heralded truly a new era.

Coincident with this spiritual change the outer forms also changed, a thing which may be observed in all lands, proving the fact that forms of art, clothing, utensils, and creations of all kinds are a necessary part of culture. In the forefront are the forms of architecture, from the dwelling up to the expression of highest political and religious thoughts. Of course, in architecture the change does not take place overnight;; a visibly longer period of time is necessary than for instance in the case of changes in clothing fashions. It is therefore especially instructive to take account of the guiding principles and the aims of such a change in architectural style.

It is of course true that changes in architecture are also in a degree subject to style, for it is not possible to look continuously at the same forms, as they wither away, like everything organic. Thus with us in Europe the architectural dress has been changed from time to time, but such changes always came slowly. Also in China this has happened repeatedly, except that the change up to now has always taken place as part of a process of organic evolution. Now however the process of fermentation may be found far back. As early as forty or fifty years age European houses in the foreign settlements were set up as examples for imitation.

In the beginning this was only child's play. Later however,

real problems arose, for instance in the building of warehouses, office buildings, and factorties, which could not be solved by the existing Chinese means. The necessity of taking on foreign forms was therefore forcefully thrust upon the Chinese. Then the new customs of those Chinese who had long been abroad were also of influence in producing a change. Finally there became effective especially after the revolution the desire for a change of form, and thus the European style was introduced into private dwellings and into the whole process of city reconstruction. These cities, with their narrow streets and their confining walls, were actually no longer adequate for the modern requirements of greater traffic and the new technology. And thus destiny had its way.

As far as the development of old Chinese cities is concerned, the development in Canton has been and still is an especially stormy one. The impression which Canton makes upon the stranger who sees these new buildings and improvements for the first time, is tremendously over-powering. I personally have just had this reaction when upon entering Canton recently I was placing my foot on Chinese territory again for the first time in tweenty-four years. You will therefore permit me to take as my example for observations about modern tendencies, our own city of Canton, which I have carefully studied in the weeks of my sojurn, with the kind and generous assistance of the Chinese officials. I shall bring before you only a few of the most essential conclusions.

It is safe already to make the following statement: The reconstruction and enlargement of Canton belong to the greatest accomplishments in the field of city architecture. This is true not only in view of the future and its problems, but actually in view of the fact that the great city with its teeming millions was hopelessly contracted because of the narrow streets, and the limitations resulting from the city wall and from the river in the south. The thorough radicalism of the young, victorious revolution quickly made a complete job of tearing down the walls and building broad streets in their place and along the main thoroughfares. Friends of the old art will naturally regret the going of a romantic past, but the practical person of vitality will recognize that it is of primary importance to live, and of only secondary importance to guard the values of the past. In the first years creation continued according to the early standards, and in the last eight years the great plans have matured. These included a complete network of streets, a new broad street along the riverside, the Canton Bund, residential and business sections according to an exact plan, development of suburbs, shore-line and harbor construction, industrial centres, imposing building layouts for the government of the city and province, for new universities and for an outer harbor in Whampoo. The energy displayed in the execution deserves admiration because we and especially the technicians know the endless troubles involved in the opening of streets. We have in Europe hardly a single example of such reconstruction of a densely-populated city. Neither does America have such an example, as far as I know. We in Europe and America incorporate new sections into the city, but limit changes within the city to the minimum. The ability and energy of your leaders are deserving of much admiration.

What is the nature, however, of this new architectural appearance of the city? About this a great deal might deal might be said.

It is evident that for the time being a very bewildering picture presents itself. The architecture even on newly completed streets is extremely uneven and most unhappily mixed, partly with old Chinese but mainly with foreign European motives. The latter unfortunately all taken from an insignificant degenerated so-called colonial style, which as a result of a few protoytpes was developed here in the Far East and which slyly found its way into the new buildings of Canton. Furthermore Canton's architectural development has been quite irregular. The ruins of buildings half torn down and temporary structures are also very evident and certainly do not help to improve the picture of city. One must in justice consider all this as an evidence of the period of transition. It is certainly ture that especially the main street will soon show greater unity as a result of improved economic conditions. But no effort should be spared to avoid the earlier mistakes and to require the building of more presentable houses by Landowners.

A good start has in my opinion, already been made in the development of a new style, which must, even in the frame of European forms of architecture, of course by thoroughly Chinese. There are a number of streets with private homes and a few with single villas placed in gardens which have in a most fortunate manner developed old Chinese forms and brought them into harmony with modern conditions. The entrances of such homes in a number of cases remind us of old Cantonese motives and in this way carry over the architechtural tradition. Here is the direction in which also a new style will be found for stores and arboured walks, two predominant aspects of the city today. Here we have proof that it is entirely possible to combine old of new in an absolutely practical way. I myself have seen in a few examples that it is possible to use successfully the beautiful, agreeable, and refined forms of old China in the interiors of homes and business houses. It is not necessary to fit everything in the interior to the meaningless surfaces and lines of the so-called modern style. In Europe we have now abandoned this view.

A special chapter could be written about the monumental buildings in conscious imitation of the old Chinese palace architecture. Lingnan University has several good examples, especially in its most recently completed hall. This building exhibits in a most masterful way the play of the old Chinese lines, surfaces, and colors. The Canton government is responsible for an even larger number of beautiful buildings in this style. I should like to mention especially the Sun Yat Sen Memorial Hall, the library on the Goddess of Mercy Hill, the Gun Yam shan, various schools, and the recently completed city library. This library is a very jewel in form, in the shape of the ground plan, and in its most practical arrangement. The practice of the Canton government should be hailed. We are obtaining by it models to which the great buildings of the future will in part have to conform. The great city hall and several halls of the New Sun Yat Sen University will be significant monuments in this direction. We should not be deceived about the fact that constant imitations of the old masterpieces of Peking belong to that field of historical architecture which we in Europe kept in vogue for a long time, but then abandoned because the too exact repetition of old forms leaves no room for freedom of artistic expression in the spirit of a new age. The organic development of the old type of architecture is taken care of by the new building program being followed for instance for the himself to the given conditions.

In connection with this I should like to say a word about the preservation of the buildings of an earlier age. It if of course true that Canton did not have an especially large number of such buildings in the classic style. Many ef these have already been sacrificed in the rebuilding of the city. Yat quite a large number are still standing. There seems to me to be an urgent necessity for preserving these transmitted pieces of architecture partly because they have accompanied the development of the new generation. If we should not preserve them the historical forms of the public buildings in the old tradition would, after the complete disappearance of the old architecture, lose their meaning, a comparison with the old protoypes would no longer be possible, and the continuous line of development would be broken. All cities protect their heritage. Peking in this respect is in the lead and now because of its art treasures has become one of the main places of interest in the world for sight-seers attracting many travellers, and directly receiving a large income from this careful protection of its historical art monuments.

Canton could adopt a similar course. Here the White Cloud Mountains. (Bak wan shan) still rise up in the north as of old, the Goddess of Mercy Hill with its Five Story Pagoda (Ng Chang lau) and the new attractive memorial column for Sun Yat Sen still is at the head of the city, the pagodas and several large temples are standing as in the past. You deserve to be proud of your great past and should perserve these monuments. In this way you will prepare for a greater future clothed in new form in which however the old China with its culture of several thousand years will still be recognizable. I congratulate Canton for having taken hold of the task with such great energy. May the new architecture also be in good hands so that the living germ plasm of a great past may bear in its new soil good fruits a thousand-fold, and give strength to the eternally youthful Chinese people!

NEW EQUIPMENT FOR NIGHT AVIATION.

By Dipl.-Ing. A. Wagner and Dipl.-Ing. M. J. Luber,
Siemens-Schuckert, Industry Department.

THE provision of special navigation signs for the hours of darkness is quite as necessary in the case of aviation as it is for shipping. For a number of years past the conditions to be fulfilled by such equipment have been the object of much observation and study, as the result of which considerable progress has recently been achieved. Following on a report published elsewhere, it is proposed in the present article to describe several new types of Siemens-Schuckert apparatus and to explain their action in conjunction with their underlying physical principles.

In April 1930 the sub-committee for aviation lighting of the International Commission on Illumination met in Berlin. During this session, which was attended by the representatives of 13 States, the following equipment was recommended for the the illumination of aerodromes and the nomenclature of the appartus decided.

Air-port beacons, preferably in the form of high-power rotating searchlights

Boundary lights, for the demarcation of the landing ground

Landing lights, for the provision of as far as possible uniform and glare-free lighting of the landing groung

Obstruction lights, to reveal the position and magnitude of obstructions,

Wind direction indicators, to indicate the direction and as far as possible the strength of the wind.

For all these purposes Siemens-Schuckert have developed new types of apparatus.

Air-Port Beacons for the Aerodrome.

It is essential that a beacon be erected in the vicinity of the aerodrome to indicate its direction and location to the pilot. For this purpose the same type of rotating searchlight is generally employed as is used on the airways (Fig. 1). The optical equipment, which consists of an accurately ground parabolic glass reflector, in the focus of which is a filament lamp, is maintained in constant rotation by a small electric motor. A second lamp is provided as a stand-by and is automatically brought into circuit in the focus of the mirror in the event of the first lamp burning out.

Fig. 1. Rotating searchlight Type G1 60-60 with automatic lamp-changing device (Air-port beacon).

As the result of experiment and observation extending over a number of years in Holland and Germany it was found that the distance at which a beacon is visible is not only dependent on its luminosity but on the duration of the impression of the light on the eye. The rating of a beacon, also known as its "light value", is thus a function of the luminosity and the duration of the flash. The presumptive characteristic of this function is illustrated in Fig. 2; more

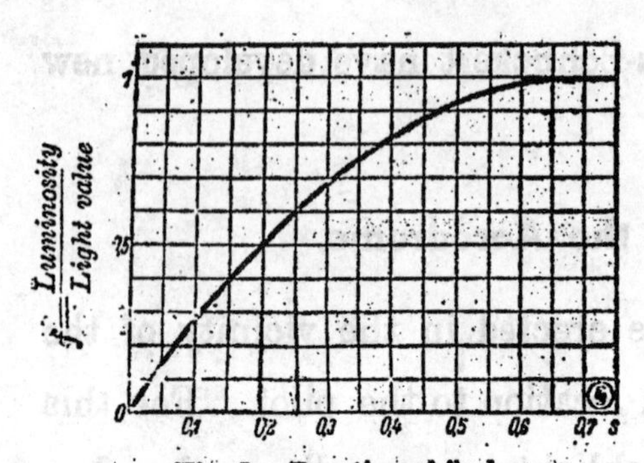

Fig. 2. Duration of flash

Fig. 2. Light value of a beacon as a function of the duration of the flash (presumptive characteristic).

exact results may be expected as soon as the extensive obseration data collected by the German Administration of Coastal Signals have been evaluated.

An easy means of increasing the light value for a given constan, value of the luminous flux was available in the shape of the ground horizontal diffusor. This diffusor is fitted to the beacon in place of the front glass; its effect on the characteristic curve will be seen by reference to Fig. 3. A comparison with the curve in Fig. 2 shows in this particular instance an increase of the light value of about 5%.

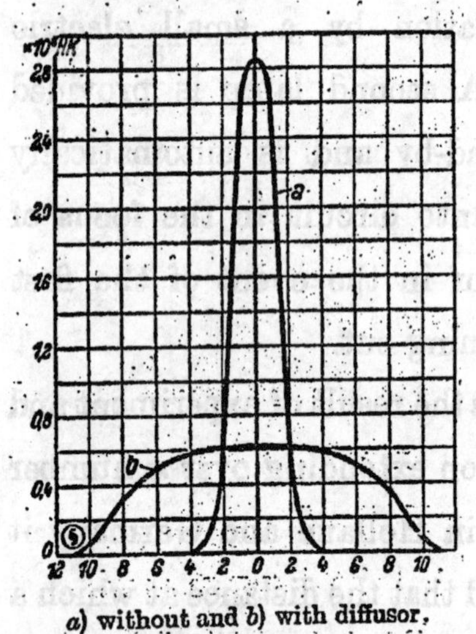

a) without and *b*) with diffusor.

Fig. 3.

Horizontal light distribution of the rotary air-port beacon Gl 60/30

It should be remarked at this point that very good figure for long distance visibility were also obtained with beacons having a relatively short flash. Navigation has been found possible from a distance of over 60 miles in fact not infrequently from over 100 miles.

The beam of the beacon can be directed upwards to an angle auf 90° and downwards to 20°, thus rendering the beacon serviceable for all general purposes such as illuminating obstacles and for signalling.

25102

Boundary Lights for Aerodromes.

A very essential requirement is that the boundaries of the flying ground should be clearly demarcated. In this respect the lead was taken by Germany, which equipped the Tempelhof aerodrome in Berlin in examplary fashion with Neon-tube boundary lights in 1926. The present form of boundary light is illustrated in Fig. 4; the cast-iron box contains the transformer for the luminous tube, an automatic cut-out, as well as a terminal board for the eight connections of the incoming and outgoing cables. The white enamelled reflector increases the brilliancy of the tubes considerably.

Fig. 4. Neon boundary light Type UFN.

A simpler and less expensive type of boundary light is obtained by the use of ordinary filament lamps. Another type of boundary light the mantle of which is enamelled yellow or red-and-white, and which can be easily picked out on the grass in daylight. The cone forming the mantle rests loosely on a low carrier ring and thus constitutes no danger in case a plane should collide with it in landing.

The boundary light imitates the outward form of the Noen light. A glass tube of approximately 6 ft. 6 in. length and coloured a light red contains a number of soffit lamps; the red-and-where reflector increases the luminosity and is also easily visible during the day. The two supports are sharply notched to ensure their immediate collapse in the event of a collision.

Obstruction Lights.

All obstructions in the danger zone of the aerodrome must be distinguished by red lights. These consist in part of filament lamp and partly of Neon lamp beacons. Particular attention must be paid to the illumination of smoke stacks for this purpose. Neon lamps have been found very serviceable; the tubes were fixed at a distance of 16—23 ft. below the top of the chimney, a ladder and platform being provided for occasional attendance.

Quite recently, the new high-capacity Neon lights which are artificially heated by electrodes, are coming into use; for currents of 1 to 20 A their luminosity is much greater than that of the lights used hitherto. Fig. 5 illustrates a type of the new light such as is used as a flash-light for a landing ground.

As an interesting novelty mention may be made of the Neon light beacon on a large gasometer in the neighbourhood of the aerodrome at Munich. In view of the inconceivably disastrous consequences of a collision with this gasometer, which contains nearly 3½ million cub. ft. of gas it was essential that the warning beacon should not be too small. The equipment therefore comprises twenty Neon tubes of over 8 ft. each in length, consuming 4000 VA in all. These lights as well as four smaller beacons on chimneys are switched in by relays from the central control station of the aerodrome.

Fig. 5. High-power Neon light Type H&F 2.

Aerodrome Floodlights.

In evolving a suitable lighting scheme for the landing ground the Engineer is faced with the following task: A great amount of light must be distributed over a horizontal surface, economic use being made of the source of light, which must not be higher than about 10 ft. above the surface to be illuminated, whilst no appreciable light must be visible above the horizontal.

Fig. 6. Portable aerodrome searchlight Type FG 90 with parabolic reflector for illuminating the landing ground.

These conditions can be met either by parabolic mirrors or by dioptric girdle lenses.

The searchlight type FG 90 (Fig. 6) has an accurately ground parabolic glass reflector of 90 cm (approximately 35·5 inches). The converging beam of light thrown by the reflector is drawn out parallel by plano-convex diffusing lenses, which simultaneously form the frontal glass of the searchlight. These lenses are likewise ground and polished with a high degree of accuracy. The total diffusion of the floodlight is 80°; by using two floodlights in combination the angle of illumination can be increased to any desired extent.

The source of light is a filament lamp of 5 kW, the filament of which forms a narrow horizontal rectangle, so as to produce a minimum vertical diffusion of the searchlight. This diffusion is accordingly about 2·5° only.

These floodlights, are frequently mounted on carriages and used at different points of the ground according to the direction of the wind (Fig. 7). It is better, however, to use a sufficient number of stationary floodlights which are connected up to an underground distribution cable. Only those floodlights should be switched on which point against the direction of the wind at any particular time, so as not to dazzle oncoming pilots. Arrangements must be made to operate the whole of the floodlights from the central control room of the aerodrome.

The arrangement of a number of stationary floodlights has the great advantage over one portable light that it is possible to follow a

Fig. 7. Aerodrome illumination by searchlight on automobile.

sudden change in the direction of the wind immediately; the bringing into position and readjustment of the portable flood light invariably occupies considerable time.

Both floodlights are preferably equipped with a lamp-changing device which switches a spare lamp into circuit immediately on the failure of the original lamp.

Fig. 8. Wind direction indicator Type WAN with Neon tubes.

Auxiliary Lights.

For days when the visibility is particularly bad it is advisable to have a powerful searchlight available. Its source of light is an arc lamp; the lighting intensity is about 25 to 250 million candle power and more, according to the diameter of the reflector.

With the aid of this powerful beam it is possible to transmit clearly visible and unmistakable signals to airmen who may have lost their bearings in the fog. It is also possible to illuminate brilliantly any unusual obstacle on the landing ground, such as a damaged airplane, so as to prevent further accidents.

Wind Direction Indicators.

Increasing use is being made of T-shaped wind vanes (Fig. 8) to indicate the direction of the wind. The rotating upper part turns into the direction of the wind and shows the pilot by its characteristic form the direction in which he must land. The vane is illuminated at night by red or blue Neon lamps or by tubular filament lamps, so called Wolfram lamps of 3 ft. length.

Recently, these wind indicators have been equipped with an anemometer with minimum contact and flicker switch. When the wind velocity drops below a certain adjustable value, a contact is closed which sets the flicker switch in operation. The lamps on the wind vane then no longer burn continuously but flicker in and out rapidly. This tells the pilot that there is very little wind on the landing ground and that he can approach from the most convenient direction, regardless of the direction of the wind.

Current Supply.

The electric apparatus of the majority of aerodromes is connected to the public supply. Since the flying services demand the hightest attainable degree of reliabitity several aerodromes have been equipped with emergency lighting plant which automatically comes into operation in the event of failure of the public supply and takes over the lighting without attendance of any kind. The prime mover is a Diesel or petrol engine. The set is electrically selfstarting, including the opening of the fuel and water cocks. The connections are such that when the emergency set is in operation the aerodrome equipment is completely disconnected from the public supply.

Plant of a similar kind, but for lower outputs, is employed for supplying beacons on the airways as well as coastal beacons. These sets work both as emergency sets, which only come into action on the failure of the main supply, or as primary supply sets for their respective beacons.

美國華盛頓大鋼橋

THE GEORGE WASHINGTON SUSPENSION BRIDGE

王叔海

美國華盛頓大鋼橋乃世界鋼橋之最巨者;其工程之設計及實施,萃集美國專門名家主持之。設計之內容頗複雜,非短篇所能盡。是篇之作,非敢研究其設計,唯着眼其數量,俾讀者感覺其偉大而已。

位置及尺寸 LOCATION & DIMENSION

斯橋橫跨哈遜河,連接紐約州及紐乍溯州(New Jersey),東端起自紐約州曼哈頓島之華盛頓壘,(Fort Washington, Menhatten Island)西端達紐乍溯州之李壘(Fort Lee. New Jersey)(參看第一圖)。

橋身長三千五百尺,兩端橋坡長約四千尺。共長七千餘尺。橋整兩座高六百五十尺。橋面(Bridge Floor)高出水面二百十五尺,最大輪船亦可在其下通過(參看第二圖第三圖)一九二七年五月動工,一九三二年十月落成,需時五年,用款六千萬。(美金)(約合大洋二萬五千萬,等於我國中央政府去歲全年收入三分之一)。開幕日有汽車六萬輛經過,而馬只有一匹,亦一有趣之事也。

文中所言尺寸皆係英尺。

第一圖

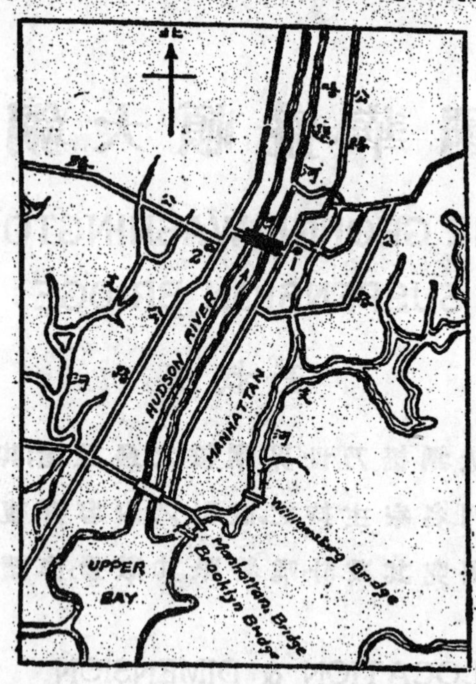

橋式　TYPE CF BRIDGE

橋爲吊橋式(Suspension Type)。兩旁各有鋼纜兩條共四條。兩橋墪上各置巨輪四。纜承輪上。纜端則繫於兩岸石穴中。兩端之間,橋纜垂空成Caternary式之曲線。全座橋身以鋼質「掛繩」Ssupende ropes數千條掛於四纜之下。橋之重量胥爲橋纜與橋墪所担負(參看第二,三,六圖)

普通鋼橋需用支柱甚多,形如蛛網既損觀瞻,又碍交通。斯橋則結構簡單對於此種支柱一掃而空之;只具橋墪,橋纜,掛繩,及橋面而已;在交通及美術上均凌上乘。

第二圖

橋墪之基礎

橋墪有二，一在東岸，一近西岸。橋墪分爲上下兩部。自橋面而下至河床磐石，稱爲橋礎（Foundation）。自橋面而上至承纜巨輪，稱爲橋塔（Bridge Tower）。橋礎之結構，下部離而爲二，將達橋面則合而爲一，成拱形。橋塔之結構亦如是，其下部（在橋面之上）離而爲二，近頂處合而爲一，成拱形。參看第二圖）

西岸橋墪建于水中，故先探其下之地質，至七十五尺深始達磐石。西岸橋礎卽由此建起，以圍堰法築之（Cofferdam Method）。先以挖泥機挖去是處浮泥，計二十日挖去砂泥七萬五千立方碼。繼以木板造成無底方堰，以鋼框緊束之，放入水中，在此堰之上又接造第二堰，如是每段從上添加，陸續放下，至河底爲止。此堰自河底磐石，上達水面，露出十餘尺，共高九十尺。堰之上部略小，下部略大；自磐石至水面之下十五尺，堰長九十三尺，寬八十九尺；自水面之下十五至水面之上十五尺，長八十四尺半，寬七十六尺。橋礎下部離而爲二，故須兩堰以爲基礎；兩堰之中線距離爲一百五十三尺半。木堰既成，卽將其內之水抽乾；塡以鋼筋三合土；凝固後卽成巨大石柱兩座，有如中流砥柱，此卽西岸橋墪之基礎也。

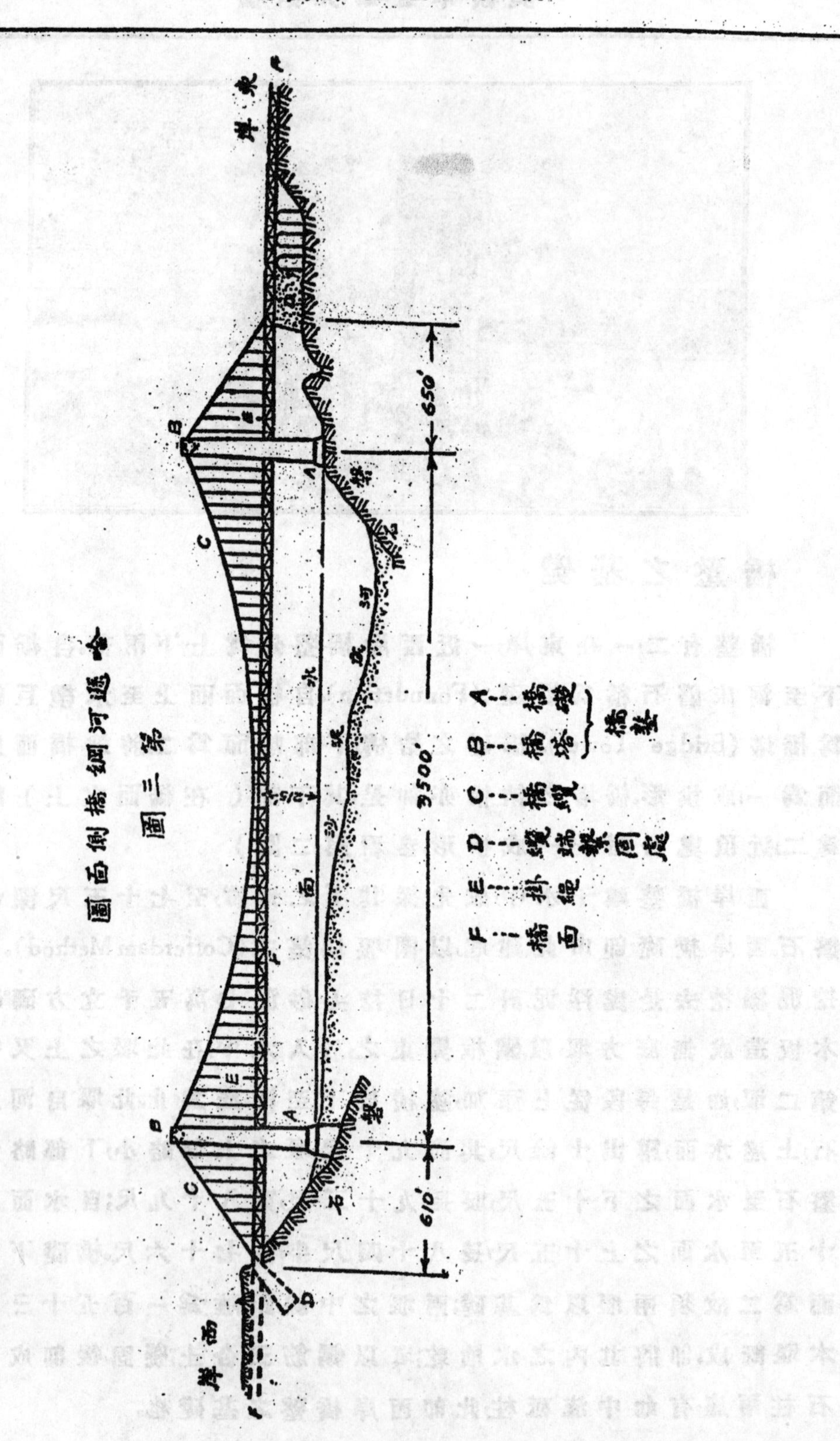

第三圖 哈遜河鋼橋側面圖

西岸橋礎下部截面

第四圖

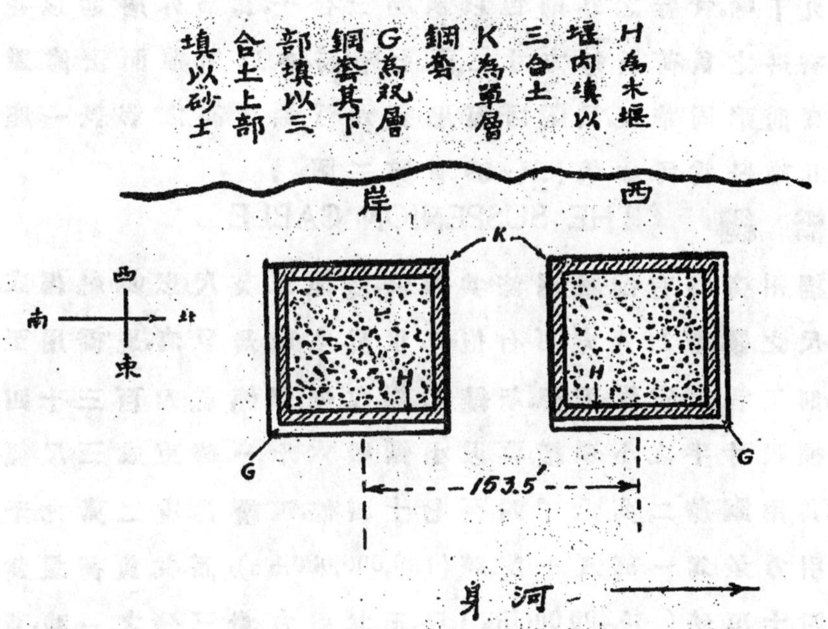

為保護橋礎勿受河水冲刮起見,木堰之外再圍以「鋼片樁」(Steel Sheet Piling)一層,下自磐石上達水面成一鋼套,套於木堰之外。此鋼套之西部,西北部,西南部均靠近岸邊,斯處水流較緩,單層鋼套已足應用。其東部,東北部,東南部均靠近河中,此處水流較急,須用夾層鋼套;夾層之內,其下部填以三合土,上部填以碎砂。橋礎得斯保護更鞏固無倫。(參看第四圖)

建造此橋礎用三合土四萬立方碼,花崗石一千五百立方碼,鋼筋四百噸。

東岸橋墪非建于水中,乃建于岸上之磐石,是以施工較易,只將石面鑿平,橋礎即築其上。(參看第三圖)

東岸橋墪之大小,形式,及結構,與西岸者相同。

橋塔 THE BRIDGE TOWER

橋塔兩座卽建兩礎之上;高六百五十尺;下部稍大,長二百一十尺,寬六十五尺;上部則畧爲收縮。以鋼架爲骨幹,重要部分以矽鋼(Silicon steel)爲之。次要部分以炭鋼(Carbon steel)爲之;共用鋼料三萬九千噸。骨幹之外則包以鋼筋三合土。其最外層則以花崗石爲表。橋塔之負荷量每方寸二萬七千磅。塔形簡單而宏偉。雖名之曰塔,實則形同華表。試閉目而思,有六百餘尺高之華表一座屹立空際其雄壯爲何如哉?!(參看第二圖)

橋纜 THE SUSPENION CABLE

橋兩旁各有橋纜兩條共四條。纜徑三英尺,以鋼絲編成。直徑三英尺之鋼纜,二人方可合抱,乃世界空前最巨者。是橋用至四條之多,則工程之宏偉可想。斯纜之編成:先以鋼線四百三十四條編成直徑四寸半之小鋼纜,再以小鋼纜六十一條束成三尺徑之大鋼纜;共用鋼線二萬六千四百七十四條。四纜共重二萬九千噸。每纜之引力量爲一萬八千萬磅(180,000,000lbs),而其負荷量爲六千五百四十萬磅(65,400,000lbs)只用其引力量三分之一強,斯則絕對安全,雖經數百年亦無危險。

纜端之繫固 ANCHORAGE OF THE SUSPENSION CABLES

每纜之負荷量爲六千五百四十萬磅,則其兩端之繫固爲極重要問題。兩岸之地形不同,故繫固之法亦稍異。西岸盤石頗高,在此鑿一斜洞,纜端埋于洞中。洞長二百五十尺,裏端六十餘尺見方,外端三十餘尺見方,成裏大外小之深穴。橋纜在穴內擺開爲數十縷,每縷編成辮形,辮端之孔則貫以鋼條。然後以三合土填滿穴中,三合土凝結則堅如岩石。纜藏其中,穩固無匹。雖有六千五百萬磅之拉力亦不能動其毫末也。(參看第五圖之一)

開鑿藏纜之斜洞及洞口之坡斜共鑿出岩石二十萬立方碼,用炸藥二十萬磅。

纜之東端亦係埋于三合土墪之內。但東岸之磐石頗低,故此座三合土墩非埋于穴中，只立于地上。先將地基之磐石故意鑿成參差不平之狀。三合土墩即築于其上。墩長二百九十尺，寬二百尺，高一百三十尺。凝結後形如小山，其重量可想；加以地基參差不平，更難動其毫末。纜端藏于其中，其方法與西端無異（參看第五圖之二）

西岸纜端繫固法

第五圖之一

C為橋纜。D為石洞。洞長二百五十尺。裏大外小。纜端藏於洞内。分為數十縫。以三合土填塞之。E為掛繩。F為橋面。

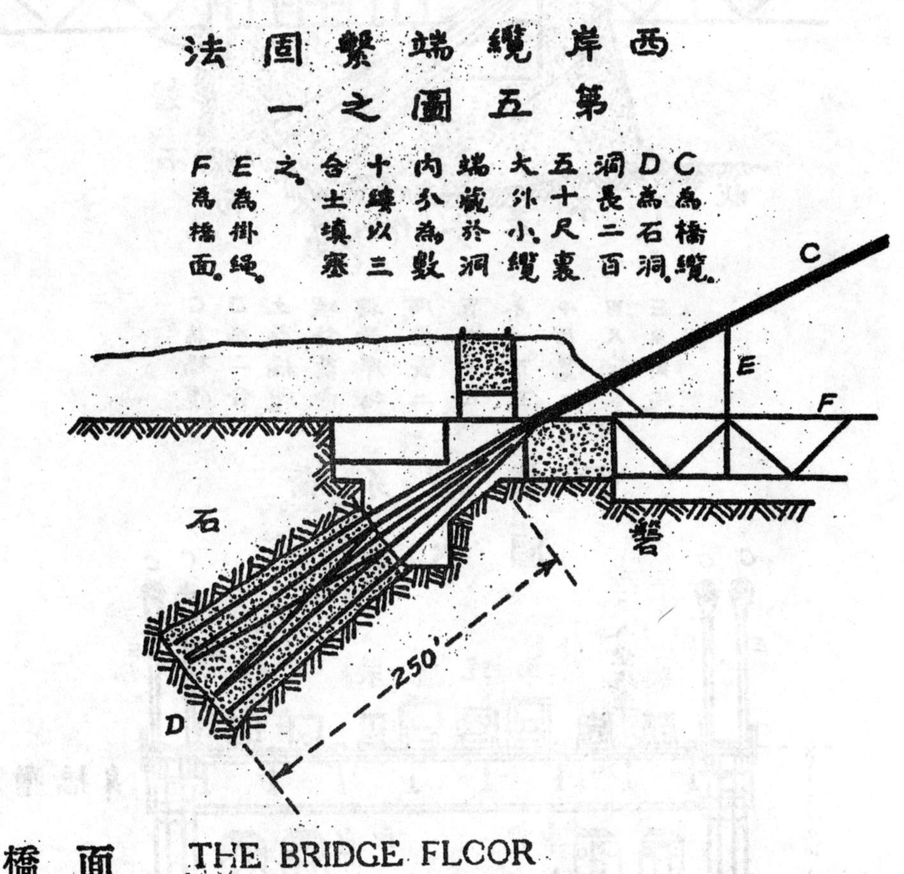

橋面 THE BRIDGE FLCOR

橋面寬一百十五尺，分上下兩層，兩層相距二十九尺。骨幹以鋼爲之。用鋼料一萬六千四百噸。最上層則舖以石子柏油。全部橋面以鋼質掛繩(Suspender ropc)懸于橋纜之下，共用掛繩七萬尺。上層橋面分爲汽車路八條，人行便道兩條；正中四條爲重載汽車用，兩旁四條爲人坐汽車用，最外兩條爲行人便道。下層橋面現尚未建，俟將來交通繁盛時，添上爲電車路之用，安設電車路四條，尚有

餘地也。（參看第六圖）

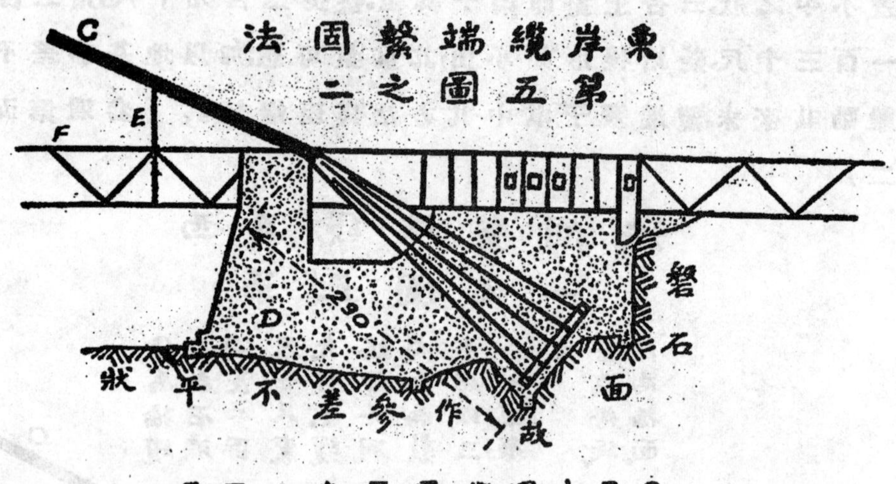

C為橋纜。D為三合土墊。橋纜埋於墊內與西岸略同。墊長二百九十尺高一百三十尺寬二百尺。E為掛繩。

橋身橫截面

第六圖

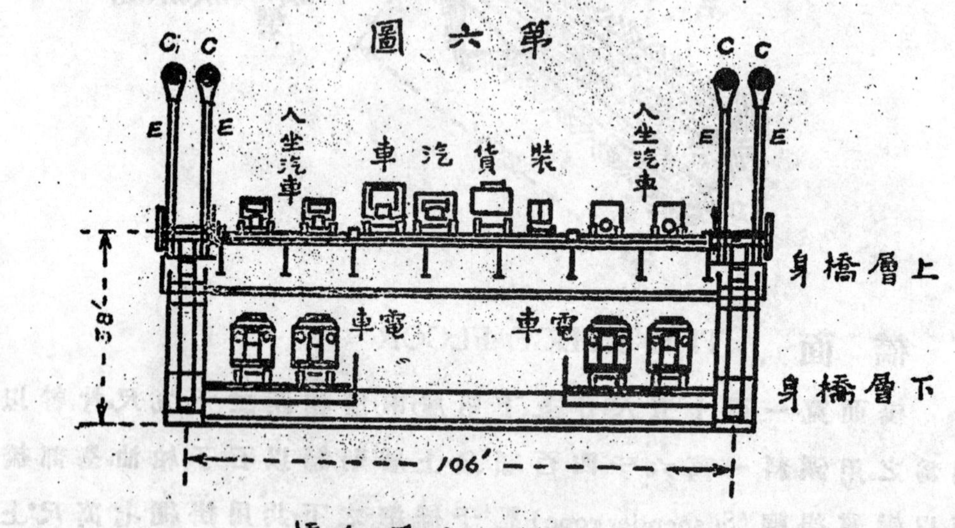

C為橋纜、E為掛繩、全座橋身用掛繩數千條掛於橋纜之下

懸橋鋼纜略論

劉 鏞

工 緒言

在世界橋樑建築中,其發展最速者,莫若懸橋(Suspen io nrBiidge)。考懸橋最初之建築,其適合于工程原理者,實始于1741年所完成之德斯河橋。是橋位於英格蘭(England)之德斯河(River Tees)上,跨度長70尺。當時用以懸吊橋重者,非鋼纜(Steel Cable)而爲鋼鍊(Chain),及至1816年美國非勒特非亞省(Philadelphia)所完成之蘇爾喬爾橋(Schuylkill Falls Suspen i n Bridge)始用鋼纜。其後所築成之懸橋,有用鋼纜者,亦有用鋼鏈者。二者之中,而以鋼纜爲優勝。其理由有七:—

(一)鋼纜之強度,勝于鋼鍊。在懸橋設計上,鋼鏈之容許設計應力(allowable design stress),每方吋爲45,000磅至50,000磅。而鋼纜之容許應力,則增至84,000磅。

(二)鋼纜比鋼鍊輕。在相等強度之鋼纜與鋼鍊中,則鋼鍊之靜重,每每比鋼纜之靜重大$2\frac{1}{2}$倍。故對于購料及建築兩方面,鋼鍊皆比鋼纜需費較多。

(三)鋼纜比鋼鍊易于安置。蓋當安置鋼纜之時,鋼纜能支持己重,不需別物支撐。

(四)鋼纜若以外皮包護,可以抵抗任何侵蝕。

(五)鋼纜懸橋所受之載重,雖超過設計時所預定之載重甚多,亦不至于崩塌。

（六）因鋼鏈在相等強度時，比鋼纜爲重。故鋼鏈祗適宜于跨度短小之建築，而鋼纜則宜於長跨度之建築。現今世界跨度最長之懸橋，爲3500尺，乃用鋼纜者。至於用鋼鏈之懸橋，其最長之跨度，亦不過700尺而已。

（七）懸橋所用之鋼鏈，爲集合多數鋼桿互相御接而成。是故各鋼桿所受之應力，常不均勻。且有甚高之副應力（Sec ndary Stresses）發生。而鋼纜則無此弊。

有以上七端，故近世之懸橋建築，多趨用鋼纜。其用鋼鏈者，雖間或有之，然亦不過施於短小之輕橋耳，至於較長之橋，多用鋼纜。今將最近二世紀所完成之鋼纜懸橋，列表如下：——

橋名	地名	跨尺度	完成之年
蘇爾喬爾瀑布橋	非勒特非亞	408	1816
格拉施爾橋	英格蘭	112	1816
京士宋道橋	英格蘭德斯河	110	1817
伏司橋	瑞士日內瓦	132	1823
布開爾橋	法蘭西羅尼河	394	1828
桃爾多尼橋	法蘭西阿根達	344	1828
淞尼橋	法蘭西里昂	335	1828
羅尼汀橋	法蘭西互倫西	384	1828
西里爾司橋	法蘭西羅尼河	332	1829
不拉舒麥恩橋	法蘭西	250	1832
路易腓力橋	巴黎塞納河	231	1833
福力堡橋	瑞士	870	1834
洛施伯爾那得橋	法蘭西	650	1836
里恩橋	柏林	57	1837

橋名	地名	跨尺度	完成之年
君士坦丁橋	巴黎窩恩河	328	1837
麥恩橋	法蘭西	344	1833
加爾里橋	法蘭西	635	1839
古布沙橋	法蘭西	360	1839
淞尼鉄路橋	法蘭西	137	1840
哥得龍橋	瑞士	746	1840
里沙德勒士橋	法蘭西	500	1840
查爾倫德橋	法蘭西	295	1841
綵爾喬爾河橋	美國腓力特非亞	358	1842
巴黎市橋	法蘭西	207	1842
斗羅河橋	葡萄牙	557	1842
聖比利橋	法蘭西	295	1845
亞基特橋	丕芝波路	162	1845
加龍納河橋	法蘭西凡爾登	500	1846
福德橋	拿加拉河	770	1847
聖基利士都非橋	法蘭西	604	1847
士密肥里特街橋	丕芝波路	188	1847
介安德橋	西維爾京那	450	1848
威嶺橋	西維爾京那	1010	1848
那破崙橋	法蘭西里昂	274	1849
密德橋	法蘭西里昂	398	1849
拿加拉橋	紐約省	1040	1850
厄耳北河橋	查利士頓	473	1852
聖約翰橋	紐不倫瑞克	628	1852

橋　名	地　名	跨尺度	完成之年
方打橋	紐約省	556	1853
拿加拉鐵道橋	拿加拉河	821	1854
摩根市	西維爾京那	608	1855
明尼亞波利斯橋	密西西皮河	620	1855
亞里根尼河橋	丕芝波路	344	1857
區砵恩可嵛馬橋	加利福尼亞省	258	1862
林白斯橋	倫敦	280	1863
俄亥俄河橋	俄亥俄	1057	1867
巴蘇橋	日爾曼	246	1869
克利福頓橋	拿加拉瀑布	1268	1869
華口橋	得撒省	470	1870
康內的告特河橋	麻沙朱失芝省	452	1870
華倫橋	賓西爾瓦尼省	470	1871
亞利橋	瑞士	160	1875
油城橋	亞利根尼河	500	1877
明尼亞波利斯第二橋	密西西皮河	675	1877
晉凡尼斯橋	蘇格蘭	173	1877
聖以利伯斯橋	法蘭西	232	1879
彼祿克連橋	紐約	1595	1883
林毛司	法蘭西	377	1884
以利克河第二橋	查利士頓	273	1884
桀得梭閘橋	康內的告特河	500	1884
亞維諾橋	法蘭西	282	1888
淞尼河橋	法蘭西	261	1888

橋名	地名	跨度尺	完成之年
廣恩橋	赤道國	275	1888
李察們得橋	印第安那	150	1889
華利男順橋	白水河	498	1889
吉林斯	紐約省	380	1890
伏爾特	法蘭西	560	1891
悉尼橋	澳大利亞	500	1891
依士得河橋	留尼江島	475	1893
哥加河橋	哥崙比亞	940	1894
東利物浦橋	俄亥俄河	705	1896
落徹斯得爾橋	俄亥俄河	800	1896
蘭根傘根橋	日爾曼	236	1898
拉加華新橋	紐約省	135	1898
遼威斯頓橋	傘加拉河	800	1899
米安皮米橋	墨西哥	1030	1900
以士頓橋	里希河	279	1900
康尼斯以克勞斯橋	法蘭西	902	1901
凡爾傘順橋	法西蘭	764	1902
加比頓橋	西維京那亞	510	1903
威廉斯堡橋	紐約	1600	1903
得躬尼橋	華得爾維利	400	1904
彭鴻未橋	法蘭西	525	1904
士彫賓維利橋	俄亥俄河	700	1904
得士金比亞橋	密蘇里河	627	1905
東利物浦第二橋	俄亥俄河	750	1905

橋名	地名	跨尺度	完成之年
求利次河橋	美國華盛頓省	300	1905
維利法蘭斯橋	法蘭西	512	1906
爪拉伯橋	墨西哥	184	1908
巴拿馬運河橋	巴拿馬	600	1909
牛布利扑橋	美國麻撒朱天芝省	244	1909
螢喀坦橋	紐約	1470	1909
馬仙拿橋	紐約	400	1910
印河橋	瑞士	550	1911
連溪橋	密蘇里	525	1911
布列恩橋	哥倫比亞	331	1913
拜爾司橋	美國得撒省	568	1914
温那茨河橋	美國華盛頓省	190	1159
拍克爾士堡橋	西維京那亞	775	1916
得爾羅橋	美國得撒省	450	1917
里奥支里基橋	巴拿馬	410	1917
卑比橋	哥崙比亞河	632	1919
螢那華都河橋	紐絲蘭	632	1919
金比爾蘭河橋	拿施威利	540	1919
不列芝模橋	奥加拉河馬	600	1921
隆陡特橋	美國紐約省	705	1922
亞林溪橋	俄亥俄		
亞打康尼華橋	奥加拉何馬	400	1922
熊山橋	里特順河	1632	1924
奴康那橋	得撒省	700	1924

25122

橋名	地名	跨尺度	完成之年
特拉華河橋	肥勒特非亞	1750	1926
拜恩芬尼橋	得撒省	400	1927
亞連尼橋	得撒省	700	1927
喬治亞佛里達橋	當奴得順維利	600	1927
米得亞康奴亞橋	哥崙比亞	380	1927
撲次茅橋	俄亥俄河	700	1927
門得珍橋	法蘭西	302	1927
斯彭賓維利橋	俄亥俄河	689	1928
蘇維斯崖橋	與加拉何馬	500	1928
羅馬橋	得撒省	630	1928
希打爾高橋	得撒省	350	1928
地斯亞克橋	阿康撒斯	650	1928
安那開羅橋	哥崙比亞	417	1928
搬那到阿蘭高橋	哥崙比亞	623	1929
孟得何皮橋	來得島	1200	1929
地多勞哀橋	美利堅與加拿大間	1850	1929
格蘭米利橋	圭培克	948	1929
中黑得順橋	紐約省	1500	1930
黑得順河橋	紐約	3500	1932

觀察上表,足見近代懸橋建築之突飛猛進,誠屬驚人,言跨度,則由57尺增至35 00尺.迺最近所策畫建築者,有長至45 00尺,約.8 52英里之遙,其工程之大,可想而知.至於載重方面,亦有長足之進展.昔日之懸橋,祗用以渡行人及小車輛而已.今之懸橋,不特可以渡行人及其他各種車輛,且可支持數千噸之火車,奔馳其上.上表

所列祇限于鋼纜懸橋,致於鋼鏈懸橋,尚未列入。然懸橋之所以能適宜於長跨度之建築者,蓋鋼鉄抵抗牽力之能力,較任何建築材料爲大,而懸橋之成功亦全在能利用鋼鉄此特殊之品質,將之造成鋼鏈或鋼纜,以支持各種靜動載重故也。

懸橋之發展,已如上述。懸橋之中,普通分爲加固懸橋(Stiffened Suspension Bridge)與不加固懸橋(Unstiffened Suspension Bridge)二種。不加固懸橋乃將路面懸掛于鋼纜之上,使橋面各種靜動載重,直接傳達於鋼纜。此種建築,不甚堅固,故重要橋樑,甚少用之。加固懸橋則以桁架(truss)支持路面,再將桁架掛于鋼纜之上,如前篇第三圖。此種建築,甚爲堅固。各種不對稱(Unsymmetrical)之集中載重(Concentrated Loading)及均等載重(Uniform Loading),皆能支持。故跨度長載重大之懸橋多採用之。此兩種懸橋之結構雖不同,然二者皆以鋼纜支持路面之靜動載重則同。故鋼纜實爲懸橋中最重要之部分,亦爲設計上所最應注意者也。

II 鋼纜之形狀

鋼纜之形狀,與鋼纜之長短及其所受之應力之大小,有密切之關係。而鋼纜所受應力之大小,每爲其設計上之唯一標準。其長度亦爲鋼纜估價不可少之條件。故研究懸橋之鋼纜,必須先觀察其所成之各種形狀。

鋼纜之形狀,每隨其所受之載重不同而異。試取一繩,兩端用力拉之,使之平直。若繫重量于其上,則繩不復平直,必向懸重量之處下垂,成一角度。若懸均等重量于其上,則此繩必成一曲線形。懸橋之鋼纜,常不平直而成種種曲線形者,亦同此理。大抵鋼纜因其所受之重量不同而成之形狀,可分爲垂曲線形及拋物線形二種。今將之分別論之。

(A) 垂曲線形 Catenary.

豎立懸橋之第一步工作,爲將鋼纜曳過江面,固定其兩端于橋塔之頂。此時鋼纜所受之載重，祇其本身之重量而已，全無其他外力加之其上。設鋼纜本身之重量爲每單位長度w磅。若在此鋼纜上截取一小段ds，如第一圖(b)所示，則此小段兩端截面所受之引力，必不相同。設此二引力各爲T及T-dT，及此二引力之縱分力

第一圖

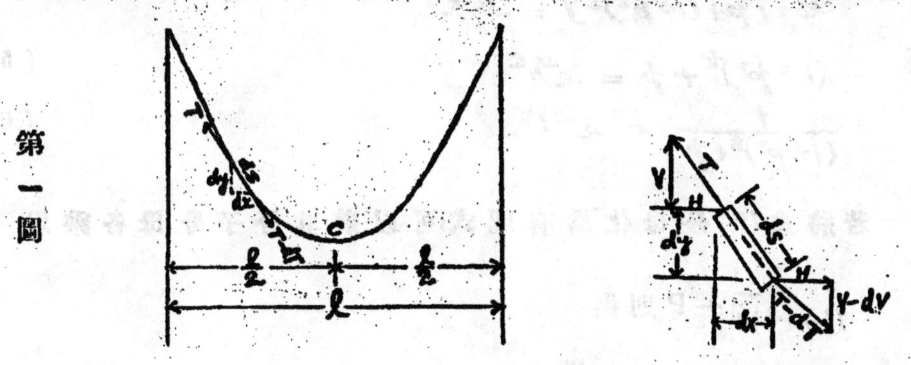

各爲V與V-dV

則 $V - wds - (V - dV) = 0$

$$dV = wds \tag{1}$$

將(1)兩節除以dx則

$$\frac{dV}{dx} = \frac{wds}{dx}$$

因 $V = H\frac{dy}{dx}$ $\quad \frac{dV}{dx} = H\frac{d^2y}{dx^2} = \frac{wds}{dx}$

$$\frac{dy^2}{dx^2} = \frac{wds}{Hdx} \tag{2}$$

$$ds = \sqrt{dx^2+dy^2} = \sqrt{1+\left(\frac{dy}{dx}\right)^2}\,dx$$

將式(3)代入式(2) $\quad \frac{d^2y}{dx^2} = \frac{w\left[1+\left(\frac{dy}{dx}\right)^2\right]^{\frac{1}{2}}}{H}$

設 $p = \frac{dy}{dx}$ 則 $\frac{dp}{dx} = \frac{d^2y}{dx^2}$

$$\frac{dp}{dx} = \frac{w(1+p^2)^{\frac{1}{2}}}{H}$$

$$\frac{dx}{H} = \frac{dp}{w(1+p^2)^{\frac{1}{2}}}$$

$$\frac{x}{H} = \int\frac{dp}{w(1+p^2)^{\frac{1}{2}}} = \frac{1}{w}\log\left[p+(1+p^2)^{\frac{1}{2}}\right]+K_1$$

若以C爲原點(Origin),則在C點時

X＝O P＝O 所以K_1＝O

$$\frac{x}{H}=\frac{1}{w}\log\left[p+(1+p^2)^{\frac{1}{2}}\right] \tag{4}$$

$$\log\left[p+(1+p^2)^{\frac{1}{2}}\right]=\frac{wx}{H}$$

$$(1+p^2)^{\frac{1}{2}}+p=e^{\frac{wx}{H}} \tag{5}$$

$$\frac{1}{(1+p^2)^{\frac{1}{2}}+p}=e^{-\frac{wx}{H}} \tag{6}$$

若將式(6)分母化爲有理式,可以將其分子分母各乘以

$(1+P^2)^{\frac{1}{2}}-P$ 則得

$$(1+P^2)^{\frac{1}{2}}-P=e^{-\frac{wx}{H}} \tag{7}$$

由式(5)減式(7)則得 $2P=e^{\frac{wx}{H}}-e^{-\frac{wx}{H}}$

$$p=\frac{1}{2}\left[e^{\frac{wx}{H}}-e^{-\frac{wx}{H}}\right]$$

$$\frac{dy}{dx}=\frac{1}{2}\left[e^{\frac{wx}{H}}-e^{-\frac{wx}{H}}\right]$$

$$y=\int\frac{1}{2}\left[e^{\frac{wx}{H}}-e^{-\frac{wx}{H}}\right]$$

$$=\frac{H}{2w}e^{\frac{wx}{H}}+\frac{H}{2w}e^{-\frac{wx}{H}}+K_2 \tag{8}$$

當 $x=0$ $y=0$ $K_2=-\frac{H}{w}$

式(8)變爲 $y=\frac{H}{2W}\left(e^{\frac{wx}{H}}+e^{-\frac{wx}{H}}-2\right)$ (9)

式(9)爲一垂曲線之方程式,由此可知鋼纜因其本身重量所成之形狀,必爲一垂曲線。設 $C=\frac{w}{H}$ 則上式變爲

$$y=\frac{1}{2c}\left(e^{cx}+e^{-cx}-2\right) \tag{10}$$

若用雙曲線函數表之,則爲

$$y=\frac{1}{c}\left(\frac{e^{cx}+e^{-cx}}{2}-1\right)=\frac{1}{c}\left[(\cosh cx)-1\right] \tag{11}$$

(B) 拋物線形

鋼纜既固定于橋塔之頂，第二步工作爲將支持路面之桁架，懸于其上。因桁架之靜重，與橋面之活重，比鋼纜本身重量，大數千倍。故鋼纜之形狀，每爲前者所定，不復爲其本身重量所影響。圖二示一鋼纜 AB，上懸每單位長度 w 磅之均等重量。設於纜上截取

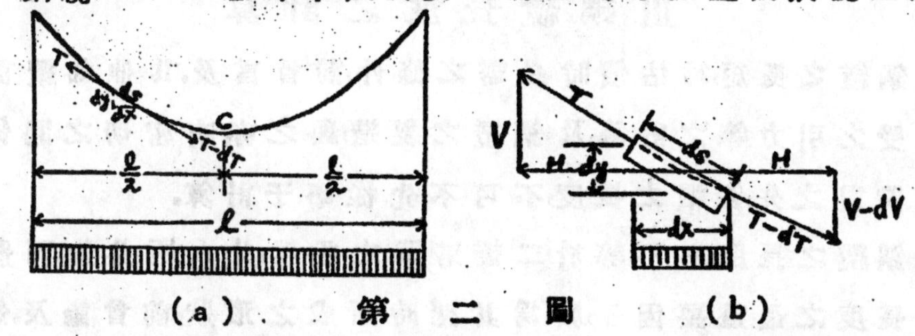

(a) 第二圖 (b)

任何小段 ds 如圖二(b)。設 T 及 T + dT 爲此小段兩端所受之引力而 wdx 爲此小段所受之載重，則

$$V = H\tan\alpha = H\frac{dy}{dx}$$

$$\frac{d^2y}{dx^2} = \frac{w}{H}$$

$$\frac{dy}{dx} = \int\frac{w}{H} = \frac{wx}{H} + K$$

若以C點爲原點，當 $X = O$ $\frac{dy}{dx}$ $O =$ 及 $K_1 = O$

則 $\frac{dy}{dx} = \frac{wx}{H}$

$$y = \int\frac{wx}{H} = \frac{wx^2}{2H} + K_2$$

當 $X = O$ $y = O$ 所以 $K2 = O$

$$y = \frac{wx^2}{2H} \tag{12}$$

此式爲二次式，則知鋼纜在懸掛路面之後，所成之曲線，爲拋物線無疑矣。但當 $x = \frac{c}{2}$，時 $y = f$，則式(12)變爲

$$f=\frac{wl^2}{8H} \quad H=\frac{wl^2}{8f} \tag{12a}$$

若將此H之值,代入式(12)則得

$$y=\frac{4fx^2}{l^2} \tag{13}$$

Ⅲ 鋼纜長度之計算

鋼纜之長短,爲估價時必需之條件,前曾言及。其他如纜重橋塔所受之引力等之計算及鋼纜之製造,與之亦有密切之關係。故在未設計之先,鋼纜之長度,不可不先從事于計算。

鋼纜之長度,必不等於二橋塔間之距離,此人所共知。而影響鋼纜長度之最重要因子,厥爲其懸時所成之形狀,前曾論及。然計算其長度之公式,則多以其兩端之距離及其中點之垂度(sag)表之。今將其公式之求法,分錄於下:

(A) 垂曲線長度公式之求法

第壹圖(a)爲一垂曲線形。設 l 爲其兩端之距離,L 爲其長度,d_s 爲小段之長度。則

$$L=2\int_0^{\frac{l}{2}}ds=2\int_0^{\frac{l}{2}}\left[1+\left(\frac{dy}{dx}\right)^2\right]^{\frac{1}{2}}dx \tag{14}$$

式(10) $y=\frac{1}{2c}(e^{cx}+e^{-cx}-2)$

$$\frac{dy}{dx}=\frac{1}{2}(e^{cx}-e^{-cx})$$

$$\left(\frac{dy}{dx}\right)^2=\frac{1}{4}(e^{2cx}+e^{-2cx}-2) \tag{15}$$

將式(15)代入式(14)

$$L=2\int_0^{\frac{l}{2}}\left[1+\frac{1}{4}(e^{2cx}+e^{-2cx}-2)\right]^{\frac{1}{2}}dx$$

$$=2\int_0^{\frac{l}{2}}\frac{1}{2}(e^{cx}+e^{-cx})dx$$

$$=\frac{1}{c}(e^{cx}-e^{-cx})\Big]_0^{\frac{l}{2}}$$

$$=\frac{1}{c}\left(e^{\frac{cl}{2}}-e^{-\frac{cl}{2}}\right) \tag{16}$$

若用双曲線函數表之，

則 $\frac{e^{\frac{cl}{2}}-e^{-\frac{cl}{2}}}{2}=\sinh\frac{cl}{2}$

式(16)變爲 $L=\frac{2}{c}\left[\frac{e^{\frac{cl}{2}}-e^{-\frac{cl}{2}}}{2}\right]$

$$=\frac{2}{c}\sinh\frac{cl}{2} \tag{17}$$

(B)拋物線長度公式之求法

第二圖(a)爲一拋物線形，設 l 爲其兩端之距離，L 爲其長度，d_s 爲任何一小段之長度一如前式。

$$L=2\int_0^{\frac{l}{2}}\left[1+\left(\frac{dy}{dx}\right)^2\right]^{\frac{1}{2}}dx \tag{14}$$

由式(13) $y=\frac{4fx^2}{l^2}$ (13)

$$\frac{dy}{dx}=\frac{8fx}{l^2}$$

$$\left(\frac{dy}{dx}\right)^2=\frac{64f^2x^2}{l^4} \tag{18}$$

$$\frac{ds}{dx}=\left(1+\frac{64f^2x^2}{l^4}\right) \tag{18a}$$

將式(18)入代式(14)則得

$$L=2\int_0^{\frac{l}{2}}\left[1+\frac{64f^2x^2}{l^4}\right]^{\frac{1}{2}}dx$$

$$=2\int_0^{\frac{l}{2}}\frac{8f}{l^2}\sqrt{\frac{l^4}{64f^2}+x^2}\,dx$$

$$=\frac{16f}{l^2}\left[\frac{x}{2}\sqrt{x^2+\frac{l^4}{64f^2}}+\frac{l^4}{128f^2}\log\left(x+\sqrt{x^2+\frac{l^4}{64f^2}}\right)\right]_0^{\frac{l}{2}}$$

$$=\frac{16f}{l^2}\left[\frac{l}{4}\sqrt{\frac{l^2}{4}+\frac{l^4}{64f^2}}+\frac{l^4}{128f^2}\log\left(\frac{l}{2}+\sqrt{\frac{l^2}{4}+\frac{l^4}{64f^2}}\right)\right]-\frac{16f}{l^2}\left(\frac{l^4}{128f^2}\log\frac{l^2}{8f}\right)$$

$$=\frac{16f}{l^2}\left\{\frac{l^2}{8}\sqrt{1+\frac{l^2}{16f^2}}+\frac{l^4}{128f^2}\left[\log\left(\frac{l}{2}+\frac{l}{2}\sqrt{1+\frac{l^2}{16f^2}}\right)-\log\frac{l^2}{8f}\right]\right\}$$

$$=2f\sqrt{1+\frac{l^2}{16f^2}}+\frac{l^2}{8f}\left[\log\left(\frac{l}{2}+\frac{l}{2}\sqrt{1+\frac{l^2}{16f^2}}\right)-\log\frac{l^2}{8f}\right]$$

$$=\frac{l}{2}\sqrt{1+\frac{16f^2}{l^2}}+\frac{l^2}{8f}\log\left[\frac{4f}{l}+\sqrt{\frac{16f^2}{l^2}+1}\right]$$

$$=\frac{l}{2}\left[\sqrt{1+\frac{16f^2}{l^2}}+\frac{l}{4f}\log_e\left(\frac{4f}{l}+\sqrt{1+\frac{16f^2}{l^2}}\right)\right]$$

設 $n=f/l$

$$L=\frac{l}{2}\left[\sqrt{1+16n^2}+\frac{1}{4n}\log_e\left(4n+\sqrt{1+16n^2}\right)\right] \tag{20}$$

由式(20)所得之值,為鋼纜之準確長度,若欲得其概數,可用次式原理,將式(19)之 $\left(1+\frac{16f^2x^2}{l^4}\right)^{\frac{1}{2}}$ 一項展開,然後將之求積分便得。

IV 鋼纜所受應力之計算

(A) 垂曲線形鋼纜

鋼纜所受之應力之大小,恆視其外力之大小,及其所成之形狀及長度而定。設 T 為鋼纜上任何點所受之引力,H 為此引力之橫分力,由第三圖(b)則得

$$T=H\frac{ds}{dx} \tag{21}$$

由式(3)及式(15)

則得 $T=H\left[1+\frac{1}{4}\left(e^{2cx}+e^{-2cx}-2\right)\right]^{\frac{1}{2}}$

$=\frac{H}{2}\left(e^{cx}+e^{-cx}\right)$

因 $H=\frac{w}{c}$ $\quad T=\frac{w}{2c}\left(e^{cx}+e^{-cx}\right)$ (22)

如第壹圖所示,因以 C 為原點,則在鋼纜跨度中點時,

$x=0$ $\quad T=H=\frac{w}{c}$ 若在兩端支點之時,則

$x=\frac{l}{2}$, $T=\frac{w}{2c}\left(e^{cx}+e^{-cx}\right)$ 為全鋼纜所受應力之最大者也。

(B) 拋物線形鋼纜

因 $T=H\frac{ds}{dx}$ (21)

$H=\frac{wl^2}{8f}$ (12a)

$\frac{ds}{dx}=\left[1+\frac{64f^2x^2}{l^4}\right]^{\frac{1}{2}}$ (18a)

則 $T=\frac{wl^2}{8f}\left[1+\frac{64f^2x^2}{l^4}\right]^{\frac{1}{2}}$ (23)

鋼纜之最大應力為在兩端支點處，

$$T=\frac{wl^2}{8f}\left(1+\frac{64f^2}{l^2}\right)^{\frac{1}{2}} \tag{24}$$

$$=wl\left(1+\frac{l^2}{64f^2}\right)^{\frac{1}{2}} \tag{25}$$

V 鋼纜長度之改變

物體遇熱則漲，冷則縮，此人所共知者也。而此種改變以金屬最顯著。懸橋之鋼纜，皆以鋼為之。雖經多年研究，發明各種膨脹係數最少之合金，以充此任，然其受溫度之影響，仍不可免。設 t 為溫度增加之度數，L 為鋼纜之長度，a 為製造鋼纜所用材料之直線膨脹係數，et 為鋼纜因溫度之增而延長之長度，則得下式

$$e_t=\alpha tL \tag{26}$$

在垂曲線時 $L=\frac{1}{c}\left(e^{\frac{cl}{2}}-e^{-\frac{cl}{2}}\right)$

則式(26)變為 $e_t=\frac{\alpha t}{c}\left(e^{\frac{cl}{2}}-e^{-\frac{cl}{2}}\right)$ (27)

在拋物線 $L=\frac{l}{2}\left[(1+16n^2)^{\frac{1}{2}}+\frac{1}{4n}\log_e\left(4n+\sqrt{1+16n^2}\right)\right]$

則式(26)變為 $e_t=\frac{\alpha tl}{2}\left[(1+16n^2)^{\frac{1}{2}}+\frac{1}{4n}\log_e\left(4n+\sqrt{1+16n^2}\right)\right]$ (28)

物體之形狀，非獨受溫度昇降之影響，而始改變，其所受外力之性質亦與焉。物體所受之外力，若為壓力，則必縮小，若為牽力，則必延長。今懸橋鋼纜所受之應力，全為牽力，前已言之，則其長度之改變必為延長，可無疑矣。由材料力學得知物體之變形，與其所受之應力，成正比例。今設 d 為單位變形，E 為此物體之彈性係數，s 為此物體所受之單位應力，則 $d=\frac{s}{E}$ (29)

若A為鋼纜之橫截面，T為鋼纜上任何一點所受之牽力

$$d=\frac{T}{AE} \tag{30}$$

長度ds之延長 $= \frac{T}{AE} ds$

全鋼纜之總延長 $= e_s = 2\int_0^{\frac{l}{2}} \frac{T ds}{AE}$ (31)

在垂曲線時 $T = \frac{w}{2c}\left(e^{\frac{cl}{2}} + e^{-\frac{cl}{2}}\right)$

則
$$
\begin{aligned}
e_s &= 2\int_0^{\frac{l}{2}} \frac{w}{2AEC}\left(e^{cx} + e^{-cx}\right) ds \\
&= \frac{w}{2CAE}\int_0^{\frac{l}{2}} \left(e^{cx} + e^{-cx}\right)\left(e^{cx} + e^{-cx}\right) dx \\
&= \frac{w}{2CAE}\int_0^{\frac{l}{2}} \left(e^{2cx} + e^{-2cx} + 2\right) dx \\
&= \frac{w}{2CAE}\left(\frac{1}{2c}e^{2cx} - \frac{1}{2c}e^{-2cx} + 2x\right)\Big]_0^{\frac{l}{2}} \\
&= \frac{w}{4CAE}\left(e^{2cx} - e^{-2cx} + 2cl\right)
\end{aligned}
$$
(32)

在拋物線時 $T = \frac{wl^2}{8f}\left[1 + \frac{64f^2x^2}{l^4}\right]^{\frac{1}{2}}$

則
$$
\begin{aligned}
e_s &= 2\int_0^{\frac{l}{2}} \frac{wl^2}{8fAE}\left[1 + \frac{64f^2x^2}{l^4}\right]^{\frac{1}{2}} ds \\
&= \frac{wl^2}{4fAE}\int_0^{\frac{l}{2}} \left[1 + \frac{64f^2x^2}{l^4}\right]^{\frac{1}{2}}\left[1 + \frac{64f^2x^2}{l^4}\right] dx \\
&= \frac{wl^2}{4fAE}\int_0^{\frac{l}{2}} \left(1 + \frac{64f^2x^2}{l^4}\right) dx \\
&= \frac{wl^2}{4fAE}\int_0^{\frac{l}{2}} \left(1 + \frac{64f^2x^2}{l^4} + \cdots\cdots\right) dx \\
&= \frac{wl^2}{4fAE}\left(x + \frac{64f^2x^3}{3l^4} + \cdots\right)\Big]_0^{\frac{l}{2}} \\
&= \frac{wl^2}{4fAE}\left(\frac{l}{2} + \frac{64f^2l^3}{24l^4} + \cdots\right) \\
&= \frac{wl^3}{8fAE}\left(1 + \frac{16f^2}{3l^2} + \cdots\cdots\right) \\
&= \frac{wl^2}{8AE}\left(n + \frac{16n^3}{3} + \cdots\right)
\end{aligned}
$$
(32)

設dL為鋼纜長度改變之總數

則在垂曲線時

$$dL=\frac{\alpha t}{c}\left(e^{\frac{cl}{2}}-e^{-\frac{cl}{2}}\right)+\frac{w}{4c^{2}AE}\left(e^{cl}-e^{-cl}+2cl\right) \quad (34)$$

則在抛物線時

$$dL=\frac{\alpha tl}{2}\left[(1+16n^{2})^{\frac{1}{2}}+\frac{1}{4n}\log_{e}\left(4n+\sqrt{1+16n^{2}}\right)\right]+\frac{wl^{2}}{8AE}\left(n+\frac{16n^{3}}{3}\right) \quad (35)$$

Ⅵ 鋼纜之製造

(A) 材料

懸橋之鋼纜，多用鍍鋅鋼線造成，而其所以鍍鋅者，蓋鋅能抵抗空氣之侵蝕故也。鍍鋅之法有三：一曰熱浸法（Hot Dip Process）乃將所需鍍鋅之鋼鐵，先以酸液去其銹汚，再以清水洗之，然後將之浸于已鎔之鋅中，或將之在已鎔之鋅中徐徐曳過亦可。但包裹鋼鐵之鋅層，不可過薄，過薄則抵抗侵蝕之能力弱，亦不可過厚，過厚則易於破裂，反爲不美。二曰電鍍法，（Electroplating）乃將鋅片及需鍍鋅之鋼鐵，置于電鍍池中，然後通以電流，則一層薄鋅層在所置之鋼鐵上造成。三曰甑鍍法（Sherardizing）此法乃將所需鍍鋅之鋼鐵，與鋅片同置于曲頸甑中熱之，使甑內之鋅片氣化，然後將之凝結于鋼鐵之上。用此法所鍍之鋅層，與鋼鐵之結合力甚大。惜其費用較他法爲昂，故不普遍應用。

(B) 鋼纜之格式

普通所用之鋼纜，可分爲二種，一爲集合多數平行鋼絲（Parallel steel wires）所成，一爲絞鋼線而爲鋼索（Steel Rope）再集多數鋼索而成。而後者復分爲鋼心與麻心二種。尋常較大之橋，多用平行鋼線造成者，以其強度大而在載重下長度之變遷少故也。集合鋼索而成之鋼纜，在相同直徑之時，其之強度遜於平行橋線鋼纜，且易於延長。但安置上之便利，則遠勝於前者，故跨度短小之橋，多採用之。至於鋼心與麻心二種之分別，則鋼纜心鋼之強度較大，而麻心者則易于彎曲而已。普通鍍鋅鋼線造成之鋼纜，其最大強度爲每方英寸215,000至230,000磅,而鋼索造成者則每方英寸爲80,000 X(直徑)^{2}磅。

四 結論

以上所論,乃就懸橋之一小部分——橋纜大概情形而言,至於橋纜本身之設計,及其在橋之兩端所成之形狀,尙未論列。其他關於橋塔,桁架等之設計與安置,則更不能顧及。他日有暇,將另篇述之。

候氏桁架之固重應力 DEAD LOAD STRESS OF HOWE TRUSS.計算法

陳景洪

我國鋼鐵廠之欠缺,鋼鐵材料多仰給舶來,其價值必較本地之材木貴,故於營造設計上選擇材料,能以木材代之者,較爲經濟.通常樓房屋頂之設計,多採用候氏式桁架.普通有四格(Four panels.)(圖一),六格(圖二),及八格(圖三)三種.四格之最長支距爲40呎;六格爲48呎;八格則支距可較長,蓋其結構全以鋼鐵爲之.四格與六格者,除垂直引力之肢段(Tension member)用圓鐵枝之外,餘均用木材,較省費用.

最經濟之斜度 Pitch 爲$^1/_4$,用材料最少,而有適度屋蓋下之空位.$^1/_3$斜度頗宜於北方,因可減輕雪之載重,但不宜於南方,因增加風力載重,且用材料較多故也.若斜度爲$^1/_5$,則所費更大.

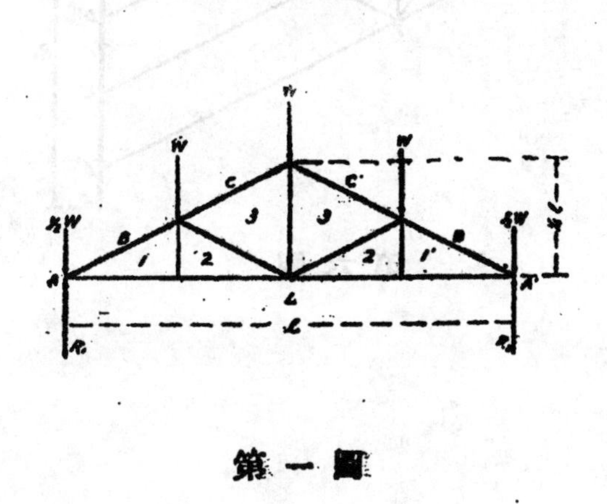

第一圖

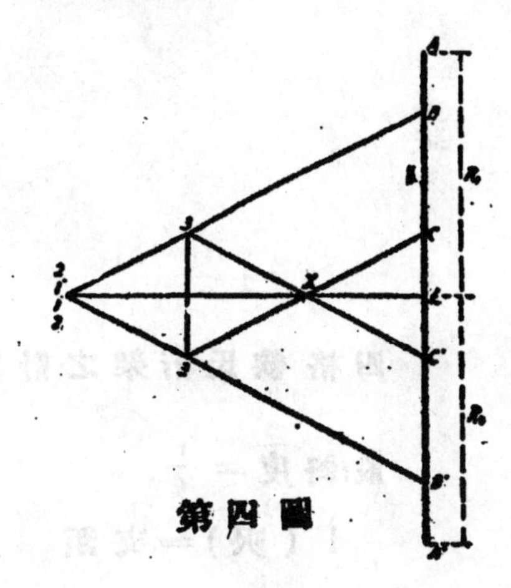

第四圖

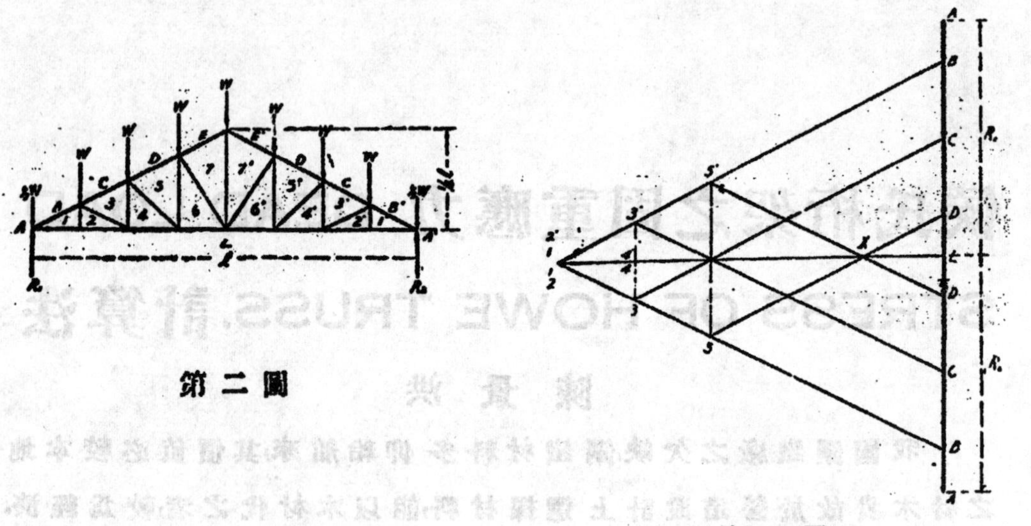

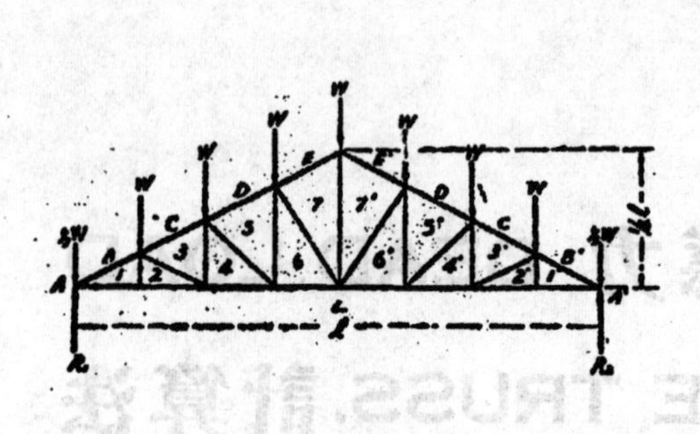

第二圖

第五圖

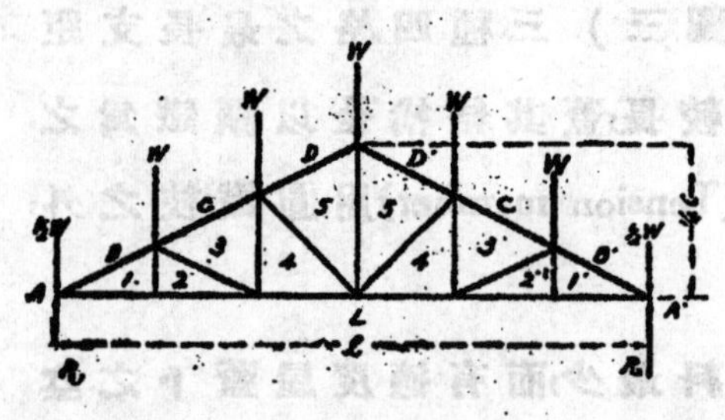

第三圖

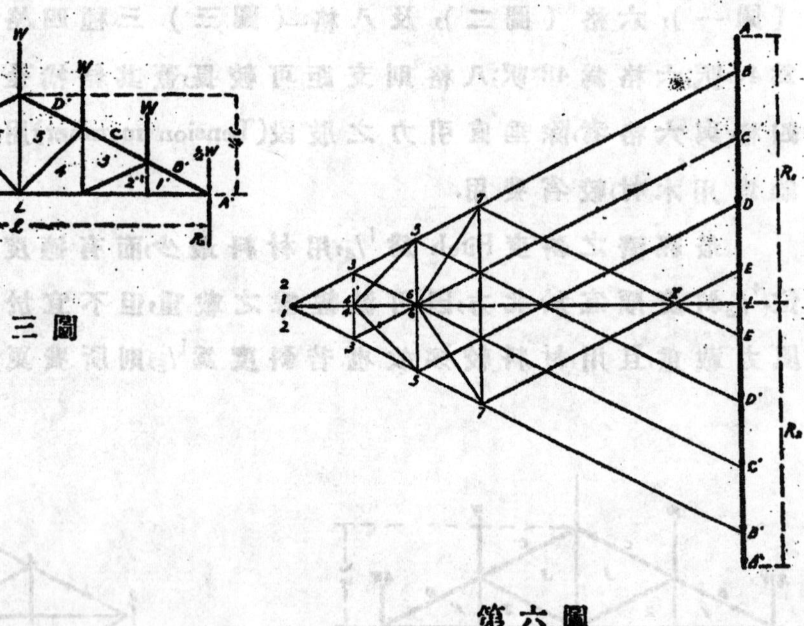

第六圖

四格候氏桁架之計算法：

設：斜度 $= \frac{1}{4}$

1（呎）= 支距

w（磅）＝每格所載之固重(Dead panel load)（包含瓦面，桁，樑，及候氏桁架本身之重量）

每肢段所受之應力，可繪圖計算之，如（圖二）．從圖二可得簡易之綫段關係，每三角形中之橫綫必等於兩倍垂直綫．

即 $\overline{L1} = \overline{L2} = 2 \times \overline{BL} = 2 \times 1\frac{1}{2}W = 3W.$

$\overline{33'} = \overline{CC'} = W.$

$\overline{B1} = \overline{L1} \sec\alpha = 3W \sec\alpha.$

$\overline{C3} = 2W \sec\alpha$

$\overline{23} = W \sec\alpha.$

若其斜度為 $^1/_3$ 或 $^1/_5$ 時，可用一定量之數(Constant)乘之，即得其值因．節段之長短與受應力之大小適成正比例．斜度 $^1/_3$ 或 $^1/_5$ 之節段與斜度 $^1/_4$ 之節段關係如下．參看圖七．

其他肢段，可照此法算之．從（圖二，四，六）所得各肢段之固重應力，列表于後．表只列一半肢段，他半互相對稱，故從畧．

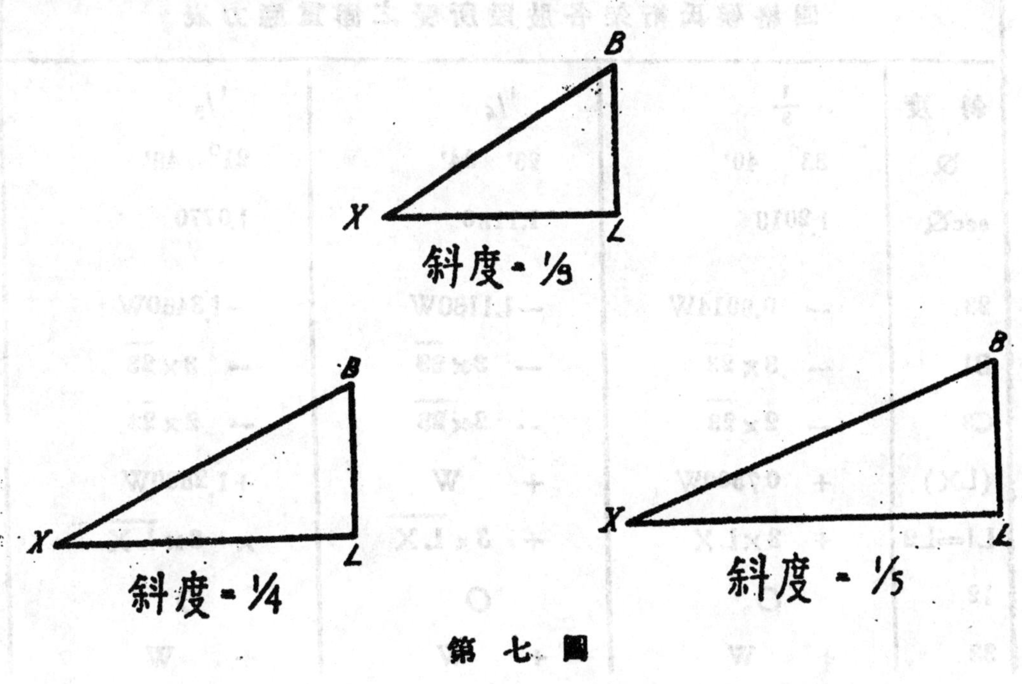

第七圖

設 $\overline{BL}=1$（長度）

則斜度 $^1/_3$ 之 $\overline{LX}=1\,^1/_2=1.5$

斜度 $^1/_4$ 之 $\overline{LX}=2$

斜度 $^1/_5$ 之 $\overline{LX}=2\frac{1}{2}=2.5$

$$\frac{^1/_3\text{斜度之}\overline{LX}\text{應力}}{^1/_4\text{斜度之}\overline{LX}\text{應力}}=\frac{^1/_3\text{斜度}\overline{LX}\text{之長}}{^1/_4\text{斜度}\overline{LX}\text{之長}}=\frac{1.5}{2}$$

$\therefore$ $^1/_3$ 斜度之 $\overline{LX}$ 應力 $=\frac{1.5}{2}\times\overline{LX}=.75\overline{LX}$.

$^1/_3$ 斜度之 $\overline{BX}$ 應力 $=.75\ \overline{LX}\ \sec 33^\circ\ 40'=.9014\ \overline{LI}$.

$^1/_5$ 斜度之 $\overline{LX}$ 應力 $=\frac{2.5}{2}\ \overline{LX}=1.25\ \overline{LX}$.

$^1/_5$ 斜度之 $\overline{BX}$ 應力 $=1.25\ \overline{LX}\ \sec 21^\circ 48'=1.346\ \overline{LX}$.

其他肢段，可照此法算之。從（圖二，四，六）所得各肢段之固重應力，列表于後。表只列一半肢段，他半互相對稱，故從畧。

四格候氏桁架各肢段所受之固重應力表

斜度	$\frac{1}{3}$	$^1/_4$	$^1/_5$
α	33° 40'	26° 34'	21° 48'
secα	1.2018	1.1180	1.0770
23	− 0.9014W	−1.1180W	−1.3460W
B1	− $3\times\overline{23}$	− $3\times\overline{23}$	− $3\times\overline{23}$
C3	− $2\times\overline{23}$	− $2\times\overline{23}$	− $2\times\overline{23}$
(LX)	+ 0.7500W	+ W	+1.2500W
L1=L2	+ $3\times\overline{LX}$	+ $3\times\overline{LX}$	× $3\times\overline{LX}$
12	O	O	O
33	+ W	+ W	+ W

六格候氏桁架各肢段所受之固重應力表

23	$-0.9014W$	$-1.1180W$	$-1.3460W$
B1	$-5\times\overline{23}$	$-5\times\overline{23}$	$-5\times\overline{23}$
C3	$-4\times\overline{23}$	$-4\times\overline{23}$	$-4\times\overline{23}$
D5	$-3\times\overline{23}$	$-3\times\overline{23}$	$-3\times\overline{23}$
(LX)	$0.7500W$	W	$1.2500W$
L1=L2	$+5\times\overline{LX}$	$+5\times\overline{LX}$	$+5\times\overline{LX}$
L4	$+4\times\overline{LX}$	$+4\times\overline{LX}$	$+4\times\overline{LX}$
12	O	O	O
34	$+\frac{1}{2}W$	$+{}^{1}/_{2}W$	$+{}^{1}/_{2}W$
55'	$+2W$	$+2W$	$+2W$
45	$-1.2500W$	$-1.4142W$	$-1.6008W$

八格候氏桁架各肢段所受之固重應力表

23	$-0.9014W$	$-1.1180W$	$-1.3460W$
B1	$-7\times\overline{23}$	$-7\times\overline{23}$	$-7\times\overline{23}$
C3	$-6\times\overline{23}$	$-6\times\overline{23}$	$-6\times\overline{23}$
D5	$-5\times\overline{23}$	$-5\times\overline{23}$	$-5\times\overline{23}$
E7	-4×23	$-4\times\overline{23}$	-4×23
(LX)	$0.7500W$	W	$1.2500W$
L1=L2	$+7\times\overline{LX}$	$+7\times\overline{LX}$	$+7\times\overline{LX}$
4L	$+6\times\overline{LX}$	$+6\times\overline{LX}$	$+6\times\overline{LX}$
L6	$+5\times\overline{LX}$	$+5\times\overline{LX}$	$+5\times\overline{LX}$
12	O	O	O

34	+ $\frac{1}{2}$W	+ $\frac{1}{2}$W	+ $\frac{1}{2}$W
56	+ W	+ W	+ W
77'	+ 3W	+ 3W	+ 3W
45	− 1.2500W	− 1.4142W	− 1.6008W
67	− 1.6770W	− 1.8027W	− 1.9516W

細察上列之表,乃由固重應力圖得來;除肢段$\overline{45}$及$\overline{67}$較爲複雜外;其他節段關係至簡祇須計肢段$\overline{23}$及$\overline{LX}$之應力;其他肢段乃此二數之倍數而已。垂直之引力肢段,由(圖二,四,六)可直接看出

肢段$\overline{45}$應力之計算法:(參看圖三,及圖六)

從圖三取出△4,及圖六之△456.作圖八.

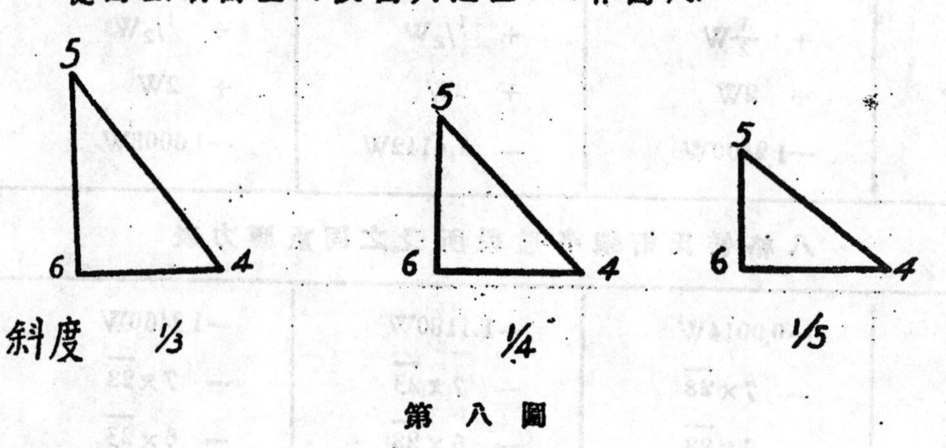

第八圖

(圖八)

每三角形底邊$\overline{64}$之長度皆等于$\frac{1}{8}$l,$\overline{56}$應力＝W

斜度$\frac{1}{3}$,圖八a:

$\overline{64}=\frac{1}{8}l$,　　$\overline{56}=\frac{1}{6}l$,(圖六)

$\overline{64}:\overline{56}=\frac{1}{8}:\frac{1}{6}=3:4$　(長度之比)

$\therefore \overline{45}=\sqrt{3^2+4^2}=5$

$\overline{45}$應力：$\overline{56}$應力＝5：4

$\therefore \overline{45}$應力＝$\frac{5}{4}\times\overline{56}$＝1.25 W

斜度；$\frac{1}{4}$圖八b：

$\overline{46}=\frac{1}{8}l$，$\overline{56}=\frac{1}{8}l$，

$\overline{46}$：$\overline{56}=\frac{1}{8}l:\frac{1}{8}l=1:1$　　（長度之比）

$\therefore \overline{45}=\sqrt{1^2+1^2}=\sqrt{2}$.

$\overline{45}$應力：$\overline{56}$應力＝$\sqrt{2}$：1

$\therefore \overline{45}$應力＝$\sqrt{2}\ \overline{56}$＝1.4142 W.

斜度圖八C

$\overline{46}=\frac{1}{8}l$.　$\overline{56}=\frac{1}{10}l$

$\overline{46}$：$\overline{56}=\frac{1}{8}l:\frac{1}{10}l=5:4$.　（長度之比）

$\therefore \overline{45}=\sqrt{5^2+4^2}=\sqrt{41}$

$\overline{45}$應力：$\overline{56}$應力＝$\sqrt{41}$：4

$\therefore \overline{45}$應力＝$\frac{\sqrt{41}}{4}\times\overline{56}$＝1.6008 W

肢段$\overline{67}$可依此法算之.

根據以上所列之表，可以作圖表之.因各肢段應力之方程式皆爲一次方，故必爲一直線圖.今作圖二種，其用法相同.但必先計算每格所載之固重W，則可肢段所受之固重應力，立即可從圖中得之.

舉例：今有六格候氏桁架，斜度＝$\frac{1}{4}$，w＝1800磅.

(一)各肢段應力之確值可從應力表算之.

$\overline{23}$＝－1.1180 W＝－1.1180 × 1800＝－2012磅.

$\overline{B1}$＝－5 × $\overline{23}$＝－5 × 2012＝－10060磅.

25141

$\overline{C3} = -4 \times \overline{23} = -4 \times 2012 = -8048$ 磅

$\overline{D5} = -3 \times \overline{23} = -3 \times 2012 = -6036$ 磅

$(\overline{LX}) = W = 1800$

$\overline{L1} = \overline{C2} = +5 \times \overline{CX} = +5 \times 1800 = +9000$ 磅

$\overline{C4} = +4 \times \overline{CX} = +4 \times 1800 = +7200$ 磅

$\overline{12} = 0$

$\overline{34} = +\frac{1}{2} W = +\frac{1}{2} \times 1800 = +900$ 磅

$\overline{55'} = +2W = +2 \times 1800 = +3600$ 磅

$\overline{45} = -1.4142\,W = -1.4142 \times 1800 = 2546$ 磅

(b)各肢段應力之約值,可從圖十得之.(因製版關係,迫得將圖縮小,至難得各肢段應力之確值)

先尋得斜線W=18者與六格之各肢段(垂直線)相交之點,而後看橫線,即得各肢段之應力(S),再乘100,即應力之值.

$\overline{34} = 9.00$

$\overline{23} = 2.$

$\overline{45} = 25.5$

$\overline{55} = 36.0$

$\overline{D5} = 60.2$

$\overline{L4} = 72.0$

$\overline{C3} = 80.6$

$\overline{L1} = 90.0$

$\overline{B1} = 100.6$

(c)或從圖十三得之.先尋得垂直線W=1.8.圖中第一行之注字爲四格之肢段;第二行爲八格;第三行爲八格.此爲六格桁架故從第二行注字之斜線與W=1.8之垂直線相交之點,而後看橫線,即得支肢段之應力(S),再乘

1000,卽得其值.

$\overline{34} = .095$

$\overline{23} = .20$

$\overline{45} = .26$

$\overline{55} = .36$

$\overline{D5} = .601$

$\overline{L4} = .72$

$\overline{C3} = .80$

$\overline{L1} = .90$

$\overline{B1} = 1.00$

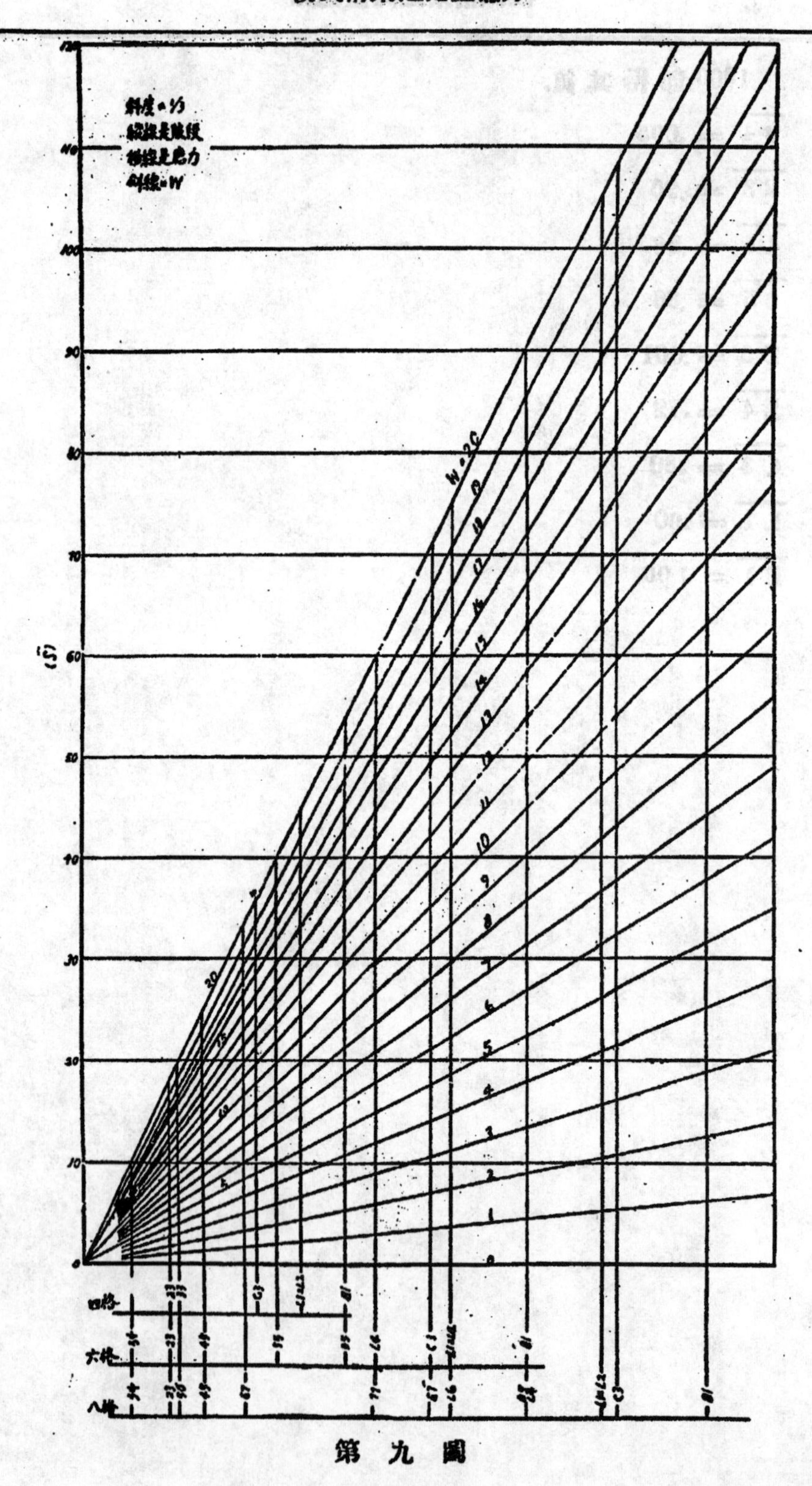

第九圖

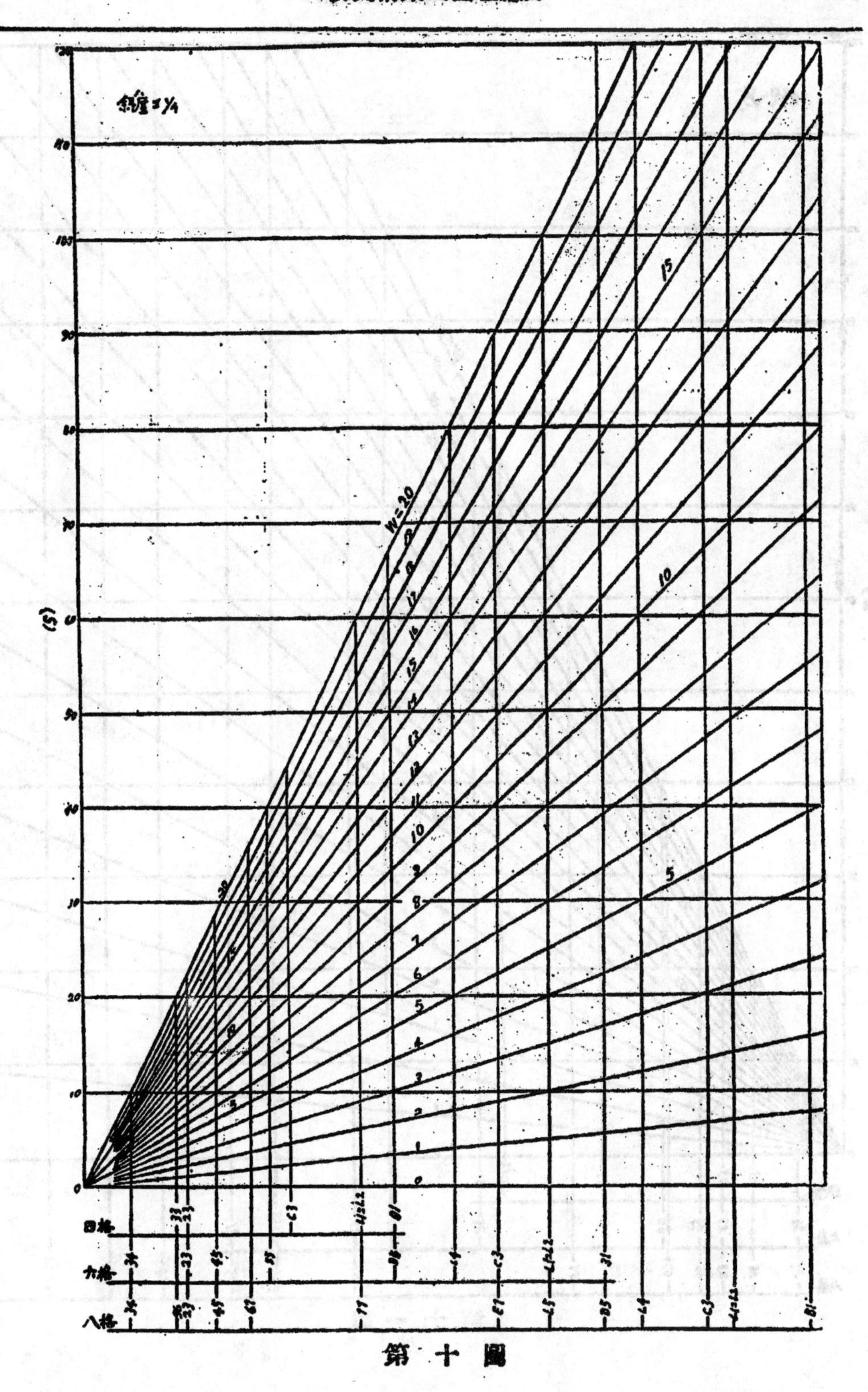

第十圖

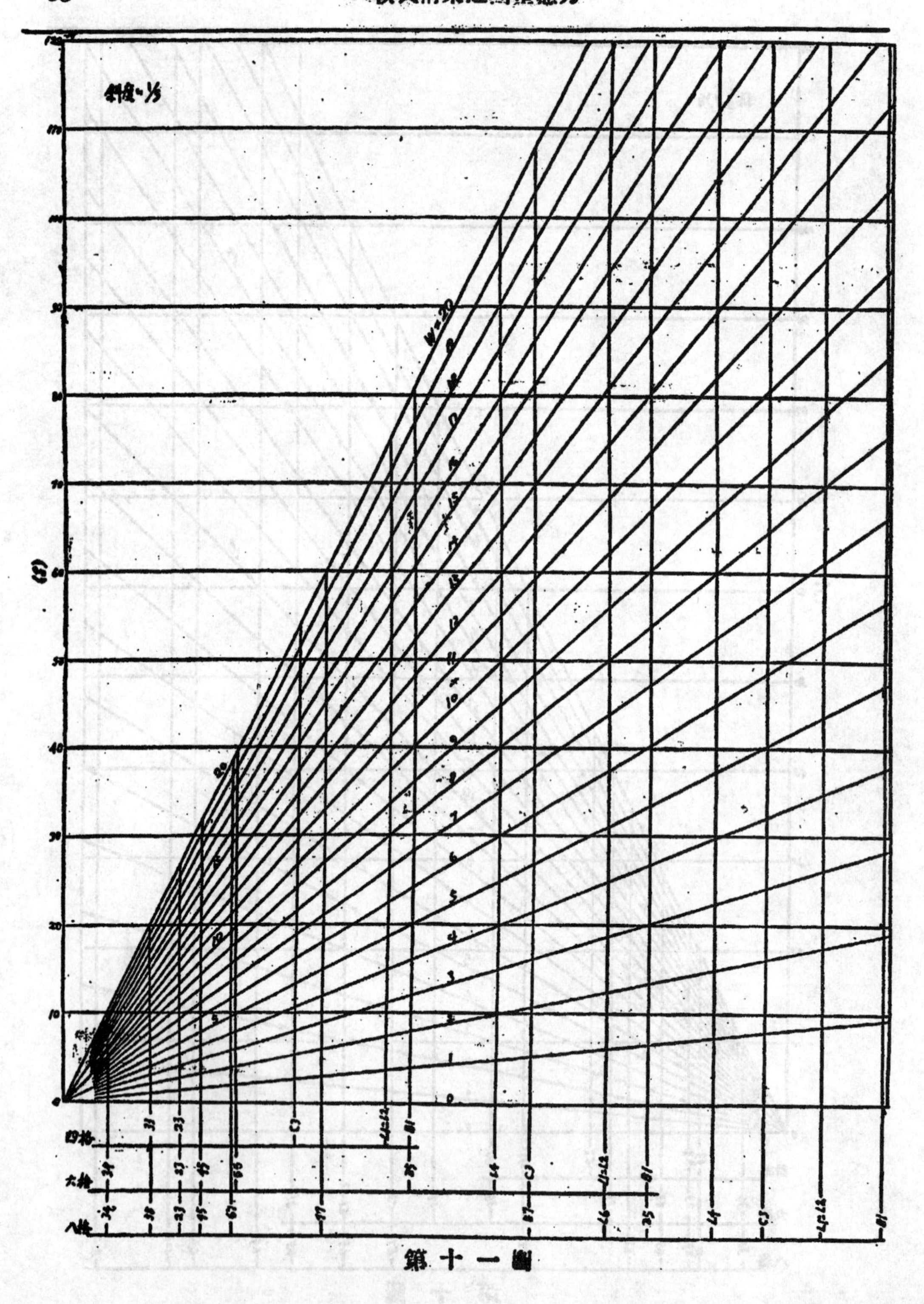

第十一圖

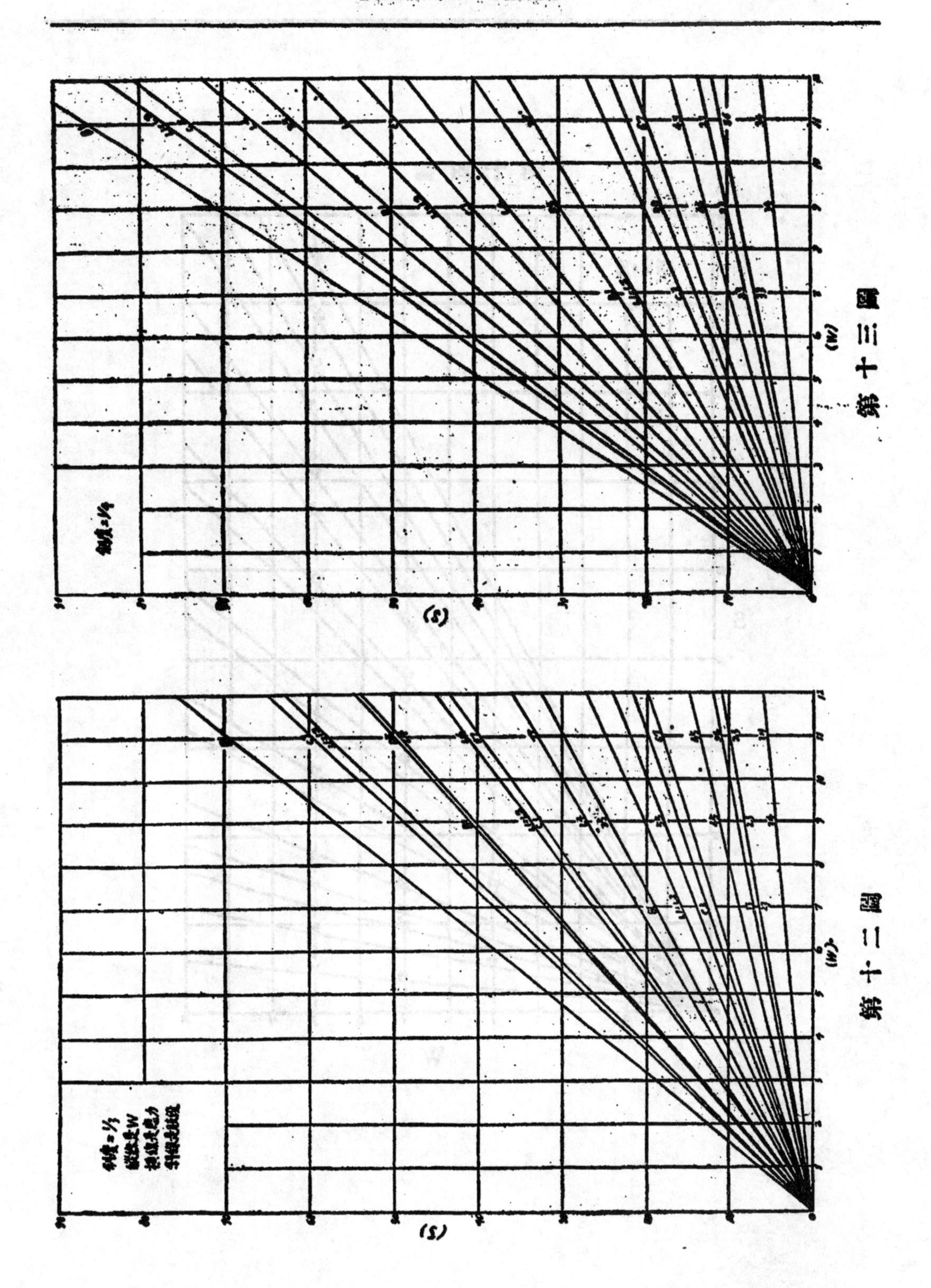

第十三圖

第十二圖

第十四圖

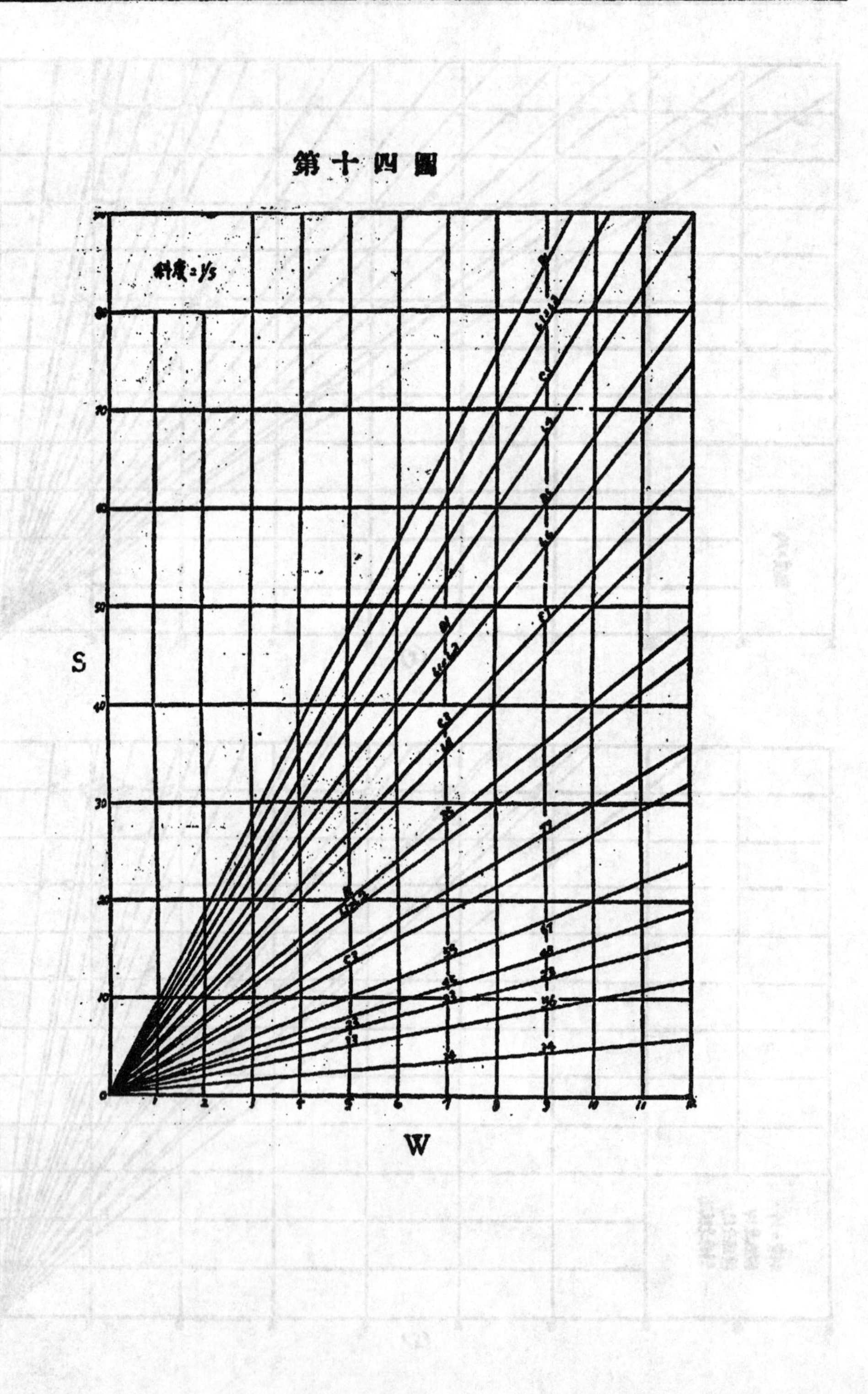

平面圖形重心之求法

李 文 泰

I·緒言　在各種工程設計中時有求平面圖形之重心之需要。然課本中論及者多以代數法求之。但普通幾何平面圖形之重心以圖解法求之更爲簡便。故特舉數種普通平面圖形之重心之圖解法於此。學雖膚淺，但或能供實用之參考。

重心的意思就是說地球對于該物体各部份引力的合力的作用點。若將此物體分爲無數小部份，而地球對于各小部份的引力都是平行的。故一不規則之平面，或由多種幾何圖形組合而成之平面，皆可先求各部份之重心，然後求各重力，即一組之平行力，之合力之作用線。再假設此引力作用於他一方向，求此方向之合力之作用線。此先後求得之作用線之交點即爲此平面之重心。如圖一，圖二，圖三所示者是也。圖一之平面爲三矩形A. B. C. 組合而成。先假設引力在 fa, fb, fc, 之方向作用 fa, fb, fc, 之長度，是與各平面A, B, C, 爲比例的。用圖解法求得R之作用線 × — ×。（此圖解法在各力學課本中均有詳論，故不贅）又假設引力在 f'a fb', fc, 之方向作用，如前求得合力R, 之作用線Y — Y。× — × 與Y — Y之交點 g, 即爲此個平面之重心也。

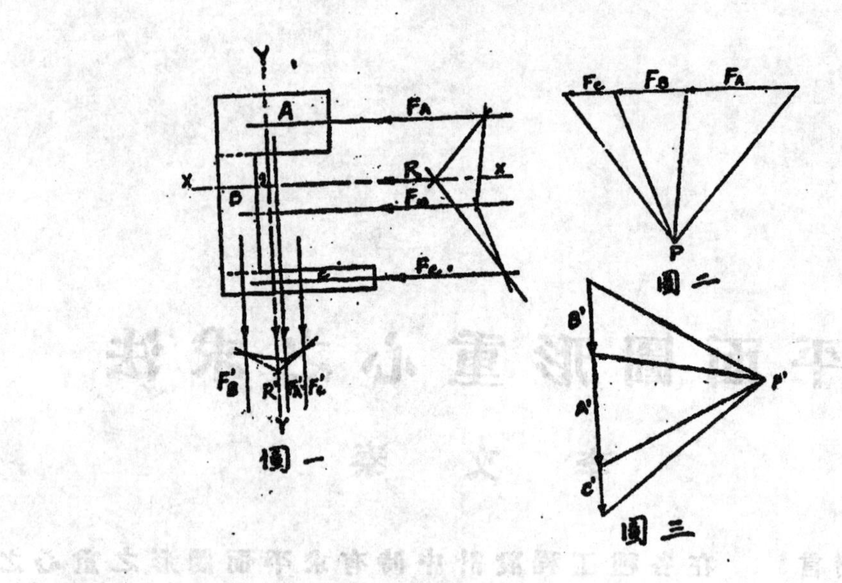

圖一　　圖二　　圖三

但下列各法均與此法稍異其趣。

Ⅱ　折線之重心

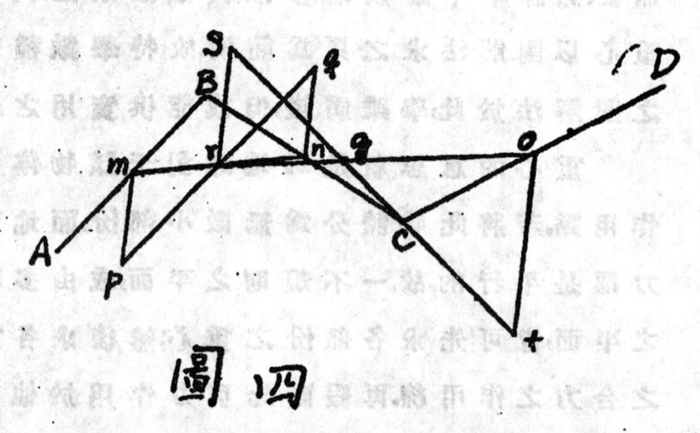

圖四

作法如圖四。聯合A-B之中點m和B-C之中點r。在m，n上畫二平行線m-p。和n-q使nq：mp=AB：BC p-q和m-n之交點r。r便是折線A—B—C之重心。又在r與C—D之中點作二平行線 r—S與 ot使ot。：rs＝AB十BC：CD。St與or之交點 g便是折線A—B—C—D之重心。此作法甚易證明。A—B之重心當在其中點m。BC之重心當在其中點N。則折線A—B—C之重心r必為m—n中之一點。由力學知r必分m—n為二段與AB BC，為反比。即mr：rn＝BC：AB。在圖四△mpr與△rnq由平面幾何知mp：ng＝mr：rn。由作圖mp：nq＝BC：AB。故mr：rn＝BC；AB。因此r即為折線A—B—C之重心。同理rg：go—CD：AB＋BC。故g為折線A—C—B—D之重心明矣。

圓弧之重心 作法如圖五

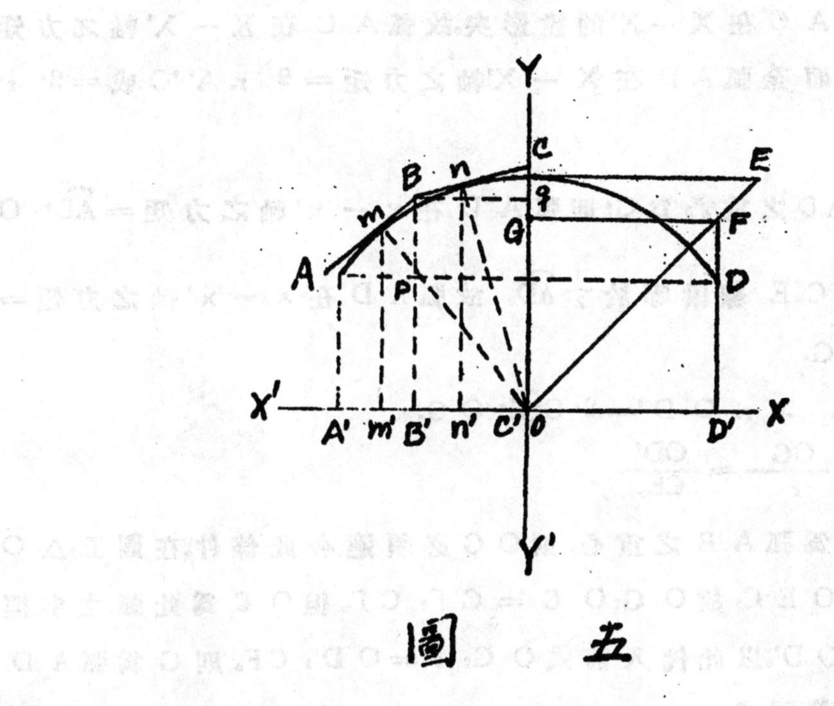

圖 五

由弧AD(以後寫作$\widehat{AD}$)之圓心。作$\widehat{AD}$之平分線Y—Y'。再由o點作Y—Y'之垂直線X—X'。在C點作切線CF，並量出CE＝$\frac{1}{2}\widehat{AD}$＝$\widehat{CD}$聯合OF。在D點作一線與X—X'垂直，交OE於F。由F作FG垂直於Y—Y'。則G點為AD之重心。

此法之證明在圖五之左邊。A—B—C為$\widehat{AG}$之外切多邊形。M, N是AB, BC之各中點。AB線段在X—X'軸之力矩＝AB. m'm.由此圖的幾何，△APB∽△Omm'.故AB: AP＝Om:mm'，則AB·mm＝Om·AP。但是Om是這弧的半徑。AB·mm'＝r. AP。AB線段在X—X¹軸之力矩＝r.AP。同理BC線段在X—X'軸之力矩r＝.Bc。折線A—B—C在X—X'軸之力矩＝r·AP＋rBq＝r(AP＋Bq)·但AP＋Bq為折線A—B—C在X—X'之投影＝A'C'。故折線B—B—C在X—X'

25151

軸之力矩 $= r \cdot A'C'$. 倘若這外切多邊形的邊數增加至無限多，則此多邊形與弧AC接近，因此，此外切多邊形在X－X'的投影亦即是弧AC在X－X'的投影矣。故弧AC在X－X'軸之力矩 $= r \cdot A'C'$。而全弧AD在X－X'軸之力矩 $= 2 \cdot r \cdot A''C$ 或 $= 2 \cdot r \cdot OD'$。

設弧AD之重心為G，則弧AD在x－x'軸之力矩 $= \widehat{AD} \cdot OG$。由作圖CE線段等於 $\frac{1}{2}\widehat{AD}$。故弧AD在x－x'軸之力矩 $= 2 \cdot CE \cdot OG$.

所以　$2 \cdot r \cdot OD' = 2 \cdot CE \cdot OG$.

$$\frac{OG}{r} = \frac{OD'}{CE}.$$

若G為弧AB之重心，則OG必須適合此條件。在圖五：△OFG ∽ △OEC，故 $OG : OC = GF : CF$。但OC為此弧之半徑r，$GF = OD'$。以此代入前式 $OG : r = OD' : CE$。則G為弧AD之重心已證明矣。

三角形之重心。作法聯合A點及其對邊BC之中點n，得中線An又聯合B點及其對邊AC之中點o，得中線Bo。此兩中線之交點g，即為此△ABC之重心。證明若將△ABC分為無數小片平行於其任一邊，如ST為其中之任一小片平行於AB，則中線Cm必平分ST易言之，Cm必含有ST之重心。△ABC可以說是由此種小片平行於AB，組合而成，而每一小片之重心皆在中線Cm之內則中線Cm必含有△ABC之重心。別一中線BO以同一之理也含有△ABC之重心。所以二中線之

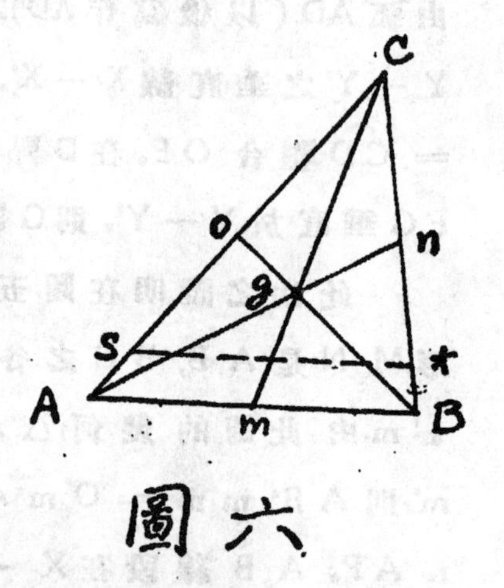

圖六

交點即爲△ABC之重心也.由平面幾何知OG＝$\frac{1}{3}$OB,mg＝$\frac{1}{3}$mC,ng＝$\frac{1}{3}$Ar。故三角形之重心可以說是其任一中線之第三點.

V 四邊形之重心作對角線AC,BD.平分AC得中點c。在BD作Bf＝DK.ef之第三點g,eg＝$\frac{1}{3}$ef,g便是四邊形A－B－C－D之重心.

證明 在圖七裡和是g和g是△ACD和△ACB之重心.並由前段知eg_1＝$\frac{1}{3}$eD;eg_2＝$\frac{1}{3}$eB.

又由力學知g點必須分g_1－g爲下列之比例g_1－g:g－g_2＝△ABC:△ABC又由平面幾何知△ABC:△ADC＝Bk_2

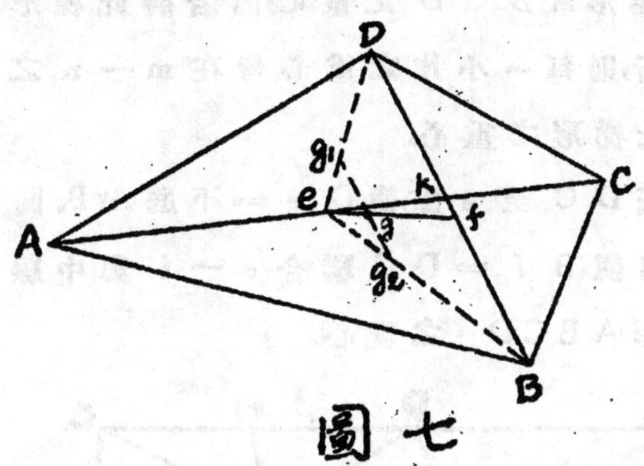

圖七

Dk.但由作圖Bf＝Dk;和Df＝Bk.代入前式,△ABC:△ADC＝Df。Bf在△eBD裡,g_1－g:g－g_2＝Df:Bf。故g_1－g:g g_2＝△ABC:△ADC。g已適合其重心之幾何條件矣.又因$\frac{eg}{ef}$＝$\frac{eg_1}{eD}$＝$\frac{eg}{eB}$＝$\frac{1}{3}$故 eg＝$\frac{1}{3}$ef。

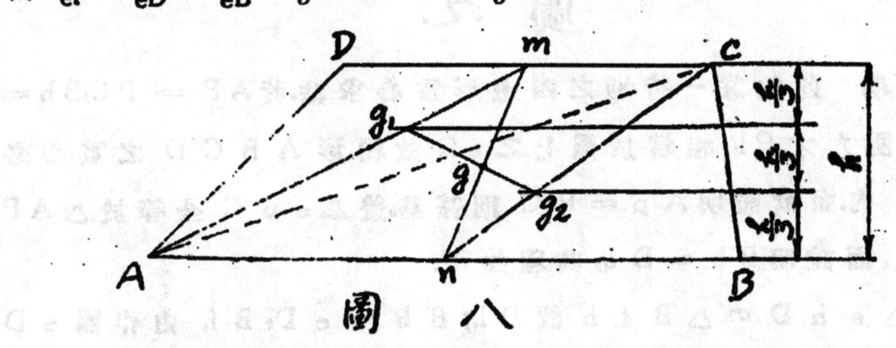

圖八

梯形之重心

梯形之重心之求法有二:(甲)平分上底DC及下底AB得中點m,n.聯合m,n.聯合Am及Cn.求得Am及Cn之各第三點g_1和g_2.聯合g_1—g_2.g_1—g_2與m—n之交點g便是梯形ABCD之重心.

此法證明甚爲簡易.作對角線AC,把梯形ABCD分爲兩個三角形ACD和ABC.(圖八)Am是△ACD之中線,則g_1便是△ACD之重心矣.同樣Cn是△ABC之中線,而g_2是△ABC之重心了.故梯形ABCD之重心必在g—g_2之內.同時mn也必含有梯形ABCD之重心,因若將此梯形分爲無數小片與其底邊平行,則每一小片之重心皆在m—n之內也.故此二線之交點卽爲此梯形之重心.

(乙)引長上底DC至e點,使De=下底AB.同樣引長下底AB至相反方向使Bf=DC聯合e—f與中線m—n相交於g.g便是梯形ABCD之重心.

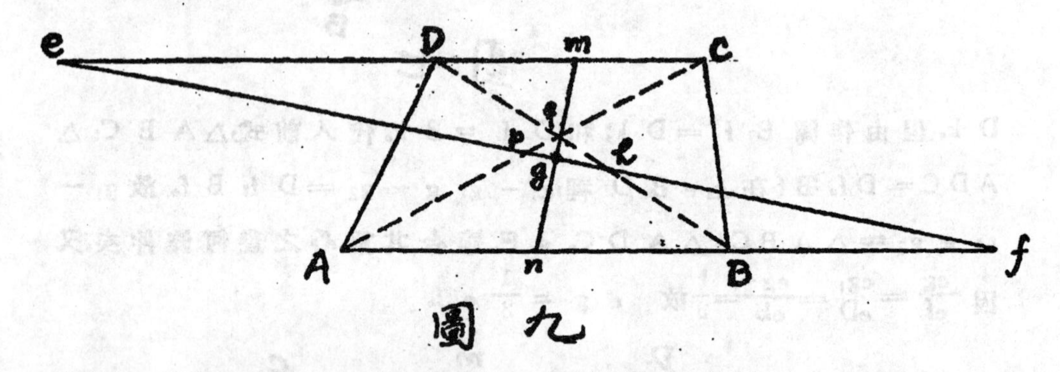

圖九

證明 此法爲一特別之四邊形重心求法.若AP=PC,Bh=Dq則圖九之Ph相當於圖七之ef.故梯形ABCD之重心必在Ph內.而欲證明Ap=PC則甚易.蓋△epC全等於△APf故也.而證明Bh=Dq,其理如下.

△ehD∽△Bfh故Dh:Bh=eD:Bf.由作圖eD

$=AB\ Bf=DC$ 故 $Dh:Bh=AB:DC$。

$\triangle ABq \backsim \triangle DqC$ 故 $Bq:Dq=AB:DC$。代入前式

$Dh:Bh=Bq:Dq$。或寫作 $\frac{Dh}{Bh}=\frac{Bq}{Dq}$。加每邊之分母於其分子得 $\frac{Dh+Bh}{Bh}=\frac{Bq+Dq}{Dq}$。$Dh+Bh$ 和 $Bq+Dq$ 皆等於 DB 故 $Bh=Dc$。

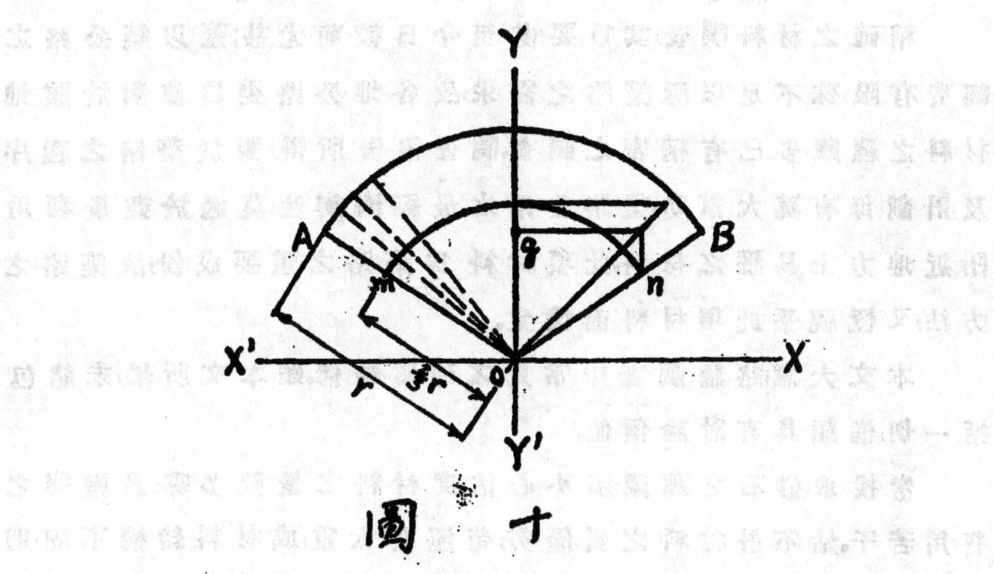

圖十

(1)扇形之重心.以該扇形之半徑之三份二作一弧.以第三段之法求此圓弧之重心亦即該扇形之重心.因此扇形可分爲無數小三角形(如圖十之右)而每一三角形之重心與圓心之距離爲 $\frac{2}{3}r$。此扇形之面積可看作集中于此弧,而此弧爲一等密度之線.故該弧之重心亦即此扇形之重心矣.

(注意:以上各圖之虛線皆用以爲証明者與作法無關.)

翻　譯

築路材料調查與地質結構之關係

倫拿 D. G. Rrrner作　八嘯譯

精確之材料調查，其重要處至今日較前尤甚；蓋以築公路之經費有限，殊不足以應築路之需求，故各地公路委員會對於該地材料之蘊藏，多已有精密之調查。調查報告所得，對於築路之程序及計劃，每有莫大幫助。築路費用之最經濟辦法，莫過於盡量利用附近地方上易獲之材料。此項材料爲築路之重要成份，故築路之方法，又恆視乎此項材料而後定。

本文大意，略論調查中常見之地質結構。雖本文所舉，未能包括一切，惟頗具有討論價值。

當找求砂石之蘊藏，須小心估度材料之量數多寡，及搬運之費用若干。姑不計材料之質優劣，苟運費太重，或材料結構不固，則雖用就地材料，究毫無利益可言。例如冰川流過之地面，其砂石厚度不同，故其搬運用，殊難預算。第一圖表示磐石上之砂石，爲冰川

第　一　圖

冰川構成之磐石上砂石層橫斷圖

所構成。砂石之厚度因各地高低不齊，故雖距離甚近，其厚薄亦不等，因是而砂石之量數及其搬運費用，全無標準。磐石反懸浮砂石，分界甚清，因地面浮土，被冰川磨擦，屢被遷移，祇留下磐石為砂石之基礎，故冰川流過之地，須加意查察，以測定其大約厚度，即附近地方之地質，亦須留意觀察，藉以帮助材料之調查。

斷層為地殼上之破裂，各地每有發見。為地質結構所常見。此種斷層，常令開掘發生困難。第二圖乃一尋常斷層。當一平面層斷

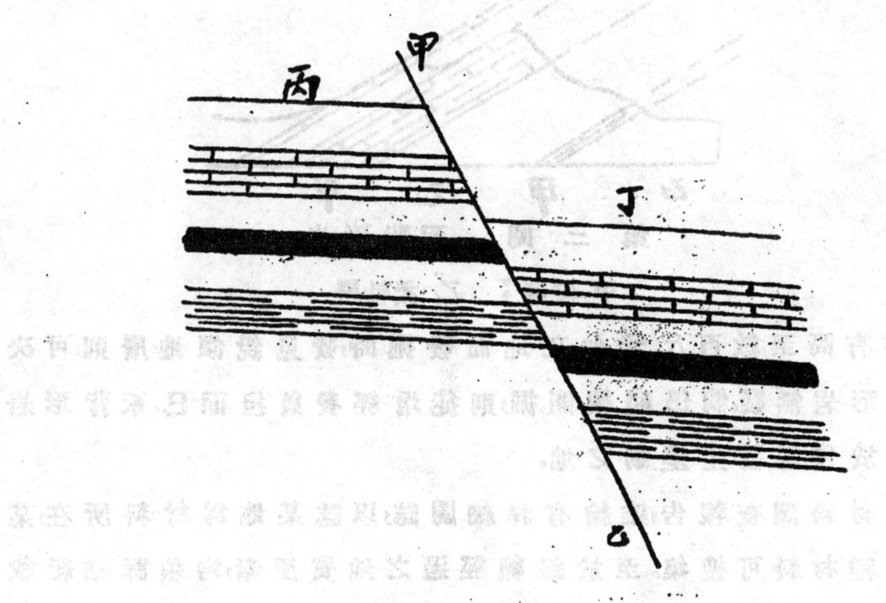

第二圖　平面斷層圖

甲一乙 斷層界；　丙 原地；　丁 下沉層

離則一便往往向低滑下數寸或至數百尺，此種轉變，常由於地殼之伸張力，壓力，或扭力有以致之。當開掘地面石岩，如遇斷層，則石層每因而中止，或一不同性質之石層在此斷層之傍，此種轉變，對於搬運費用，影响殊大，總而言之，在斷層地方多有下列數點標識，

1. 斷層界兩傍地層，有顯明之變動。

2. 同層再次發見

3. 溪澗

4.地層中止

5.斷層常令數層不同抵抗風化力之地層貼合,其結果則地層不堅者,因風化侵蝕而漸次消失,地層堅固者,其抵抗力較大,故高出其他地層而成爲高脊。

第三圖論及別一種開掘困難之地質構造圖示傾斜愈大,則抵抗風化作用較其傍之岩石愈強,此種結構,謂之豕背形岩,(hog-

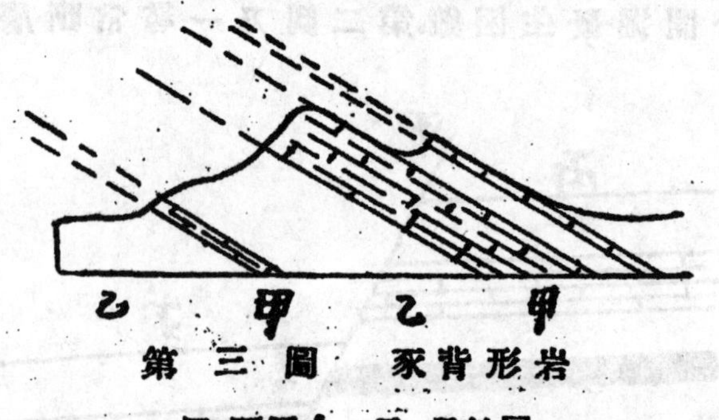

第 三 圖 豕背形岩

甲硬層; 乙柔軟層

back)常有高至數百尺者。如在地面發掘時,發見銳傾地層則可决爲豕背形岩無疑。倘仍繼續開掘,則徒增經費負担而已。豕背形岩常發見於曾經發生變動之地。

做材料調査報告,雖繪有詳細圖誌,以誌某地爲材料所在,某地有某種材料可搜集。至於路線經過之地質原素,均須詳細紀載此種圖誌,雖非絕對須要,惟對於築路人員,及築路策劃,有絕大幫助。

直面曲線之圖表解法

Robert N. Abbctt作

煦　譯

設計工程師之任職於道路或都市機關者,時須計算無限量之直面曲線(Verticale Curves)若能用下列之圖表,則對于計算上之工作與時間,皆有極大之減省。

尋常拋物線形之直面曲線,其在交點(P.I.)處由切線坡度至曲線之縱坐標,多用$Y_o=\frac{LG}{800}$一式求之。式中之L爲曲線之橫投影之長度;而G爲兩切線斜度之代數和。在普通長度之曲線,此式可以圖表解之。

在曲點(P.C.)或切點(P.T.)與交點之間,由切線坡度至曲線上任何一點之縱坐標爲$Y_x=\frac{Y_o X^2}{\left(\frac{L}{2}\right)^2}$。下表所列者,爲曲線在各種長度時$\frac{x^2}{\left(\frac{L}{2}\right)^2}$之值,如欲求$Y_o X$之值時,將表中所列X之值乘$Y_o$便得.

下列之圖,可用插入法(Interpolation)將L之値增加,因Y_o與L之值,恰成正比例。表中之值,雖祇限于L＝300,然亦可用以解L較大之曲線,蓋當x之值相似時,各種長度之曲線其$\frac{x^2}{\left(\frac{L}{2}\right)^2}$之

值皆同故也。(100 尺曲線 X 等于 10 尺則 200 尺曲線 × 必等于 20 尺)

此圖示各種長度之直面曲線在各種斜度不同時正中縱坐標之值

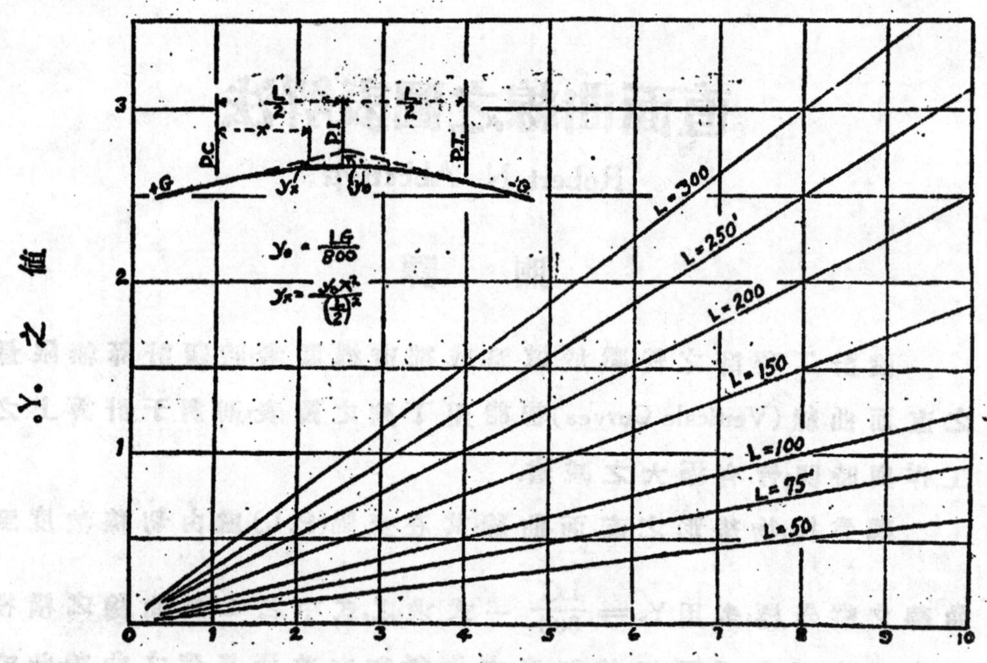

下表所示為公式 $Y_x = \dfrac{Y_o X^2}{\left(\frac{L}{2}\right)}$ 中 $\dfrac{X^2}{\left(\frac{L}{2}\right)}$ 之值，而 Y_x 為直面曲線上曲點與交點間任何一點之縱坐標.

x	L=75	L=100	L=150	L=200	L=250	L=300	x	L=150	L=200	L=250	L=300	x	L=250	L=300
1	0.001	0.000	0.000	0.000	0.000	0.000	51	0.462	0.260	0.166	0.116	101	0.653	0.453
2	0.003	0.002	0.001	0.00	0.000	0.000	52	0.481	0.270	0.173	0.120	102	0.666	0.462
3	0.006	0.004	0.002	0.001	0.001	0.000	53	0.499	0.281	0.180	0.125	103	0.679	0.472
4	0.012	0.006	0.003	0.002	0.001	0.001	54	0.518	0.292	0.187	0.130	104	0.692	0.481
5	0.018	0.010	0.004	0.003	0.002	0.001	55	0.538	0.303	0.194	0.135	105	0.706	0.490
6	0.026	0.014	0.006	0.004	0.002	0.002	56	0.557	0.314	0.201	0.140	106	0.719	0.499
7	0.035	0.020	0.009	0.005	0.003	0.002	57	0.578	0.325	0.208	0.145	107	0.733	0.509
8	0.045	0.026	0.012	0.006	0.004	0.003	58	0.598	0.336	0.215	0.150	108	0.746	0.518
9	0.058	0.032	0.014	0.008	0.005	0.004	59	0.619	0.348	0.223	0.155	109	0.760	0.520
10	0.071	0.040	0.018	0.010	0.006	0.004	60	0.640	0.360	0.230	0.160	110	0.774	0.538
11	0.086	0.048	0.022	0.012	0.008	0.005	61	0.662	0.372	0.238	0.165	111	0.789	0.548
12	0.102	0.058	0.026	0.014	0.009	0.006	62	0.683	0.384	0.246	0.171	112	0.803	0.557
13	0.120	0.068	0.030	0.017	0.011	0.008	63	0.706	0.397	0.254	0.176	113	0.817	0.567
14	0.140	0.078	0.035	0.020	0.013	0.009	64	0.728	0.410	0.262	0.182	114	0.832	0.578
15	0.160	0.090	0.040	0.023	0.015	0.010	65	0.751	0.423	0.271	0.188	115	0.846	0.588
16	0.182	0.102	0.045	0.026	0.017	0.012	66	0.774	0.436	0.279	0.194	116	0.861	0.598
17	0.206	0.116	0.052	0.029	0.019	0.013	67	0.798	0.449	0.287	0.200	117	0.876	0.608
18	0.230	0.130	0.058	0.032	0.021	0.014	68	0.822	0.462	0.296	0.206	118	0.891	0.619
19	0.257	0.144	0.064	0.036	0.023	0.016	69	0.846	0.476	0.305	0.212	119	0.906	0.629
20	0.284	0.160	0.071	0.040	0.026	0.011	70	0.871	0.490	0.314	0.218	120	0.922	0.640
21	0.314	0.176	0.078	0.044	0.028	0.020	71	0.896	0.504	0.323	0.224	121	0.937	0.651
22	0.344	0.194	0.086	0.048	0.031	0.022	72	0.921	0.518	0.332	0.230	122	0.953	0.662
23	0.376	0.212	0.094	0.053	0.034	0.024	73	0.947	0.533	0.341	0.237	123	0.968	0.672
24	0.410	0.230	0.102	0.058	0.037	0.026	74	0.973	0.548	0.351	0.244	124	0.984	0.683
25	0.444	0.250	0.111	0.063	0.040	0.028	75	1.000	0.563	0.360	0.250	125	1.000	0.694
26	0.481	0.270	0.120	0.068	0.044	0.030	76		0.578	0.370	0.257	126		0.706
27	0.518	0.292	0.130	0.073	0.047	0.032	77		0.593	0.380	0.264	127		0.717
28	0.557	0.314	0.140	0.078	0.050	0.035	78		0.608	0.389	0.270	128		0.728
29	0.598	0.336	0.150	0.084	0.054	0.037	79		0.624	0.399	0.277	129		0.740
30	0.640	0.360	0.160	0.090	0.058	0.040	80		0.640	0.410	0.284	130		0.751
31	0.683	0.384	0.171	0.096	0.061	0.043	81		0.656	0.420	0.292	131		0.763
32	0.728	0.410	0.182	0.102	0.065	0.045	82		0.672	0.430	0.299	132		0.774
33	0.774	0.436	0.194	0.109	0.070	0.048	83		0.689	0.441	0.306	133		0.786
34	0.822	0.462	0.206	0.116	0.074	0.052	84		0.706	0.452	0.314	134		0.798
35	0.871	0.490	0.218	0.123	0.079	0.055	85		0.723	0.463	0.321	135		0.810
36	0.921	0.518	0.230	0.130	0.083	0.058	86		0.740	0.474	0.329	136		0.822
37	0.973	0.548	0.244	0.137	0.088	0.061	87		0.757	0.484	0.336	137		0.834
38		0.578	0.257	0.144	0.092	0.064	88		0.774	0.495	0.344	138		0.846
39		0.608	0.270	0.152	0.097	0.068	89		0.792	0.507	0.352	139		0.859
40		0.640	0.284	0.160	0.102	0.071	90		0.810	0.518	0.360	140		0.871
41		0.672	0.299	0.168	0.108	0.075	91		0.828	0.530	0.368	141		0.884
42		0.706	0.314	0.176	0.113	0.078	92		0.840	0.541	0.376	142		0.896
43		0.740	0.329	0.185	0.118	0.082	93		0.865	0.554	0.384	143		0.909
44		0.774	0.344	0.194	0.124	0.086	94		0.884	0.566	0.393	144		0.921
45		0.810	0.360	0.203	0.130	0.090	95		0.903	0.578	0.401	145		0.934
46		0.846	0.376	0.212	0.136	0.094	96		0.922	0.590	0.410	146		0.947
47		0.884	0.393	0.221	0.141	0.098	97		0.941	0.602	0.418	147		0.960
48		0.922	0.410	0.230	0.147	0.102	98		0.960	0.614	0.427	148		0.973
49		0.960	0.427	0.240	0.154	0.107	99		0.980	0.627	0.436	149		0.987
50		1.000	0.444	0.250	0.160	0.511	100		1.000	0.640	0.444	150		1.000

四種路面價格之比較

李文泰註譯

(By W.W. Zass.)

各種路面價格之比較問題是時常發生的。雖然這個問題的直接解答是不可能,但我們可以由每種路面每英里之最初投資及改進後之壽命等,算出每一種路面之極價(Ultimate Cost)則每一種路面之經濟可由此比較矣。或將此極價化爲現值(PresentWorth)因此種將來支給之數須折扣爲今日之值也。

此種比較性質能否準確是很難說得定的,須視各地之特別情形而定;如初價(First cost)須依物料及人工價錢而定;年中之維持費須依運輸之密度及路面之壽命等而定。其他如氣候,土壤,和排水情形等,皆爲影響各種費用之要素。利率之高低又須視該地之金融情形而定。

下列各圖表爲各種路面每英里之建築費及維持費包含路面之鋪砌而已。其他各種工作之費用,如路基的建築,排水,號誌等費用,無論路面是用何種材料此等費用都是一樣,故不包連。

年中之維持費在高等鋪路中包含中線誌號,接縫之修理及小面積之塡補等。在沙礫路中之維持費包含刮平(blading)所需之機械購置及運用費等。在地瀝青路面之維持費則只包含小面積之塡補。材料因長時間之運輸而損壞,須重行補置者則不在維持費之內。如改進後之壽命,假設爲三年,八年,十年,則每年須重行補還之材料及人工費用爲初築費之$\frac{1}{3}$，$\frac{1}{8}$，或$\frac{1}{10}$如此則路面初築時之值可得繼續維持矣。

表中各曲線是代表最低和最高情形的。最低初價和最低維持費合爲一類。最高初價和最高維持費合爲一類。若情形變換，則所繪出之曲線當在此二者之間。極價之現值可由$P=(1+r)^{-n}$公式得之 P＝一元之現值。r＝利率。n＝年數若r爲百分之五則$(1+r)^{-n}$之值如下：

5年	10年	15年	20年	25年
0.7835	0.6139	0.4810	0.3769	0.2953

極價之現值減去廢值（Salvage Value）之現值，其差可以爲價格比較之基礎矣。

例　今引高等鋪路與地瀝青路面之比較作爲例。設高等鋪路二十年後之極價爲$56,500 (1)此數爲最高情形和最低情形之平均數。設二十後之廢值爲初價之百分二十五，(2)可得一平均之廢值爲$4,750。(3)極價之現值爲$21,295元廢值之現值爲$1790元其差約爲每英里 $19,505元此數可代表高等鋪路中各種費用之比較的價值。

二十年後沙礫路地瀝青面之極價爲$57,000。(4)此數爲三年及八年壽命之最高及最低之平均值。其平均廢值爲$7,500。(5)極價之現值爲$21,483，廢值之現值爲$2,827。其差爲$18,656。此數可代表沙礫路地瀝青面各種費用之比較的價值矣。

在此引證中是照實際支用之金額而言。此二種路面的比較的價值是不相上下的。第一種需要很高的最初投資，但運用費少。第二種之最初投資低，但運用費大。在此種或別種之比較中我們還須顧及車輛在此兩種不同路面上之運用費，及旅行之舒適等最後還須視有無充裕之基金爲最初之建築費用否。

各種路面價格比較表

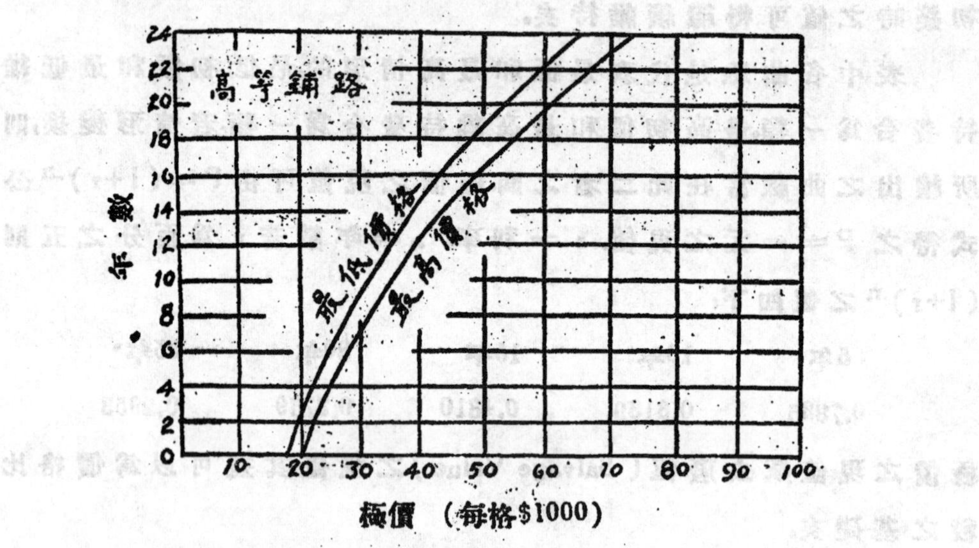

改良後之壽命由15年至25年。在壽命期內沒有補設

每英里之鋪砌面積爲 10560 sg.yd.

	最低	最高
每英里之初價	$18,000	$20,000
每英里年中之維持費	150	200

表 二

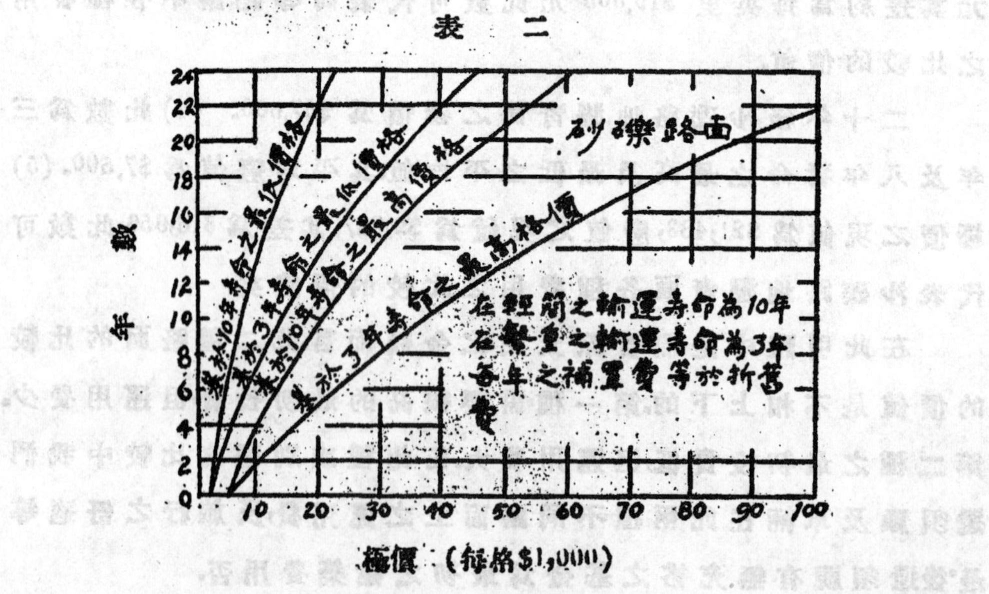

每英里之砂礫路面材料爲 1,760 cu.yd.

	最低	最高
每英里之初價	$2,000	$5,000
每英里年中之維持費	150	450
每英里年中之補置費（3年壽命）	667	1,667
每英里年中之補置費（10年壽命）	200	500

表　三

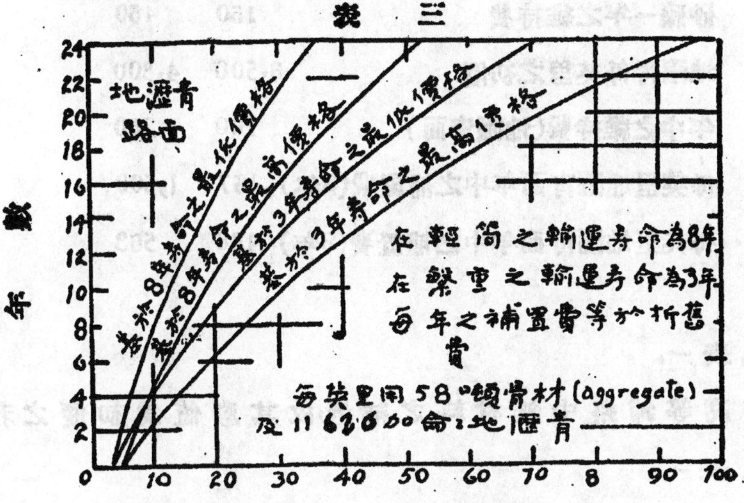

極價　（每格$1,000）

	最低	最高
每英里之初價	$3,500	$4,500
每英里年中之維持費	100	250
每英里年中之補置費（3年壽命）	1,167	1,500
每英里年中之補置費（8年壽命）	438	563

表　四

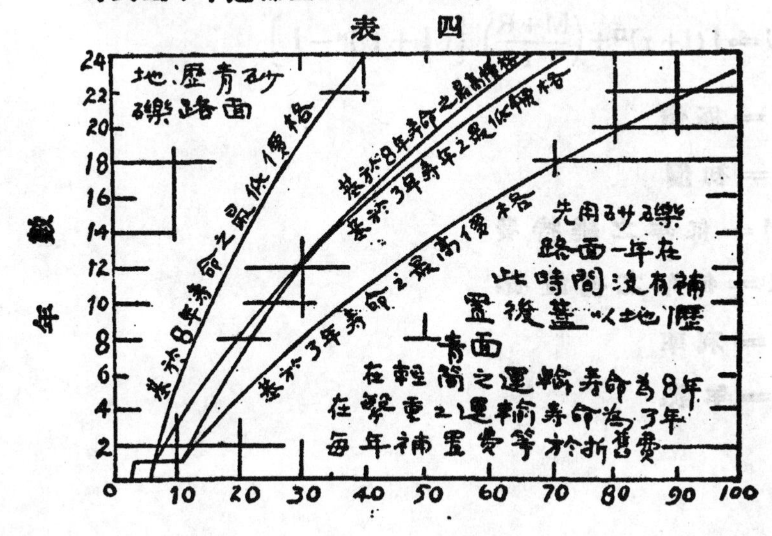

極價　(每格$1,000)

每英里之砂礫路面材料爲 1,760 cu.yd.

每英里用58噸骨材及 11,618 加崙之瀝青

	最低	最高
每英里砂礫之初價	$2,000	$5,000
砂礫一年之維持費	150	450
地瀝青每英里之初價	3,500	4,500
年中之維持費(地瀝青面)	100	250
每英里地瀝青面年中之補置費(3年)	167	1,500
每英里地瀝青面年中之補置費(8年)	438	563

註釋:

(1)見表一。

(2)因高等鋪路中無材料之補置,故其廢值爲初價之若干%

(3)見表一。

(4)見表四。

(5)爲砂礫及地瀝青每英里元最高初價和最低初價之平均數。見表四。

(6)此等曲線之普通公式約爲:

$$U = I(1+r)^n + \left(\frac{M+R}{r}\right)\left[(1+r)^N - 1\right]$$

U＝極價

I＝初價

M＝每年之維持費

R＝每年之補置費。

n＝利率

r＝年數。

25166

記事

南山測量記

梁卓芹

(1) 起因

依據嶺南工學院的課程,我們二年級和三年級要有四星期的野外普通測量及野外鉄路測量實習,是在暑假期中舉行.這兩個野外測量實習,我們只要顧名思義便知道是做什麼工作,在去年的經驗中——這也許是南大工學院首次的野外測量實習——因爲沒有確實的地點和工作,故隨便指定一個適宜於我們生活的地點,假設一件事去做,結果我們是嘗不到野外的生活,更不十分信任我們的工作是可靠,同時又因爲這是假設的工作,也許會隨意一點.今年適值獨立第二師肅清南山共匪,進行移民墾殖工作,由南山移墾委員會呈諸西南政務會派本校同學前往測量,一則我們是有個野外實習的好地點,二則政府也省却了不少的費用,結果西南政務會會議通,並撥津貼費三千元.這大概是我們往南山測量的起因.

(2) 南山之地勢及歷史

未說我們到南山之前,我們該要首先明白南山形勢及其歷史.南山位於汕頭之西南,由潮陽縣城西行約六十里可達其邊境之兩英墟,北爲普寧縣,西南毗惠來縣城,再南可直達海濱,縱橫約百餘里,實跨潮陽普寧惠來三縣.入內則山脈縱橫,形勢險要,惟地土肥沃,盛產菓類.從前村落百餘,居民十餘萬糧食頗足自給,因交通不便之故,山內居民甚鮮外出,洵一世外桃園.

南山本其險要地勢,豐富物產,故早爲共黨涎涎,于民國十二年(?)共黨勢力初到廣東,即重視該地,並同時進行各種工作,迨清共時,所有共黨都竄匿該地,以後勢力日漸澎漲.內地居民,其稍爲富有或不甘附共者,均逃往潮陽,普寧,惠來三縣,聞逃出者,固七八年不能歸,所帶出些少現金,早已用盡,彼等旣無戚屬故舊,又無工作可做,故餓死者十之七八.其貧窮不能逃走或甘心附共者,均被共黨赤化.民國廿年共匪勢力已及潮陽,惠來,普寧,海豐,陸豐,揭陽,等縣,汕頭市亦曾一度幾瀕於險.政府此時方注意及之,乃調集大軍圍剿,閱二年始將共匪盡數驅回山內,然尙未能澈底肅清匪巢.迨去年改調獨立第二師張瑞貴部隊由潮陽正面進攻會同各軍包圍南山,計費時一年零八個月始攻破共黨老巢,然軍隊方面損失想亦不少.計共黨赤化南山凡十餘年,其中不甘附共者均逃之四方,故山內完全赤化.至肅清共匪時山內十室十空,屋宇盡遭焚毀,其最完整者亦祗四壁危墻而已.山內縱橫百餘里,除軍隊駐紮外,了無人煙,滿目荒凉,目不忍睹,誠浩刼也.

(3) 出發情形

本來我們決定在七月四日早車往港,隨即轉輪赴汕,各同學早已準備行裝,有先行往港者,但在三日晚,因政府欵項未到,未能

成行,故四日早有不知消息的同學,直赴車站等候,可憐的還是依依不捨的送別人,白到車站跑了一趟.

六日欵項是領到了,是晚由本校庶務先將各同學行李及儀器先行送往廣九站,七日早我們始由學校乘電船轉赴車站,計此行共三十三人,內二年級同學十二人,三年級同學十五人,由王淑

海和羅石麟二位教授率領,另學校派出工人四名同行.

到港後,本擬即日轉輪赴汕,惟該輪因事改期,故在港逗留一天,八日下午始離港,九日下午一時抵汕,不幸因消息不靈之故,師部派來接船的專員,誤會我們九日始來,故于是日十二時折返潮陽,致兩不相會.我們在沒有辦法的當兒,只得先到旅店.惟因汕市素以搶匪著名,故抵步時由全體同學包圍所有行李,幸無絲毫損失.十日午,得師部派來副官一名,領導我們首途,下午一時由汕乘小輪赴潮陽,再由潮陽乘長途汽車逕南山麓之兩英墟,計抵步時已下午八時許矣.該南山移墾委員會亦設于此地,當晚並蒙殷勤設宴招待.計此行除王,羅二教授向學校及外界負責外,並由同學舉出陳銘珊,梁卓芹,崔兆鼎三君,負責辦理一切事務.

(4) 工作情形

我們固爲工作關係,要隨時遷移的,不過此次因住宿不便問題,故祇遷居一次,然每日赴工作地點,及收工回來,要跑不少路程.

我們工作的分配,就是將全隊分爲十小隊,均依其工作種類,以定人數之多少.最重要的一隊,就是「幹線隊」,(main lineParty),此隊派同學二人,及學校派同來之測佚二人,此隊之工作須比其餘各隊先一二日工作,故常比各隊深入七八里,其次就是「水準隊」(Level Party)及「覆水準隊」(Check Level Party),這兩隊每隊都是派同學二人,專隨着「幹線隊」(Main line Party)做水準工作(Leveling).再其次就是「地形測量隊」(Topo Party),這種工作大概每天要有六七隊出發,每隊或三四人不等.其餘就要留下二三同學專司繪圖工作.每天先行之幹線隊將其所測幹線(Main line)之方向,距離,及水準隊所得各點之高低(Eelevation)交繪圖同學先行繪圖,然後大概依地形情勢分段交各地測量隊於翌日負責測繪,大概幹線(Main line)及水準(leve)兩隊,每天可測一英哩至二英里.地形測量隊則將幹線分爲左右翼,長約二三千呎,左右直深入約五千呎,纍計地形測量隊每日六隊出發,可測得兩方哩左右.

我們每天因爲要跑十里八里不等才到測量地點,且儀器笨重,要是完全自己拿的話,每天吃苦不少,幸得移墾委員會代顧得挑伕二十名,每隊派一二名,幫同携帶笨重儀器,其餘精細及輕便儀器,還是要我們自己拿.同時又得獨立第二師部派出工兵十餘人幫助,彼等經我們訓練數日,堪稱良好測伕,對於我們幫助很大.再其次就是我們工作時的治安問題,要知深入這無人烟的匪窟,共黨大隊雖然被擊散,然流離山內者尙屬不少,治安當然不甚安全,故每天由獨立第二師派出武裝軍隊數十名,每隊分撥數名,以資隨行保護.所以每隊裏面雖然只得二三同學,然總計每隊人數亦總至十人左右.

我們每天早上在六時半至七時早膳,早膳後隨着出發去,午餐約至下午一時,是由挑伕分隊担來,下午六時收工,回來已七時許.晚飯後趕着到小河裡洗澡,各種工作做完後已是時九時許.因工作困倦之故,大都比較早一點睡覺,大概每天工作約十一二小時.這時我們每天的生活情形.

在這四星期中的工作和生活情形,可依據我們所住宿的地點分爲兩個時期.第一星期居住在兩英墟,爲第一期,由第二星期遷居林招,直至完工,爲第二期.

第一期:——在這期內我們是住在兩英墟,此地離南山約四五里,爲入山孔道,該移墾委員會亦設立于此,全墟數百戶,間亦有遭共黨焚毁者蓋此地曾一度陷於共匪手上.居民約二千餘,亦有小商店數十,我們居此尙不感十分痛苦.

到此地的第二天,由移墾委員會派員協同兩位教授率領一部份同學往觀察所應測地帶.其餘留守大本營.正午得張鏡澄副師長過訪,談及該師部女宣傳隊隊員,爲前共黨宣傳隊及重要份子,自新後,在師部担任宣傳工作,藉資將功贖罪.我們極欲一睹彼等面目,詢問彼等身世及各種情形,蒙張副師長即傳令彼等到見.計該隊共約十餘人,今天到者只七人,彼等皆潮普惠附近鄉人,彼等說其所以加入共黨原因,大都爲受着双重家庭禮教壓迫,適共黨以自由,解放,婚姻自主爲口號,故樂意加入,迨熟察裡面情形,又覺其只有名無實,斯時又不能脫離,及後革命軍至,彼等始冒險逃出自新,自後願盡忠黨國等語,彼等皆説湖州話,幸得同學某君頗諳該種方言,乃爲我們繙譯然我們所聽到的祇十分之五六而已.

第三天派出幹線一隊,水準兩隊,先行測量幹線,另派出兩隊專從事測量該地之經緯度(Observations for Longitude and Latitude)及其準確方向(Observations for Azimuth)其餘同學則任整理儀器(adjustment of Instrument)工作.

因爲移墾委員會預備分田及種墾工作,所以我們所測繪的圖要精細詳密,雖田界之微亦須給在圖內,我們用的此例尺是百份一,所以工作特別來得慢,同時又因爲樹林菓木極多,對於我們工作上發生不少的防碍.

在這星期內,我們每天都是機械地工作着.每天都在暑炎的太陽下汗流夾背的工作着,雖然比較辛苦,不過我們却感覺到其中別有興趣.這幾天的測量區,多是無主菓園,一切的菓品自然是任我們吃,有一兩位同學吃至作嘔.在這裡間中仍可見三數兩英墟居民到來摘取菓品——據說被等因有軍隊隨同保護我們,故敢冒險入——,因爲菓樹上的菓太多之故,我們每天各隊都很少得會面的機會,有時雖聞聲,而總不能會面.如果有會面的機會我們都很快樂的暢談一會,多是彼此誇耀自已區域範圍內的菓品怎樣精美茂盛,不錯這裏完全是我們的天地.

第二期:——爲着方便我們往返起見,我們在第二星期一便遷居林招,此地爲一大村落駐有警衛隊數百名,但所有房屋均遭焚燬,危牆兀立,瓦礫滿地,我們的住所是在一所燒毁了的祠堂,上面是臨時用葵蓋的.這裏離兩英墟約二十里,除軍隊之外全無居民,蕭條不堪.

在這裏可以說是我們最苦生活的開始,入山愈深,人跡全絕,危險性也愈大,晚上到村外小河洗澡也須派軍隊保護同往,最不幸的就是到這裏第一天的晚上給我們一個極不好的印象,在遷居的這天傾盆大雨,所有行李床鋪,完全濕濕,但我們今天仍照常工作故也變了落湯雞,回來看見到完全濕透了的東西,已很厭悶,同時天公更不造美是晚仍傾盆大雨,四處滴漏,各同學都說在水裏睡了一夜.

到了這裏的第三天,更發生令我們恐怖的事,下午忽聞槍聲數响,驚破這荒野的沈寂,我們立即收隊到適當地點集合,以防意外,隨即查悉某隊同學在其測區內之大山石洞裏,發覺襤褸可疑的人三名在這個荒凉區域內,而又伏在大山的石洞中,雖或不是共黨,然亦萬分可疑,隨同保護的軍隊.一則因爲自已實力過于薄弱,二則不知彼方虛實如何,所以只離遠開槍射擊,結果逃去二人,拿獲中年婦人一名.因爲言語不通之故,我們不能詢向彼是何樣人,及後當局只宜稱彼乃一癲婦而已.然我們終是疑團莫釋,認爲危險性甚大,結果當局答應加派軍隊隨行保護.並另派軍隊往來附近巡查,我們始較爲安心工作.

我們到林招後感覺生活最苦,在這裡因爲交通不便,糧食也是每三天往兩英購買一次.在毫無人烟的荒野,雖火柴一物之微也要預先購備,尤令我們不快就是通訊非常困難,所有來往信件都是由軍隊的糧食購買員替我們作一站的郵差.

我們認爲最不幸就是到這裡後,有三位同學患病,在這裡不特沒有醫生,除了我們的攜帶的平常葯品外,連藥也沒辦法找,一位折返汕頭某同學家裡調理，幾天後隨即回來照常工作，有兩位大概因為利害一點，故要返廣州醫治。

本來在這種工作着下雨，是很不妥的，可是我們却常在歡迎着下雨，因為光着讓雨淋下，一消暑熱的辛苦，却別有風味，不過体質稍弱的，恐怕很容易生病。

兩星期都是這樣過去，在最後的幾天裡，不幸的危險的消息又來臨，我們只看到軍用電話不通，和軍隊調動紛紛，便知不大妥當，多方查詢，始知附近發現共黨一隊，約五六十人，惟槍械犀利，曾於日前在牛角拗村開會。我們快測到牛角拗了，該地是離林招約十餘里，我們只要三四天的工作便可連牛角拗也測完，同時也是我們工作的結束，不幸的却發見了那裏有共黨，所以欲罷不能，欲進不敢，結果得軍隊方面答應每于上午四時先派軍隊到該山搜索，並加派步哨，在各處山顛守望，所以比較安穩一點。不過我們仍不大放心，都以早點離開這危險地帶較為上算一點，後來我們全數讚成每天提早出發，延遲收工，並且各人須趕速工作，俾得早日結束，幸得各同學努力，在兩天把工作趕完，不過在兩天中却辛苦萬分，每天四時半起床，五時早膳，天色微亮便即出發，晚上七時才收工，回到住宿處的時候，已時八時許，故每天都要携帶電燈隨同出發。

七月卅一日，我們便在晨光熹微、大雨如注之下和林招告別。蒼天却是有情，在我們來的時候，下雨以迎，當我們去的時候，也流着淚送了三程。正午抵兩英墟，乃乘長途汽車往潮陽，轉輪往汕，旋於八月二日由汕乘輪返港。

在這期中，還有一部份同學因為要負繪圖之責，故留在林招繪圖，記者因為沒有繪圖，同時這繪圖的幾天中也無甚可記，該部份同學直至八月七日始工作完竣，隨即赴汕返港。

(5)最後的雜記

本來我們預計在四星期內可以測得英里一百平方里,可是一則因比例尺的限制,二則受着詳細的拘束,所以工作上遲緩了幾倍.同時又因爲儀器不十分充足,時間上除了星期日和雨天外實在工作的時間約十六七天.加之以危險性大,即有時間亦不敢再行深入,結果離我們的預計很遠,總計我們的幹線約七英哩,(約合廿二華里),兩翼深入共約一哩半(約合華里五里),面積共約十平方哩(Io Sq Mil)約合華里一百方里,其中山脈,河流,大路,小路,田園,屋宇,橋樑等,俱在圖內,明瞭詳晰.

在生活上,我們此次可以說是嘗到眞實的野外生活,我們數星期內都生活于荒野的深山中,城市好生活自然全無,即使鄉村的生活也嘗不到,可以說是軍人在出征時或是野人生活吧!一切用具,除了輕便的自已帶來的外,別樣用具是完全沒有的,我們吃飯却是蹲在地上,一月來洗面也沒有用過盆,其他可想而知.

這荒凉滿目的南山,自從給我們的歌聲和笑聲打破了這荒山的沉寂後,我們更帶給她無限的希望.事情就是這樣,我們入山以後,原來逃了往別處的鄉民,也因爲我們入了山,才有三數人壯着胆子隨我們回到他們一別數年的荒凉的故鄉.自然他們須先經過手續向政府(南山移墾委員會)取得良民証方得入內,否則以共黨論,以資識別.聞說我們離開此地後,鄉民回去的也逐漸增加起來,這隱病垂危的南山,也逐漸得着生氣了,我們希望她逐漸強壯起來,繁榮起來,盡她一點對民族對國家的責任.然而不知她將來亦會忘記曾爲她犧牲了不少的生命,及我們三十多人一月的事跡不?

在四星期中我們感覺到合作的偉大,我們嘗到大自然的生活,我們當然不勝感謝南山移墾委員會的職員對我們的幫忙和指導，我們更感謝師部派出的十多個工兵替我們做了不少的最苦的工作。

雖然得到政府的補助費，可是除了膳食及川資外，我們自已也用了不少的費用。同時在這短短的時期中，我們到過一處人烟絕跡和常人不能到的境地，我們嘗到野外的真正生活，同時我們也得到不少的實際工作的經驗，雖然受了四星期的苦，也還算值得。

廿二年十一月卅日於爪哇堂

珠江河水化驗結果

由1930年十一月至1931年一月

日期	水	氣味	色澤	濁度	沉澱	含硫總量	燃燒損失	暫時硬度	永久硬度	硬度總量	鈣	鎂	鐵	遊離二氧化碳	溶解氧	硫化氯	氧之成分	遊離硫精	亞硝酸鹽	硝酸鹽	氯
十九年十一月十一日	S	無	5	50	甚少	149	22.6	47.5	10.0	57.5			1.7		5.01	0.2555	1.1				24.0
	B	杉腐味	10	100	少	164	26.9	50.0	12.5	62.5			2.0		5.25	0.1440	1.5				24.0
	C	微腐味	5	450	多	169	59.7	56.0	10.0	66.0			1.6			0.1700	1.2				25.0
十九年十一月十八日	S	無	5	30	甚少	43.1	13.4	52.5	2.5	55.0					5.74	0.383	1.9				30.0
	B	杉腐味	12	100	少	143.1	23.1	52.5	7.5	60.0					5.50	0.296	2.7				31.0
	C	無	5	80	多	96.2	56.1	52.5	2.5	55.0						0.289	3.2				30.0
十九年十一月廿四日	S	無	5	100	甚少	59.4	16.5	55.0	——	55.0	2.10	0.60	1.7		5.25	0.341	3.0	0.00	0.000	0.000	24.0
	B	微腐味	5	200	少	153.1	17.9	55.0	0.0	55.0	2.10	0.60	2.0		5.24	0.256	3.3	0.00	0.000	0.000	24.0
	C	杉腐味	10	350	多	145.3	21.3	57.5	0.0	57.5	22.0	0.60	1.5			0.213	3.6	1.00	0.000	0.000	24.0
十九年十二月四日	S	無	20	50	甚少	89.7	52.8	55.0	2.5	57.5	19.0	2.4		4.0	5.45	0.139	2.5	0.00	0.000	0.000	24.0
	B	微草味	12	30	甚少	159.2	12.2	55.0	5.0	60.0	18.0	3.6			5.16	0.129	2.9	0.00	0.000	0.000	24.0
	C	微草味	12	30	甚少	156.2	49.7	55.0	5.0	60.0	18.0	3.6		5.0		0.186	2.2	0.00	0.000	0.000	22.0
十九年十二月十日	S	無	10	80	甚少	139.1	53.6	55.0	5.0	60.0	21.0	1.8		4.0	5.50	0.202	2.6	0.00	0.000	0.000	27.0
	B	微草味	20	150	少	146.6	22.2	60.0	0.0	60.0	21.0	1.8			5.95	0.129	2.7				23.0
	C	微草味	20	150	少	109.9	31.7	60.0	0.0	60.0	21.0	1.8		10.0		0.107	3.5	1.40	0.000	0.000	26.0
十九年十二月十六日	S	無	10	30	甚少	92.7	58.6	55.0	5.0	60.0	20.0	2.4		6.0	5.45	0.170		0.10	0.1000	0.000	30.0
	B	坭味	10	70	甚少	153.8	32.5	53.0	5.0	58.0	19.0	2.4			5.35	0.152					30.0
	C	坭味	10	70	甚少	294.3	166.5	53.0	5.0	58.0	19.0	2.4		6.0		0.152		0.20	0.1000	0.000	28.0
十九年十二月廿三日	S	無	5	40	無	54.8	52.6	55.0	0.0	55.0	16.0	3.6	1.5	6.0	5.80	0.170	2.2	0.00	0.066	0.000	36.0
	B	汽油味	15	150	少	154.5	73.3	60.0	0.0	60.0	18.0	3.0	1.3		5.83	0.130	2.4				34.0
	C	汽油味	15	150	少	235.0	138.5	60.0	0.0	60.0	18.0	3.6	1.6	9.5		0.130	3.8	1.20	0.107	0.000	34.0
廿年一月十三日	S	無	10	25	甚少	132.0	94.2	57.5	0.0	57.5	19.0	2.4	1.0	3.0	6.25	0.214	2.4	0.20	0.074	1.990	28.0
	C	無	10	30	甚少	122.0	83.0	57.5	0.0	57.5	19.0	2.4	1.0	4.0	6.40	0.193	2.2	0.30	0082	2.210	26.0

廿　年	S	無	10	40	甚少	168.7	100.9	60.0	0.0	60.0	21.0	1.8		7.0		0.214	2.3	0.00	0.049	3.320	22.0
一　月																					
廿　日	C	無	10	30	甚少	246.1	158.6	60.0	0.0	60.0	21.0	1.8		7.0		0.085	2.3	0.20	0.066	3.090	21.0
廿　年	S	無	12	20	無	134.4	34.6	57.0	0.00	57.0	14.8	4.9		7.5		0.170	1.8	0.20	0.250	3.320	20.0
一　月																					
廿七日	C	汽油味	22	50	少	162.9	20.9	75.0	0.00	75.0	20.0	6.1		10.0		0.085	4.9	3.20	0.300	4.870	22.0
廿　年	S	無	5	20	少	223.4	79.9	60.0	17.0	77.0	26.8	2.4		5.5	5.47	0.174	2.2	0.00	0.300	1.330	65.4
二　月	B	泥　味	5	120	少	230.7	44.6	60.0	17.0	77.0	26.8	2.4		5.0	56.0	0.174	2.0	0.00	0.300	1.330	67.3
五　日	C	汽油味	10	250	多	162.1	30.1	65.0	7.5	72.5	25.0	2.4		19.0		0.130	3.7	1.80	1.600	3.000	41.0
廿　年	S	無	10	100	無	276.0	30.0	57.0	15.0	72.0	20.8	4.9	1.4	6.0	8.40	0.266	2.0	0.00	0.006	0.027	75.0
二　月	B	無	10	100	無	255.0	86.0	57.0	15.0	72.0	20.0	4.9	1.6	7.0	8.20	0.006	1.7	0.10	0.006	0.017	75.0
十二日	C	無	10	100	無	342.0	135.0	57.0	15.0	72.0	20.8	4.9	1.4	7.5	8.20	0.006	1.3	0.10	0.005	0.027	46.0
廿　年	S	無	10	130	甚少	365.0	148.0	60.0	32.0	92.0	34.8	1.2	0.7	6.0			1.5	0.00	0.002	0.010	12.3
二　月	B	無	10	200	少	248.0	104.0	60.0	32.0	92.0	34.8	1.2	0.8	6.0			2.2	0.10	0.002	0.009	11.5
十六日	C	汽油味	13	130	甚少	337.0	36.0	60.0	32.0	92.0	34.8	1.2	0.7	7.0			2.1	0.00	0.004	0.018	7.1
廿　年	S	無	10	100	甚少	169.0	64.0	55.0	00.0	55.0	18.0	2.4	0.5	7.5	6.90	0.380	1.7	0.00	0.033	0.070	14.0
三　月	B	無	10	150	少	161.0	81.0	55.0	00.0	55.0	18.0	2.4	0.70	7.5	6.30	0.400	1.3	0.00	0.033	0.058	16.0
八　日	C	植物味	20	300	多	330.0	106.0	52.0	20.0	82.0	28.8	2.4	0.52	16.0	6.10	0.400	5.9	2.80	0.031	0.199	52.0
廿　年	S	魚鯹味	20	400	少	155.0	56.0	42.0	00.0	42.0	15.6	0.7	2.8	8.2	8.13	0.270	2.4	0.00	0.021	0.035	6.0
四　月	B	魚　味	15	600	多	210.0	52.0	42.0	00.0	42.0	15.6	0.7	3.2	9.0	7.50	0.000	2.7	0.40	0.008	0.044	6.0
十五日	C	魚　味	15	400	少	167.0	58.0	42.0	00.0	42.0	15.6	0.7	2.8	12.0	7.96	0.340	2.8	0.20	0.016	0.044	6.0
廿　年	S	泥　味	25	600	多	204.0	60.0	38.0	10.0	48.0	15.2	2.4	11.4	9.9	7.31	0.340	1.7	0.20	0.033	0.018	5.0
四　月	B	泥　味	25	800	多	233.0	40.0	38.0	10.0	48.0	15.2	2.4	8.0	10.0	7.02	0.340	1.8	0.20	0.082	0.035	5.0
廿　日	C	泥　味	25	850	多	248.0	43.0	33.0	00.0	33.0	9.2	2.4	11.0	9.0	7.18	0.410	2.1	0.40	0.033	0.018	5.0
廿　年	S	無	30	600	少	188.0	34.0	35.0	00.0	35.0	10.0	2.4	6.6	12.0	59.0	0.408	2.5	0.60	0.033	0.044	5.0
四　月	B	無	20	800	多	254.0	52.0	35.0	00.0	35.0	10.0	2.4	8.8	13.0	61.5	0.306	2.7	0.60	0.066	0.035	5.0
廿七日	C	泥　味	20	700	少	262.0	57.0	40.0	00.0	40.0	12.0	2.4	8.8	12.0	62.5	0.272	1.9	0.40	0.066	0.013	5.0

廿 年	S	無	15	300	少	132.0	53.0	50.0	00.0	50.0			5.6	8.0	53.0		2.4	0.40	0.130	0.023	2.5
五 月	B	無	20	350	適中	131.0	40.0	50.0	00.0	50.0			6.4	8.0	58.0		2.1	0.40	0.021	0.031	2.5
四 日	C	無	15	300	少	135.0	42.0	50.0	00.0	50.0			8.0	10.0	64.0		2.6	0.45	0.014	0.021	3.0
廿 年	S	無	15	250	甚少	121.0	31.0	36.0	00.0	31.0	7.6	2.9	7.6	5.0	5.8		2.5	0.40	0.016	0.753	3.0
五 月	B	無	25	300	少	158.0	59.0	37.0	00.0	36.0	7.2	4.4	6.0	6.0	5.8		2.6	0.40	0.016	0.885	3.0
十一日	C	無	15	250	甚少	125.0	34.0	38.0	00.0	33.0	10.0	1.9	4.9	7.0	6.3		2.2	0.45	0.008	0.753	3.0
廿 年	S	無	15	350	甚少	164.0	59.0	30.0	——	30.0	11.2	0.5	4.0	6.0	4.40	0.260	3.1	0.20	0.027	0.292	2.0
五 月	B	無	35	900	多	252.0	83.0	30.0	——	30.0	11.2	0.5	7.6	10.0	4.40	0.200	5.2	0.20	0.066	0.243	2.5
十八日	C	植物味	15	350	少	172.0	61.0	30.0	——	30.0	11.2	0.5	4.4	8.0	4.30	0.200	2.7	0.40	0.021	0.257	2.3
廿 年	S	無	15	150	甚少	115.0	35.0	40.0	——	40.0	11.2	2.9	3.4	5.0	4.5	0.273	1.1	0.20	0.0004	0.310	2.0
五 月	B	微泥味	20	180	甚少	118.0	42.0	40.0	——	40.0	11.2	2.9	3.2	5.0	4.9	0.239	2.2	0.20	0.0003	0.530	2.0
廿六日	C	微泥味	20	150	甚少	114.0	37.0	38.0	——	38.0	10.4	2.9	3.0	6.0	5.2	0.239	1.9	0.20	0.0002	0.400	2.0
廿 年	S	微草味	20	150	甚少	120.0	35.0	39.0	2.0	41.0	12.0	2.7	9.8	6.0	7.6	0.239	0.6	0.20	0.0246	0.354	2.5
六 月	B	微草味	20	200	多	160.0	54.0	39.0	1.0	40.0	12.0	2.0	5.9	7.0	7.0	0.168	1.6	0.20	0.0739	0.841	2.5
廿一日	C	微草味	15	100	甚少	110.0	38.0	40.0	18.0	58.0	19.2	2.4	5.0	8.0	7.2	0.237	0.6	0.20	0.0452	0.398	2.5
廿 年	S	無	25	200	甚少	120.0	30.0	44.0	25.0	69.5	25.5	1.5	5.9	6.0	5.1	0.000	0.8	0.20	0.000	0.532	1.0
七 月	B	魚 味	20	400	多	183.0	62.0	44.0	34.0	78.5	34.5	2.3	13.0	6.0	5.7	0.051	2.0	0.20	0.0041	0.620	1.0
廿 日	C	無	20	250	少	134.0	37.0	46.0	26.5	72.5	26.5	1.6	6.1	8.0	5.8	0.000	0.7	0.40	0.0082	0.620	1.0
廿 年	S	無	15	150	——	97.0	23.0	45.5	27.5	73.0	28.3	0.48	4.3	8.0	9.5	0.273	1.65	0.00	0.164	0.354	3.5
九 月																					
廿九日	C	泥 味	10	100	000	96.0	20.0	44.0	31.5	75.5	29.3	0.61	5.5	13.0	8.2	0.273	1.85	0.00	0.131	0.265	3.0
廿 年	S	植物味	12	80	少	113.0	28.0	50.0	25.0	75.0	28.4	1.0	1.4	4.5	5.3	0.068	1.45		0.029	0.443	3.0
十 月																					
十三日	C	植物味	10	65	甚少	105.0	30.0	49.0	18.5	67.5	24.2	1.7	2.0	3.0	6.1	0.034	1.85		0.062	0.576	3.0
廿 年	S	無	5	60	00	100.0	30.0	55.0	80.0	63.0	32.0	1.2	1.2	2.0	5.1	0.100	1.00		0.000	0.000	12.0
十一月																					
廿七日	C	無	10	80.0	甚少	110.0	27.0	55.0	80.0	63.0	22.0	1.5	1.5	4.0	5.4	0.120	1.20		0.000	0.000	11.0

S……河面所取之水

B……河底所取之水

C……小涌所取之水

附　　錄

廣州市西南鐵橋之概要

廣州自珠江鉄橋完成後，一時兩河南北交通，極稱便利。玆當局復以粵漢廣三兩鐵路有聯運之必要，乃築黄沙鉄橋，以利兩路之接駁。惟是該橋横跨珠江，如遇河水高漲，對於來往船隻不無阻碍，且該橋附雙軌鉄道，坡度不能過高，遂用旋轉式開合之設計。

(一)橋身之長寬度及橋底至水面高度

西南鉄橋全橋共分兩段，一段由廣州之黄沙至牛牯沙，計長六百五十六呎。餘一段由江心之牛牯沙跨江而直達對岸之石圍塘，長九百八十四呎。全橋共長一千六百四十呎。為中國西南部之唯一大橋，故曰西南橋。全座寬度共四十二呎。橋底至最高潮水水面為拾四呎。

(二)橋身及橋躉之載重

橋身載重之限制亟宜審慎。而載重之決定，恆視其運輸之性質與情形而定。今西南鉄橋不獨用以接駁粵漢廣三兩鐵路，且用以利便廣州與花地之汽車輸運與行人。故其運輸之性質，非單純一種而兼三者而有之。現西南鉄橋所採取橋身之載重，爲古柏氏E－35，蓋中國目下鐵路運輸情形其載重鮮有超過E－40者故也。而橋躉則用E－50，以備他日之擴充。

(三)橋身之構造

該橋之構造,純爲鋼料所造成。除兩岸均築橋躉外由黄沙至牛牯沙一段,在河中築橋躉三座。另旋轉機橋躉一座。由牛牯沙至石圍塘一段,有橋躉五座。共有橋躉八座,旋轉機橋躉一座。每座距離一百六十四呎,旋轉機橋躉位于第二第三橋躉之中央,每邊各距離八十二呎。橋面用三合土造成,上鋪雙軌鉄路。

(四)交通

此橋建築之目的,除在接駁粤漢廣三兩鉄路,以便聯絡此兩路之運輸外,並兼以利便珠江兩岸之汽車之交通及行人之來往。故橋身之中部,鋪築三合土路面,敷設鉄軌,以便火車及汽車通行外,復于兩傍另附寬約四呎之人行路使徒步行人,亦得相當便利。

黄埔闢港第一期工程計劃

治河會自接又開闢黄埔爲商港以來,經已積極籌築。茲查關於黄埔築港計劃自審定後,該會隨飭技士李文邦妥擬第一期工程計劃及預算,以便請欵着手實施。李氏奉諭,即經悉心擘劃將該項計劃圖則擬就呈覆各常委核示辦理。並探錄李氏簽呈中所擬就之第一期工程計劃及第一期工程預算如下:

工程計劃

第一期工程計劃包含下列各項(一)建築鋼樁,堤岸四千二百英尺,用賴生第五號鋼樁,長六一二一英尺,堤岸位置離河岸約一百英尺至三百英尺。現在河邊圍基高度爲一〇七.四七,將來堤岸高度擬定爲一〇八.一〇,高出于盛潮高水度七英尺,高出于盛潮低水度一五.一九英尺,高出於規測所得最高水度二六二五英尺。即水漲至最高時亦無汎濫之虞。現在河邊第一層之土爲稀薄之浮泥及坭漿,深約十一英尺,不甚適宜於貨倉地台之用,而泥漿橫向之壓力甚大,將來建築時須將樁邊之泥漿移至岸上,而填以魚頭石及沙。魚頭石橫向之壓力比浮坭及坭漿之壓力小，宜塡近樁邊，以減少壓力。堤岸之位置，現根據盛潮低水度下十八英尺等深淺酌量而定。因河床在十八英尺等深處，卽不見浮泥，（請參照挖泥圖）可省却移去浮泥一層手續。及因流速關係，河流深度在此處以外，易於保存。鋼樁由此打入河床內三十英尺可保其堅固不致搖動。在此深度之河床，高於挖掘石河床七英尺。以防在此深度有不規則之深處，以免河底距樁底少於預定之數，而致樁底向外崩潰。

（二）塡高堤岸內之地至標高一〇八.一〇以免水患。計由規定之鋼樁堤邊至現在之河邊，共塡地九〇，七九四平方公尺，卽一〇八.二〇華畝（排錢尺）。塡泥處之平均河水深度約一二.五五英尺（照一九〇七年河序深度測量圖塡泥約四一〇六二六.五方公尺，塡魚頭石一〇四，〇〇〇立方公尺，塡沙一八六〇〇〇立方公尺，所塡沙泥，均可由挖掘河床之沙泥移來塡築。

（三）挖去第一沙（First Bar）第一沙在鰲魚洲之北岸對開，橫梗河中，其深度由十英尺至十八英尺，阻碍航行。須挖至盛潮低水度下二十五英尺，使與鋼樁堤邊港深相符合。計挖泥約二六，一四五立方公尺，卽九，二三二，〇〇〇立方公尺，用爲塡築堤內地方之用。

（四）挖深鋼樁堤岸對開處之河床。鋼樁堤岸對開處卽龍船沙北便對開之沙，現設浮燈處水甚淺，必須挖深之，然後輪船方能有空位週轉。計約挖坭一四九，一七六立方米突，卽五，二七〇〇〇〇立方英尺，用爲塡築堤內地方之用。

（五）建築貨倉六座。貨倉每座寬一百二十英尺，長六百英尺。大洋船之長度約六百英尺，每貨倉之前，可泊船一艘。貨倉六座，同時可供六隻船之用。

（六）展築中山公路。由蟹山至堤岸一段，其路面擬增寬爲原路面之一倍，計長四四〇公尺，即一，四四〇英尺，寬四十六英尺。

（七）建築黃埔公路。由東堤起，經職德，貝村，程界，棠下，東圃，而至魚珠，約長五二，五〇〇英尺。

工程預算

共分七項計（一）鋼樁堤岸，（二）填築堤岸內至河邊之地，（三）挖泥，（四）建築貨倉，（五）展築中山公路（六）建築黃埔公路，以上六項，估計需費六，三七五，八六〇元，連監視工程及管理費用三〇〇，〇〇〇元，合計全部工程費總數爲六，六七五，八六〇元

黃浦港分區施政計劃

治河委員會自將黃埔商港區域範圍審定後，旋以都市之發展，必須有一種規定，將來方能井然有條理，故特採行分區制度，將黃埔商港分爲十二區，因地制宜，每區限制其用途及建築以求一致，其所擬各區名稱及用途如下：（一）港業區，區內有貨倉轉運塲碼頭船澳，船塢修船廠煤棧及露天貯物塲設備（二）港市行政區，區內爲港市行政機關所在地，將行政辦公樓宇聚于一區，所以利便辦公，且集敞壯宏麗之樓宇於一地，以成港市之中區，必能獲人民之敬仰，而辦事亦易得民衆之同情也，（三）笨重工業區，此區專爲大工廠而設，蓋笨重工業所在之地煤煙惡氣散佈，機聲嘈雜妨擾市民，其工業帶有危險性者，對於市民之安全，尤爲妨害故須另爲一區擇其地價低廉運輸利便之地而置之，（四）輕便工業區，輕便工業，乃指手工業而言，廣州爲手工業有名之城市，黃埔將來之手工業，亦必隨之而興，手工業雖無笨重工業區之危險，唯嘈什之聲與不潔之氣，勢所難免，故此區宜島近笨重工業區，不宜與住宅

25182

區或商業區混什也，(五)大商業區，大商業區爲發行商店而設，凡貨物之批發所在焉，此區最宜介於碼頭貨倉與小商業或輕便工業區之間，(六)小商業區，小商業區，即普通之商業區，平常市民之貿易在焉，此區最宜介於大商業與住宅之間，(七)甲種住宅區，爲住宅區中之最優美者，即田園住宅區是也，位于郊外，擇平坦山崗之地以爲之，所以求其清靜優雅遠離塵囂，區內多闢花園，路旁夾植樹木，一家式住宅之樓房，屹立於花叢菓林之中，乃其特色，(八)乙種住宅區，爲最普通之住宅區，兩家式之住宅，准建于此，此區不若甲種住宅之華美，而其適合于居住之衛生不稍遜也，故不宜與大商業區及工業區相近，俾能脫離城市穢濁之氣味，及車輛之煩擾，(九)丙種住宅區，此區專爲利便工商人等居住而設，位于城市之內工商區附近，俾工商人等就近往工廠或商店工作，可省來往之時間，(十)農林帶，自豪華濕工 Howa d 提倡田園市先後實施於列窩市，Otchr oi h 及威爾文 Welwgo 之後世界，人乃知農林帶對於新城市之重要，蓋必於市郊保留一農林地域，不准房屋建築其間，方能與城市以一種田園景緻及清新空氣也，(十一)公園及遊樂場，此地乃市民游憩休息之所，藉以怡養性情調劑生活者也，如城市有充份遊息之所，則市民可免沉淪于邪穢之區，及作不正當之娛樂，故現代之市政家，多注意于公園及遊樂場之建設，誠有益于市民之生活康健，及社會之風化者也，(十二)公共墳場，我國舊城市罕有公共墳場，而市民之埋葬，亦鮮有限制，其結果致近郊形勢優美之山崗，散佈窀穴，形狀不同，大小不一，旣廣田畝，復失觀瞻，故必由市劃定地區開闢墳場，以爲市民西方極樂之地，設員管理之，務使成爲公園化。

市政府今年施政計劃

分整理建設兩部進行

廣州市政府自訂定三年施政計劃後,關于工務公用教育土地財政社會衛生等項,業已依照計畫積極進行,成績極有可覩,茲者二十二年份已告終結,第一年計劃所未完成者繼續辦理而第二年計畫亦待籌劃進行,以促市政建設之發展,茲將第二年施政計劃探錄如下:

整理之部

(工務事項)(一)改良路面,1.郊外路一律鋪石碎或掃臘青,2.市心次要馬路改鋪半寸厚臘青,(二)私人自開街之整理,1.私人自開街一律改良路面,(三)整理全市內街圖,1.全市內街總圖,2規定全市各內街退縮,(四)整理全市水準點,1.完成全市水準點,2製全市水準點圖,(五)實施改善公園圖案,1.越秀公園,2.中山公園(六)美化原有各公園,1.加種花卉樹木,2.增加建亭台池沼,3.改善東山公園4.增加播音機,(七)公開試驗,1.士敏士2.臘青(八)試驗各種建築材料1.磚,2.木,3鋼鐵,4.石,(九)改善郊外木橋,1.西村路木橋改造爲三合土橋,(十)確定清理全市渠道計劃,1.籌劃清渠經常費用,2.確定全市渠道,輪流清理辦法,(十一)馬路渠之清理,1.增加夫役,2.計劃加設自動冲洗水閘,(十二)勵行公尺制度,(十三)增設建築美術審查會,(十四)改良計算書負責人問題,(公用事項)(一)繼續擴充整理自來水廠,(二)繼續擴充整理電力事業,(三)繼續擴充整理自動電話及長途電話,(四)改良電燈藏地綫,(五)增設馬路電燈,(六)繼續整飭廣告場位,(七)敷設全市自來水街喉及公共龍頭,(八)增設長途汽車路綫,及改善車輛式様,(下畧)

建設之部

(工務事項)(一)完成全市道路系統測量,(二)繼續開闢市郊馬路幹線,(三)完成全市三角網測量,(四)開闢河南新橋市心,(五)開始濬深珠江前後航線,(六)完成粵秀山至蟠龍崗大隧道,(七)完成東濠,(八)建築西濠,(九)建設內街大渠道,(十)繼續開闢市心馬路,(十一)繼續開闢內街,(十二)續建黃沙鐵橋,(十三)續建西堤鐵橋,(十四)完成增步橋,(十五)完成小港橋,(十六)續填內港坦地,(十七)續建河南堤,(十八)續建碼頭,(十九)續建貨倉,(二十)完成試驗塲,(廿一)續建馬路廣塲,(廿二)籌建黃沙堤,(廿三)規劃河南尾堤,(廿四)開闢東沙公園,(廿五)開闢小港公園,(廿六)開闢南石頭公園,(廿七)開闢漱珠公園,(廿八)繼續建築市立各小學,廿九)完成勷勤大學,(三十)完成公共墳塲,(卅一)繼續建築市塲,卅二)完成市府合署第二期工程,(卅三)完成大運動塲,(卅四)完成娛樂塲,(卅五)與築農村試驗塲,(卅六)完成國貨陳列館,(卅七)完成山谷游泳塲,(卅八)完成賽馬望台,(卅九)建築民用飛機塲,(四十)建築市立大戲院,(四一)完成市立賓館,(四二)完成展覽會塲,(公用事項)(一)繼續籌辦市營事業,(二)計畫與築從化江水利,(三)繼續完成煤汽廠,(四)繼續完成市內無軌電車,(五)繼續辦理燃用火油渣汽車,(六)繼續辦理計程汽車。(下畧)

編輯後話

編者在編輯本刋之初,便抱定兩個宗旨,第一就是鼓勵同學在課餘之時,從事著作和譯述,第二便是增加本刋中文篇幅。所以本期稿件,學生作品和中文寫成的文章都比較前期爲多。

本期內容頗爲充實,分論著翻譯記事附錄四欄。論著翻譯兩欄所載俱是應時之作,對於中國目下建設,不無有相當補助。附錄所載,亦爲有價值之參考材料。編者本不應自嘆花香,本刋實質如何,還是最好讓讀者自已來批評。

編者自感學識淺陋,力量稀微,謬誤之處,恐不能免。尚希碩學鴻儒,有以教之。

Lam Construction Co.,

HONG NAME (WAH YICK)

18 LUEN FAT STREET. HONG KONG. TEL No. 26125.

SHAH KI, CANTON. TEL. No 12304.

香港

華益建築公司

啓者本公司承接建造樓房屋宇各欵洋樓地盆碼頭堤壆橋樑所有海陸一切工程連工包料并代計劃繪圖快捷妥當如蒙光顧祈為留意

香港聯發街十八號

電話弍陸壹弍伍號

廣州市沙基中約橫弍巷十六號

電話壹弍叁零肆號

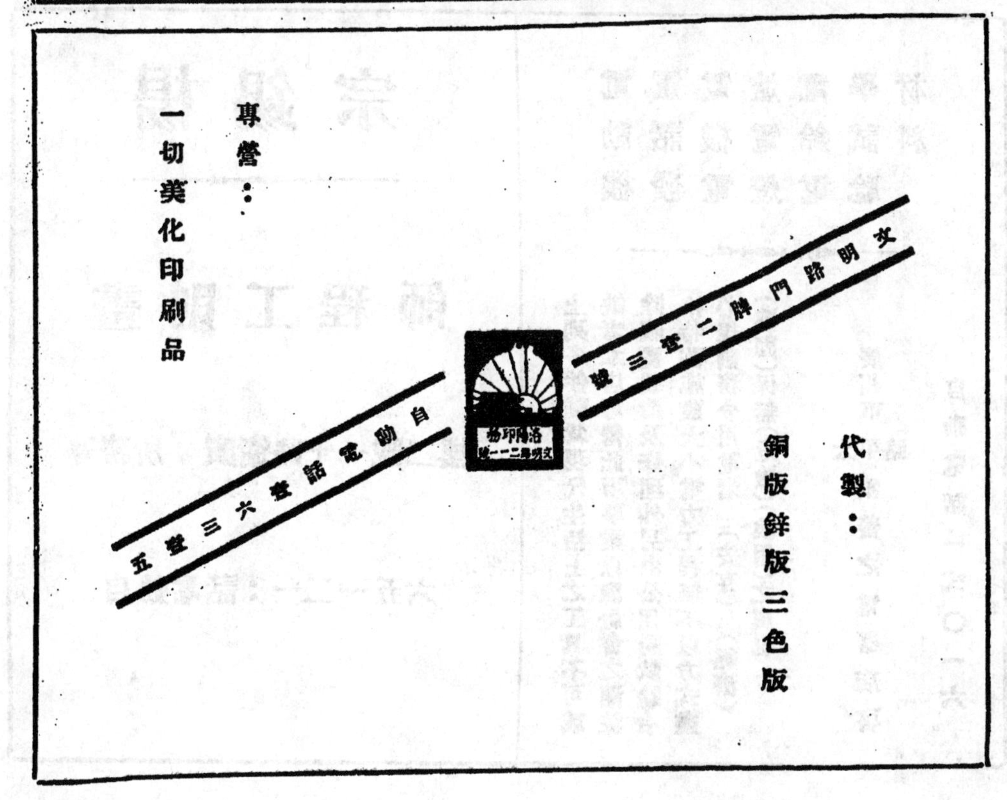

THE JOURNAL
OF
THE LINGNAN ENGINEERING ASSOCIATION

VOL. 1 No. 2

Editor-in-chief Liau Tsoi Woh

Associate Editors: Chan King Hung, Leung Cherkan, Lee Man Tai

Advertising Manager George Lau

Proofreader Lee Chuk Kit

Circulating Manager Liu Kin Man

中華民國二十三年二月十日出版

南大工程

第二卷 第二期

(每册四角) 郵費··本市二分··國內五分··國外三角

編輯者……………………劉載和

出版者……………………南大工程學會

發行者……………………南大工程學會

總發售處……………………南大書局

分售處……………………各大書局

南大工程

第三卷第一期　嶺南大學工程學會發行　民國廿四年六月

THE JOURNAL OF THE LINGNAN ENGINEERING ASSOCIATION

VOL. 3, NO. 1.　JUNE, 1935.

英商怡和機器有限公司（即渣甸洋行）

經理各國名廠專家出品

發電採礦 航空航海 農墾水利 紡織救火 鐵路印刷 各種機械及製冰機冷氣機升降機潔淨具軍用品等諸君惠顧無任歡迎

香港 必打街十四號

廣州 沙面英租界

SIXTY-FIVE OF THE **LEADING**

ENGINEERING MANUFACTURERS

OF THE WORLD EACH

SPECIALISING IN A PARTICULARB RANCH

OF ENGINEERING

ARE REPRESENTED

IN CHINA

By

THE JARDINE ENGINEERING CORPORATION

LIMITED.

(Incorporated under the Companies Ordinances of Hong Kong)

HONG KONG, 14, Pedder St.

CANTON, Shameen.

THE NATIONAL GAS ENGINE Co. LTD
ASHTON-UNDER LYNE.

吳翹記 批盪專家

廣州惠愛西路五十四號 · 電話一六二六六

華南最偉大建築物之批盪工程均為本號所營造

介紹意大利藝術吳翹記批盪專家

現代文明都市其偉大建築物與紀念物莫不競尚意大利批盪蓋其堅耐美化雅麗堂煌光艷奪目能發人美感現本市有吳瑞翹君者研究此種批盪歷有年所其藝術之湛深工料之精選洵非時下庸才所及且能就圖案而造意因光線而設色故其成績斐然大觀不愧當局品題名流贊許邇來吳君為精事研求特禮聘美術專家化學藥師工程技師等共事改良故成績比前更勝一步如本市中山紀念堂牌亭及市立中山圖書館市府合署中山大學電器機械科農科文學法學理學等院胥為吳君所經營批盪其成績可見一班其他如舖櫃地臺碑墓花盆浴盆用具等均能巧奪天工精美絕倫 弟等為本市美化計願為吳君介紹俾各界知所問津焉

介紹人建築工程師 黃謙益 郭振聲 鄭校之 卓文揚 鄭成佑 江宗漢 譚勵宸 李文邦 陳榮枝 李卓 郭秉琦 林克明 同啟

全省偉大建築工程本號批盪成績之一班

1. 鄧番先生題詞
2. 蕭佛成先生題詞
3. 劉紀文先生題詞
4. 金曾澄先生題詞
5. 吳鐵仙先生題詞
6. 中大文學院
7. 羅樂之先生題詞
8. 余定邦先生題詞
9. 廣州市府合署
10. 中央銀行
11. 中大機械科學院
12. 中大農科學院
13. 廣州展覽會獎狀
14. 中山紀念堂牌亭
15. 李烈士紀念碑
16. 東山電話分所
17. 總統大商店銀盾
18. 番禺縣政府
19. 嶺南大學女生宿舍
20. 市立中山圖書館

南大工程

嶺南大學工程學會會刊

第三卷第一期目錄

25195

投稿簡章

本刊歡迎投稿且鑒於外國文稿件,已爲國內僅有之二三工程刊物所摒棄在作者方面,每苦於專門名詞未有標準之迻譯,恆習慣直接用外文撰稿,本刊在中央編譯館未將各標準名詞公佈以前,外文稿件亦樂爲登載。爰錄簡章數則於後:

(一) 投寄之稿,自撰或翻譯,均以還於工程者爲限,文體文言白話不拘。

(二) 中國文稿件暫以英德法三文爲限。

(三) 來稿請繕寫清楚,如有附圖附表,尤須書寫清潔外文稿件打字寄來尤佳。

(四) 如係譯稿,請書明原著書名篇目出版地點日期。

(五) 稿末請注明姓名住址,以便通訊,至揭載時如何署名,聽投稿者自便,作者能將詳細履歷敘明尤佳。

(六) 投寄之稿在二千字以上者,如不揭載,編輯部於每期出版後按址郵寄回。

(七) 來稿經揭載後,酌酬本刊是期多本。

(八) 來稿編輯部或酌量增删之,但投稿人不願他人增删者,請預先聲明。

(九) 來稿請逕寄廣州嶺南大學工程學會出版部收。

南大工程學會職員
民國廿四年六月

南大工程學會出版部職員
民國廿四年六月

卷首語

鄒煥新

本刊內容雖未臻充實，然刊中登載之稿件，則頗屬珍貴，如本校教授羅石麟先生著之"Analysis of Continuous Arches on Elastic Piers"，爲羅先生領受碩士學位之畢業論文之一部，堪稱難得。林榮棟君之磚凝鋼根，"The Reinforced Brick" 乃此刊中最足引人注意之著作，蓋此文論着新頴，取材豐富，且歷一年之研究及實驗，撮其概要而成此文。所論爲工程界最新發明之磚凝鋼根，此項工程美國工程界已極注意，務求其能適於應用。按磚凝鋼根之效能，雖稍遜於三合土凝鋼根，然其價格則僅約值三合土凝鋼根三分之二，故彼極適於規模較小之工程，如小樓房之橫樑，及短距離之橋樑等。其價格之低廉也，原因凡三：(一)材料包含磚，鋼根，與小量之士敏土及沙，惟大部份之材料屬磚，而磚之價格不高。(二)製造手續簡單，所費工資不昂，(三)製時所需之支撑極少，遠不如三合土凝鋼根之多。由是觀之，此種工程極適合於我國之需求及應用，蓋我國工程未臻興隆，規模多屬不大，且財源短絀，經濟落後，更無力擔負昂貴之工程費用，再者，我國無大量士敏土之出產，且士敏土之價格亦高，若處缺乏士敏土供給之區域，欲稍興工土木，殊非易易，此困難之解決惟有待於磚凝鋼根之應用焉。林君有見及此，特悉心研究，並作種種試驗，歷時一載，所獲效果，頗稱滿意。據林君言，其試驗已達相當成功，每樑之試驗之結果，皆可從計算預測之，屢驗無訛，林君又云此乃嘗試中初步之成功，僅奠此工程之基礎，現仍繼續研究，務達完滿之成功云。黃文煒教授之

"A Study of the Characteristics of the Pearl River Water."研究及試驗珠江之水之特質，黃教授對於此項工作，已致力有年，成績卓著，茲得黃教授之許可，將其試驗與研究之結果轉載此刊，以供諸學者之參考。黃玉瑜建築師之 "How to Appreciate Good Architecture" 乃表揚建築學之眞義而偏重於欣賞之作，並引世界有名之建築物爲例証。彭奠原君之"廣州市土雜杉木之檢討"分述廣州市市面應用之木材之來源及性質，此種論著亦屬可貴，對於工程界尤有莫大貢獻。當今政府提倡應用國產木材聲中，亟需此等論著，以供工程界之參考，以應社會之要求。茲因時間所限，未克完成其工作，祇將其著作前部先行付梓，餘待續刋於下卷，閱下部全屬試驗之紀錄云。

廿四，六，十二。

Analysis of Continuous Arches on Elastic Piers

羅 石 麟

Shih Lin Lo

INTRODUCTION: In this paper the design of a three span hingeless reinforced concrete arch bridge supported on comparatively slender piers is presented. The method of analysis used, was devised by Ernst Pichl as published in "Der durchgehende gelenklose Bogen auf elastischen stuetzen." The method is wholly analytical and as such will appeal to most engineers who have occasion to design such structures.

The general characteristics of a continuous arch bridge differ from those of an ordinary fixed end arch, in that, each span cannot be considered as an element by itself. The ends of the various arches are continuous over the supports and intimately connected to the piers, so that, any movement of the latter will cause a corresponding movement of the adjacent springings. Now, if a load is applied on any of the arches the span of the loaded arch is increased, and reactions are developed on the piers. This increase in span of the loaded arch will tend to relieve part of the thrust and transmitting it over to other spans. Through the yielding of the piers and the rotation of the joints a greater positive moment at the crown and negative moments at the springings are developed in the loaded span as compared with a fixed-end arch of the same dimensions. The effect of the yielding of the piers decreases as we go farther and farther away from the loaded span, and for usual cases, where the piers are not exceedingly tall and slender, only three spans need be considered at one time.

The method here used may be outlined as follows. The arches are first assumed to be intimately connected with each other and the piers, but imaginary restraints which permit rotation only without any displacement, are placed at all pier heads. The system is analyzed under this assumption, and later on the restraints are removed, and the displacement of the pier tops is taken up. The algebraic sum of the stresses obtained in both cases gives us the desired results for the actual structure.

The procedure may be classified under the following headings.

(1) The continuous arch system is first separated into various units — arches and piers — and the required elastic constants of each unit is determined.

(2) These units are then put together to form the original structure, and the mutual effects of each component unit upon the rest of the structure in consequence of continuity are calculated. These mutual effects are expressed in terms of degrees of restraint, distribution factors and carry-over factors.

(3) Imaginary supports, which permit rotation unaccompanied by any horizontal or vertical displacement, are inserted at all pier heads. The continuous structure is analyzed for all conditions of loading desired under this assumption, and stresses and reactions obtained.

(4) The imaginary supports used in step (3) are removed by applying forces equal to, but opposite in direction to the reactions obtained in step (3), and the desired stresses are calculated. This process is exactly equivalent to releasing the pier heads i. e. permitting them to displace under loads.

(5) Add algebraically the moments, thrusts and reactions of steps (3) and (4).

(6) Calculate maximum fibre stresses and checked cross sections adopted in preliminary design.

ELASTIC PROPERTIES OF ARCHES AND PIERS: In a previous paper entitled "Analysis of Restraint Arches" published in the first issue of this journal, the writer developed general formulae for constants of any shape of restrained arches based on the work of Pichl. As constant reference will be made to that paper those readers who are interested in this subject may profit to some extent by reviewing that article. To save trouble in printing the following symbols have been changed.

α, β and γ changed to X, Y and Z respectively

φ, ψ and Δ " " s, t and D "

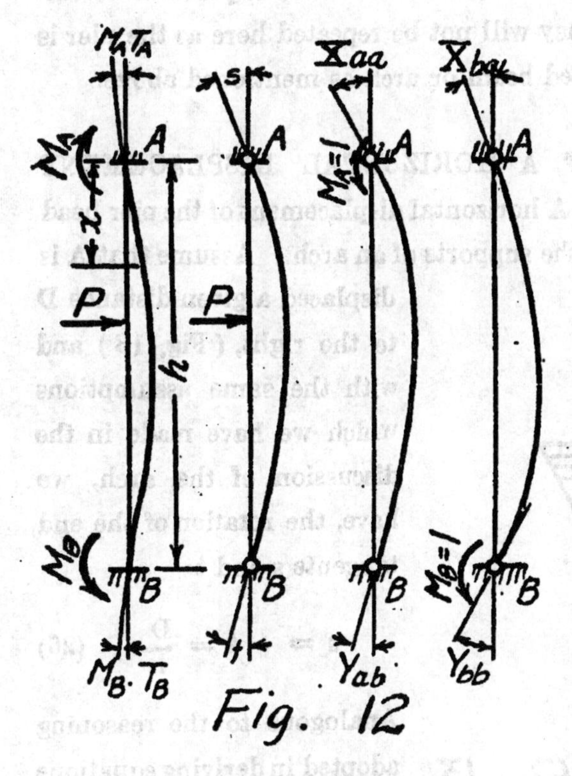

Fig. 12

THE ANGLES OF ROTATION X, Y AND Z OF THE END TANGENTS OF A TWO HINGED COLUMN DUE TO UNIT END MOMENTS (FIG: 12):

The angle of rotation, X of the end tangent at A due to unit moment $M_A = 1$ applied at that end, Y, that at B due $M_B = 1$ applied at B, and Z, that at A due $M_B = 1$ or that at B due to $M_A = 1$, will follow from Eq. 8 by putting $y = 0$.

$$X = \frac{1}{E} \cdot \frac{1}{h^2} \int (h-x)^2 dw \qquad 24a)$$

$$Y = \frac{1}{E} \cdot \frac{1}{h^2} \int x^2 dw \qquad 24b)$$

$$Z = \frac{1}{E} \cdot \frac{1}{h^2} \int x(h-x) dw \qquad 24c)$$

dw being as usual the elastic weight $\frac{ds}{I}$ of any element. Similarly, the angles of rotation of the end tangents, s and t, due to the application of external loads, may be had from Equations 9 (ibid. page 12) by putting y=0, giving

$$s = \frac{1}{E}, \frac{1}{h} M_O \int (h-x)\, dw \qquad 25a)$$

$$t = \frac{1}{E}, \frac{1}{h} M_O \int x\, dw \qquad 25b)$$

For other useful relations, the readers are referred to the previous discussion on restrained arches. They will not be repeated here as the pier is simply a special case of a curved beam or arch as mentioned above.

THE INFLUENCE OF A HORIZONTAL DISPLACEMENT OF THE PIER-HEAD:- A horizontal displacement of the pier head corresponds to the sinking of the supports of an arch. Assume that A is displaced a given distance D to the right, (Fig. 13) and with the same assumptions which we have made in the discussion of the arch, we have, the rotation of the end tangents s and t

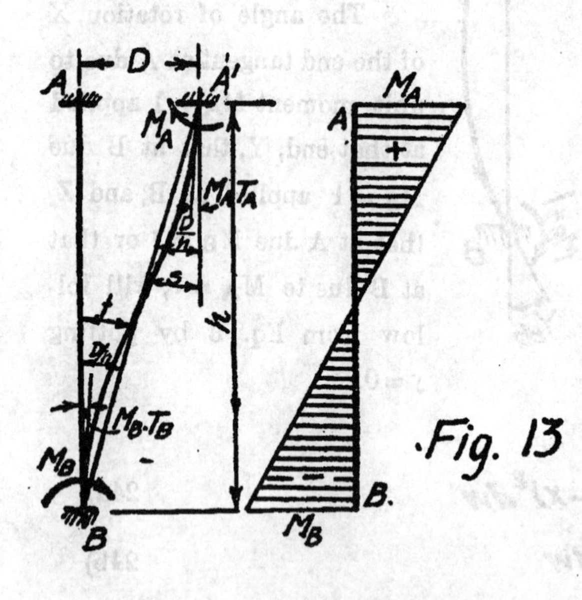

Fig. 13

$$s = -t = \frac{D}{h} \qquad (26)$$

Analogous to the reasoning adopted in deriving equations 2 and 4 p 7-8 in the previous discussion, we have the following elastic equations

$$M_{AD} X + M_{BD} Z + \frac{D}{h} = - M_A T_A$$

$$M_{AA} Z + M_{BD} Y - \frac{D}{h} = - M_B T_B$$

and $$M_{AD} = -(C_3 + C_1)\frac{D}{h} \qquad 27a)$$

$$M_{BD} = +(C_3 + C_2)\frac{D}{h} \qquad 27b)$$

The horizontal reactions will be

$$A_D = -B_D = -\frac{1}{h}(M_{AD} + M_{BD})$$

A_O and B_O of Eq.12 are zero since there are no external loads. D is considered positive, when it is a displacement from left to right, while a moment which causes tension on the right hand side of the pier is also taken as pisitive.

COMBINATION OF ARCHES AND PIERS:- If we apply a unit moment to the end of any straight or curved member) say e g. an arch or pier detached from the rest of the system), and denote the resulting rotation of that end as e, then, in the case of a beam which has unit rotation at its end, there will be developed at that end of the beam a resisting moment $M = \frac{1}{e}$, or, for any other angle of rotation say q the mement will be $M = \frac{1}{e}q$. In the case of an absolutely rigid joint, when the joint rotates a certain amount, all the members will rotate the same amount. Now to distribute the moment, which causes the rotation of the joint, among the members, we will have to know what the result will be for each individual member with a unit moment acting on it, when the whole joint is disconnected and each member acting freely.

In order to arrive at the value of the angle of rotation of the arch under external forces, let us separate the arch A B (Fig. 14) from the rest of the arches, and assume, that in A we insert a hinge while B remains in the same condition as if the arch is not isolated from the rest of the arches. For the present we will consider the degree of restraint at B, T_B, or amount of rotation of the end tangent at B due to unit moment

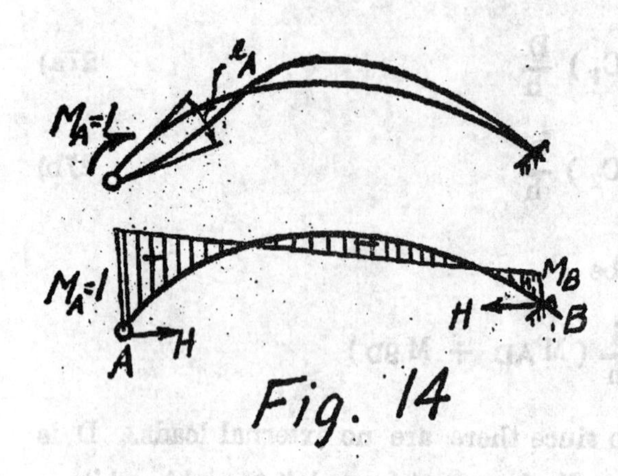

Fig. 14

$M_B = 1$ as known. We now have under our consideration an arch with a hinge at one end A, and the other end B partially fixed, and we shall find the effects of applying a unit moment $M_A = 1$ at A. Applying this unit moment at A through a lever mechanism or imaginary device, there will be developed at B an unknown moment M_B due to its end restraint. Similar to the derivation of Eq. 2, we will have for the end B

$M_A Z + M_B Y = -M_B T_B$ the expression for Y being given by Eq. 8b, that for Z by 8c. Since $M_A = 1$ therefore $M_B = m_A =$

$$-\frac{Z}{Y+T_B} \qquad (28)$$

which gives us the value of the moment at B when there is a unit moment at A. It will be seen that m_A depends only on the arch dimensions and the degree of end restraint at B. With the help of Eq. 10. the thrust developed in the arch, under this loading of $M_A = 1$, will be

$$H = M_A B_A + M_B B_b = B_a + M_A B_b \qquad (29)$$

The moment developed at any section in the arch, following Eq. 11, will be

$$M_X = \frac{L-x}{L} \cdot M_A + \frac{x}{L} M_B - Hy$$

or
$$M_X = \frac{L-x}{L} \cdot M_A + M_A \frac{x}{L} - (B_a + m_A B_b)\, y \qquad (30)$$

The angle of rotation e_A of the end tangent at A follows from Eq. 6

$$e_A = \int \frac{M_X M_a\, ds}{EI} + \int \frac{N_X N_a\, ds}{AE}$$

As in the equation given at the top of page 11, $M_a = \frac{1}{L}(L-x) - B_a y$, and with some approximation, we may put $N_a = B_a$ and $N_X = H$.(Eq. 29) Substituting all these values in the equation for e_A, and, with due consideration of those already derived, equations 8a and 8c. we obtain

$$e_A = X + m_A Z = X - \frac{Z^2}{Y+T_B} \qquad (31\ a)$$

This value of e_A will follow directly had we utilized our base system the two hinged arch with $M_A = 1$ and $M_B = m_A$

$$e_A = M_A X + M_B Z = X + m_A Z$$

If the degree of restraint at A is given as T_A we have, similarly,

$$e_B = Y + m_B Z = - \frac{Z^2}{X+T_A} \qquad 31\ b)$$

and the moment ratio of the arch for the springing A

$$m_B = - \frac{Z}{X+T_A} \qquad 31\ c)$$

thus, giving us the value of the moment at A, M_A when $M_B = 1$. These moment ratios m_A and m_B will be known as carry over factors.

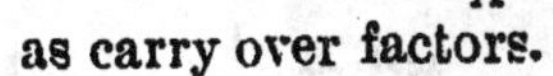

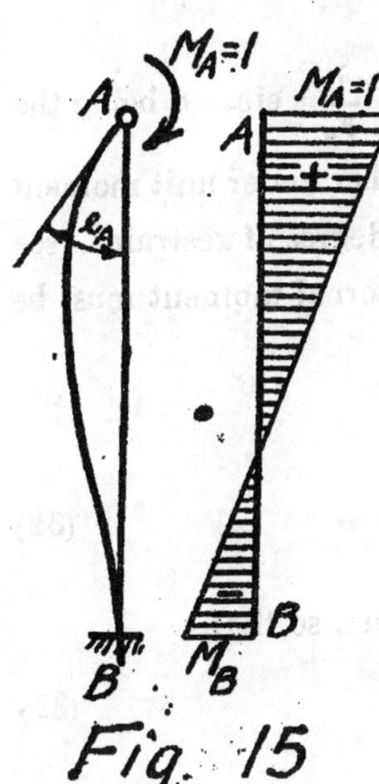

Fig. 15

ANGLES OF ROTATION OF THE PIER ENDS:- (Fig. 15) As was mentioned previously, the pier acts in the same manner as the arch, except, that all values for y are equal to zero. Equations 29 and 31 are applicable for the calculation of the angles of rotation of the pier ends, only the corresponding values X, Y and Z for the pier should be used instead of those for the arch.

DETERMINATION OF THE ANGLE OF ROTATION OF THE WHOLE JOINT:- (Fig. 16) Having thus determined the behaviour of the individual members as regards to their resistance to rotation under external loads, we will now proceed to find the effects produced when the members are put together to form a joint. We will assume, that the rotation of the whole joint under the influence of a unit moment applied at the joint as T, and that the joint is absolutely rigid as far as the relative position of the individual members is concerned, i. e. the angles between the various members remain unaltered even under loads. It follows, therefore, that all the members will rotate the same amount T. Now if the joint has a unit rotation it will correspond to a moment of $M = \frac{1}{T}$ acting at that joint. The ends of all the members meeting at the joint will also have the same unit rotation, and the moments developed at their ends will be

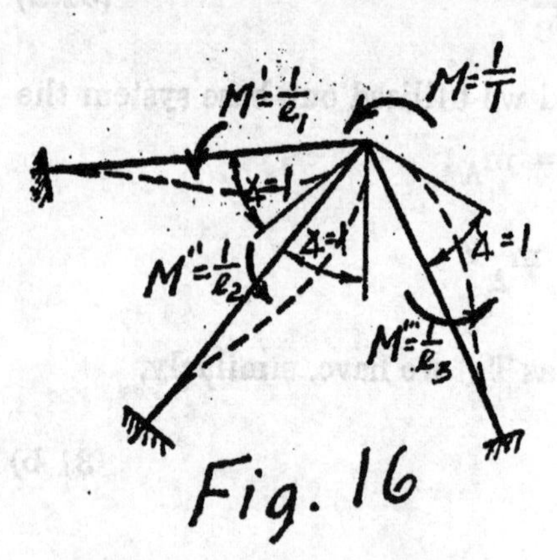

Fig. 16

$$M' = \frac{1}{e_1} \qquad M'' = \frac{1}{e_2} \qquad M''' = \frac{1}{e_3}$$ etc. e being the amount of rotation at the hinged end of the member under unit moment at that end while the far end is under its actual degree of restraint. (see pp. 5-6.) As the joint is in equilibrium the external moment must be equal to the total internal moment, so that

$$M = M' + M'' + M''' + \text{etc.}$$

or

$$\frac{1}{T} = \frac{1}{e_1} + \frac{1}{e_2} + \frac{1}{e_3} + \cdots \cdots\cdots\cdots \tag{32}$$

For our case we have to consider only two members, so that,

$$T = \frac{e_1 e_2}{e_1 + e_2} \tag{33}$$

Having determined the angle of rotation of a joint due to unit mo-

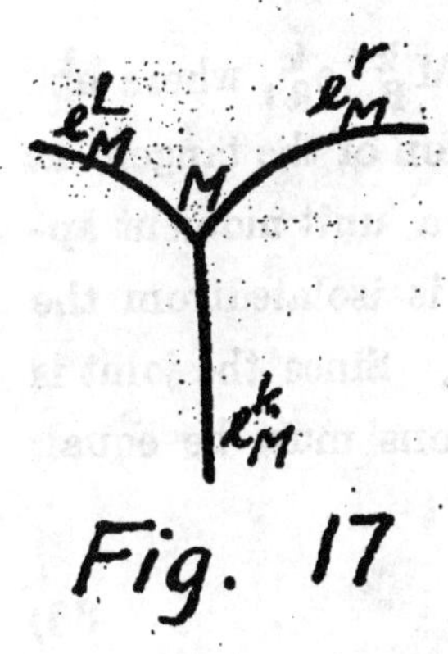

Fig. 17

ment we will take a specific joint say M. Taking e_M^L as the angle of rotation of the tangent at the right end of the left arch, under a unit moment at that end, when that arch is isolated from the rest; (Figure 17). e_M^r the angle of rotation of the left end of the arch (Figure) and e_M^k the angle of rotation of the top of the pier: we have

$$T_M^L = \frac{e_M^r \cdot e_M^k}{e_M^r + e_M^k}$$ as the degree of restraint of the right end of the le left arch.

$$T_M^r = \frac{e_M^l \cdot e_M^k}{e_M^l + e_M^k}$$ for the left end of the right span and (34)

$$T_k = \frac{e_M^r \cdot e_M^l}{e_M^r + e_M^l}$$ for the pier top. (35)

In a continuous arch system, the condition of the extreme ends of the system as well as the pier foundations, will be determined by the construction methods and the soil condition. If the supports are absolutely rigid, i. e. hingeless and unyielding, $T=0$. T is the angle of rotation of the end tangent due to a unit moment applied at that end. For a hinged end, $T=$ infinity, and the corresponding moment at the springing disappears.

DISTRIBUTION OF MOMENT AT A PIER-HEAD:- In the joint R shown in figure 18, let there be a moment M at the right end of the left arch then the angle of rotation of the whole joint as well as all the members will be MT_R^L, in which, T_R^L is the angle of rotation of that springing due to unit moment applied there. The moment developed in any member, say the pier is M_R^k. The angle of rotation of

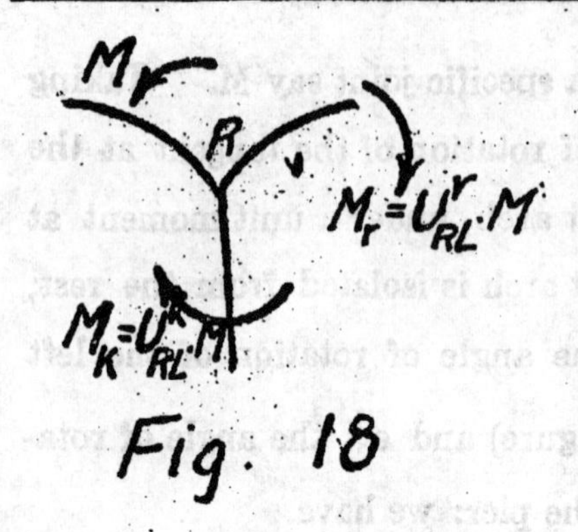

Fig. 18

that body will be $M^k_R \, e^k_R$, where e^k_R is the angle of rotation of the tangent at the pier top due to a unit moment applied when the pier is isolatedfrom the rest of the structure. Since the joint is rigid the two rotations must be equal; so that

$$MT^L_R = M^k_R \, e^k_R \tag{36}$$

or $M^k_R = M\frac{T^L_R}{e^k_R} = U^k_{RL}\,M$, and if $M = 1$, we have $M^k_R =$

$$U^k_{RL} = \frac{T^L_R}{e^k_R} \tag{37}$$

wherein R represents any joint R, U^k_{RL} represents the distribution coefficient in joint R for a moment coming from the left of the pier to the pier. Likewise, $M^r_R = U^r_{RL}\,M$, is the moment developed at the springing of the right adjacent arch. U^r_{RL} is the distribution coefficient in joint R for a moment coming from the left of joint R to the left springing of the right adjacent arch. By statics $M = M^r_R + M^k_R$ and $U^r_{RL} + U^k_{RL} = 1$. The same process may be applied to the rotation of the joint due to any moment which may exist at other springing.

Thus, we are able to determing by the distribution ot moment over the pier head, the mutual influence of the rigid connection at the pier head, and the degree of end restraint of each individual member meeting at a joint.

CONTINUOUS ARCHES ON ROTATING CUT NON-DISPLACEABLE PIERS:- With the treatment of the component parts which make up a continuous arch system before us, we are now in a position to discuss the system proper, as being made up of individual arches resting on and rigidly connected to the elastic piers. For convenience in analysis, we will consider the effects of rotation and lateral displacement separately.

Considering first the case of rotation, each individual span of an arch series may be considered as resting on partially fixed supports, whose degree of end restraint is determined by the conditions of the foundation, and the elastic properties of the arch and the rest of the structure. In other words, insofar as the loaded span is concerned, the problem simply winds up into determinations of end restraints. With the end restraints known, both moments and thrusts can be easily found for any loading by methods given in the previous paper. The stresses

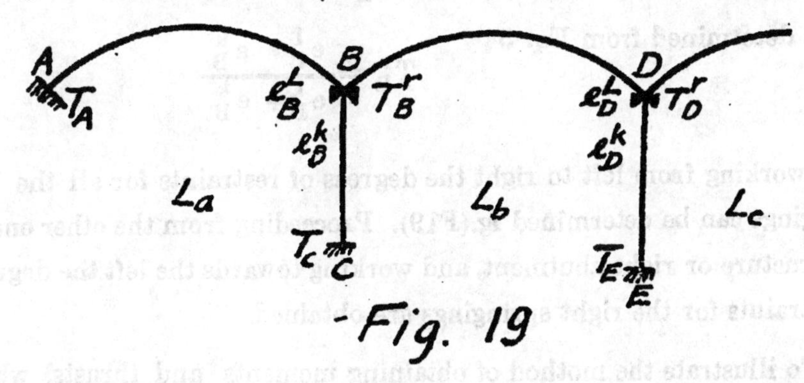

Fig. 19

in the adjacent spans and in the piers will follow from the distribution and carry-over factors. (Refer to Fig. 20) The readers should bear in mind however, that the method is applicable only to loading one span at a time.

To return to our subject of determination of end restraints, and, coincidentally, the distribution and carry-over factors, let us consider the case of Fig. 19. The dimensions of the arches and piers, or X, Y, Z and X' Y' Z' are assumed as known. The conditions at the abutment A and the pier bases C, and E will be considered as perfectly fixed i. e. $T_A = T_C = T_E = 0$. From Equation 31c the carry over ratio from

B to A is $m_B^L = \dfrac{-Z}{Y+T_A^r}$ or $= \dfrac{-Z}{X+T_A}$

From Equation 31b. $e_B^L = Y + m_B^L Z$

Similarly for the pier BC

$$m_B^k = \frac{-Z'}{Y' + T_C} \quad \text{and} \quad e_C^k = X' + m_B^k Z'$$

T_A and T_C are determined from the foundation conditions, zero for perfectly fixed conditions, and infinity for hinged ends. X, Y, and Z are determined from the arch dimensions AB, while X', Y' and Z' from the pier dimensions BC.

Finally, the degree of restraint T_B^r for the left springing of Arch BD is determined from Eq. 34

$$T_B^r = \frac{e_B^L \cdot e_B^k}{e_B^L + e_B^k}$$

Thus working from left to right the degrees of restraints for all the left springings can be determined ig.(F19). Proceeding from the other end of the structure or right abutment, and working towards the left the degrees of restraints for the right springings are obtained.

To illustrate the mothod of obtaining moments and thrusts, when only pier-head rotation is considered, let us take the example of Fig. 20.

The span L_L supports a vertical load P at a distance x to the left of joint K. Assuming, that the end restraints T_K^r and T_M^L, as well as the distribution, and carry-over factors of all the members have been calculated by methods just outlined, the moments at the left and right springings or M_K^r and M_M^L respectively and thrust H_L for the span KM can be

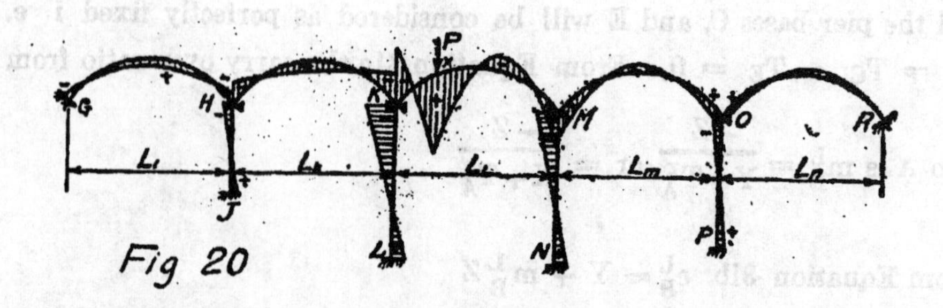

Fig 20

determined by processes given in the previous paper. Using the distribution factors U^{k}_{Kr} and U^{L}_{Kr}, the distribution factors of the left arch, and the pier respectively at joink K for a moment M^{r}_{K} coming from the right, we obtain the moment at the pier head K $\quad M^{k}_{K} = U^{k}_{Kr} M^{r}_{K}$.

the moment at the right springing of the adjacent arch $M^{L}_{K} = U^{L}_{Kr} M^{r}_{K}$

The sum of these two moments must, of course, equal to M^{r}_{K},

With a moment M^{k}_{K} at pier-head K, there exists simultaneously a moment M^{f}_{L} at the pier base L

$$M^{f}_{L} = m^{k}_{K} \cdot M^{k}_{K} = m^{k}_{K} U^{k}_{K} \cdot M^{r}_{K}$$

The moment diagram for the pier is a straight line, as there is no external load acting on the pier. The signs of the moment at the pier-head and pier base are opposite in sense.

Similarly the moment at the left springing of span HK is

$$M^{r}_{H} = m^{L}_{K} M^{L}_{K} = m^{L}_{K} U^{L}_{Kr} M^{r}_{K}.$$

The signs of the two springing moment are the same. The moment diagram for this span may be expressed by the equation

$$M_{X} = M^{r}_{H} \frac{L-x}{L} + M^{L}_{K} \frac{x}{L} - Hy \qquad \text{in which H by Eq. 10 is}$$

$$H = M^{r}_{H} B^{r}_{H} + M^{L}_{K} B^{L}_{K}$$

Thi process is carried torwards the left until the end of the sturcture, abutment G is reached. In a similar way, with M^{L}_{M} as a starting point, we can work from joint M towards the right and towards the base until all the springing and pier moments are obtained.

In practical designing it is generally useless to go beyond one span on each side of the loaded span. In fact, for piers fairly stiff the whole subject of pier-head rotation may be completely ignored.

INFLUENCE OF DISPLACEMENT OF THE PIERS:- In the foregoing discussion, we have assumed that the system has piers which permit rotation at the top but no displacement. When the arches are loaded with external forces, there will be developed at the supports horizontal forces, which are equal to the algebraic sum of the horizontal thrusts existing in the two adjacent arches and the shear in the pier. The sign adopted to signify the direction of this horizontal force in the pier will be positive for a force acting from left to right. If we release the supports these horizontal forces at the pier heads will cause the joints to displace, until the horizontal forces so adjust themselves as to reach equilibrium through the changes in span lengths of the arches and the bending of the piers. The process of taking out the horizontal reactions at the pier heads caused by external loads, when the joints are restricted from any movement, is equivalent to the addition of equal and opposite

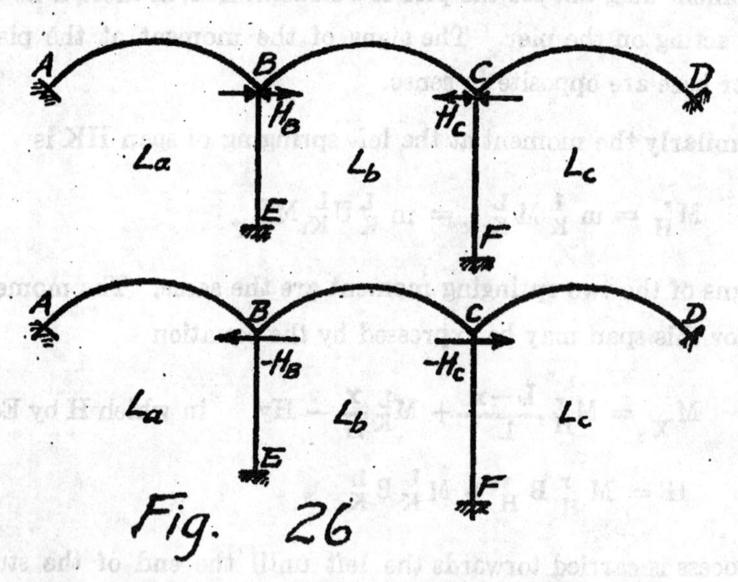

Fig. 26

external forces.(See Fig. 26) This does not mean that there are no horizontal forces at the joints. Of course, ther are, but these internal horizontal forces balance each other, when the piers are permitted to displace. We will, therefore, investigate the influence of a horizontal force say H upon the whole system, when displacement of the pier-heads is permitted. This we will show in the following approximate method.

We will apply at one of the pier-heads a unit horizontal force and note its actual displacement in the original structure. Under this loading, the pier will bend, and the whole system will be involved in consequence of continuity. To achieve the final results in a simple way we will divide the resulting movement into individual parts. Proceeding with the system we have adopted, i. e. continuous arches with rotating but non-displaceable pier-heads, we shall release each individual joint one after the other, and determine the influence of the horizontal forces acting at those joints.

With this purpose in mind, we will proceed to let joint M be released and displaced a distance D to M^1 under the influence of a certain horizontal force (Fig. 21). The arches L_m and L_n connected to this joint

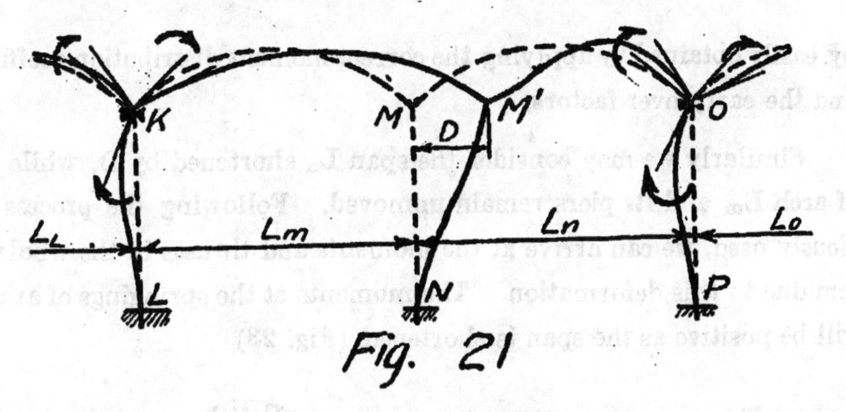

Fig. 21

will be directly involved through this displacement, and every member in this system will set up resisting forces. To determine their numerical values we will subdivide the displacement of the joint into three portions corresponding to the number of individual members connected to the joint, investigate them separately and add the results thus obtained.

Continuing the imposed condition of the immovableness of the remaining pier heads we will increase the span of arch L_m by D (Fig. 22) assuming the span L_n and the position of its supports unchanged. As a consequence of this change of span in arch L_m, internal forces-moments and thrusts-will be developed in this arch, the expression for which can be obtained from paragraph 11 of the previous paper. The moments developed at the springings will be negative. The moments and thrusts at other sections of the structure, as well as the horizontal reactions and moments at the bases of the piers under this specific deformation, may

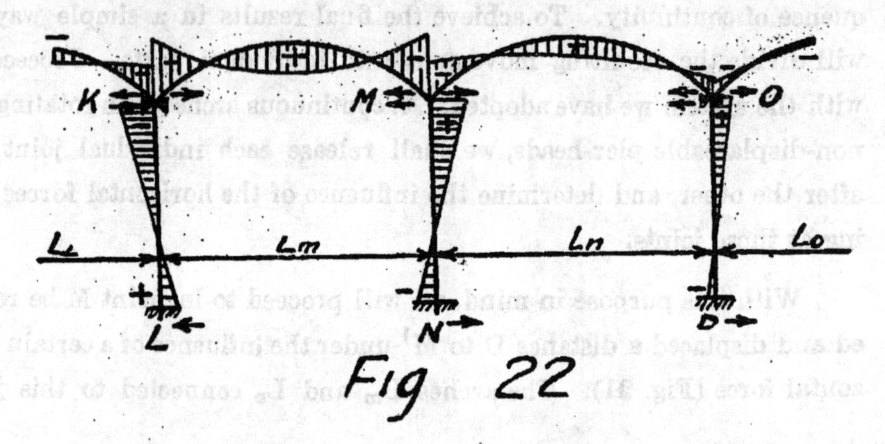

Fig 22

by easily obtained by applying the correct moment distribution coefficients and the carry over factors.

Similarly we may consider the span L_n shortened by D, while that of arch L_m and its piers remain unmoved. Following the process previously used, we can arrive at the moments and thrusts in the whole system due to this deformation. The moments at the springings of arch L_n will be positive as the span is shortened. (Fig. 23)

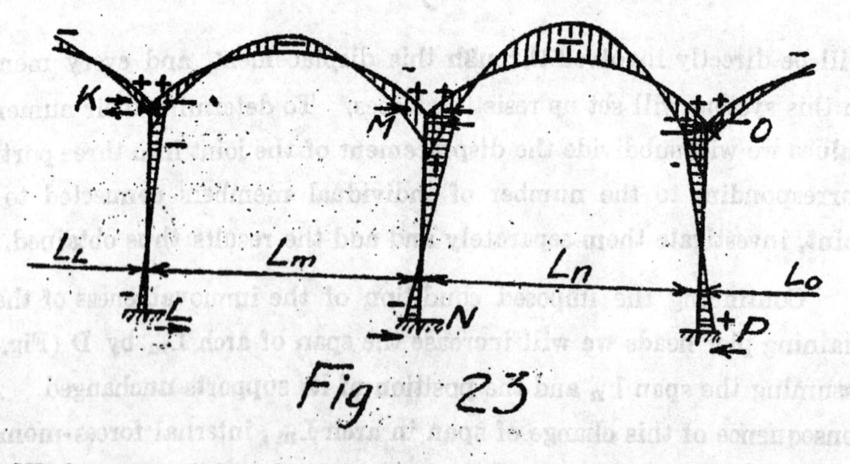

Fig 23

We have now left only the action of the displacement of the pier, and we can determine from pp. 2-3 of this article the moments existing at the pier ends when it is displaced a distance D, with the span lengths of the adjacent arches assumed unchanged. The effects of this portion of the displacement on the rest of the structure will be obtained by the utilization of the distribution coefficients and the carry-over factors.(Fig.24)

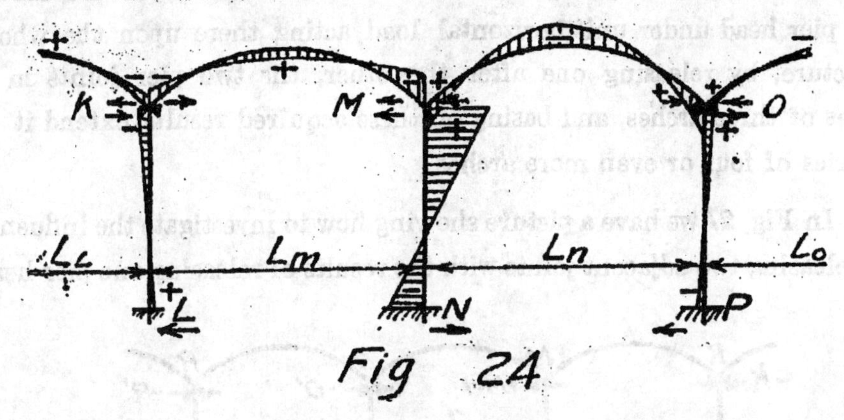

Fig 24

Adding all these results together, we will have determined all the inner forces and reactions developed through the displacement D of joint M (Fig. 25). The resultant of all the horizontal forces at joint M is the "Force" M acting in the sense of the displacement. At other pier heads as K and O are the "Reactions" K, O, etc. acting in a sense opposite to "Force" M. The influence of a unit load at the released joint with all the other joints restricted from movement, may be obtained by dividing the results by the value of M: the amount of displacement of the joint by this unit load will be $\frac{D}{M}$.

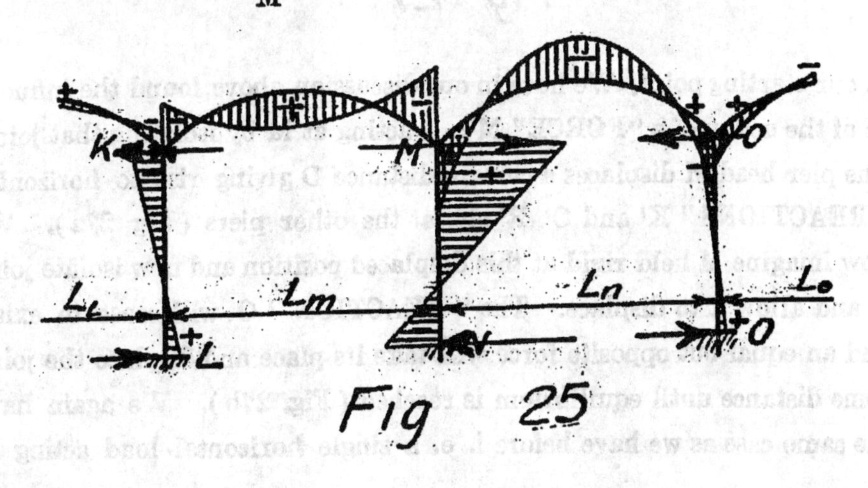

Fig 25

The same process of applying a unit load at one of the joints and allowing it to displace while the rest are held rigid is to be applied to all the joints of the structure.

From this investigation we can obtain the effect of the movement of a pier head under unit horizontal load acting there upon the whole structure, by releasing one after the other, the two pier joints in a series of three arches, and basing on these acquired results, extend it to a series of four or even more arches.

In Fig. 27 we have a picture showing how to investigate the influence of releasing two adjacent joints with the results of releasing one pier head

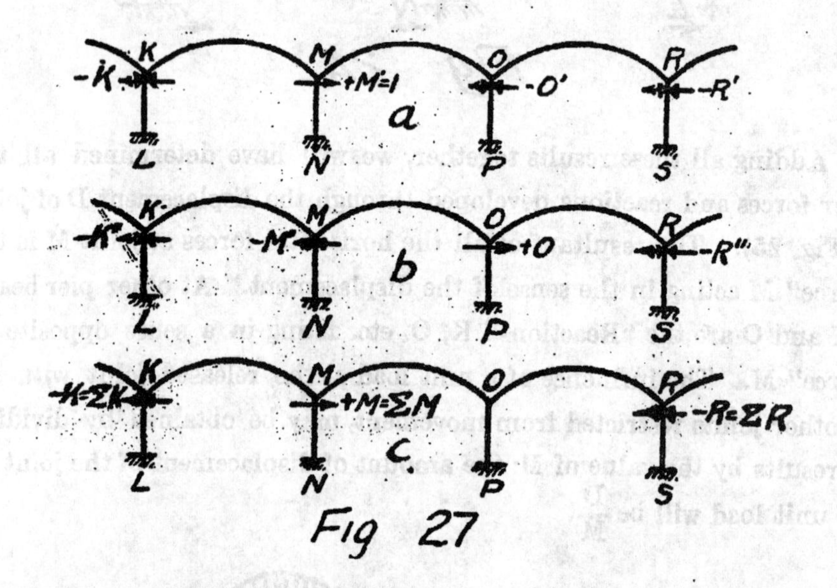

Fig 27

as our starting point. We have in our discussion above, found the influence of the action of a "FORCE" M' = 1 acting at M by isolating that joint. The pier head M displaces a certain distance D giving rise to horizontal "REACTIONS" K' and O', R' etc. at the other piers (Fig. 27a). We now imagine M held rigid at this displaced position and now isolate joint O and allow it to displace. The "REACTION" O' will cease to exist, and an equal but opposite force will take its place and displace the joint some distance until equilibrium is reached (Fig. 27b). We again have the same case as we have before i. e. a single horizontal load acting at the released joint O while all other joints are held rigid, thus developing horizontal " REACTIONS " at all the other joints, which in addition to those due to M' will give us the resulting " REACTIONS " as a consequence of the displacement D of joint M through a certain " FORCE " with the pier heads M and O released. Likewise, we can add the interior forces produced in these two cases and obtain the state of stress for this case of loading. The " FORCE " developed through the displacement of M with the two joints M and O released is M=1-M" where M" represents the horizontal reaction at M caused by the application of +O' at O. Dividing all the values by 1 - M" will give us the influence of a unit load at M, with 2 neighbouring joints released. The example of releasing two adjacent joints should be extended to the whole arch system if there are more tham three spans. In Fig. 28 is shown

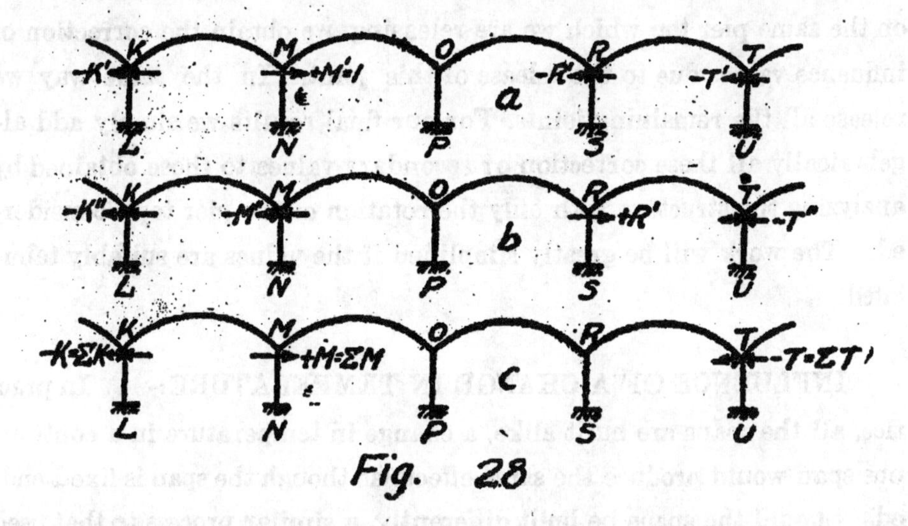

Fig. 28

the method of releasing 3 adjacent joints under the action of "FORCE" +M= M at pier-head M with the three consecutive pier-heads M, O and R unlocked for displacement.

We have now approached a point in which a general coordination of our treatment is necessary. As mentioned previously under the assumption that the pier heads are not allowed to displace but permitted to rotate, horizontal forces are developed at the supports. Following the above discussion of finding the influence of a unit horizontal load acting upon the joint which is allowed to displace, we need to multiply the horizontal forces developed by the effects of the unit load and add the results. This may be taken as a correction for the displacement of the piers, and, if we add this to the values obtained by analyzing the system with immovable piers, we get the results of analyzing the structure as it is-multiple arches on elastic piers. To obtain values for influence lines, we, as before, determine the influence of a unit horizontal load acting on any and each of the pier heads. Then proceeding with the influence values for thrust of the arches with immovable piers, and adding algebraically the horizontal pier reaction at the joint, we determine the influence values for the horizontal forces acting at the pier heads. By multiplying these influence values of the horizontal forces by the corresponding numerical values of the influence of a unit horizontal load on the same pier top which we are releasing, we obtain the correction of influence values due to the release of this joint. In the same way we release all the remaining joints. For our final results we simply add algebraically all these correction or secondary values to those obtained by analyzing the structure with only the rotation of the pier tops considered. The work will be greatly simplified if the values are suitably tabulated.

INFLUENCE OF A CHANGE IN TEMPERATURE:- In practice, all the spans are built alike, a change in temperature in a continuous span would produce the same effects as though the span is fixed-ended. Should the spans be built differently, a similar process to that used in the analysis for vertical loads may be used. The first step is to find the effects of this change of temperature on each individual arch. We then analyze our system of arches for the effects of rotation of the pierheads, and, by the use of the distribution coefficients and the carry-over factors we can obtain the interior forces developed in the various parts of the structure, as well as the horizontal forces acting at the pier heads. With these horizontal forces, and using the same process of releasing the joints, the correction for the displacement of the piers is obtained, which, in addition to the forces developed when the springings are assumed to be immovable, will give us the final results due to a change in temperature on the whole arch system The effects of a change in height of the piers due to a change in temperature is in usual cases negilibly small, but, if it is to be taken account of, the following discussion will lend some help.

EFFECTS OF SINKING OF PIERS:- The sinking o piers has not been explained in detail, as such sinking in all cases should be prevented or else the design is carried out so that the whole structure will sink an equal amount during loading, which in words, can be easily expressed but in actual cases is very difficult to carry out especially for live loads. In all usual cases which occur in practice, it is seldom that a continuous structure is built on yielding soils, which explains why sinking of piers are not usually considered. Further-more, the deformation of the piers itself is small, so the stresses in them are small.

We will consider that one of the piers sinks a distance D, while the rest remaining unchanged. For the case of rotation of pier-heads, this will affect directly only the two arches adjacent to the pier. The pier itself suffers no deformation if the effect of direct stress is neglected. The idea is similar to that of side displacement. In this case we may separate the deformation into two portions for the two individual arches. We investigate each individual arch for the sinking of one of its supports, and determe the effects on other parts of the structure. This will develop horizontal forces at the pier heads which may be utilized to calculate the correction values for the displacement of the piers, as we did in the case of vertical loads by releasing all the joints.

CONCLUSION:- The treatment given above of the analysis of multiple arches on elastic piers, is hoped to be simple enough as to warrant its use in actual disign work. The practical considerations of the choice of the number of spans, the length of each individual span, the amount of rise, the height of the piers as well as the arch form can not be taken up in any theoretical treatment, since for reasons of economy, aesthetics and suitability, every case is a study by itself, and wise judgment obtained through long and varied experiences will tell us what to use.

Example of A Three-span Continuous Arch Bridge

The individual arches of this example have been taken from an analysis of a two-span continuous arch system by Whitney in T. A. S. C. E Vol. 90 pp. 1094-1146. The first part of the calculations is similar to that given in a previous paper, and detailed explanation of the steps will not be given.

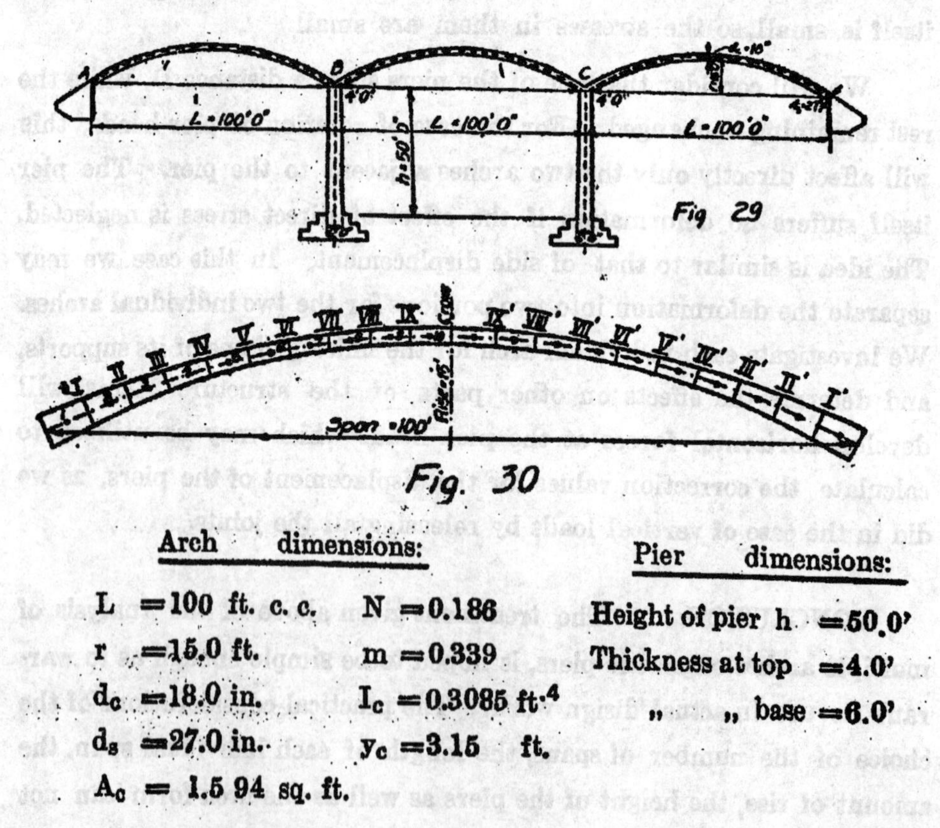

Fig 29

Fig. 30

Arch dimensions:

$L = 100$ ft. c. c.

$r = 15.0$ ft.

$d_c = 18.0$ in.

$d_s = 27.0$ in.

$A_c = 1.594$ sq. ft.

$N = 0.186$

$m = 0.339$

$I_c = 0.3085$ ft.^{4}

$y_c = 3.15$ ft.

Pier dimensions:

Height of pier $h = 50.0'$

Thickness at top $= 4.0'$

„ „ base $= 6.0'$

TABLE I.

Pt.	$w=\frac{ds}{I}$	x	$\frac{w}{L}(L-x)$	y	wy
1	5.87	2.5	5.72	1.85	10.9
2	6.90	7.5	6.38	5.18	35.7
3	8.01	12.5	7.01	7.81	62.6
4	8.93	17.5	7.37	9.89	88.3
5	9.87	22.5	7.65	11.51	113.7
6	11.17	27.5	8.10	12.75	142.3
7	12.72	32.5	8.59	13.67	173.9
8	13.78	37.5	8.61	14.32	197.6
9	14.82	42.5	8.52	14.74	218.5
10	15.75	47.5	8.27	14.98	235.8
10'	15.75	52.5	7.48		1279.2
9'	14.82	57.5	6.30		
8'	13.78	62.5	5.16		
7'	12.72	67.5	4.13		
6'	11.17	72.5	3.07		
5'	9.87	77.5	2.22		
4'	8.93	82.5	1.56		
3'	8.01	87.5	1.00		
2'	6.90	92.5	0.52		
1'	5.87	97.5	0.15		
	216.24				

$E\int dw = \frac{L}{I_c}\cdot\frac{1+m}{2} = 217.02$

$E\int y\,dw = (r-y_c)E\int dw = 216.8(15-3.15) = 2568$

$E\int y_0^2\,dw = \frac{C L r^2}{I_c} = \frac{0.04223\times100\times\overline{15}^2}{0.3085} = 3082$

$E\int y^2\,dw = E\int y_0^2\,dw + (r-y_c)^2 E\int dw$
$= 3082 + (15-3.15)^2\,216.8 = 33580$

$\int dv = \frac{L}{A_c C'_m} = \frac{100}{1.113\times1593} = 56.3$

$E\int x(L-x)\,dw = \frac{L^3}{I_c}\cdot\frac{5+3m}{48} = 40683$

$E\int x^2\,dw = \frac{L^3}{I_c}\cdot\frac{7+9m}{48} = 678,910$

$E\int xy\,dw = \frac{L}{2}(r-y_c)E\int dw = \frac{100}{2}(15\text{-}3.15)216.8 = 128660$

In Table II is given the calculations for the influence line for H_0 - horizontal thrust of the two-hinged base system. The value $\frac{1}{2}\int dw$ $=1279.2$ used is obtained through a summation process instead of the more correct value $2568/2=1284$ obtained by integration, so as to be consistent with our calculations.

The following constants of the arch, the equations of which had been given in the previous paper are necessary. The various integrals used are calculated by formulas either given by, or may be derived from those given by Whitney. (See Table I)

$$B = \frac{L}{L} = \frac{\Sigma xyw}{(\Sigma wy^2+\Sigma v)} = \frac{128700}{100(33580+56.3)} = 0.0383$$

$$EX = EY = \frac{1}{L^2}\sum wx^2 - B^2\left(\sum wy^2 + \sum v\right)$$

$$= \frac{678200}{100^2} - 0.0383\,(33580 + 56.3) = 18.65$$

$$EZ = \frac{1}{L^2}\sum wx(L-x) - B^2\left(\sum wy^2 + \sum v\right)$$

$$= \frac{40633}{100^2} - 0.0383^2\,(33580 + 56.3) = -8.605$$

TABLE II. INFLUENCE LINE VALUES FOR HORIZONTAL THRUST H_O OF THE TWO-HINGED ARCH BASE SYSTEM.

Section	Moment of ydw	$H_O = \frac{\int M_O y dw}{\int y^2 dw + \int dv}$
Springing		0
I	1279.2×2.5+1268.3×2.5= 6369	0.189
II	6369+1268.3×2.5+1232.6×2.5=12622	0.375
III	12622+1232.6×2.5+1170.0×2 5=18628	0.553
IV	18628+1170.0×2.5+1081.7×2.5=24257	0.721
V	24257+1081.7×2.5+ 968.0×2.5=29381	0.872
VI	29381+ 968.0×2.5+ 825.7×2.5=33865	1.007
VII	33865+ 825.7×2.5+ 651.8×2.5=37558	1.116
VIII	37558+ 651.8×2.5+ 454.2×2.5=40324	1.199
IX	40324+ 454.2×2.5+ 235.7×2.5=42048	1.249
Crown	42048+ 235.7×2.5 =42637	1.268

The inflnence lines for s aud t (Table III) or rotations of the end tangents of the 2-hinged base system are obtained from the equation

$$S = \frac{1}{E}\left[\frac{1}{L}\sum M_o(L-x)w - H_o B\left(\sum wy^2 + \sum v\right)\right]$$

in which the expression $\frac{1}{L}\sum M_O(L-x)\,w$, for the case of a unity load, is none other than the moment diagram of a simple beam loaded with a variable load $w\frac{(L-x)}{L}$. In finding this moment diagram we again use the method of finding the area under the shear curve.

TABLE III. INFLUENCE LINES FOR X AND Y.

Sect.	Moment of $\frac{w(L-x)}{L}$	$\frac{-H_0B}{(\sum wy^2+\sum v)}$	EX	EY
Spring		0	0	0
I	+67.22×2.5+61.50×2.5= 321.79			
II	321.79+61.50×2.5+55.12×2.5= 613.33	− 482.92	+130.41	− 78.26
III	613.33+55.12×2.5+48.11×2.5= 871.40			
IV	871.40+48.11×2.5+40.74×2.5=1093.51	− 928.08	+165.43	−135.59
V	1093.50+40.74×2.5+33.09×2.5=1278.07			
VI	1278.07+33.09×2.5+24.99×2.5=1423.26	−1295.68	+127.58	−153.89
VII	1423.26+24.99×2.5+16.40×2.5=1526.73			
VIII	1526.73+16.40×2.5+ 7.79×2.5=1587.20	−1542.81	+ 44.39	−124.17
IX	1587.20+ 7.79×2.5− 0.73×2.5=1604.84			
Crown	1604.84− 0.73×2.5− 9.00×2.5=1580.50	−1631.30	− 50.80	− 50.80
IX'	1580.50− 9.00×2.5−16.48×2.5=1516.79			
VIII'	1516.79−16.48×2.5−22.78×2.5=1418.64	−1542.81	−124.17	+ 44.39
VII'	1418.64−22.78×2.5−27.94×2.5=1291.83			
VI'	1291.83−27.94×2.5−32.07×2.5=1141.79	−1295.68	−153.89	+127.58
V'	1141.79−32.07×2.5−35.14×2.5= 973.75			
IV'	973.75−35.14×2.5−37.36×2.5= 792.49	− 928.08	−135.59	+165.43
III'	792.49−37.36×2.5−38.92×2.5= 601.78			
II'	601.78−38.92×2.5−39.92×2.5= 404.66	− 482.92	− 78.26	+130.41
I'	404.66−39.92×2.5−40.44×2.5= 203.75			
Spring.	203.75−40.44×2.5−40.59×2.5 ∼ 0	0	0	0

The left reaction of this imaginary beam so loaded is

$$R_L = \frac{1}{L^2}\sum (L-x)^2 w = \frac{1}{L^2}\sum wx^2 = \frac{672180}{100 \times 100} = 67.218$$ (obtained

by summation as against 67.820 obtained by integration.

Assume the following dimensions for the pier:

Thickness at top $t_B = 4.0$ ft.

" " base $t_E = 6.0$ ft.

height of pier $h = 50.$ ft.

Width of pier = Width of rib.

$k = t_E / t_B = 6/4 = 1.5$

Incfeasing the moment of inertia of the concrete 10 % to allow for reinforcement.

$$I_B = 1.1 \times \frac{4^3}{12} = 5.87 \text{ ft.}^4$$

From T.A.S.C.E Vol. 90 P. 1114

$$E\int dw = 0.5555 \times \frac{50}{5.87} = 4.73 \qquad E\int y\,dw = 0.3333 \times \frac{50^2}{5.87} = 141.9$$

$$E\int y^2 dw = 0.2437 \times \frac{50^3}{5.87} = 5180$$

$$E\int y(h-y)\,dw = \frac{h^3}{I_B} \times \frac{1}{2k} - \int y^2 dw = \frac{50^3}{5.87 \times 3} - 5180 = 1920$$

$$E\int (h-y)^2 dw = h^2 \int dw - 2h\int y\,dw + \int y^2 dw$$

$$= 50^2 \times 4.73 - 2 \times 50 \times 141.9 + 5180 = 2815$$

The rotations of the end tangents of the pier due to unit moments are

$$EX' = \frac{1}{h^2}\int y^2 dw = 0.2437 \times \frac{50}{5.87} = 2.072$$

$$EY' = \frac{1}{h^2}\int (h-y)^2 dw = \frac{2815}{50^2} = 1.127$$

$$EZ' = \frac{1}{h^2}\int y(h-y)\,dw = \frac{1920}{50^2} = 0.768$$

The abutments and the pier bases are assumed as perfectly fixed in this example. With all the elastic properties of the arches and the piers before us, we can proceed to calculate (a) the degrees of restraint of the individual members T, (b) the distribution ratios U and (c) the carry-over ratios m.

According to Eq. 34 the carry-over ratio for the pier is

$$m_B^k = \frac{-Z^1}{Y^1 + T_E^f} = \frac{-0.768}{2.072 + 0} = -0.371 = m_c^k \text{ by symmetry.}$$

The angles of rotation of the piers are

$$e_B^k = e_C^k = X' + m_B^k Z' = \frac{2.072}{E} - 0.371 \times \frac{0.768}{E} = 1.787/E$$

Similarly for the 2 exterior symmetrical arches with T_A^r and T_D^L equal to zero by assumption

$$m_B^L = m_C^r = \frac{-Z}{X + T_A^r} = \frac{8.605}{18.65} = 0.461$$

$$e_B^L = e_C^r = Y + m_B^L Z = \frac{1}{E}[18.65 + 0.461(-8.605)] = \frac{14.68}{E}$$

$$T_B^r = T_C^L = \frac{e_B^L \cdot e_B^k}{e_B^L + e_B^k} = \frac{14.68 \times 1.787}{E\,14.68 + 1.787)} = \frac{1.592}{E}$$

The distribution factors at joint B with any span to the right of B loaded are $U_{Br}^L = \frac{T_B^r}{e_B^L} = \frac{1.592}{14.68} = 0.109$ for the left exterior arch

$$U_{Br}^k = \frac{T_B^r}{e_B^k} = \frac{1.592}{1.787} = 0.891 \quad \text{for the left pier}$$

At joint C through symmetry

$$U_{CL}^r = 0.1085 \quad \text{for the right exterior arch}$$

$$U_{CL}^k = 0.8915 \quad \text{for the right pier}$$

when any span to the left of C is loaded.

Proceeding thus, we obtain

$$m_B^r = m_C^L = \frac{-Z}{X + T_B^r} = \frac{8.605}{18.65 + 1.592} = 0.425$$

$$e_B^r = e_C^L = X + m_B^r Z = \frac{18.65}{E} - \frac{8.605}{E} = \frac{14.99}{E}$$

$$T_B^L = T_C^r = \frac{e_B^r \cdot e_B^k}{e_B^r + e_B^k} = \frac{14.99 \times 1.787}{E(14.99 + 1.787)} = \frac{1.597}{E}$$

The distribution factors at joint B when any span to the left of C is loaded are

$$U_{BL}^r = \frac{T_B^L}{e_B^r} = \frac{1.597}{14.99} = 0.1066 \quad \text{for the right arch or intermediate arch}$$

$$U_{BL}^k = \frac{T_B^L}{e_B^k} = \frac{1.597}{1.787} = 0.8934 \quad \text{for the left pier}$$

The degree of restraint of the pier head through its rigid connection with the two arches is determined by the corresponding arch rotation angles

$$T_B^k = T_C^k = \frac{e_B^L \cdot e_B^r}{e_B^L + e_B^r} = \frac{1.468 \times 14.99}{E(14.68 + 14.99)} = \frac{7.41}{E}$$

The distribution factors of the arches for moments existing at the pier head are

$$U_{Bk}^L = U_{Ck}^r = \frac{T_B^k}{e_B^L} = \frac{7.41}{14.68} = 0.505 \quad \text{for the 2 exterior arches.}$$

$$U_{Bk}^r = U_{Ck}^L = \frac{T_B^k}{e_B^r} = \frac{7.41}{14.99} = 0.495 \quad \text{for the middle arch.}$$

From the above calculations it may be noted that U_{Br}^L and U_{BL}^r are around 0.11 or 11% so that the effect of the rotation of the pier heads is by no means negligible. Any approximate method which considers only the horizontal displacement of the piers may thus incur considerable error for the proportions used in our example. In passing it may be mentioned that if the pier thicknesses were doubled and even if the pier height be increased to 60 ft. these values of U are only 0.016. In the latter case the effect of rotation of the pier heads may safely be neglected.

The moment constants for the two exterior arches are according to Eq. 5 in the previous paper

$$C_1 = \frac{Y + T_B^L}{(X + T_A^r)(Y + T_B^L) - Z^2}$$

$$= \frac{(18.65 + 1.597)\,E}{(18.65 + 0)(18.65 + 1.597) - 8.605^2} = \frac{20.25E}{303.6} = 0.0666E$$

$$C_2 = \frac{X + T_A^r}{(X + T_A^r)(Y + T_B^L) - Z^2} = \frac{(18.65 + 0)\,E}{303.6} = 0.614E$$

$$C_3 = \frac{Z}{(X + T_A^r)(Y + T_B^L) - Z^2} = \frac{-8.605\,E}{303.6} = -0.614\,E$$

In Table IV is shown the calculations of influence line values for the springing moments and thrusts of the exterior arch. In this table only the values for the loaded span are given. The fundamental equations used were derived in the previous paper. They are;

$$M_A = C_3 t - C_1 s$$

$$M_B = C_3 s - C_2 t \qquad \text{and } H = H_0 + (M_A + M_B)B.$$

For other sections of the structure use may be made of the distribution and carry-over factors already calculated. The results are tabulated in Tables Va and Vb following. In all these cases rib shortening has been taken care of.

For the branch of influence line for sections not situated in the loaded span discussed above it may be obtained by making use of the distribution and carry-over factors.

For example, let it be required to find the moments for the sections enumerated below for a unit load placed at section VI of Arch I or left exterior arch.

TABLE IV.- INFLUENCE VALUES FOR STRESSES IN ARCH I. — Unit load on Arch I

Load at Sect.	C_3t	$-C_1s$	M_A^r	C_3s	$-C_2t$	M_B^L	M_A+M_B	$B(M_A+M_B)$	H_o	H_{al}
Spring.			0			0			0	0
II	+2.20	− 8.69	−6.49	−3.70	+ 4.81	+1.11	−5.38	−0.206	0.375	0.169
IV	+3.82	−11.02	−7.20	−4.69	+ 8.33	+3.64	−3.56	−0.136	0.721	0.585
VI	+4.32	− 8.50	−4.18	−3.60	+ 9.44	+5.82	+1.64	+0.063	1.007	1.070
VIII	+3.50	− 2.96	+0.54	−1.26	+ 7.63	+6.37	+6.91	+0.264	1.199	1.463
Crown	+1.43	+ 3.38	+4.81	+1.44	+ 3.12	+4.56	+9.37	+0.358	1.268	1.626
VIII'	−1.25	+ 8.27	+7.02	+3.52	− 2.73	+0.79	+7.81	+0.299	1.199	1.498
VI'	−3.60	+10.24	+6.64	+4.37	− 7.83	−3.46	+3.18	+0.122	1.007	1.129
IV'	−4.66	+ 9.03	+4.37	+3.84	−10.17	−6.33	−1.96	−0.075	0.721	0.646
II'	−3.68	+ 5.21	+1.53	+2.22	− 8.01	−5.79	−4.26	−0.163	0.375	0.212
Spring.			0			0			0	0

From Table IV $M_A^r = -4.18\#$ $\qquad M_B^L = +5.82\#$

$M_B^r = U_{BL}^r M_B^L = 0.107 \times 5.82 = +0.621\#$ $\qquad M_B^k = U_{BL}^k M_B^L = 0.893 \times 5.82 = -5.20^o\#$

$M_C^L = m_B^r M_B^r = 0.425 \times 0.631 = +0.264\#$ $\qquad M_E^f = m_B^k M_B^k = -0.371(-5.20) = +1.93\#$

$M_C^r = U_{CL}^r M_C^L = 0.109 \times 0.264 = +0.0286\#$ $\qquad M_C^k = U_{CL}^k M_C^L = 0.892(0.264) = -0.235^o\#$

$M_D^L = M_C^r M_C^r = 0.461 \times 0.0286 = +0.0132\#$ $\qquad M_F^f = m_C^k M_C^k = -0.371(-0.236) = +0.0875\#$

o These moments change sign because of the convention of signs chosen positive for tension on right side of pier and at intrados of arch. In thus follows, that if M_B^L and M_C^L are positive, M_B^k and m_B^k must be negative. In later calculations the negative sign will be placed before U_{BL}^k and U_{CL}^k

The horizontal forces accompanying these moments are:

Horizontal thrust in Arch I $\quad H_{a1} = +1.070\#$

" " " " II $\quad H_{a2} = (M_B^r + M_C^L) B = (0.631 + 0.264) \times 0.0383 = +0.0343\#$

" " " " III $\quad H_{a3} = (M_C^r + M_D^L) B + (0.0286 + 0.0132) \times 0.0383 = +0.0016\#$

" shearing force at Pier Head 1. $H_{p1} = \frac{1}{h}(-M_B^k + M_E^f) = \frac{1}{50}(5.20 + 1.93) = +0.143\#$

TABLE Va:- MOMENTS IN UNLOADED SPANS FOR UNIT LOAD ON ARCH I.

Load at Sect.	M_B^L Table IV	$M_B^r = 0.167 M_B^L$	$M_C^L = 0.425 M_B^r$	$M_C^r = 0.109 M_C^L$	$M_D^L = 0.461 M_C^r$	$M_B^k = -0.894 M_B^L$	$M_E^f = -0.371 M_B^k$	$M_C^k = 0.892 M_C^L$
II	+1.11	+0.118	+0.050	+0.005	+0.003	−0.99	+0.37	−0.045
IV	+3.64	+0.388	+0.165	+0.018	+0.008	−3.25	+1.21	−0.147
VI	+5.82	+0.621	+0.264	+0.029	+0.013	−5.20	+1.93	−0.235
VIII	+6.37	+0.679	+0.289	+0.031	+0.014	−5.69	+2.11	−0.258
Crown	+4.56	+0.486	+0.206	+0.022	+0.010	−4.07	+1.51	−0.184
VIII'	+0.79	+0.084	+0.036	+0.004	+0.002	−0.71	+0.26	−0.032
VI'	−3.46	−0.369	−0.157	−0.017	−0.008	+3.09	−1.15	+0.140
IV'	−6.33	−0.675	−0.287	−0.031	−0.014	+5.66	−2.10	+0.256
II'	−5.79	−0.617	−0.262	−0.028	−0.013	+5.17	−1.92	+0.234

TABLE Vb:- M AND H IN UNLOADED SPANS FOR UNIT LOAD ON ARCH I.

Load at Sect.	$M_F^f = 0.371 M_C^k$	$H_{a2} = 0.0383 (M_B^r + M_C^L)$	$H_{a3} = 0.0383 (M_C^r + M_B^L)$	$H_{P1} = \frac{1}{50} (M_E^f - M_B^k)$	$H_{P2} = \frac{1}{50} (M_F^f - M_C^k)$	$H_B = H_{a1} - H_{a2} + H_{P1}$	$H_C = H_{a2} - H_{a3} + H_{P2}$
II	+0.017	+0.006	+0.000	+0.027	+0.001	+0.190	+0.007
IV	+0.055	+0.021	+0.001	+0.089	+0.004	+0.653	+0.024
VI	+0.087	+0.034	+0.002	+0.143	+0.006	+1.179	+0.038
VIII	+0.096	+0.037	+0.002	+0.154	+0.007	+1.580	+0.042
Crown	+0.068	+0.026	+0.001	+0.112	+0.005	+1.712	+0.030
VIII'	+0.012	+0.005	+0.000	+0.019	+0.001	+1.512	+0.006
VI'	−0.052	−0.020	−0.001	−0.085	−0.004	+1.064	−0.023
IV'	−0.095	−0.037	−0.002	−0.155	−0.007	+0.528	−0.042
II'	−0.087	−0.034	−0.002	−0.142	−0.006	+0.104	−0.038

25232

Hor. shearing force at Pier Head 2. $H_{p2} = \frac{1}{h}(-M_C^k + M_F^f) =$

$$\frac{1}{50}(0.236+0.0875) = +0.0045$$

Horizontal forces are positive if their directions are from left to right. For arches, thrust is positive and tension is negative.

At the top of pier I or joint B in which we have placed an imaginary support which permits rotation only, there will exist a horizontal reaction caused by the thrusts of Arches I and II and the shear due to the moments of the pier ends. This horizontal reaction will be the algedraical sum of the following items.

Hor. Thrust of Arch I	$H_{a1} = +1.070$	acting	from	left to right
,, ,, ,, ,, II	$H_{a2} = -0.0343$	,,	,,	right to left
,, Shear at top of pier 1	$H_{p1} = +0.143$	,,	,,	left to right
Total hor. force acting at support B	$H_B = +1.179$	,,	,,	left to right

The reaction at the imaginary restraining support will of course act from right to left. Owing to the fact that when this imaginary restraint is removed the joint B will displace from left to right, we will depict this force H_B and all horizontal forces or actions (not reactions) acting from left to right as positive.

Accordingly, through this loading, i. e. unit load at Sect VI of Arch I, there will also appear at joint C a horizontal force H_C, which tends to displace that joint towards the right.

$$H_C = H_{a2} - H_{a3} + H_{p2} = 0.034 - 0.002 + 0.006 = +0.038.$$

In this way Tables Va and Vb are filled up, and we have a branch of the influence line for each of the stresses enumerated in the column headings, for unit load placed over Arch I.

If the moments at any section in Arch I for a unit load at a distance x from the left support is desired, we may utilize the equation

$$M_x = M_o + M_a \frac{L-x}{L} + M_b \frac{x}{L} - Hy.$$

Using the same process of calculation as above, the influence line values for moments, thrusts and horizontal forces for load placed on Arch II are compiled in Tables VI, VII. No calculations need be made for loads placed on Arch III as the stresses may be easily obtained from Tables IV and Va and b owing to the symmetrical nature of the whole structure.

The constants required for calculating influence line values for moments at the springings of Arch II (Table VI) or intermediate arch are:

$$C_1 = C_2 = \frac{X+T^r_B}{(X+T^r_B)^2 - Z^2} = \frac{18.65\ E}{(18.65+1.592)^2 - 8.605}$$

$$= \frac{18.65\ E}{335.69} = 0.0556\ E$$

$$C_3 = \frac{EZ}{335.69} = \frac{-8.605E}{335.69} = -0.02563E$$

We will now proceed to remove the imaginary restraints which we placed at the pier head B and C in our previous calculations, We will loosen that placed at B first, leaving the imaginary restraint at C undisturbed. Assuming that B is displaced from left to right a distance ED=1000 then the support of each of the 3 members Arch I and II and pier B E will be displaced the same amount. To simplify our calculations, we consider the displacement of each individual member separately and sum up the results.

Considering Arch I as lengthened by the amount ED=1000, while the span of Arch II and the pier head of BE imagined as being undisturbed, then according to Eq. 14 the thrust in our two-hinged base system is

$$H = \frac{ED}{\Sigma wy^2 + \Sigma v} = \frac{1000}{33640} = -0.0297\#\ (\text{tension})$$

TABLE VI:- INFLUENCE VALUES FOR STRESSES IN ARCH II.

Unit load on Arch II

Load at Sect.	C_3 t	$-C_1$ s	M^r_B	M^L_C	(M_B +M_C)	(M_B +M_B)	H_o	H_{a2}
Spring.	0	0	0	0	0	0	0	0
II	+2.00	−7.25	−5.25	+1.00	−4.25	−0.163	0.375	0.212
IV	+3.48	−9.20	−5.72	+3.28	−2.44	−0.094	0.721	0.627
VI	+3.94	−7.08	−3.14	+5.28	+2.14	+0.082	1.007	1.089
VIII	+3.18	−2.47	+0.71	+5.76	+6.47	+0.248	1.199	1.447
Crown	+1.30	+2.82	+4.12	+4.12	+8.24	+0.316	1.268	1.584
VIII'	−1.14	+6.90	+5.76	+0.71	......	... ...		1.447
VI'	−3.27	+8.55	+5.28	−3.14	... ...	... ...	......	1.089
IV'	−4.25	+7.53	+3.28	−5.72	......	... ...	......	0.627
II'	−3.35	+4.35	+1 00	−5.25	... ...	... ...	......	0.212
Spring.	0	0	0	0			0	0

TABLE VII:- MOMENTS AND HORZONTAL FORCES IN UNLOADED SPANS FOR UNIT LOAD ON ARCH II.

Load at Sect.	M^r_B (Table VI)	M^L_B =0.109 M^r_B	M^r_A =0.461 M^L_B	M^k_B =0.892 M^r_B	M^f_E =-0.371 M^k_B	H_{a_1}=B (M^r_A + M^L_B)	H_p,	H_B	H_{c2}
II	−5.25	−0.569	−0.262	−4.68	+1.74	−0.032	+0.128	−0.116	+0.230
IV	−5.72	−0.621	−0.286	−5.10	+1.89	−0.035	+0.140	−0.522	+0.687
VI	−3.14	−0 341	−0.157	−2.80	+1.04	−0.019	+0.077	−1.031	+1.186
VIII	+0.71	+0.077	+0.036	+0.63	−0.23	+0.004	−0.067	−1.460	+1.553
Crown	+4.12	+0.447	+0.206	+3.68	−1.36	+0.025	−0.101	−1.660	+1.660
VIII'	+5.76	+0.625	+0.288	+5.14	−1.91	+0.035	−0.141	−1.553	+1.460
VI'	+5.28	+0.573	+0.264	+4.71	−1.75	+0.032	−0.129	−1.186	+1.031
IV'	+3.28	+0.356	+0.169	+2.92	−1.08	+0.020	−0.080	−0.687	+0.522
II	+1.00	+0.109	+0.049	+0.89	−0.33	+0.006	−0.024	−0.230	+0.116

From Eq. 15a and b of the previous paper

$$Es = Et = H_{oD}B(\Sigma wy^2 + \Sigma v) = 0.0297 \times 0.03826 \times 33640 = 38.26$$

The springing moments

$$M^r_{AD} = (C_3 - C_1)s = (-0.02838 - 0.0666)38.26 = -3.635'\#$$

$$M^L_{BD} = (C_3 - C_2)s = (-0.0284 - 0.0614)38.26 = -3.440'\#$$

(see Eq. 16a and b)

The negative thrust in the arch will be (Eq. 17)

$$H_{D1} = H_{oD} + (M^r_{AD} + M^L_{BD})B = -0.0297 + (-3.645 - 3.44)0.03826 = -0.301\#$$

Using the value of M^L_{BD} and the distribution and the carry-over factors calculated previously, we may obtain the moments at other springings and at the extremities of the piers listed in Table VIII by simple multiplication. In the lower portion of the table is given the calculation of horizontal reactions caused by these moments. Horizontal reactions acting from left to right are considered positive.

The second portion of the displacement of joint B, the shortening of Arch II, with other spans and the position of the pier heads imagined as undisturbed, will develop moments at the springings

$$M^r_{BD} = M^L_{CD} = (C_2 - C_1)s = (-0.02563 - 0.05556)(-38.3) = +3.110'\#$$

The total thrust in Arch II will be

$$H_{D2} = 0.0297 + (3.11 + 3.11)0.0383 = 0.268\#$$

For moments developed at other sections due to this shortening of Arch II see Table VIII.

The remaining component of the displacement of joint B- the moving of pier-head B a distance ED = 1000 with the arch spans remaining unchanged-will cause a rotation of the end tangents of the pier

$$s = -t = \frac{D}{H} = \frac{1000}{50} = 20 \qquad \text{Eq. 26}$$

The moment constants of the piers by Eq. 5 are

$$C_1 = \frac{Y' + Z^f_E}{(X' + T^k_B)(Y' + T^f_E) - (Z')^2} = \frac{(1.127 + 0)E}{(2.072 + 7.41)\ (1.127 + 0) - (0.768)^2}$$

$$= \frac{1.127}{10.10} = 0.1117\ E$$

$$C_2 = \frac{X' + T^k_B}{10.10} = \frac{(2.072 + 7.41)E}{10.10} = 0.939\ E$$

$$C = \frac{Z'}{10.10} = \frac{0.768E}{10.10} = 0.0761\ E$$

By Eq. 27a and b the end moments of the pier are

$$M^k_{BC} = (C_3 + C_1)\frac{D}{H} = (0.0761 + 0.1117)\frac{1000}{50} = +3.756\# \text{ tension on the right side.}$$

$$M^f_{ED} = -(C_3 + C_2)\frac{D}{H} = (0.0761 + 0.939)\frac{1000}{50} = -20.302\# \text{ tension on the left side.}$$

The horizontal shear of these moments is

$$H_{pD} = \frac{1}{H}(+M^f_{ED} - M^k_{BD}) = \frac{1}{50}(-20.302 - 3.756) = -0.4812\#$$

This reaction acts from left to right at the top of the pier so in Column B^k in Table VIII it is listed as posifive. At the bottom of the pier it must be from right to left E^f is listed as negative

The moment in B, M^k_B, of course, imparts moments and thrusts to other parts of the structure which may be obtained in a similar manner like that mentioned previously.

Adding the results of the three individual processes together we obtain the total horizontal force required at B=1.378# to displace joint B a distance ED=1000 with joint C restrained from shifting. Dividing all the stresses by 1.378 we will have the effect of applying a horizontal force at B equal to 1 acting from left to right. This unit force at B develops at joint C a horizontal reaction equal to −0.303#. In our

Table VIII — INFLUENCE OF A HORIZONTAL FORCE B=1# at JOINT B.

		A^r	B^L	B^r	B^k	C^L	C^r	C^k	D^L	E^F	E^F
MOMENT.	B Loosened and Shifted D=1000 C Restrained: Displacement of Arch I	−3.635	−3.440	−0.367	+3.073	−0.156	−0.017	−0.139	−0.008	− 1.140	−0.055
	Displacement of Arch II	+0.156	+0.338	+3.110	+2.772	+3.110	+0.338	−2.772	+0.156	− 1.028	+1.028
	Displacement of pier-head	−0.874	−1.896	+1.860	+3.756	+0.791	+0.086	−0.705	+0.040	−20.202	+0.262
	Result. Displacement thru B=1.38	−4.353	−4.998	+4.603	+9.601	+3.745	+0.407	−3.338	+0.188	−22.470	+1.238
	Displacement of B thru Force B=1#, C restrained	−3.16	−3.63	+3.34	+6.97	+2.72	+0.30	−2.42	+0.14	−16.32	+0.90
	B restrained, C loosened under Force C=0.303#	−0.04	−0.10	−0.82	−0.73	−1.01	+1.10	+2.11	+0.96	+ 0.27	−4.93
	B and C loosened, B acted upon by Force B=0.909#	−3.20	−3.72	+2.52	+6.24	+1.71	+1.40	−0.31	+1.10	−16.05	−4.03
	Influence of Force B=1# upon whole structure	−3.51	−4.09	+2.76	+6.85	+1.86	+1.53	−0.33	+1.20	−17.60	−4.44
HORIZONTAL FORCES.	B Loosened acd Shifted ED=1000 C Restrained: Displacement of Arch I	−0.301	+0.301	−0.020	+0.084	+0.020	−0.001	+0.004	−0.001	−0.084	−0.004
	Displacement of Arch II	+0.019	−0.019	+0.268	+0.076	−0.268	+0.019	−0.076	−0.019	−0.076	+0.076
	Displacement of Pier-head	−0.106	+0.106	+0.101	+0.481	−0.101	+0.005	−0.019	−0.005	−0.481	+0.019
	Result. Displacement thru B=1.38 #	−0.388	+0.388	+0.349	+0.641	−0.349	+0.023	−0.091	−0.023	−0.641	+0.091
				+1.378			−0.417				
	Displacement of B=1#, C restrained	−0.282		+1.000			−0.303		−0.016	−0.465	+0.066
	B restrained, C loosened under C=0.303 # as action force	−0.005		−0.091			+0.303		−0.086	−0.020	−0.141
	B and C loosened under B=0.909#	−0.287		+0.909			....		−0.102	−0.445	−0.075
	Influence of B=1# upon whole structure	−0.314		+1.000			....		−0.112	−0.490	−0.083

original structure both joints B and C are free to move so we must relieve the reaction at C also. For this purpose we shall need the results obtained above through the action of the unit load B=1#. We imagine that joint B is restrained in the displaced position and remove the imaginary supports at C so that the resulting value of force C=0.303# may come into action. We have now, the same case as before–B released and C restrained–; only this time we have at C the force C= +0.303# acting with joint B restrained at its displaced position in symmetrical contrast to the preceeding case. As the whole arch system is symmetrical, the influence of force C=0.303# may be obtained by the multiplication of the corresponding symmetrically opposite values caused by the unit load at B.

Assembling these two component results together, i. e. the effect of the displacement of B under the unit load with C restrained, and the subsequent release of joint C of its imaginary restraint through the application as "Action" of the reaction C=0.303# caused by the former displacement–we have, the effect on the whole system of the force B=0.909# acting at B.

By dividing all the values by this amount there is finally obtained the influence of the horizontal unit load at B upon the whole system under the original and actual conditions of support.

The various details of the calculation are compiled in Table VIII.

The thrusts of the individual arches developed, as a consequence of the displacement caused by the force B=1#, are;

$H_{a1} = A_r = -0.314\#$ tension

$H_{a2} = D_L + F^F = 0.112 + 0.083 = +0.195\#$ compression

$H_{a3} = D_L = +0.112\#$ compression

On account of the symmetry of the whole structure the influence of a unit horizontal load at joint C will be in the relationship of a mirror image of that acting at B.

TABLE IX INFLUENCE LINE VALUES OF SECONDARY EFFECTS OF H_B FOR UNIT LOAD ON ARCH I.

Sect.	H^1_B Table V_b	$M^r_A = -3.51\, H^1_B$	$M^L_B = -4.09\, H^1_B$	$M^r_B = +2.76\, H^1_B$	$M^k_B = +6.85\, H^1_B$	$M^L_C = +1.86\, H^1_B$	$M^r_C = +1.53\, H^1_B$	$M^k_C = -0.33\, H^1_B$	$M^L_D = +1.20\, H^1_B$	$M^F_E = -17.60\, H^1_B$	$M^F_F = -4.44\, H^1_B$	$H^I_a = -0.314\, H^1_B$	$H^{II}_a = +0.195\, H^1_B$	$H^{III}_a = +0.112\, H^1_B$
II	+0.190	−0.67	−0.77	+0.52	+1.30	+0.35	+0.29	−0.06	+0.23	−3.34	−0.84	−0.060	+0.037	+0.021
IV	+0.653	−2.29	−2.67	+1.80	+4.47	+1.21	+1.00	−0.22	+0.78	-11.49	−2.90	−0.205	+0.127	+0.073
VI	+1.179	−4.13	−4.81	+3.25	+8.07	+2.19	+1.80	−0.39	+1.41	-20.72	−5.23	−0.370	+0.230	+0.132
VIII	+1.580	−5.55	−6.46	+4.36	+10.82	+2.94	+2.42	−0.52	+1.90	-27.82	−7.01	−0.496	+0.308	+0.177
Crown	+1.712	−6.01	−7.00	+4.72	+11.72	+3.18	+2.62	−0.57	+2.06	-30.15	−7.60	−0.537	+0.334	+0.192
VIII'	+1.512	−5.31	−6.19	+4.17	+10.36	+2.82	+2.31	−0.50	+1.82	-26.63	−6.71	−0.475	+0.295	+0.169
VI'	+1.064	−3.74	−4.25	+2.94	+7.29	+1.98	+1.63	−0.35	+1.28	-18.72	−4.72	−0.334	+0.208	+0.119
IV'	+0.528	−1.85	−2.16	+1.46	+3.62	+0.98	+0.82	−0.17	+0.63	−9.29	−2.34	−0.166	+0.103	+0.059
II'	+0.104	−0.36	−0.43	+0.29	+0.71	+0.19	+0.16	−0.03	+0.12	−1.83	−0.46	−0.033	+0.002	+0.012

TABLE X INFLUENCE LINE VALUES OF SECONDARY EFFECTS OF H_C FOR UNIT LOAD ON ARCH I

Sect.	H^I_C	$M^r_A = -1.20\, H^1_C$	$M^L_B = -1.53\, H^1_C$	$M^r_B = -1.86\, H^1_C$	$M^L_C = -2.76\, H^1_C$	$M^r_C = +4.09\, H^1_C$	$M^L_D = +3.51\, H^C_1$	$M^k_B = -0.31\, H^1_C$	$M^k_C = -6.85\, H^C_1$	$M^F_E = -4.44\, H^1_C$	$M^F_F = -17.60\, H^1_C$	$H^I_a = -0.112\, H^I_C$	$H^{II}_a = -0.195\, H^1_C$	$H^{III}_a = +0.314\, H^I_C$
II	+0.007	−0.01	−0.01	−0.01	−0.02	-0.03	−0.03	−0.00	-0.05	-0.03	−0.12	−0.001	−0.001	+0.002
IV	+0.024	−0.03	−0.04	−0.04	−0.07	-0.10	−0.08	−0.01	-0.16	-0.11	−0.42	−0.003	−0.005	+0.008
VI	+0.038	−0.04	−0.06	−0.07	−0.11	-0.15	−0.13	−0.01	-0.26	0.17	−0.67	−0.004	−0.007	+0.012
VIII	+0.042	−0.05	−0.06	−0.08	−0.12	-0.17	−0.15	−0.01	-0.29	-0.19	−0.74	−0.004	−0.008	+0.003
Crown	+0.030	−0.04	−0.05	−0.06	−0.08	-0.12	−0.11	−0.01	-0.21	-0.13	−0.53	−0.003	−0.006	+0.009
VIII'	+0.006	−0.01	−0.01	−0.01	−0.02	-0.02	−0.02	−0.00	-0.04	-0.03	−0.11	−0.001	−0.001	+0.002
VI'	−0.023	+0.03	+0.04	+0.04	+0.06	+0.09	+0.08	+0.01	+0.17	+0.10	+0.41	+0.003	+0.004	−0.007
IV'	−0.042	+0.05	+0.06	+0.08	+0.12	+0.17	+0.15	+0.01	+0.29	+0.19	+0.74	+0.005	+0.008	−0.013
II'	−0.038	+0.05	+0.06	+0.07	+0.10	+0.16	+0.13	+0.01	+0.26	+0.17	+0.67	+0.004	+0.007	−0.012

TABLE XI INFLUENCE LINE VALUES FOR SECONDARY EFFECTS OF H_B FOR UNIT LOAD ON ARCH II

Sect.	H_B^{II} Table VII	$\frac{M_A^r}{H_B^{II}}=-3.51$	$\frac{M_B^L}{H_B^{II}}=-4.09$	$\frac{M_B^r}{H_B^{II}}=+2.76$	$\frac{M_C^L}{H_B^{II}}=+1.86$	$\frac{M_C^r}{H_B^{II}}=+1.53$	$\frac{M_D^L}{H_B^{II}}=+1.20$	$\frac{M_E^F}{H_B^{II}}=-17.60$	$\frac{H_a^I}{H_B^{II}}=0.314$	$\frac{H_a^{II}}{H_B^{II}}=+0.195$	$\frac{H_a^{III}}{H_B^{II}}=+0.112$	$\frac{M_F^F}{H_B^{II}}=-4.44$	$\frac{M_B^k}{H_B^{II}}=+6.85$	$\frac{M_C^k}{H_B^{II}}=-0.33$
II	−0. 116	+0. 41	+0. 47	−0. 32	−0. 22	−0. 18	−0. 14	+2. 04	+0.036	−0.023	−0.013	+0. 52	−0. 79	+0. 04
IV	−0. 522	+1. 83	+2. 14	−1. 44	−0. 97	−0. 80	−0. 63	+9. 19	+0.164	−0.118	−0.059	+2. 32	−3. 58	+0. 17
VI	−1. 031	+3. 62	+4. 21	−2. 84	−1. 92	−1. 58	−1. 24	+18. 13	+0.324	−0.201	−0.115	+4. 57	−7. 06	+0. 34
VIII	−1. 460	+5. 12	+5. 97	−4. 03	−2. 72	−2. 24	−1. 75	+25. 68	+0.459	−0.285	−0.163	+6. 48	−10. 00	+0. 48
Crown	−1. 660	+5. 82	+6. 79	−4. 58	−3. 09	−2. 54	−1. 99	+29. 22	+0.521	−0.324	−0.186	+7. 37	−10. 92	+0. 55
VIII'	−1. 553	+5. 45	+6. 35	−4. 29	−2. 89	−2. 38	−1. 86	+27. 37	+0.488	−0.303	−0.174	+6. 90	−10. 64	+0. 51
VI'	−1. 186	+4. 17	+4. 85	−3. 28	−2. 21	−1. 82	−1. 42	+20. 87	+0.372	−0.231	−0.133	+5. 26	−8. 12	+ .39
IV'	−0. 687	+2. 41	+2. 81	−1. 90	−1. 28	−1. 05	−0. 83	+12. 08	+0.216	−0.134	−0.077	+3. 05	−4. 71	+0. 23
II'	−0. 230	+0. 81	+0. 94	−0. 63	−0. 43	−0. 35	−0. 27	+4. 05	+0.073	−0.045	−0.026	+1. 02	−1. 58	+0. 07

As mentioned previously when the arch is loaded there are developed at the imaginary supports horizontal reactions, which are listed in Tables Vb and VI b. Upon the removal of these imaginary restraints, these horizontal reactions will doubtless disappear, but the act of removing these restraints is none other than the application of equal but opposite forces to the reactions removed, thus causing stresses or secondary effects at all sections of the structure.

With a unit load at Sect. VI of Arch I e. g. there are developed at the imaginary supports B and C according to Table V b

$H_{B1}=1.179.$ $H_{C1}=0.038$

The value of the moment at the section just to the left of joint B is

M_B^L for restrained pier-heads (Table IV) ... ... $=+5.82$

M_B^L in consequence of releasing $B=-4.09\ H_{B1}$

(or by Tab. VIII) $=-4.81$

M_B^L in consequence of releasing $C=-1.53\ H_{C1}$

(or by Tab. VIII) $=-0.06$

Final influence line value of M_B^L ... ... $=+1.05$

In a similar way all the secondary values of influence line values of moments and thrusts are calculated and are listed in Tables IX, X and XI. The former two tables are presumably correct for loads placed on Arch I. but they may be used for loads placed on Arch III by utilizing the idea of mirror image. Table XI are values for loads placed on Arch II or the middle arch.

The final influence line ordinates, which are obtained by adding the corresponding secondary values to the values obtained when the pier-heads were restricted from lateral movements, are tabulated in Tables XII to XVI. The influence line ordinates for moments at the crown sections of Arches I and II are given in Tables XVII and XVIII respectively. To give a general idea of the loading conditions which may have to be considered in designing continuous arches, these tables are platted as Figures 31 and 32

TABLE XII:- INFLUENCE LINE VALUES FOR MOMENT AT LEFT ABUTMENT - M^r_A

	ARCH I				ARCH II				ARCH III			
Load at Sect.	M^r_A Tab.IV	M^r_A Tab.IX	M^r_A Tab. X	M^r_A	M^r_A Table VII	M^r_A Table XI	$-M^L_D$ reversed. Table XI	M^r_A	M^L_D reversed Tab. Va	M^L_D Tab. X Reversed	$-M^L_D$ Tab. X Reversed	M^r_A
II	−6.49	−0.67	−0.01	−7. 7	−0.26	+0.41	−0.27	−0.12	−0.013	+0.12	−0.13	−0.02
IV	−7.20	−2.29	−0.03	−9.52	−0.29	+1.83	−0.83	+0.71	−0.014	+0.63	−0.15	+0.47
VI	−4.18	−4.13	−0.05	−8.36	−0.16	+3.62	−1.42	+2.04	−0.008	+1.28	−0.08	+1.19
VIII	+0.54	−5.55	−0.05	−5.06	+0.04	+5.12	−1.86	+3.30	+0.002	+1.82	+0.02	+1.84
Crown	+4.81	−6.01	−0.04	−1.24	+0.20	+5.83	−1.99	+4.04	+0.010	+2.06	+0.11	+2.17
VIII'	+7.02	−5.34	−0.01	+1.70	+0.29	+5.46	−1.75	+4.00	+0.014	+1.90	+0.15	+2.06
VI'	+6.64	−3.74	+0.03	+2.93	+0.26	+4.17	−1.21	+3.19	+0.013	+1.41	+0.13	+1.56
IV'	+4.37	−1.85	+0.05	+2.57	+0.17	+2.41	−0.63	+1.95	+0.008	+0.78	+0.08	+0.88
II'	+1.53	−0.37	+0.05	+1.20	+0.05	+0.81	−0.14	+0.72	+0.003	+0.23	+0.03	+0.29

TABLE XIII:• INFLUENCE LINE VALUES FOR MOMENT M_B^L

	ARCH I				ARCH II				ARCH III			
Load at Sect	M_B^L Table IV	-4.09 H_B or Table IX	-1.53 H_C or Table X	M_B^L	M_B^L Tab. VII	M_B^L Tab. XI	M_C^r Reversed Tab. XI	M_B^L	M_C^r Reversed Tab. Va	M_C^r Reversed Tab. IX	M_C^r Reversed Tab. X	M_B^L
II	+1.11	−0.77	−0.01	+0.33	−0.57	+0.47	−0.35	−0.45	+0.01	+0.16	+0.16	+0.33
IV	+3.64	−2.67	−0.04	+0.91	−0.62	+2.14	−1.05	+0.47	+0.02	+0.82	+0.17	+1.01
VI	+5.82	−4.71	−0.06	+1.05	−0.34	+4.21	−1.82	+2.05	+0.03	+1.63	+0.09	+1.75
VIII	+6.37	−6.46	−0.06	−0.15	+0.08	+5.97	−2.38	+3.51	+0.03	+2.31	−0.02	+2.32
Crown	+4.56	−7.01	−0.05	−2.50	+0.45	+6.79	−2.54	+4.70	+0.02	+2.62	−0.12	+2.52
VIII'	+0.79	−6.19	−0.01	5.41	+0.63	+6.35	−2.24	+4.74	+0.00	+2.42	−0.17	+2.25
VI'	−3.46	−4.36	+0.04	−7.78	+0.57	+4.85	−1.58	+3.84	−0.02	+1.80	−0.16	+1.62
IV'	−6.33	−2.16	+0.06	−8.43	+0.35	+2.81	−0.80	+2.36	−0.03	+1.00	0.10	+0.87
II'	−5.79	−0.43	+0.05	−6.16	+0.02	+0.94	−0.18	+0.78	−0.03	+0.31	−0.03	+0.25

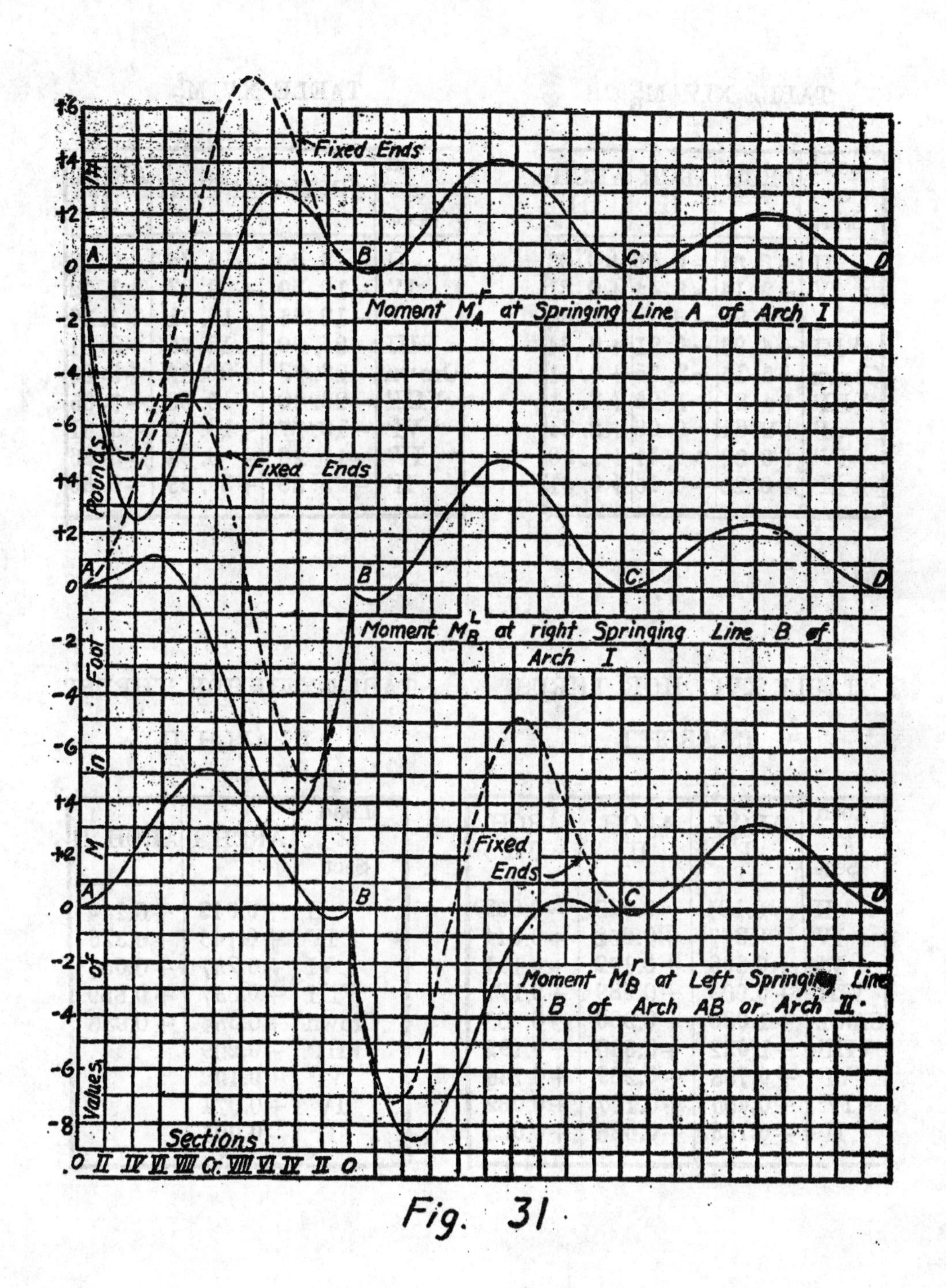
+6
+4
+2
0
-2
-4
-6
+4
+2
0
-2
-4
-6
+4
+2
0
-2
-4
-6
-8
Values of M in Foot Pounds
Fixed Ends
A
B
C
D
Moment M_A^r at Springing Line A of Arch I
Fixed Ends
Moment M_B^L at right Springing Line B of Arch I
Fixed Ends
Moment M_B^r at Left Springing Line B of Arch AB or Arch II
Sections
O II IV VI VIII Cr. VIII VI IV II O

Fig. 31

TABLE XIV M^r_B

Load at Sect.	ARCH I	ARCH II	ARCH III
II	+0.64	-6.00	+0.03
IV	+2.14	-8.44	+0.81
VI	+3.80	-8.19	+1.88
VIII	+4.96	-6.21	+2.84
Crown	+5.15	-3.55	+3.31
VIII'	+4.24	-1.25	+3.11
VI'	+2.61	+0.08	+2.34
IV'	+0.86	+0.41	+1.30
II'	-0.26	+0.15	+0.41

TAELE XV M^f_E

Load at Sect.	ARCH I	ARCH II	ARCH III
II	- 3.00	+ 2.76	-0.12
IV	-10.39	+ 8.03	+1.72
VI	-18.96	+13.91	+4.36
VIII	-25.90	+18.55	+6.81
Crown	-28.77	+20.49	+8.06
VIII'	-26.40	+18.98	+7.65
VI'	-19.77	+13.55	+5.81
IV'	-11.20	+ 8.68	+3.26
II'	- 3.58	+3 .25	+0.80

TABLE XVI HOR. THRUST IN ARCH I

Load at Sect	ARCH I	ARCH II	ARCH III
II	+0.104	-0.022	-0.002
IV	+0.377	+0.052	+0.044
VI	+0.696	+0.172	+0.111
VIII	+0.963	+0.289	+0.171
Crown	+1.086	+0.360	+0.202
VIII'	+1.022	+0.360	+0.192
VI'	+0.798	+0.289	+0.146
IV'	+0.485	+0.177	+0.082
II'	+0.183	+0.066	+0.025

TABLE XVII HOR. THRUST IN ARCH II

Load at Sect	ARCH I	ARCH II
II	+0.042	+0.144
IV	+0.143	+0.375
VI	+0.257	+0.657
VIII	+0.337	+0.859
Crown	+0.354	+0.936
VIII'	+0.299	
VI'	+0.192	
IV'	+0.074	
II'	-0.025	

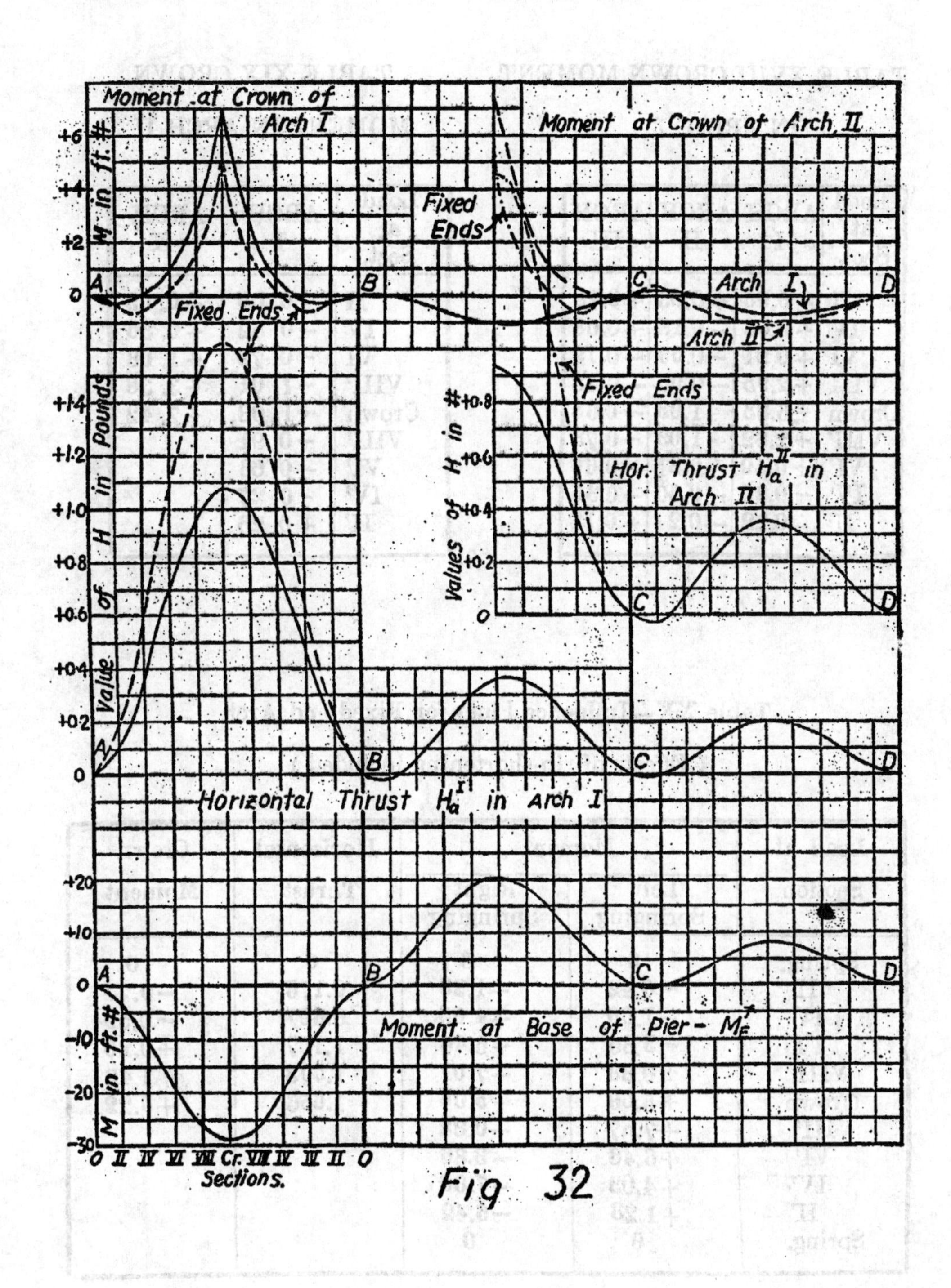

Moment at Crown of Arch I
Moment at Crown of Arch II
Fixed Ends
Fixed Ends
Fixed Ends
Arch I
Arch II
M in ft. #
Value of H in Pounds
Values of H in #
Hor. Thrust H_a^{II} in Arch II
Horizontal Thrust H_a^{I} in Arch I
Moment at Base of Pier - M_F'
M in ft. #
A
B
C
D
Sections.

Fig 32

TABLE XVIII CROWN MOMENT IN ARCH I

Load at Sect.	ARCH I	ARCH II	ARCH III
II	−0.06	+0.03	+0.18
IV	+0.03	−0.18	+0.08
VI	+0.91	−0.53	−0.19
VIII	+2.95	−0.93	−0.48
Crown	+6.83	−1.03	−0.68
VIII'	+2.82	−1.03	−0.73
VI'	+0.59	−0.81	−0.60
IV'	−0.20	−0.50	−0.36
II'	−0.22	−0.24	−0.12

TABLE XIX CROWN MOMENT IN ARCH II

Load at Sect.	ARCH I	ARCH II
II	−0. 09	−0. 09
IV	−0. 52	+0. 36
VI	−0. 79	+1. 08
VIII'	−1. 01	+3. 38
Crown	−1. 08	+7. 42
VIII'	−0. 91	
VI'	−0. 63	
IV'	−0. 27	
II'	+0. 26	

Table XX—Influence Lines for Fixed-end Arch

(Effects of rib-shortening included)

Load at Section	Moment		Horizontal Thrust	Crown Moment
	Left Springing	Right Springing		
Spring.	0	0	0	0
II	−6.42	+1.23	0.176	−0.24
IV	−7.00	+4.03	0.607	−0.59
VI	−3.86	+6.46	1.107	+0.20
VIII	+0.88	+7.07	1.503	+1.43
Crown	+5.06	+5.06	1.656	+5.22
VIII'	+7.07	+0.88		
VI'	+6.46	−3.86		
IV'	+4.03	−7.00		
II'	+1.23	−6.42		
Spring.	0	0		

Reinforced Brick

By

Lam Wing Tung

林 榮 棟

Introduction

Although records of the use of reinforced brick mansonry date back to the building of a tunnel under the Thames by Burnel in 1825, comparatively little is reported concerning this type of construction until 1923 in America. The work was also done as a thesis by three civil engineering seniors, Messrs L. E. Angoli, L. I. Krasin and B. F. Ludowise in spring of 1932 in America. They deserved much credit for their work in testing and in supervising the work of the mason who frabricated the beams.

It seems to me that neither this problem had been taken up by any body nor records were reported.

This problem is suggested by Professor Y. Y. Wong, and the work is done under the direction of Professors C. Y. Leung and Y. M. Wong and the aid of Mr. T. K. Mok and the class of material testing of sophomore.

Before the actual testing, a lot of things must be born in mind, such as kind of brick to be used, various factors to be assumed, correction of the deflectometer, and the adjustment of the testing machine, etc.

The capacity of the testing machine is 50,000 pounds. The deflectometer used in this testing is a thread wound -dial type as shown in the later figures. In order to make sure about this deflectometer, two of 4 × 6-in. × 12 ft-0 in and two of 4 × 6-in × 9 ft- 0in. wooden beams were tested. The average modulus of elasticity found from those data is about 1,500,000 lbs per sq. in. and this value is close enough to those given in books and in other data. This resutle verifies that the deflectometer is workable.

I dare not say that this thesis is valuable. But it is the aim of this thesis to present problem such that all of our people's attention will be focussed on the practical application of the work of the reinforeed brick.

Materials

Services of the mason and materials are provided by the College of Civil Engineering.

Brick:-

The brick used throughout this test is the first grade o the Nankong (南崗上明金). The average dimenrion is $2\times 3\frac{3}{4}\times 8\frac{1}{2}$—in. while the average woight is 100 pounds per cubic foot of brick masonry.

The Modulus of Elasticity, the unit stress and the stress-strain diagram of brick are shown as follows:

Brick specimen No. 1.

Let P = total load applied, in pounds

A = cross-sectional area of the specimen

$f_b = \frac{P}{A}$ = unit ustress, pounds per square inch.

D = total deformation, in inches.

L = length of the specimen, in inches.

$S = \frac{D}{L}$ = unit deformation, inches per inch.

$E_b = \frac{f_b}{S}$ = Modulus of Elasticity of brick pounds per square inch.

Trial 1.

$P = 1{,}000\#$; $f_b = \frac{P}{A} = \frac{1{,}000}{4.25} = 236\#/\text{in.}^2$

$S = \frac{D}{L} = \frac{0.0004}{1.75} = 0.00023$ in./in.

$E_b = \frac{f_b}{S} = \frac{236}{0.00023} = 1{,}037{,}000\#/\text{sq.}''$

TABLE I

No. 1 Section, $2\frac{1}{4} \times 2\frac{3}{6}$ – in.; Length, $L = 1\frac{7}{8}$ – in.; Area, A = 5.34 sq.in.

Applied Iood		Deformation		Modulus of Elasticity #/in.² Eb	Remark
Total #. P	Unit Stress #/in.² F	In Gage Length In. D	Unit per In. Length S		
1	2	3	4	5	
2,000	374	0.0005	0.00027	1,405,000	
4,000	750	0.0015	0.00080	950,000	
6,000	1,124	0.0020	0.00106	1,057,000	
8,000	1,498	0.0025	0.00133	1,127,000	
10,000	1,873	0.0030	0.00160	1,176,000	
12,000	2,248	0.0040	0.00213	1,055,000	
14,000	2,625	0.0050	0.00266	987,000	
16,000	3,000	0.0065	0.00346	867,000	
18,090	3,374	0.0080	0.00426	781,000	
20,000	3,744	0.0100	0.00531	706,000	
22,000	4,120	0.0120	0.00639	645,000	
24,000	4,500	0.0130	0.00681	660,000	
26,000	4,870	0.0140	0.00745	653,000	
28,000	5,245	0.0160	0.00851	616,000	Initial
30,000	5,620	0.0170	0.00905	621,000	crack

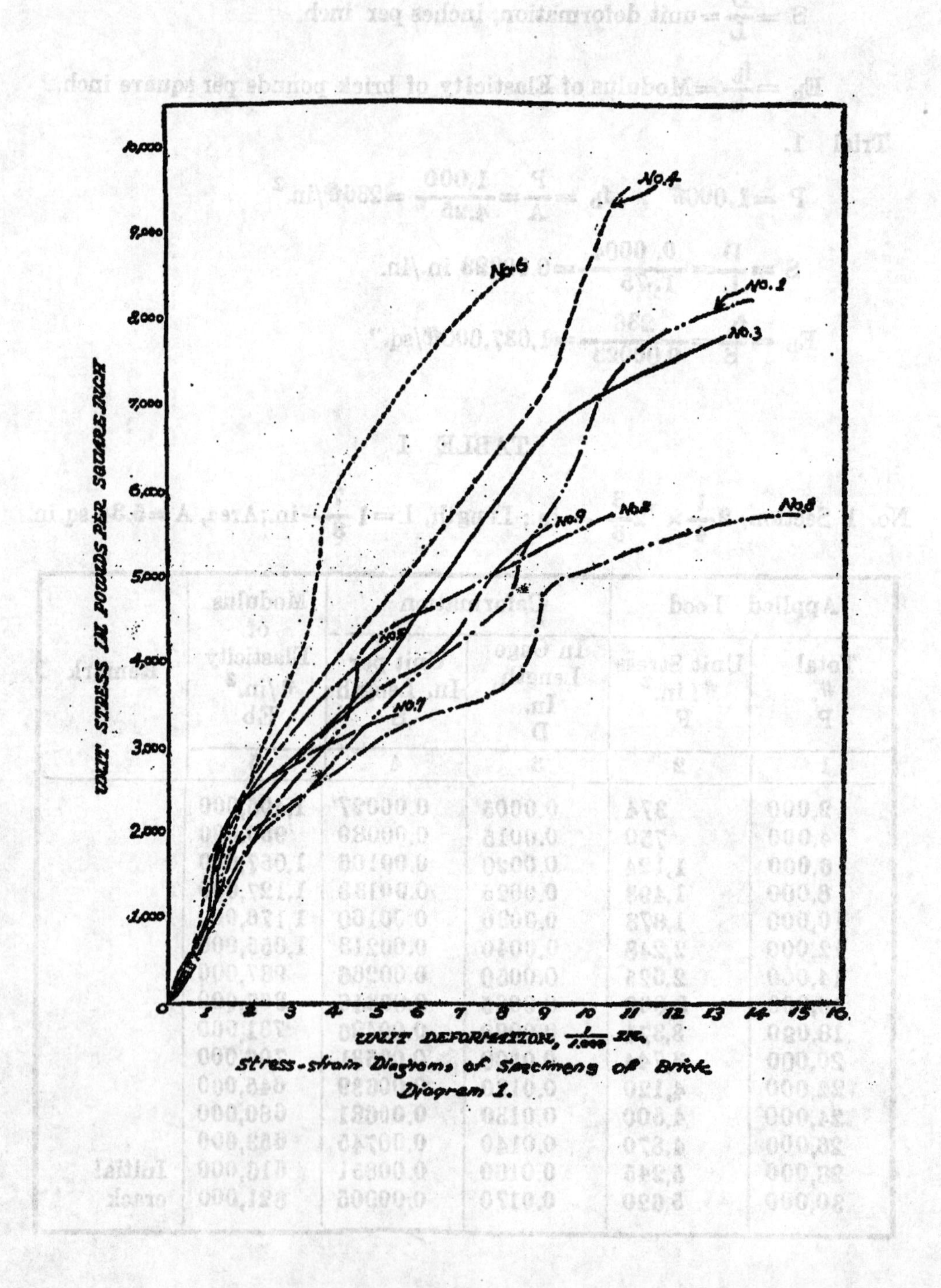

Stress-strain Diagrams of Specimens of Brick.
Diagram 1.

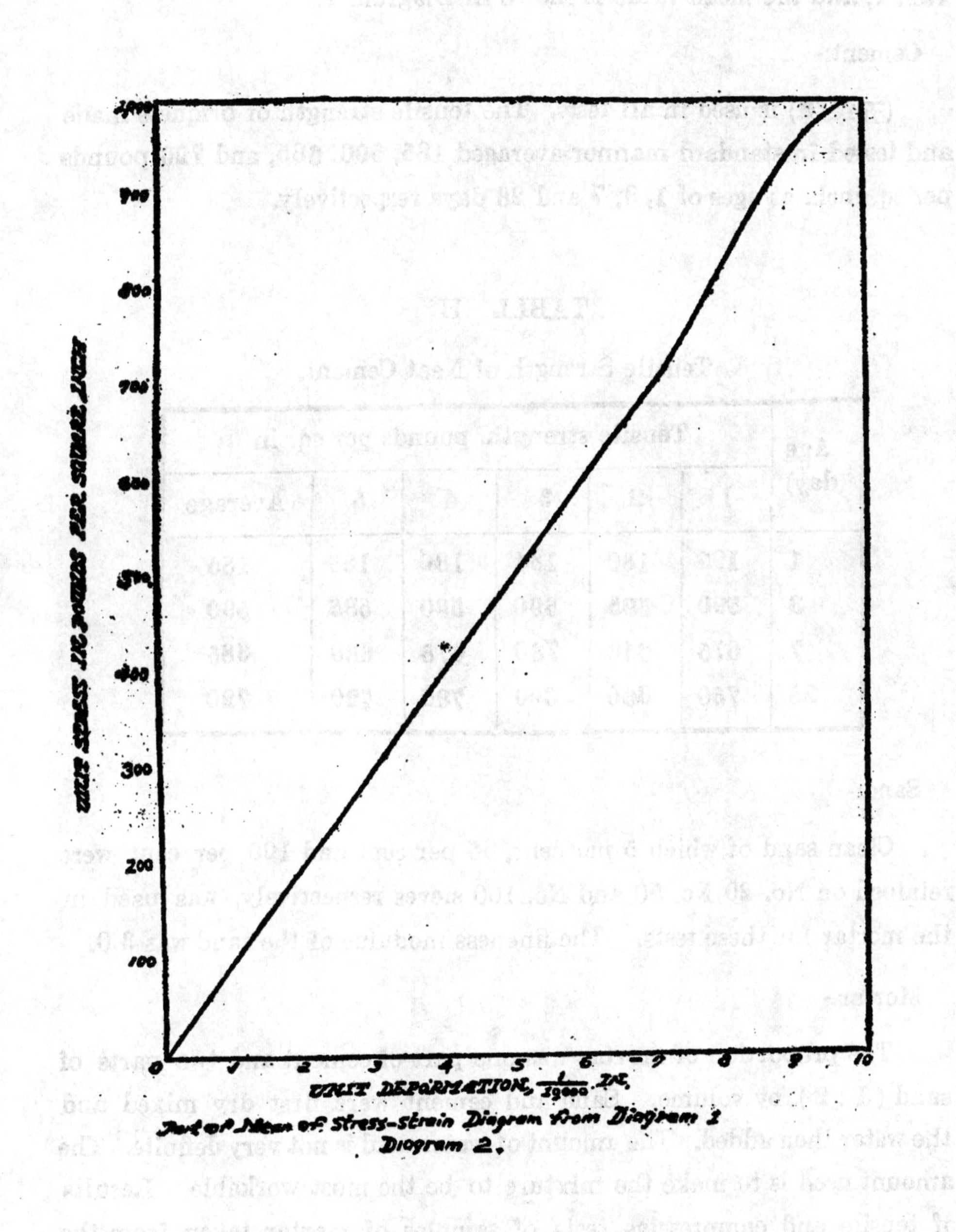

Part of Mean of Stress-strain Diagram from Diagram 1

Diagram 2.

The rest of the results of the brick specimens are shown in Diagram 1; and the mean value is shown in Diagram 2.

Cement:-

(五羊牌) is used in all tests. The tensile strength of briquets made and tested in standard manner averaged 185, 590, 685, and 720 pounds per sq. inch. at ages of 1, 3, 7 and 28 days respectively.

TABLL II

Tensile Strength of Neat Cement.

Age (day)	Tensile strength, pounds per sq. in					
	1	2	3	4	5	Average
1	195	180	184	180	185	185
3	590	595	590	590	585	590
7	675	613	780	675	680	685
28	750	660	690	780	720	720

Sand:-

Clean sand of which 5 per cent, 65 per cent and 100 per cent were retained on No. 20 No. 50 and No. 100 sieves respectively, was used in the mortar for these tests. The fineness modulus of the sand was 3.0.

Mortar.-

The proportion of mortar was one part of cement and two parts of sand (1 : 2), by volume. Sand and cement were first dry mixed and the water then added. The amount of water used is not very definite. The amount used is to make the mixture to be the most workable. Results of tensile and compressive tests of samples of mortar taken from the mason's batches are given in Table III.

TAB E III

Strength of 1—2 Mortar from Mason's Batches. Proportions were: 1 cement, 2 sand; $\frac{2}{3}$ to $\frac{3}{4}$ parts of water by Volume.

	Tensile Strength #/in.^{2}			Compressive Strength #/in^2.							
				Cube, 2×2×2—in.				Cylinder, 2—in.dia. ×4—in.			
					Max. Load #	Sec. tion, in.^{2}	Unit strest #/in.^{2}		Max. Lood, #	Sec. tion, in^2	Unit stress #/in.^{2}
1	430	539	495	14	26,500	4	6,625	28	11,000	3.14	3,500
2	395	575	490	28	30,000	„	7,500	28	11,000	„	3,500
3	330	395	500	28	25,000	„	6,250	42	10,000	„	3,200
4	360	520	510	42	26,000	„	6,500	42	10,500	„	3,350
5	375	470	530	42	28,000	„	7,000	60	9,490	„	3,035
Average	378	500	505				6,775				3,330

Steel:-

Excepting the $\frac{1}{4}$—in. and $\frac{3}{8}$—in. rods, all reinforcement consisted of deformed bars of the type indicated in TABLE VI and with the yield point shown in Diagram 3. The Modulus of Elasticity found from the tests of steel specimens is shown in TABLES IV & V. Percentage of steel in beams are baed on actual area of rods.

Specimen 1.

Load applied, P = 500#.

Area of section, A = 0.196 in.^{2}

Unit stress, $f = \frac{P}{A} = \frac{500}{.196} = 2{,}550$ #/in.^{2}.

Total Elongation, $D = 0.3 \times \frac{1}{10{,}000}$ —in.

Gage Length, l = 2.0—in.

$$\text{Unit elongation, e} = \frac{D}{l} = \frac{0.3}{2.0 \times 10{,}000} = \frac{0.15}{10{,}000}$$

$$\text{Modulus of Elasticity} = \frac{f}{e} = \frac{2{,}550 \times 10{,}000}{.15} = 17{,}000{,}000\ \#/\text{in}^2$$

The rset are in table form.

TABLE IV

Tension Test of Steel

Diameter =0.5 – in.; Area, A = 0.196 sq in. Gage length, l = 2.0 in.

Applied Load, lb.		Elongation, Inches.		Modulus, of Elasticity #lin.²	Remark
Total	Der sg. in	In gage length.	Per inch length		
P	f	D *	e *	E	
500	2,550	0.30	0.15	17,000,000	
1,510	7,710	3.80	1 90	40,500,000	The
2,270	11,600	6.00	3.00	38,700,000	diameter of
3,100	15,800	8.90	4.45	35,500,000	rodis
3,860	19,700	10.30	5.15	38,200,000	different
4,400	22,500	11.20	5.60	40,200,000	from part
4,920	25,100	13.00	6.50	38,700,000	to part
5,320	27,200	14.40	7.20	37,800,000	
5,840	29,900	16.00	8.00	37,400,000	
6,660	34,000	18.30	9.15	37,200,000	
7,190	36,600	20.10	10.05	36,600,000	
7,350	37,600	21.00	10.50	35,900,000	
7,660	39,100	22 30	11.15	35,200,000	
7,780	39,600	24.50	12.25	32,700,000	yield point

* D and e are in $\frac{1}{10{,}000}$ of an inch.

TABLE V

Tension Test of Steel

Diameter =0.5-in.; Area, A=0.196 sq. in. Gage length, 1=2.0-in.

Applied Load lb.		Elongation, Inches		Modulus of Elasticity #/in.²	Remark
Total	Per sq.in.	In gage length	Per inch length		
P	f	D *	e *	E	
500	2,550	0	0	....	
1,000	5,100	4.0	2.00	25,500,000	
1,500	7,650	6.0	3.00	25,500,000	
2,000	10,200	7.8	3.90	26,200 000	
2,500	12,750	9.8	4.90	26,100,000	As soon as the
3,000	15,300	10.2	5.10	30,000,000	rod begins to
3,500	17,850	11.9	5.95	30,000,000	neck, the elon-
4,000	20,400	13.0	6.50	31,400,000	gation increas-
4,500	22,950	15.0	7.50	30,650,000	es rapidly
5,000	25,500	16.2	8.10	31,500,000	
5,500	28,050	18.2	9.10	30,800,000	
6,000	30,600	20.0	10.00	30,600,000	
6,500	33,150	22.0	11.00	30,100,000	
7,000	35,700	24.0	12.00	29,780,000	
7,500	38,250	36.0	18.00	21,250,000	yield point
8,220	42,000	240.0	120.00	3 500,000	
10,665	54,800 — ultimate strength				

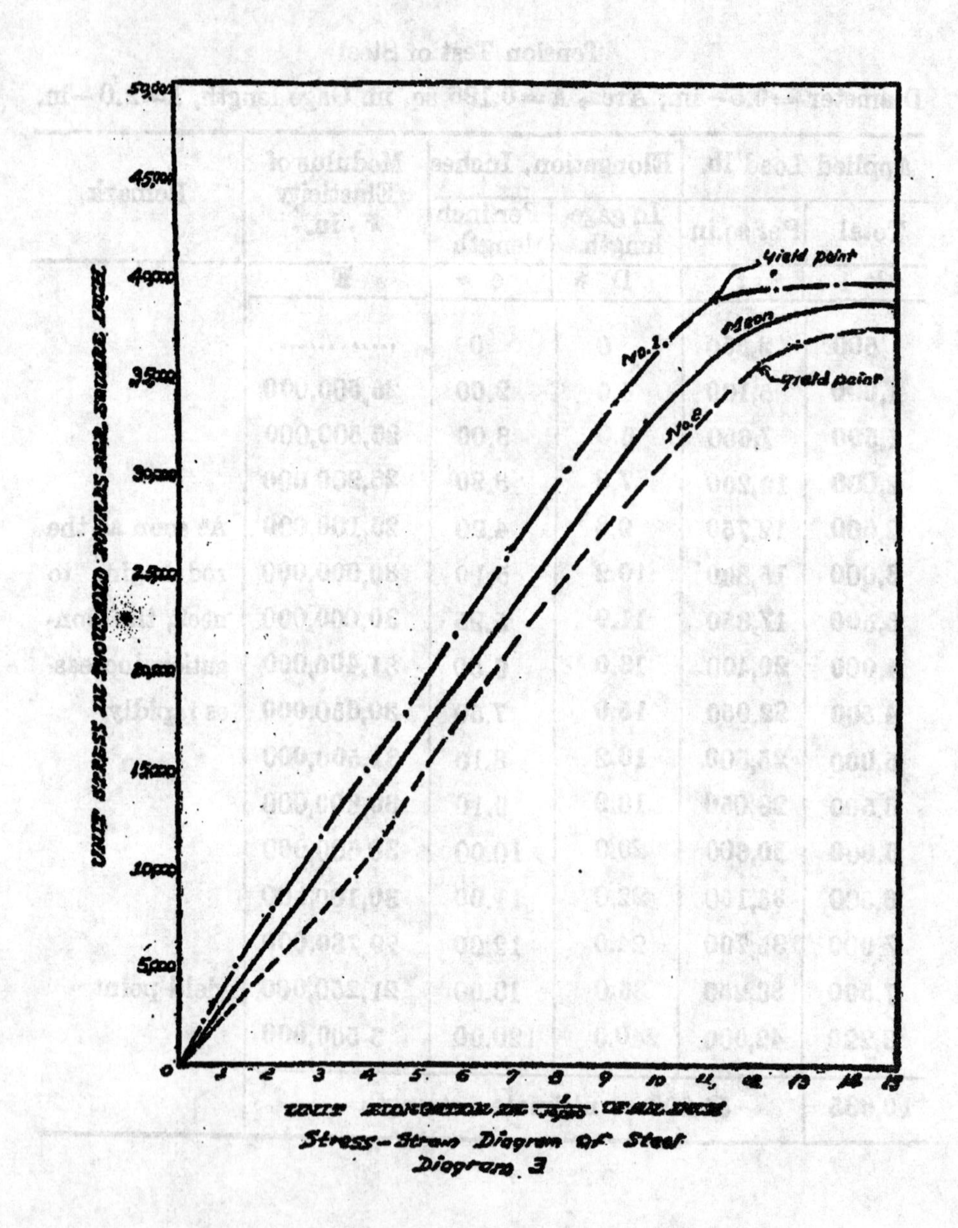

Stress-Strain Diagram of Steel
Diagram 3

Design of Beams for Various Failures

1. Notation.

E_s = modulus of elasticity of steel, pounds por square inch.

E_b = modulus of elasticity of brick, pounds per square inch.

f_s = unit stress in steel, pounds per sq. in.

f_b = unit compression in brick, pounds per square inch.

f_u = unit stress in stinups, pound per squart inch.

A_s = area of cross-scction of longitudinal steel.

A_s = area of cross-section of stirrup per spacing interval.

b = breadth of beam

d = effective depth from top of beam to center of longitudinal steel.

k = ratio of depth of neutral axis from top of beam to d.

j = $1-\frac{1}{3}k$ for straight-line formula and $1-\frac{3}{8}k$ for parabalic formula.

M_b = resisting moment in terms of stress in brick.

M_s = resisting moment in terms of stress in steel.

M = bending Moment.

n = E^s/E_b

o = perimeter of one longitudinal rod

p = A/bd

u = unit bond stress of mortar to steel rod

v = unit shear stress due to total shear, V

2. Computation of p. k and j. from the assumed data as follows.

Assuming,

E_s = 30,000,000 #/in²

E_b = 1,500,000 #/in²

f_s = 16,000 #/in², working stress

60,000 #/in², ultimate stress

f_b = 300-700 #/in², average = 500 #/in²

weight of brick masonry = 100 #/ft³

The above assumptions are useful before tests carried out.

$$n = \frac{E_s}{E_b} = \frac{30,000,000}{1,500,000} = 20$$

$$p = \frac{2/3}{\frac{f_s}{f_b}\left(\frac{f_s}{2nf_b}+1\right)}$$

$$= \frac{2/3}{\frac{16000}{500}\left(\frac{16000}{2\times20\times500}+1\right)}$$

$$= \frac{2/3}{32(.8+1)} = 0.0116$$

$$k = \sqrt{3pn+\left(\frac{3}{2}pn\right)^2} - \left(\frac{3}{2}pn\right)$$

$$= \sqrt{3\times0.0116\times20+\left(\frac{3}{2}\times0.0116\times20\right)^2} - \frac{3}{2}\times0.0116\times20$$

$$= \sqrt{.696+.1214} - .348$$

$$= 0.905 - 0.348 = .557$$

$$j = 1 - \frac{3}{8}k$$

$$= 1 - \frac{3}{8}\times.557$$

$$= 1 - .209 = .791$$

3. Design for Shear Failure.- B 1

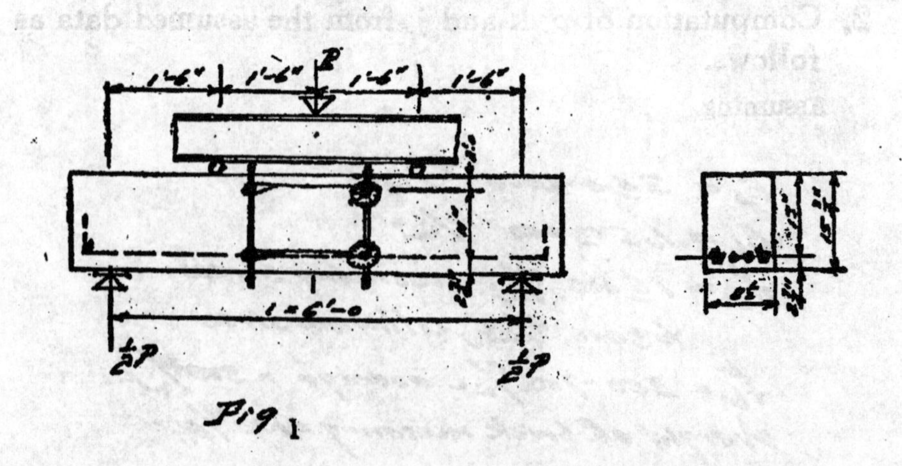

Fig 1

Dimension $8\frac{1}{2} \times 15\frac{3}{4}$-in $\times$ 8'-00in

$b \times d = 8\frac{1}{2} \times 13\frac{1}{4}$-in

Span length = 6'-0

Load applied at $\frac{1}{4}$-point as shown in Fig

$$\omega = \frac{8.5 \times 15.75}{144} \times 100 = 93 \ \#/\text{ft. of beam}$$

$$M_b = M_s = \frac{2}{3} f_b j k b d^2$$

$$= \frac{2}{3} \times 300 \times .791 \times .357 \times 8.5 \times \overline{13.25}^2$$

$$= 126{,}500 \ ''\#.$$

$$A_s = \frac{126{,}500}{16000 \times .791 \times 13} = 0.77 \ \square''$$

use $2 - \frac{5}{8}''\phi = 0.78 \ \square''$

$2 - \frac{1}{4}''\phi = \underline{0.098}$

$0.878 \ \square''$

Moment due to applied load $= M_b - \left(\frac{wl^2}{8} - \frac{wl^2}{32}\right)$

$$= M_b - \frac{3wl^2}{32}$$

$$= 126{,}500 - \frac{3 \times 93 \times 6^2 \times 12}{32}$$

$$= 126{,}500 - 3760 = 122{,}740 \ ''\#$$

$$\frac{1}{2} P = \frac{122740}{12 \times 1.5} = 6830 \ \#$$

Maximum load applied, $P = 2 \times 6830 = 13660 \#$

$$V = 6830 + \frac{1}{2} \times 93 \times 6$$

$$= 6830 + 279 = 7007 \ \#$$

$$v = \frac{7007}{.791 \times 8.5 \times 13} = 80 \ \#/\square''$$

No stirrups are used.

$$u = \frac{V}{\Sigma o j d} = \frac{7007}{2(.785+.25) .791 \times 13}$$

$$= 102 \ \#/\square''$$

Use bearing plate at supports of 6"x12"

$$\text{Bearing stress} = \frac{7087}{6 \times 8.5} = 137 \text{ \#/□"} \quad \text{O.K}$$

4. Design for Tension Failure.—B_2

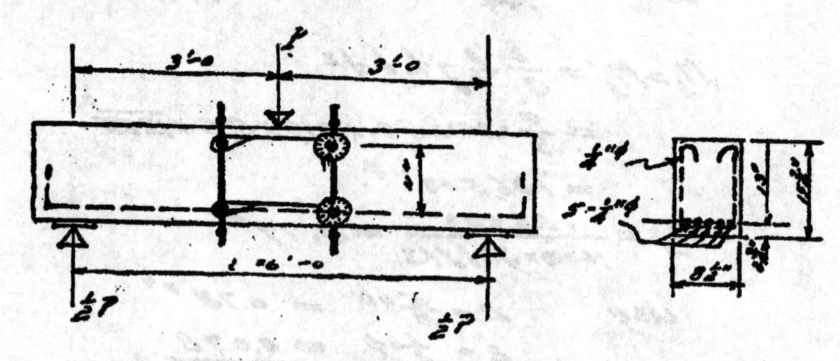

Fig 2

Dimension of beam B_2 same as B_1

Load applied at center as shown in Fig 2

$M_b = M_s = 126500$ "# (same as B_1)

$$A_s = \frac{126500}{60000 \times .791 \times 13} = 0.205 \text{ □"}$$

Use 5-$\frac{1}{4}$"ϕ = 0.245 □"

Moment due to weight of beam

$= \frac{1}{8} wl^2$

$= \frac{1}{8} \times 93 \times 6^2 \times 12 = 5{,}020$ "#

Moment due to applied load

$= M_b - \frac{1}{8} wl^2$

$= 126{,}500 - 5{,}020 = 121{,}480$ "#

$$\text{Load applied, } P = \frac{121{,}480}{\frac{6 \times 12}{4}} = 6740 \text{ \#}$$

Weight of beam = 6×93 = 558#

$V = \frac{1}{2}(6740+558) = 3,649\#$

$u = \frac{3,649}{5\times.785\times.791\times13} = 90.0\ \#/\square''$

$v = \frac{3,649}{8.5\times791\times13} = 41.7\ \#/\square''$

Use $\frac{1}{4}''\phi$ U-stirrup

Spacing of stirrup, $s = \frac{fv}{b(v-o)}$

$= \frac{16000\times2\times.049}{8.5\times41.7}$

$= 4.43$ in.

Use $s = 4\frac{1}{2}''$ "5"

5. Design for Compression Failure.—B_3

Dimension as B_1

Load applied at center, as shown in Fig 2

Assuming maximum load applied = 20000#

Reaction $= \frac{1}{2}P = \frac{20000}{2} = 10000$

Bending moment due to applied load

$= \frac{1}{4}Pl = \frac{1}{4}\times20,000\times6\times12 = 360,000''\#$

Bending moment due to the beam:

$= \frac{1}{8}wl^2 = \frac{1}{8}\times93\times6^2\times12 = 5,020''\#$

Total bending moment,

$M = 360000+5,020 = 365,020''\#$

$A_s = \frac{365,020}{16000\times.791\times13} = 2.21\ \square''$

use 5-$\frac{3}{4}$"ϕ = 2.82□"

$$f_b = \frac{M}{\frac{2}{3}kj\,bd^2}$$

$$= \frac{365{,}020}{\frac{2}{3} \times .557 \times .791 \times 8.5 \times \overline{13}^2}$$

$= 870$ #/□"

$V = 10{,}000 + 279 = 10279$ #

$$u = \frac{10279}{5 \times 2.356 \times .791 \times 13} = 85 \text{ #/□"}$$

$$v = \frac{10279}{8.5 \times .791 \times 13} = 117.2 \text{ #/□"}$$

use $\frac{3}{8}$"ϕ - Z - stirrup

$$\text{Spacing of stirrup, } s = \frac{16000 \times .11}{8.5 \times 117.2}$$

$= 1.77$"

use $s = 2\frac{1}{2}$"

6. Design for Bond Failure.—B_4

Dimension same as B_1

Load applied at $\frac{1}{4}$-point same as B_1

$M_b = M_s = 126{,}500$ "# (from B_1 P. 31)

$P = 13{,}660$ # (" " P. 32)

$A_s = 0.77$ □" (" " P. 31)

use 1 - 1"ϕ = 0.785 □"

$V = 7{,}009$ # (from B_1 P. 32)

$$u = \frac{7009}{3.142 \times .791 \times 13} = 217 \text{ #/□"}$$

$v = 80$ #/□" (from B_1, P.32)

use ⅜" φ-U-stirrups

spacing, $s = \frac{16000 \times 2 \times .049}{8.5 \times 80}$

$= 2.3"$

use $s = 2\frac{1}{2}"$ near supports

$= 5"$ at center portion

Note: Cross-section, span length and gage length of deflectometer are the same for B_1 B_2, B_3, B_4.

Design of beams B_5 — B_8 in the similar way.

Making of Beams and Testing of Beams

1. Making of Beams.-

TABLE VI

Schedule of Beam Fabricated.

Kind of Brick	Beam	Cross Section in.	Over all length ft	Span length ft	Condition of Brick	Reinforcement in Beam: Longitudinal	Per Cent	Shear
1	2	3	4	5	6	7	8	9
[illegible]	B_1	8½×15¾	8.0	6.0	Soaked throughout night	2-⅝-in. square; 2-⅝-in. round	0.78	none
	B_2	"	"	"	"	5-⅝-in. round	0.818	⅜-in. φ-U @ 4½-in.
	B_3	"	"	"	"	5-⅝-in. round	0.50	one-⅜-in. Z @ 8½-in.
	B_4	"	"	"	"	1-1-in. round	0.649	⅜-in. φ-U @ 4½-in.
	B_5	8½×10	"	"	Sprinkled 10 minutes	1-⅝-in. square	0.612	¼-in. φ-U @ 9-in.
	B_6	"	"	"	"	1-⅝-in. square	[illegible]	none
	B_7	"	"	"	"	4-⅝-in. round	0.940	¼-in. φ-U @ 9-in.
	B_8	"	"	"	"	6-⅝-in. round	1.13	1-⅜-in. φ-Z @ 3-in.

TABLE VI gives the schedule of the beams which were fabricated. The brick were given the treatment indicated in Column 6 of the table Beams were laid by an ordinary bricklayer. Brick in the bottom course were first placed flatwise on a wooden plank. The brick were so spaced

25265

that the vertical joints were obout ½ – in. Mortar was then slushed into the vertical joints and a thick layer of mortar was then spread over this course and the stirrups and longitudinal steel which were previously bound together with iron wire, were pressed into place. The bars were then covered with additional mortar and the upper courses of brick were laid. These were surmounted by a ¼ – in. mortar cap which served to cover and to level the tops of the beams. The thickness of the joint containing the longitudinal steel was about ¾ to 1 – in. and the thickness of the other joints was ½ – in. All joints were completely filled with mortar.

2. Testing of Beams:-

The set-up for testing the beams is shown in photos. Loads were opplied at forth points for testing both shear and bond failures while for testing tension and compression failures, the load was applied at center of a 6 – ft span beam. Deflection at mid span were read to 0.01 in. by an ordinary deflectometer, Deformenter readings, at 2¼ — in. from top fiber for those beams having a cross-section of 8½ × 15¾ – in. and at ½ – in. from top fiber for those beams having a cross section of 8½ × 10 – in., and in the plane of the longitudinal rods, were taken over a 24 – in. gage length by means of thread-wound-dials deformeter reading to 0.0001 in. The distance between the top deformeter and the bottom deformeeter was 11 – in. for beams having a cross-section of 8½ × 15¾ – in. and 7 – in. for beams having a cross-section of 8½ × 10 – in.

The arrangement of brick, reinforcement and point of loading are shown on photoes on pp 67 and 68 and 69 p 67 and the arrangement of cracks in Reinforced-beams are as shown on p 70.

Making of Beam

Set Up of Beam

Tension Failure

Bond Failure

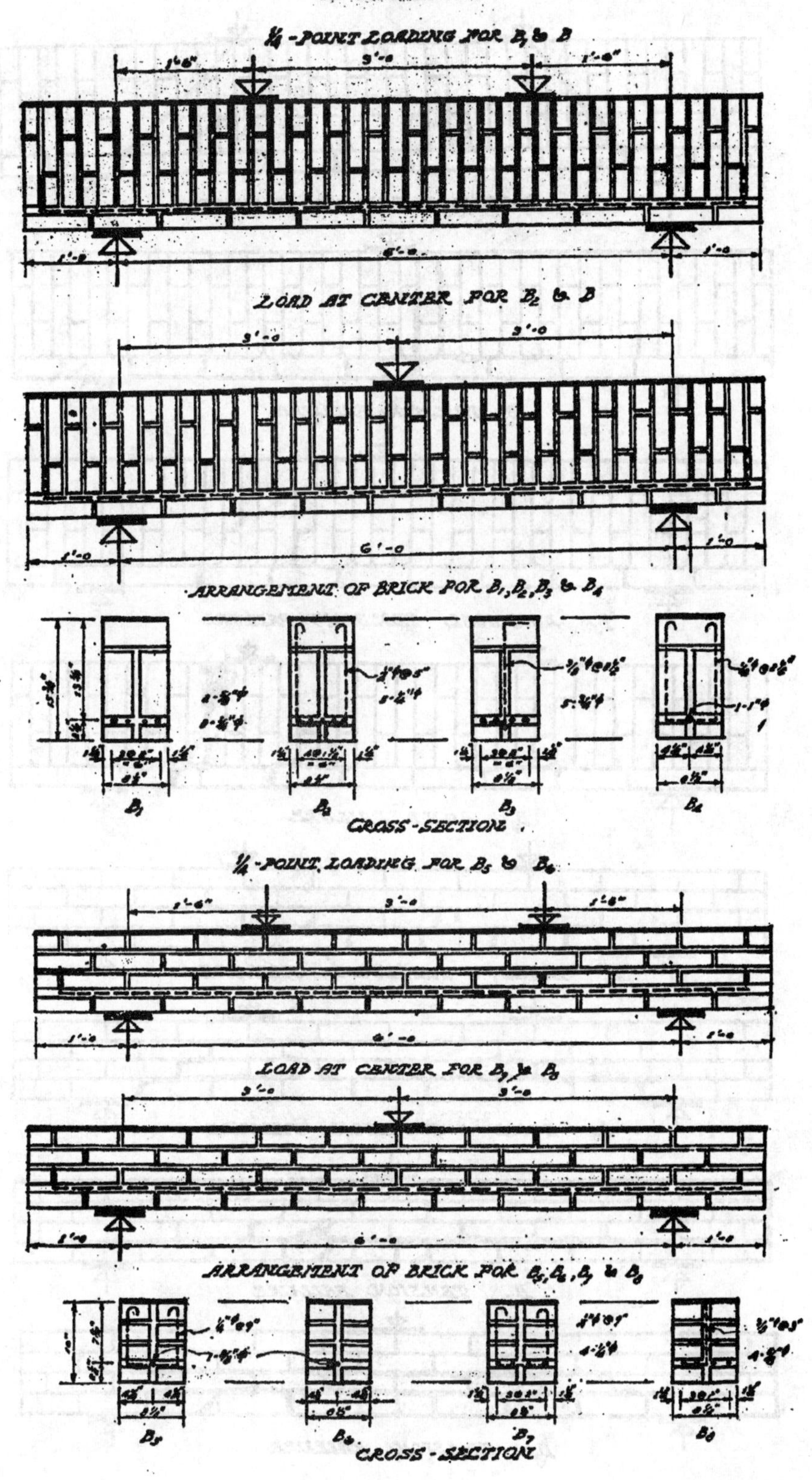
ARRANGEMENT OF BRICK REINFORCEMENT AND POINT OF LOADING.
¼-POINT LOADING FOR B1 & B
LOAD AT CENTER FOR B2 & B
ARRANGEMENT OF BRICK FOR B1, B2, B3 & B4
CROSS-SECTION
¼-POINT LOADING FOR B5 & B6
LOAD AT CENTER FOR B7 & B8
ARRANGEMENT OF BRICK FOR B5, B6, B7 & B8
CROSS-SECTION

ARRANGMENT OF CRACKS
IN REINFORCED BRICK BEAMS

B_1 DIAGONAL TENSION FAILURE

B_2 TENSION FAILURE

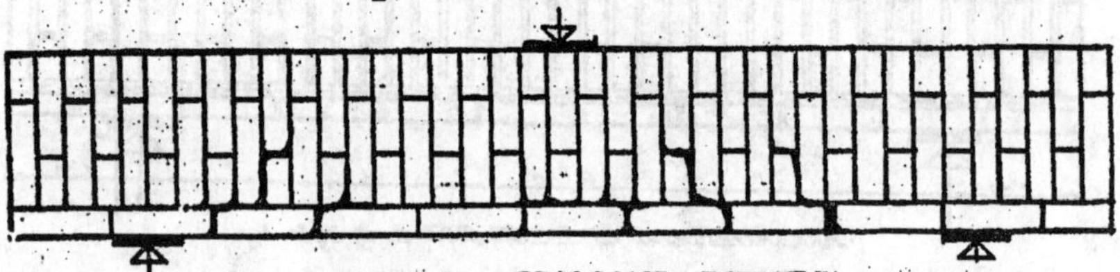

B_3 DIAGONAL TENSION FAILURE

B_4 BOND FAILURE

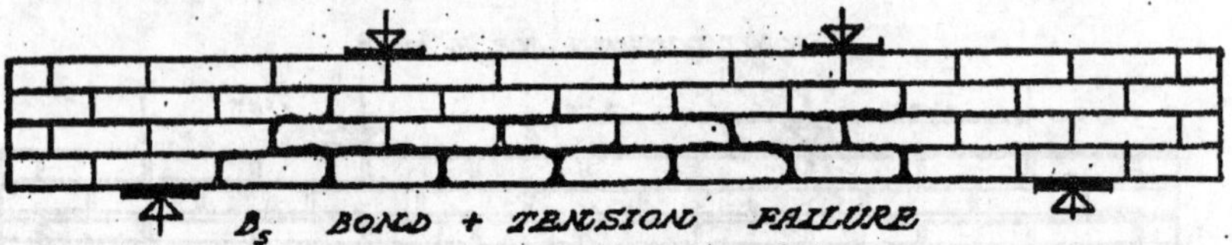

B_5 BOND + TENSION FAILURE

B_6 BOND + TENSION FAILURE

B_7 TENSION FAILURE

B_8 TENSION FAILURE

25270

Results of Tests

1. Data obtaiued from the Tests:-

Data here put in this paper are only B_1, B_4 and B_5 which are both good for computing the Modulus of Elasticity and stresses of brick masonry as shown in the Tables VII to IX on pp 72—77

2. Caoulatiou of Modulus of Elasticity and Fiber Stresses of Brick Masonry from Moments and Strains.

Notation and Formulas to be used;

General Fomulas:

$$f = \frac{My}{I}$$

$$E = \frac{f}{s}$$

in which,

f = unit fiber stress, #/in.^{2}

M = bending moment in in—lbs.

y = kd = distance from neutral aris to extreme top fiber.

I = total moment of inertia about the neutral axis,

$= I_b + A_b d_b^2 + (I_s + A_s d_s^2)(n-1)$

I_b = moment of inertia of brick masonry, $\frac{bt^3}{12}$, (t is total thickness of beam) and b is width of beam), or ($\frac{bd^3}{12}$ if it is necessary)

A_b = area of cross—section, bt or bd

d_b = distance between the center of the cross-section and the neutral aris.

I_s = moment of inertia of longitudinal reinforcement.

A_s = Area of longitudinal reforcement

d_s = distance between the center of the reinforcement and the corresponding neutral axis,

TABLE VII.

Mark	Age, (day)		Load Applied lb (P)	East Deflectometer: Top Deflectometer in +	East Deflectometer: Bottom Deflectometer in *	West Deflectometer: Top Deflectometer in +	West Deflectometer: Bottom Deflectometer in *	Average Top Deformation in +	Average Extreme Top Deformation in *	Average Bottom Deformation in *	Center Deflection in.	Remark
B	52		0	0	0	0	0	0	0	0	0	Wt. of beam =845#
		1	835	0.0002	0	0.0004	0	0.0003	.....	0	0.002	
		2	1,650	0.0003	0.0001	0.0008	0.0001	0.00055	.....	0.0001	0.005	
		3	2635	0.0007	0.0001	0.0009	0.0001	0.0008	.....	0.0001	0.010	Deformeter gage length for B_1-B_8 =24"
		4	3,630	0.0012	0.0002	0.0010	0.0002	0.0011	.....	0.0002	0.015	
		5	4,675	0.0018	0.0004	0.0016	0.0004	0.0017	0.00214	0.0004	0.020	
		6	5,545	0.0022	0.0004	0.0020	0.0005	0.0021	0.00255	0.00045	0.025	
		7	6,610	0.0025	0.0005	0.0025	0.0006	0.0025	.....	0.00055	0.030	
		8	7610	0.0031	0.0006	0.0028	0.0006	0.00295	0.0036	0.0006	0.035	
		9	8,580	0.0035	0.0006	0.0031	0.0007	0.0033	0.0040	0.00065	0.0385	
		10	9,605	0.0041	0.0008	0.0037	0.0009	0.0039	.....	0.00085	0.045	
		11	10,635	0.0047	0.0009	0.0043	0.0013	0.0045	0.0056	0.0011	0.050	
		12	11,620	0.0052	0.0010	0.0048	0.0014	0.0050	.....	0.0012	0.055	
		13	12,710	0.0054	0.0012	0.0050	0.0016	0.0051	0.00635	0.0014	0.161	

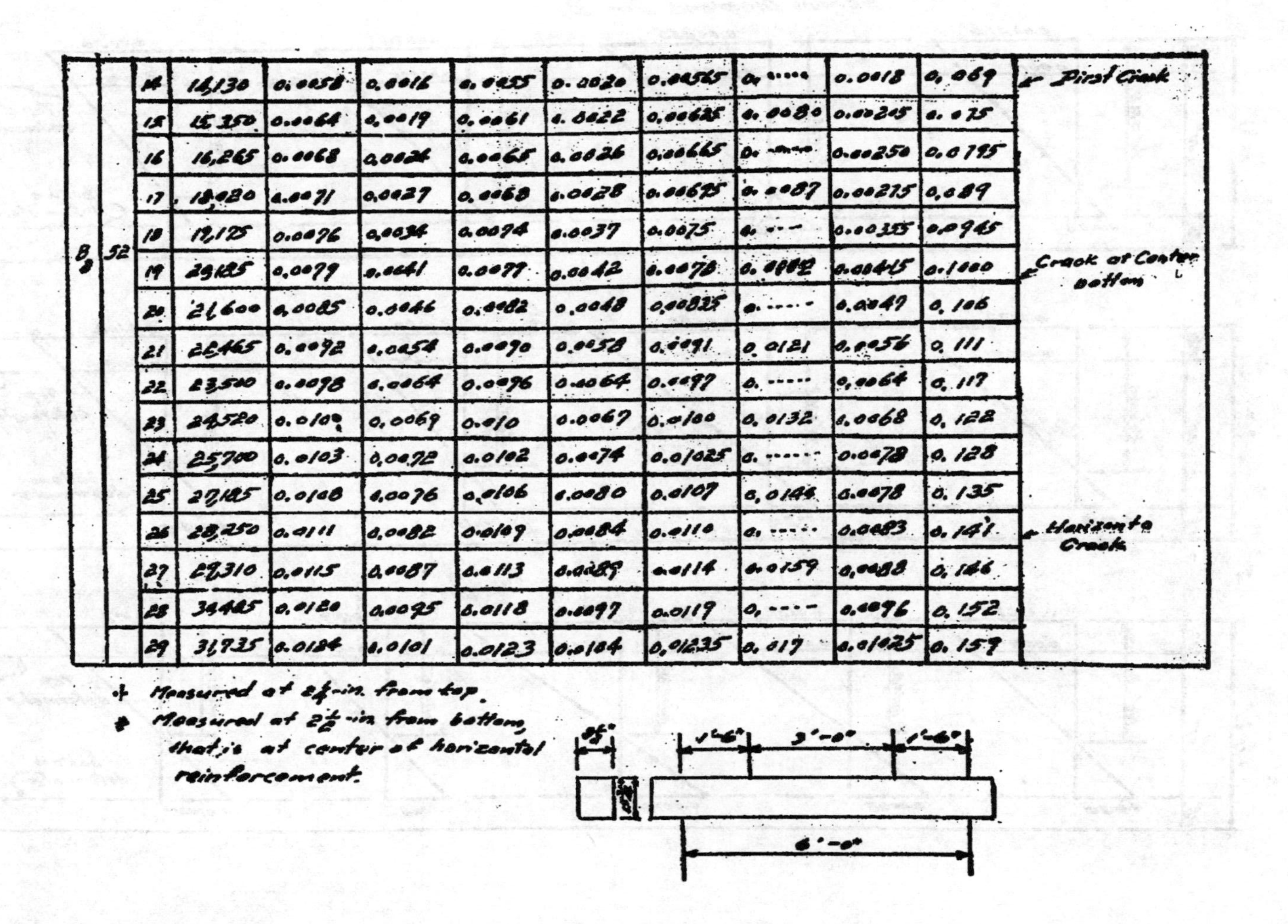

B8	52	14	16,130	0.0058	0.0016	0.0055	0.0020	0.00565	0.	0.0018	0.069	← First Crack
		15	15,350	0.0064	0.0019	0.0061	0.0022	0.00625	0.0080	0.00225	0.075	
		16	16,265	0.0068	0.0024	0.0065	0.0026	0.00665	0.	0.00250	0.0795	
		17	18,020	0.0071	0.0027	0.0068	0.0028	0.00695	0.0087	0.00275	0.089	
		18	19,175	0.0076	0.0034	0.0074	0.0037	0.0075	0.	0.00335	0.0945	
		19	20,185	0.0079	0.0041	0.0077	0.0042	0.0078	0.0092	0.00415	0.1000	← Crack at Center bottom
		20	21,600	0.0085	0.0046	0.0082	0.0048	0.00825	0.	0.0049	0.106	
		21	22,465	0.0092	0.0054	0.0090	0.0058	0.0091	0.0121	0.0056	0.111	
		22	23,500	0.0098	0.0064	0.0096	0.0064	0.0097	0.	0.0064	0.117	
		23	24,520	0.0101	0.0069	0.010	0.0067	0.0100	0.0132	0.0068	0.122	
		24	25,700	0.0103	0.0072	0.0102	0.0074	0.01025	0.	0.0078	0.128	
		25	27,125	0.0108	0.0076	0.0106	0.0080	0.0107	0.0144	0.0078	0.135	
		26	28,250	0.0111	0.0082	0.0109	0.0084	0.0110	0.	0.0083	0.141	← Horizontal Crack
		27	29,310	0.0115	0.0087	0.0113	0.0089	0.0114	0.0159	0.0088	0.146	
		28	30,485	0.0120	0.0095	0.0118	0.0097	0.0119	0.	0.0096	0.152	
		29	31,735	0.0124	0.0101	0.0123	0.0104	0.01235	0.017	0.01025	0.159	

† Measured at $2\frac{1}{4}$-in. from top.

‡ Measured at $2\frac{1}{2}$-in. from bottom, that is at center of horizontal reinforcement.

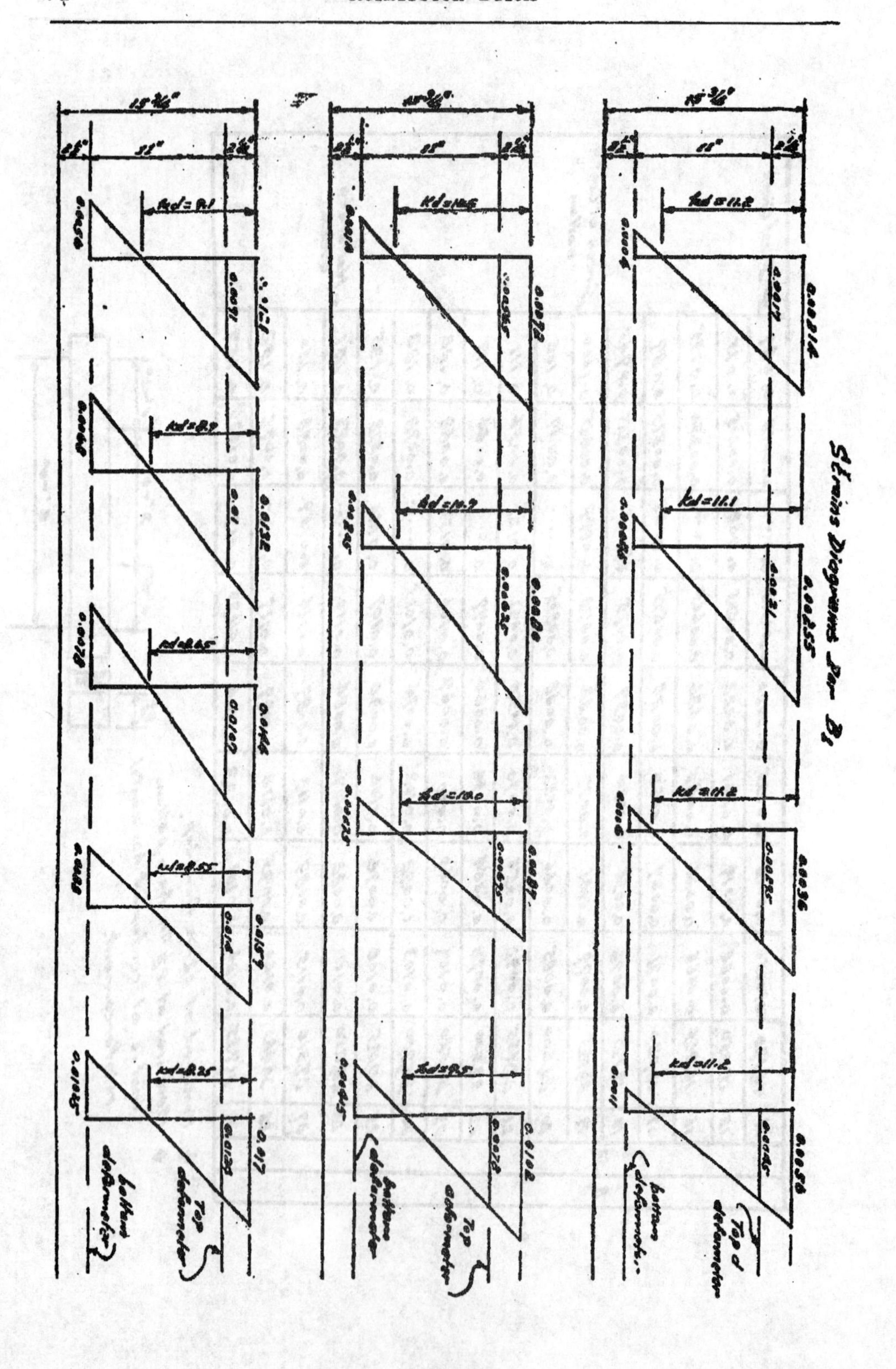

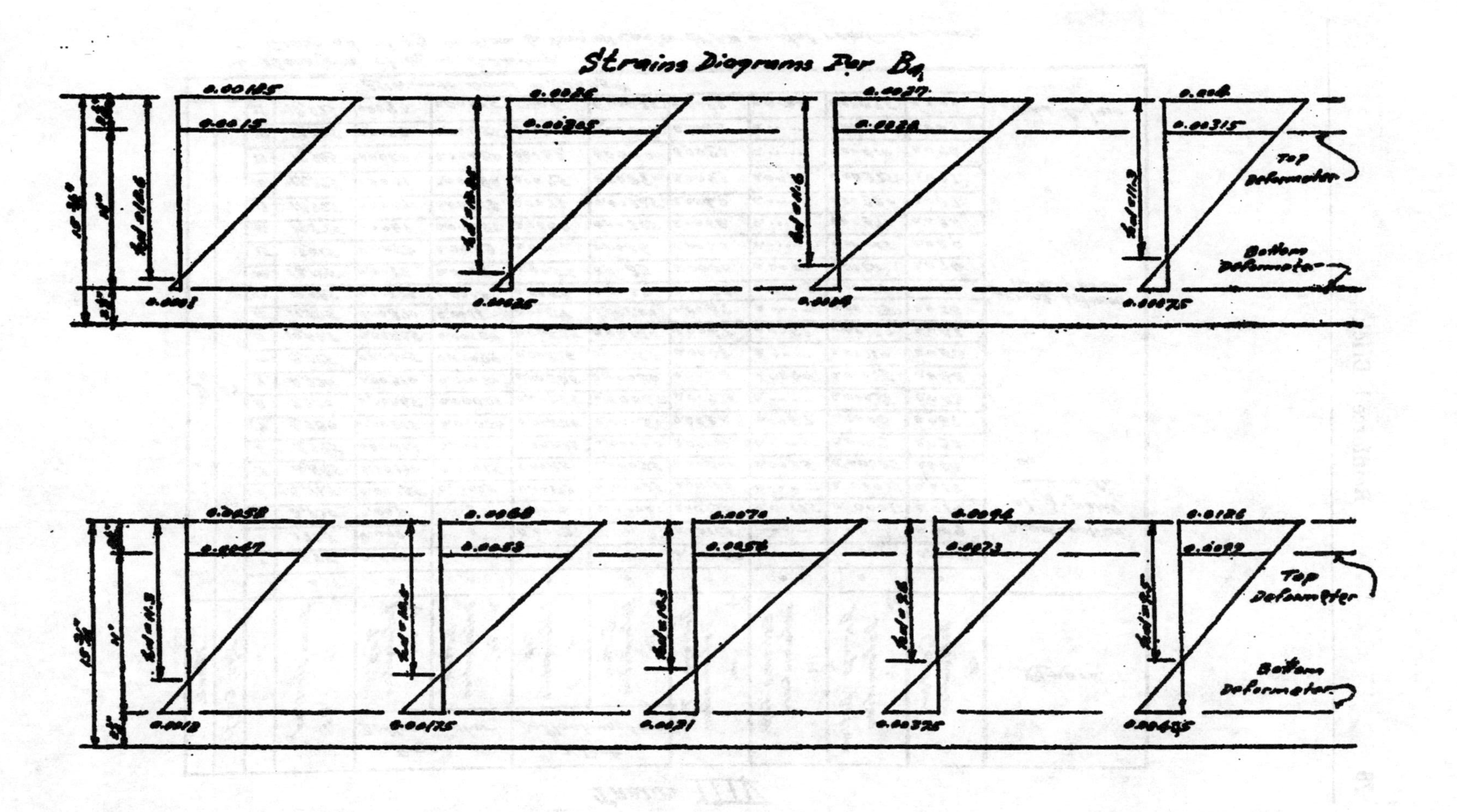

Strains Diagrams For B_{4_1}
0.00125
0.0015
0.0026
0.00205
0.0027
0.0028
0.00315
Top Deformeter
Bottom Deformeter
0.0001
0.00025
0.0006
0.00075
0.0058
0.0047
0.0068
0.0053
0.0070
0.0056
0.0096
0.0073
0.0126
0.0099
Top Deformeter
Bottom Deformeter
0.0012
0.00175
0.0031
0.00325
0.00435

TABLE IIIV

Mark	Age, (day)		Load Applied lb (P)	East Deflectometer: Top Deformation + in.	East Deflectometer: Bottom Deformation * in.	West Deflectometer: Top Deformation + in.	West Deflectometer: Bottom Deformation * in.	Average Top Deformation + in.	Extreme top Deformation in.	Average Bottom Deformation * in.	Center Deflection in.	Remark
B_4	49		0	0	0	0	0	0	0	0	0	
		1	675	0.0005	0	0.0004	0	0.00045	----	0	0.003	
		2	1,725	0.001	0	0.0012	0	0.00115	----	0	0.009	Load applied
		3	2,730	0.0014	0	0.0016	0.00020	0.00150	0.00185	0.0001	0.014	at 1/3-point,
		4	3,725	0.00165	0.0005	0.00193	0.00025	0.00180	0.----	0.00015	0.019	same as for B_1
		5	4,645	0.00190	0.00015	0.0022	0.00035	0.00205	0.0025	0.00025	0.024	
		6	5,695	0.00215	0.00020	0.00225	0.00040	0.0022	0.----	0.0003	0.032	
		7	6,780	0.00215	0.00030	0.00225	0.00050	0.0022	0.0027	0.0004	0.037	
		8	7,780	0.00265	0.00045	0.00285	0.00065	0.00275	0.----	0.00055	0.042	
		9	8,730	0.00310	0.00070	0.00320	0.00080	0.00315	0.0040	0.00075	0.043	
		10	9,825	0.00405	0.00095	0.00415	0.00125	0.0041	0.----	0.00110	0.056	
		11	10,625	0.00465	0.00100	0.00475	0.00140	0.0047	0.0058	0.00120	0.063	
		12	11,585	0.00470	0.0011	0.0051	0.00200	0.0049	0.----	0.00155	0.066	
		13	12,585	0.00520	0.0015	0.0054	0.0020	0.0053	0.0068	0.00175	0.072	crack between mortar & brick
		14	13,125	0.0056	0.0019	0.0056	0.0023	0.0056	0.0070	0.0021	0.076	
		15	13,815	0.0058	0.0023	0.0062	0.0026	0.0060	0.-----	0.00245	0.080	
		16	14,575	0.0061	0.00275	0.0067	0.00305	0.0064	0.-----	0.00290	0.084	
		17	14,925	0.0067	0.00315	0.0071	0.00325	0.0069	0.-----	0.00320	0.087	
		18	15,875	0.0071	0.00380	0.0075	0.0037	0.0073	0.0094	0.00375	0.091	
		19	16,765	0.0080	0.00420	0.0082	0.0040	0.0081	0.-----	0.00410	0.096	
		20	17,725	0.0082	0.00455	0.0090	0.00435	0.0086	0.----	0.00445	0.103	
		21	18,825	0.0090	0.00515	0.0104	0.00425	0.0097	0.0126	0.00475	0.111	Bond failure
Ultimate Load = 20,095#												

\+ Measured at 2½-in. from top.

* Measured at 2½-in. from bottom, at center of horizontal reinforcement.

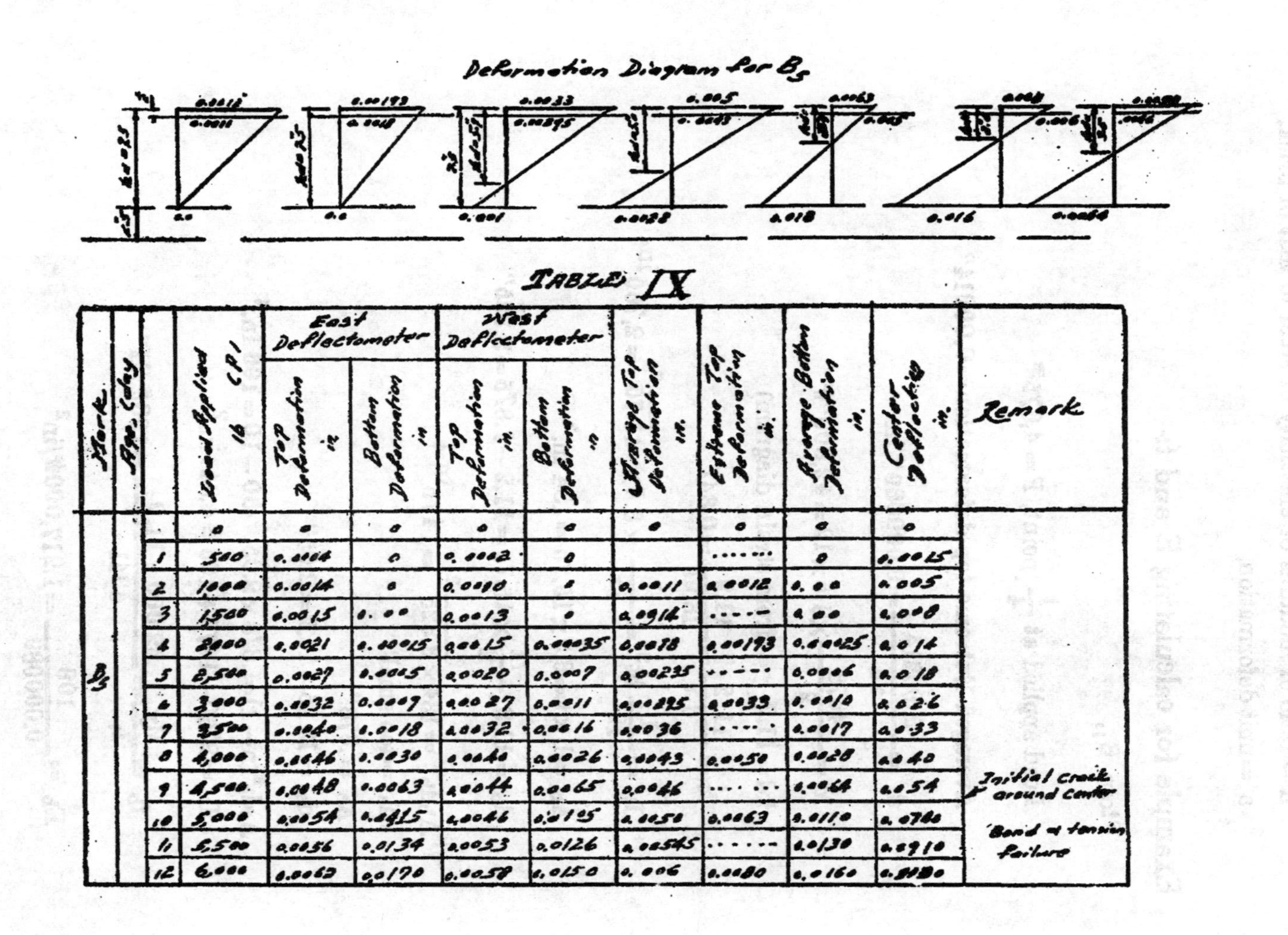

TABLE IX

Mark	Age, day		Load Applied lb CP1	East Deflectometer Top Deformation in.	East Deflectometer Bottom Deformation in.	West Deflectometer Top Deformation in.	West Deflectometer Bottom Deformation in.	Average Top Deformation in.	Extreme Top Deformation in.	Average Bottom Deformation in.	Center Deflection in.	Remark
B_3			0	0	0	0	0	0	0	0	0	
		1	500	0.0004	0	0.0002	0		......	0	0.0025	
		2	1000	0.0014	0	0.0010	0	0.0011	0.0012	0.00	0.005	
		3	1500	0.0015	0.00	0.0013	0	0.0914	----	0.00	0.008	
		4	2000	0.0021	0.00015	0.0015	0.00035	0.0018	0.00193	0.00025	0.014	
		5	2500	0.0029	0.0005	0.0020	0.0007	0.00235	----	0.0006	0.018	
		6	3000	0.0032	0.0009	0.0027	0.0011	0.00295	0.0033	0.0010	0.026	
		7	3500	0.0040	0.0018	0.0032	0.0016	0.0036	.....	0.0017	0.033	
		8	4000	0.0046	0.0030	0.0040	0.0026	0.0043	0.0050	0.0028	0.040	
		9	4500	0.0048	0.0063	0.0044	0.0065	0.0046	......	0.0064	0.054	Initial crack around center
		10	5000	0.0054	0.0415	0.0046	0.0105	0.0050	0.0063	0.0110	0.0780	Bond at tension failure
		11	5500	0.0056	0.0134	0.0053	0.0126	0.00545	......	0.0130	0.0910	
		12	6000	0.0062	0.0170	0.0058	0.0150	0.006	0.0080	0.0160	[illegible]	

n = ratio of modulus of elasticity of steel to that of brick.

a = unit deformation,

Example for calculating E and f:-

Ftom B_1,

Load applied at $\frac{1}{4}$ point, P = 4,675#

Averaged Extreme top deformation = 0.00214"

$$d = \frac{0.00214}{24} = 0.000089\text{"/in.}$$

$$M = \frac{4675}{2} \times 1.5 \times 12 = 42,000\text{"\#}$$

kd = 11.2 (from strain diagram)

$$k = \frac{11.2}{d} = \frac{11.2}{13.25} = 0.847$$

$$I_b = \frac{1}{12} bt^3 = \frac{1}{12} \times 8.5 \times \overline{15.75}^2 = 2,760^{in4}$$

$$A_b = b \times t = 8.5 \times 15.75 = 134 \text{ in.}^2$$

$$d_b = 11.2 - \frac{1}{2} \times 15.75 = 11.2 - 7.875 = 3.325\text{"}$$

$$A_b d_b^2 = 134 \times \overline{3.325}^2 = 1475^{in.4}$$

I_s = too small

A_s = .878"

d_s = 13.25 − 11.2 = 2.05"

$$A_s d_s^2 (n-1) = .878 \times \overline{2.05}^2 (30-10 = 106 \text{ in.}^4$$

$$I_t = 2760 + 1475 + 106 = 4,341 \text{ in.}^2$$

$$f_b = \frac{My}{I_t} = \frac{42,000 \times 11.2}{4341} = 108\text{\#/II"}$$

$$E_b = \frac{108}{0.000089} = 1,217,000\text{\#/in}^2$$

The others are in table form.

Caculation of Mooulus of Elasticity and Piber Stresses of Brick Masonry

Por Beams, B_1, B_2, B_3, & B_4 :

Cross-section, $b \times t = 8\frac{1}{2} \times 15\frac{3}{4}$ — in.

Area, $b \times t = 134^{\square}$"

Effective depth, $d = 13\frac{1}{4}$"

Span length $= 6 - 0$"

Gage length $= 24$"

Moment of Inortia of Cross-section $= \frac{bt^3}{12} = 2{,}760$ in.^{4}

TABLE X

B_1 — Longitudinal steel area $= 0.878^{\square}$"

Applied Load P, #	Top Deformation D, in.	Unit Deformation δ, in/in	External moment M, in-lb	kd	k	I_b in⁴	d_b in.⁴	$A_s d_b^2$ in⁴	d_s in.	$(A_s d_s^2)(n-1)$ in⁴	I_t in⁴	By Formulas, #/in² f_b	By Formulas, #/in² E_b	From tests, #/in² f_b	From tests, #/in² E_b
4,675	0.00214	0.000089	48,000	11.2	0.847	2760	3.325	1,475	2.05	106	4341	108	1,217,000	89	1,000,000
5,545	0.00255	0.000106	49,900	11.1	0.840	"	3.225	1,390	2.15	118	4268	130	1,226,000	106	"
7,616	0.00360	0.000150	68,500	11.2	0.847	"	3.325	1,475	2.05	106	4341	177	1,180,000	150	"
10,635	0.00250	0.000233	95,720	11.2	0.847	"	3.325	1,475	2.05	106	4341	249	1,060,000	233	"
14,130	0.00720	0.000300	127,000	10.6	0.800	"	2.725	990	2.65	178	3928	343	1,140,000	300	"
15,350	0.00800	0.000333	138,000	10.7	0.810	"	2.85	1,069	2.55	165	3994	369	1,138,000	333	"
18,020	0.00870	0.000363	162,180	10.0	0.755	"	2.13	605	3.25	269	3634	459	1,265,000	363	"
20,185	0.01020	0.000425	181,400	9.5	0.717	"	1.63	354	3.75	358	3472	497	1,160,000	429	"
22,465	0.01210	0.000505	202,500	9.1	0.686	"	1.23	201	4.15	438	3379	559	1,107,000	505	"
24,520	0.01320	0.000550	220,600	8.9	0.671	"	1.03	141	4.35	481	3382	580	1,065,000	550	"
27,105	0.01440	0.000600	244,000	8.65	0.653	"	0.78	79	4.60	545	3379	623	1,040,000	600	"
29,130	0.01590	0.000663	263,900	8.55	0.645	"	0.68	61	4.70	563	3384	665	1,003,000	663	"
31,735	0.01700	0.000709	285,400	8.25	0.623	"	0.28	10	5.00	637	3407	691	925,000	709	1,002,000

25279

TABLE XI

B_4: Longitudinal Steel area = 0.785 □"

Applied Load P, lb.	Top Deformation D, in.	Unit Deformation D, in/in	External Moment M, in-lbs	kd	k	I_b in⁴	d_b in.	$A_b d_b^2$ in⁴	d_s in.	$(A_s d_s^2)(n-1)$ in⁴	I_t in⁴	By Formulas, #/□" f_b	By Formulas E_b	From tests, #/□" f_b	From tests E_b
2,730	0.00185	0.000077	24,570	12.6	0.95	2960	4.825	2,745	0.65	10	5,715	54	765,000	99	1,003,000
4,645	0.00250	0.000104	41,800	12.25	0.93	"	4.375	2,546	1.00	23	5,289	97	993,000	104	"
6,780	0.00270	0.000113	61,000	11.6	0.87	"	3.725	1,860	1.65	62	4,682	151	1,341,000	112	"
8,730	0.00400	0.000167	78,000	11.3	0.85	"	3.425	1,573	1.95	87	4,420	201	1,205,000	167	"
10,695	0.00580	0.000242	96,250	11.3	0.85	"	3.425	1,573	1.95	87	4,420	249	1,022,000	242	"
12,585	0.00680	0.000284	112,800	10.5	0.79	"	2.625	925	2.75	171	3,856	306	1,083,000	284	"
13,125	0.00700	0.000292	118,100	10.3	0.77	"	2.425	790	2.95	190	3,748	325	1,115,000	292	"
15,895	0.00940	0.000371	145,800	9.6	0.72	"	1.725	378	3.65	300	3,458	404	1,038,000	371	"
18,895	0.01260	0.000525	169,700	9.5	0.71	"	1.625	355	3.75	318	3,433	490	894,000	525	"

TABLE —

B_5:
Cross section, b×t = $8\frac{1}{2}$×10 in.
Area, $A = 8\frac{1}{2} \times 10 = 85$ □"; $I_b = \frac{bt^3}{12} = 710$ in⁴
Longitudinal steel area, A_s = 0.39 □"

TABLE XII

Effective depth, d = $7\frac{1}{2}$ in.
Span Length = 6'-0
Gage Length = 24"

$A = bd = 8\frac{1}{2} \times 7\frac{1}{2}$ = [illegible] □"
$I_b = \frac{bd^3}{12} = 299$ in⁴

Applied Load P, lb.	Top Deformation D, in.	Unit Deformation D, in/in	External Moment M, in-lbs.	kd	k	I_b in⁴	d_b in.	$A_b d_b^2$ in⁴	d_s in.	$(A_s d_s^2)(n-1)$ in⁴	I_t in⁴	By Formulas, #/in² f_b	By Formulas E_b	From tests, #/in² f_b	From tests E_b
1,000	0.00120	0.00005	9,000	7.5	1	710	2.5	531	0	0	1,241	54	1,080,000	50	1,000,000
2,000	0.00193	0.00008	18,000	7.5	1	710	2.5	531	0	0	1,241	108	1,342,000	80	"
3,000	0.00330	0.00013	27,000	5.9	0.79	710	0.9	69	1.6	33	812	196	1,429,000	138	"
4,000	0.00500	0.00021	36,000	5.0	0.67	710	0.0	0	2.5	80	790	228	1,090,000	208	"
4,500	0.00580	0.00024	40,500	3.5	0.46	299	0.25	4	4.0	204	505	286	1,191,700	240	"
5,000	0.00650	0.00026	45,000	2.7	0.38	299	0.85	46	4.6	269	612	*213	819,000	260	"
6,000	0.00800	0.00033	54,000	2.6	0.34	299	1.15	69	4.9	310	676	*207	627,000	330	"

* These two values should be increased instead of decreased

3. Computation of Stresses are computed both from straight-line fomulas and parabolic gormulas.

Straight-line Formulas.

$$k=\sqrt{2pn+(pn)^2}-pn \quad \text{------- (1)}$$
$$j=1-\frac{1}{3}k \quad \text{------- (2)}$$
$$f_b=\frac{M}{\frac{1}{2}jkbd^2} \quad \text{------- (3)}$$
$$f_s=\frac{M}{pjbd^2} \quad \text{------- (4)}$$

Parabolic Formulas.

Parabolic Formulas

$$k=\sqrt{3pn+(\frac{3}{2}pn)^2}-\frac{3}{2}pn \quad \text{------- (5)}$$
$$j=1-\frac{3}{8}k \quad \text{------- (6)}$$
$$f_b=\frac{M}{\frac{2}{3}jkbd^2} \quad \text{------- (7)}$$
$$f_s=\frac{M}{pjbd^2} \quad \text{------- (8)}$$

B_1:-

Compute k, j, f_b, and f_s, from Staight-line Formulas.

p, here used, is the actual proportionate steet in beam.

p = 0.0078 n = 30

$$k=\sqrt{2\times0.0078\times30+(0.0078\times30)^2}-0.0078\times30$$
$$=\sqrt{0.468+0.0547}-0.234$$
$$=\sqrt{0.523}-0.234$$
$$=0.724-0.234=0.49$$
$$j=1-\frac{1}{3}\times0.49=.837$$
$$f_b=\frac{127,000}{\frac{1}{2}\times.837\times.49\times8.5\times13.25^2}=415\ \#/\square''$$
$$f_s=\frac{127,000}{0.0078\times.837\times8.5\times13.25^2}=13,050\ \#/\square''$$

The others are computed in the similar way

TABLE XIII

Fiber Stresses Calculated at Cracking Load
(from Straight-line Formulas)

Beam	Steel per cent p	Cracking Load P, #	External Moment M, in-lb.	From Formulas		From Strains		Stresses, #/in²				Remark
								From k_1 and j_1		From k and j		
				k_1 *	j_1	k *	j	Steel, f_s	Brick, f_b	Steel, f_s	Brick, f_b	
1	2	3	4	5	6	7	8	9	10	11	12	13
B_1	0.78	14,130	127,000	0.49	0.837	0.80	0.733	13,050	415	14,900	290	
B_2	0.218	4,370	78,700	0.39	0.878	0.00	----	2 600	310	------	----	First cracks appear at center joints
B_3	2.50	32,065	577,500	0.69	0.770	0.00	----	20,100	1460	------	----	
B_4	0.697	12,582	113,600	0.47	0.843	0.79	0.734	12,860	382	14,750	261	
B_5	0.612	4,500	40,500	0.451	0.850	0.46	0.847	16,300	440	16,300	435	
B_6	0.612	4,500	40,500	0.451	0.850	0.00	-----	16,300	440	------	----	
B_7	0.308	3,500	63,000	0.347	0.884	0.00	-----	48,570	860	----	----	
B_8	1.93	11,000	190,000	0.646	0.785	0.00	----	27,400	1630	----	----	

For f_b, values in column 10 are greater than those in column 12, that is to say, f_b obtained from straight-line formulas are greater than those from actual strains, and this is on the safe side.

* k in column 7 are greater than k_1 in column 5. This shows that part of tensile stress is taken up by brickmasonry resulting lower the neutral axis.

TABLE XIV

Fiber Stresses Calculated at Cracking Load

(from Parabolic Formulas)

Beam	Steel per cent p	Cracking Load P, #	External Moment M, in-lbs.	From Formulas k_2	From Formulas j_2	From Strain k	From Strain j	Stresses, #/in.² From k_2 and j_2 Steel, f_s	From k_2 and j_2 Brick, f_b	From k and j Steel, f_s	From k and j Brick, f_b	Remark
1	2	3	4	5	6	7	8	9	10	11	12	13
B_1	0.78	14,130	127,000	0.358	0.793	0.80	0.733	13,770	292	14,900	290	
B_2	0.218	4,370	78,700	0.356	0.866	0.00	----	27,900	256	-----	----	
B_3	2.50	32,065	577,500	0.995	0.627	0.00	----	24,800	935	-----	----	First cracks appear at center joints.
B_4	0.697	12,582	112,600	0.536	0.799	0.79	0.734	13,600	264	14,250	261	
B_5	0.612	4,500	40,500	0.515	0.807	0.46	0.847	17,150	306	16,310	435	
B_6	0.612	4,500	40,500	0.515	0.807	0.00	---	17,150	306	-----	----	
B_7	0.308	3,500	63,000	0.407	0.848	0.00	---	50,600	572	----	---	
B_8	1.93	11,000	198,000	0.707	0.735	0.00	---	29,260	1180	---	----	

Column "11" shows that values for B_1 and B_4 are almost the same as in column 12 obtained from straight-line strain variation. The values for steel in col. 9 and 11 are not different very much.

For safety of design, straight-line formulas are preferable, because f_s obtained, as shown in TABLE are greater from straight-line formulas than from strains.

TABLE XIV

Fiber Stresses Calculated at the Last Applied Load

(from Straight-line Formulas)

Beam	Steel per cent p	Last Applied Load P, #	External Moment M, in-lbs.	From Formulas		From Strains		Stresses, #/in.²				Kind of Failures
								From k_1 and j_1		From k and j		
				k_1	j_1 *	k	j	Steel, f_s	Brick, f_b	Steel, f_s	Brick, f_b	
1	2	3	4	5	6	7	8	9	10	11	12	13
B_1	0.78	31,735	285,400	0.49	0.837	0.623	0.792	29,400	1,171	30,900	770	DT (crack)
B_2	0.218	* 7,480	*134,500	0.39	0.870	----	----	47,700	551	-----	----	T
B_3	2.50	35,065	631,000	0.69	0.770	----	----	22,000	1,595	-----	----	DT
B_4	0.697	*24,065	*216,600	0.47	0.843	----	----	24,650	735	-----	----	Bond
B_5	0.612	*13,000	*117,000	0.451	0.850	----	----	47,100	1,280	-----	----	Bond+T
B_6	0.612	*14,500	*130,500	0.451	0.850	----	----	52,500	1,425	-----	----	Bond+T
B_7	0.308	* 5,000	* 90,000	0.347	0.884	----	----	62,250	1,230	-----	----	T
B_8	1.93	17,000	306,000	0.646	0.785	----	----	42,300	2,540	-----	----	T

* The maximum values

DT = Diagonal Tension

T = Tension

25284

TABLE XVI

Fiber Stresses Calculated at the Last Load Applied.

(from Parabolic Formulas)

Beam	Steel per cent p	Last Applied Load P, #	External Moment M, in-lbs.	From Formulas		From Strains		Stresses, #/in.²				Kind of Failures
								From k_2 and j_2		From k and j		
				k_2	j_2	k	j	Steel, f_s	Brick, f_b	Steel, f_s	Brick, f_b	
1	2	3	4	5	6	7	8	9	10	11	12	13
B_1	0.78	34,736	285,400	0.556	0.793	0.623	0.792	31,000	664	30,900	770	DT (crack)
B_2	0.218	* 7,480	*134,500	0.356	0.866	----	----	47,500	428	----	----	T
B_3	2.50	35,065	631,000	0.995	0.629	---	---	27,000	1,015	---	---	*
B_4	0.697	*24,065	*216,600	0.536	0.799	---	---	26,000	508	---	---	Bond
B_5	0.612	*13,000	*117,000	0.515	0.809	---	--	40,900	882	---	---	Bond + T
B_6	0.612	*14,500	*130,500	0.515	0.809	---	---	55,000	983	---	---	Bond + T
B_7	0.308	*5,000	* 90,000	0.407	0.848	---	---	72,000	820	---	---	T
B_8	1.93	17,000	306,000	0.707	0.735	---	---	45,800	1,860	---	---	T

* The maximum values

f_s and f_b in Columns 9 and 10 are quite closer to f_s and f_b in Column 11 & 13 respectively.

TABLE XVII

Shearing Stresses Developed by Reinforced Brick Beam at Cracking Load

Beam	Breadth (b) in.	Effective Depth (d) in.	(a) j_1	(i) j_2	(c) j	Cracking Load (P) #	Total vertical Shear (V) #	Moment at Cracking Load (M) in-lbs.	$\frac{M}{bd^2}$ #/in.²	Unit Shear $v=\frac{V}{bjd}$ From j_1	From j_2	From j	Kind of Failure
1	2	3	4	5	6	7	8	9	10	11	12	13	14
B_1	$8\frac{1}{2}$	13.25	0.837	0.793	0.733	14,130	7,065	127,000	53	75	80	86	DT (crack.
B_2	"	"	0.870	0.866	0.---	4,330	2,165	78,700	85	22	28	----	T
B_3	"	"	0.770	0.627	0.--	32,065	16,033	577,500	387	185	230	---	DT
B_4	"	"	0.843	0.799	0.734	12,582	6,291	112,600	76	66	70	76	Bond
B_5	"	7.5	0.850	0.807	0.847	4,500	2,250	40,500	85	38	44	42	Bond + T
B_6	"	"	0.850	0.807	0.---	4,500	2,250	40,500	85	38	44	----	Bond + T
B_7	"	"	0.884	0.848	0.--	3,500	1,750	63,000	132	31	33	----	T
B_8	"	"	0.985	0.735	0.--	11,000	5,500	198,000	415	110	118	----	T

(a) j_1 - from straight-line formulas

(i) j_2 - from parabolic formulas

(c) j - from strains

TABLE XVIII

Shearing Stresses Developed by Reinforced-brick Beam at the Last Applied Load

Beam	Breadth (b) in.	Effective Depth (d) in.	j_1	j_2	j	Last Applied Load (P) #	Total Vertical Shear (V) #	Moment at Last Applied Load (M) (in-lb).	$\frac{M}{bd^2}$ #/in.²	Unit Shear, $v=\frac{V}{bjd}$ From j_1	From j_2	From j	Kind of Failure
1	2	3	4	5	6	7	8	9	10	11	12	13	14
B_1	8.5	13.25	0.837	0.793	0.792	31,735	15,868	285,400	191	188	178	178	DT (crack)
B_2	"	"	0.870	0.866	---	* 7,480	* 3,740	* 134,500	90	38	39	----	T
B_3	"	"	0.770	0.627	---	35,065	17,533	* 631,000	424	203	280	----	DT
B_4	"	"	0.843	0.799	---	* 24,065	* 12,033	* 216,600	145	127	151	---	Bond
B_5	"	7.5	0.850	0.807	---	* 13,000	* 6,500	* 117,000	244	120	126	---	Bond + T
B_6	"	"	0.850	0.807	----	* 14,500	* 7,250	* 130,500	273	135	141	---	Bond + T
B_7	"	"	0.884	0.848	----	* 5,000	* 2,500	* 90,000	188	45	46	---	T
B_8	"	"	0.785	0.735	---	17,000	8,500	306,000	640	170	181	---	T

* The maximum values.

25287

TABLE XIX.

Bond Stresses both at Cracking Load and the last Applied Load.

Beam	Effective Depth (d) in.	Σo	j_1	j_2	At Cracking Load: Cracking Load (P) #	At Cracking Load: V #	Unit Bond Stresses $u = \frac{V}{\Sigma o j d}$: From j_1	Unit Bond Stresses: From j_2	At the Last Applied Load: Last applied Load (P) #	At the Last Applied Load: V #	$u = \frac{V}{\Sigma o j d}$: From j_1	$u = \frac{V}{\Sigma o j d}$: From j_2	Type of Failure
1	2	3	4	5	6	7	8	9	10	11	12	13	14
B_1	13.25	6.57	0.837	0.793	14,130	7,065	97	102	31,735	15,868	217	230	(Shear) (crack)
B_2	"	3.93	0.870	0.866	4,370	2,185	48	49	* 7,480	3,740	83	83	T
B_3	"	11.78	0.770	0.627	32,065	16,033	133	164	35,065	17,533	146	179	
B_4	"	3.14	0.840	0.799	12,582	6,291	180	190	*24,065	12,033	345	364	Bond
B_5	7.5	2.50	0.850	0.807	4,500	2,250	141	148	*13,000	6,500	408	430	Bond, T
B_6	"	2.50	0.850	0.807	4,500	2,250	141	148	*14,500	7,250	453	477	Bond, T
B_7	"	3.14	0.884	0.848	3,500	1,750	84	88	* 5,000	2,500	120	126	T
B_8	"	7.86	0.785	0.735	11,000	5,500	119	127	17,000	8,500	184	196	T

* Maximum Applied Loads.

k, j, f_c and f_s from Parabolic Formulas.

$p = 0.0078$. $n = 30$

$$k = \sqrt{3 \times 0.0078 \times 30 + \left(\frac{3}{2} \times 0.0078 \times 30\right)^2} - \frac{3}{2} \times 0.0078 \times 30$$

$$= \sqrt{.702 + .123} - 0.351$$

$$= \sqrt{.825} - 0.351$$

$$= .907 - .351 = 0.556$$

$$j = 1 - \frac{3}{8} \times 0.556 = 0.793$$

$$f_c' = \frac{182000}{\frac{2}{3} \times .793 \times .556 \times 8.5 \times \overline{13.25}^2} = 292 \#/in$$

$$f_s = \frac{182000}{0.0078 \times .793 \times 8.5 \times \overline{13.25}^2} = 12770 \#/in$$

The others are computed in similar ways.

Conclusion

The results of these tests show that it is possible to develop a high degree of flexural strength in reinforced brick beams. In irder to develop a desirable strength, good workmanship is very important. An ordinary mason, laying ordinary walls or other brick masonry is useally to place insufficient mortar to make a full joint of the required thickness, run the point of his trowel through the middle of the mass making an open channel with a sharp ridge of mortar on each side and then lay the brick upon the top of these two ridges, then leaving the center of the brick unsupported. But in reinforced brick masonry this practice is inapplicable. For both strength and impervious are matter of any moment, so care shoudd be taken to see that both the vertical and horizontal joints are filled solidly full of mortar, what we called "slushing the joints." From the results of these tests, the following conclusions may be drawn:

1. In order to secure a high degree of strengths particular attension must be given to see that all joints are filled solidly full of mortar.

2. With good workmanship, rich mortar and proper design of stirrup and longitudinal reinforcement, coefficients of resistance, M/bd^2, in ex-

cess of 600 lbs. per sq. in. and shear stress, v, in excess of 200 lbs per sq. in. were obtained.

3. The tests of brick specimens indicate that the straight-li neformulas used in the calculations for reinforced brick beams of fiber, shear, and bond stresses for reinforced concrete beams can with proper constants be used in like calculations for reinforced brick beams.

4. In designing the arrangement of coursing, it is preferable to have some headers as well as stretchers in order to secure good bond strength.

5. The 1 cement: 2 sand mortar used in these tests gave very satisfactory results.

6. With the mortar used in these tests shear strengths were developed without stirrups in excess of 168, 135, 163 lbs per sq. in. while with stirrups it was developed in exeess of 208 lbs per sq. in.

7. None of these beams was failed by compression although it was stressed as high as up to 2540 lbs / in.^{2}, as B_8.

8. Working stresses used for design beams in builings without any considerable impact are as follows:

For 2000-lbs concrete, working stress may be used as high as 800 lbs/in^2. for ordinayy purpose with a factor of safety 2½. If 500 lbs/in^2. for brick masonry with the maximum fiber stress of 2540 lbs/in.^{2}, we have a safety factor of 5, supposing 2½ for those conditions as for concrete and the other 2½ for poor workmanship and other unexpecte effects.

The mortar used in reinforced brick beams is richer than concrete, so the bond and shear stresses for reinforced concrete can be used for reinforced brick masonry. But because of poor workmanship and the weakness at the plane between mortar and brick at joints, lower working stresses both for bond and shear are desirable, that is 60 lbs / in.^{2} for bond and 30 lbs per in.^{2} and 80 lbs/in.^{2} for shear stress without or with web reinforcement respectively.

For the sake of clearness, they may be sum up as follows:

Unit fiber stress of brick, f_b = 500 lbs/in.^{2} for buildings
= 300 lbs/in.^{2} for structures having impact action.

Unit tensile stress of steel, f_s = 16,000 lbs/in^2

Modulus of elasticity of steel, E_s = 30,000,000 lbs/in.^{2}

Modulus of elasticity of brick, E_o = 1,000,000 lbs/in.^{2}

$$n = \frac{E_s}{E_o} = 30$$

v = 30 lbs/in.^{2} without web reinforcement and
80 lbs/in.^{2} with web reinforcement.

u = 60 lbs/in.^{2}

Tables and Diagrams for Design Retangular Beams and Slabs

TABLE I
DATA FOR DESIGN OF RECTANGULAR BEAMS

FORMULAS NEEDED

$M = Kbd^2$; or $bd^2 = \frac{M}{K}$

$A_s = pbd$

FORMULAS USED IN PREPARING TABLE:

$p = \frac{\frac{1}{2}}{\frac{f_s}{f_c}\left(\frac{f_s}{nf_c}+1\right)}$

$k = \sqrt{2pn + (pn)^2} - pn$; $j = 1 - \frac{k}{3}$.

$K = pf_sj$ or $\frac{1}{2}f_ckj$ (from formula $M = Kbd^2$)

n = 25

f_s	f_c	k	j	p	K
14,000	200	0.264	0.912	0.00190	24.2
	250	0.308	0.897	0.00275	34.5
	300	0.348	0.884	0.00374	46.2
	350	0.385	0.872	0.00490	58.8
	400	0.417	0.864	0.00596	72.1
	450	0.445	0.852	0.00718	85.7
	500	0.472	0.843	0.00845	100.0
	550	0.495	0.835	0.00974	114.0
	600	0.518	0.827	0.01110	128.2
	650	0.538	0.821	0.01250	143.7
	700	0.555	0.815	0.01390	158.2
16,000	200	0.238	0.921	0.00149	21.9
	250	0.280	0.907	0.00219	31.8
	300	0.319	0.894	0.00299	42.7
	350	0.354	0.882	0.00385	54.5
	400	0.385	0.872	0.00481	68.0
	450	0.413	0.862	0.00580	80.0
	500	0.440	0.853	0.00690	94.0
	550	0.461	0.846	0.00794	108.3
	600	0.485	0.838	0.00910	121.5
	650	0.505	0.832	0.01025	137.0
	700	0.523	0.826	0.01142	151.1
18,000	200	0.217	0.928	0.00120	20.1
	250	0.257	0.914	0.00177	27.5
	300	0.294	0.902	0.00245	38.8
	350	0.327	0.891	0.00317	51.0
	400	0.358	0.881	0.00398	63.2
	450	0.385	0.872	0.00480	75.5
	500	0.410	0.863	0.00570	88.4
	550	0.432	0.856	0.00660	101.5
	600	0.455	0.848	0.00760	115.6
	650	0.475	0.842	0.00860	130.0
	700	0.493	0.836	0.00960	144.0
20,000	200	0.200	0.933	0.00100	18.7
	250	0.238	0.921	0.00149	27.4
	300	0.272	0.909	0.00205	37.1
	350	0.304	0.899	0.00266	47.7
	400	0.334	0.889	0.00334	59.3
	450	0.360	0.880	0.00406	71.7
	500	0.385	0.872	0.00482	84.3
	550	0.407	0.864	0.00560	96.7
	600	0.430	0.857	0.00643	110.0
	650	0.449	0.850	0.00730	124.0
	700	0.467	0.844	0.00820	138.0

TABLE II
DATA FOR DESIGN OF RECTANGULAR BEAMS

$n = 30$

f_s	f_c	k	j	p	K
14,000	200	0.300	0.900	.00214	87.0
	250	0.348	0.884	.00311	49.0
	300	0.392	0.869	.00420	51.0
	350	0.429	0.857	.00536	64.1
	400	0.462	0.846	.00660	78.1
	450	0.491	0.836	.00791	92.5
	500	0.518	0.827	.00925	106.5
	550	0.540	0.820	.01060	122.0
	600	0.562	0.813	.01204	137.2
	650	0.581	0.806	.01347	151.5
	700	0.600	0.800	.01500	167.5
16,000	200	0.273	.909	.00170	24.6
	250	0.319	.894	.00249	35.6
	300	0.360	.880	.00337	47.5
	350	0.396	.868	.00443	60.0
	400	0.429	.857	.00536	73.6
	450	0.459	.847	.00645	87.5
	500	0.484	.839	.00758	101.2
	550	0.508	.831	.00870	116.0
	600	0.530	.823	.00993	131.0
	650	0.550	.817	.01115	146.0
	700	0.567	.811	.01240	161.0
18,000	200	0.250	.917	.00139	22.9
	250	0.295	.902	.00205	33.2
	300	0.334	.889	.00278	44.4
	350	0.370	.877	.00358	56.4
	400	0.400	.867	.00445	69.3
	450	0.430	.857	.00537	83.0
	500	0.455	.848	.00631	96.5
	550	0.479	.840	.00731	110.5
	600	0.500	.833	.00833	125.0
	650	0.520	.827	.00940	139.5
	700	0.538	.821	.01047	155.0
20,000	200	0.231	.923	.00115	21.2
	250	0.272	.909	.00170	30.9
	300	0.310	.897	.00234	41.8
	350	0.345	.885	.00301	53.3
	400	0.375	.875	.00375	65.9
	450	0.403	.866	.00455	78.5
	500	0.430	.857	.00536	92.0
	550	0.452	.849	.00621	105.0
	600	0.474	.842	.00710	119.5
	650	0.494	.835	.00803	134.0
	700	0.512	.829	.00898	148.0

TABLE III

DATA FOR DESIGN OF RECTANGULAR BEAMS

$n = 35$

f_s	f_b	k	j	p	K
14,000	200	0.334	0.889	0.00238	29.6
	250	0.385	0.872	0.00344	42.0
	300	0.430	0.857	0.00460	55.2
	350	0.468	0.844	0.00585	69.0
	400	0.500	0.833	0.00715	83.5
	450	0.530	0.823	0.00853	98.0
	500	0.556	0.815	0.00995	113.2
	550	0.580	0.809	0.01140	128.7
	600	0.600	0.800	0.01285	144.0
	650	0.620	0.793	0.01444	160.0
	700	0.637	0.788	0.01590	175.0
16,000	200	0.304	0.899	0.00190	27.3
	250	0.353	0.882	0.00276	34.9
	300	0.397	0.868	0.00372	51.6
	350	0.434	0.854	0.00475	65.2
	400	0.477	0.841	0.00585	79.0
	450	0.496	0.835	0.00699	93.0
	500	0.521	0.826	0.00812	107.9
	550	0.547	0.818	0.00940	122.6
	600	0.569	0.810	0.01062	138.0
	650	0.588	0.804	0.01190	154.0
	700	0.605	0.798	0.01320	168.0
18,000	200	0.280	0.907	0.00156	25.4
	250	0.327	0.891	0.00228	36.6
	300	0.369	0.877	0.00306	48.4
	350	0.405	0.865	0.00390	60.8
	400	0.438	0.854	0.00486	74.9
	450	0.465	0.845	0.00581	88.4
	500	0.494	0.835	0.00688	103.0
	550	0.518	0.827	0.00790	117.4
	600	0.539	0.820	0.00879	132.2
	650	0.558	0.814	0.01049	148.3
	700	0.576	0.808	0.01118	162.0
20,000	200	0.200	0.913	0.00130	23.7
	250	0.305	0.898	0.00191	34.2
	300	0.344	0.885	0.00268	45.7
	350	0.380	0.873	0.00332	58.2
	400	0.412	0.863	0.00412	71.4
	450	0.440	0.853	0.00475	84.7
	500	0.466	0.845	0.00584	98.7
	550	0.490	0.837	0.00673	112.5
	600	0.512	0.827	0.00770	127.5
	650	0.532	0.823	0.00865	142.2
	700	0.550	0.817	0.00964	157.5

25295

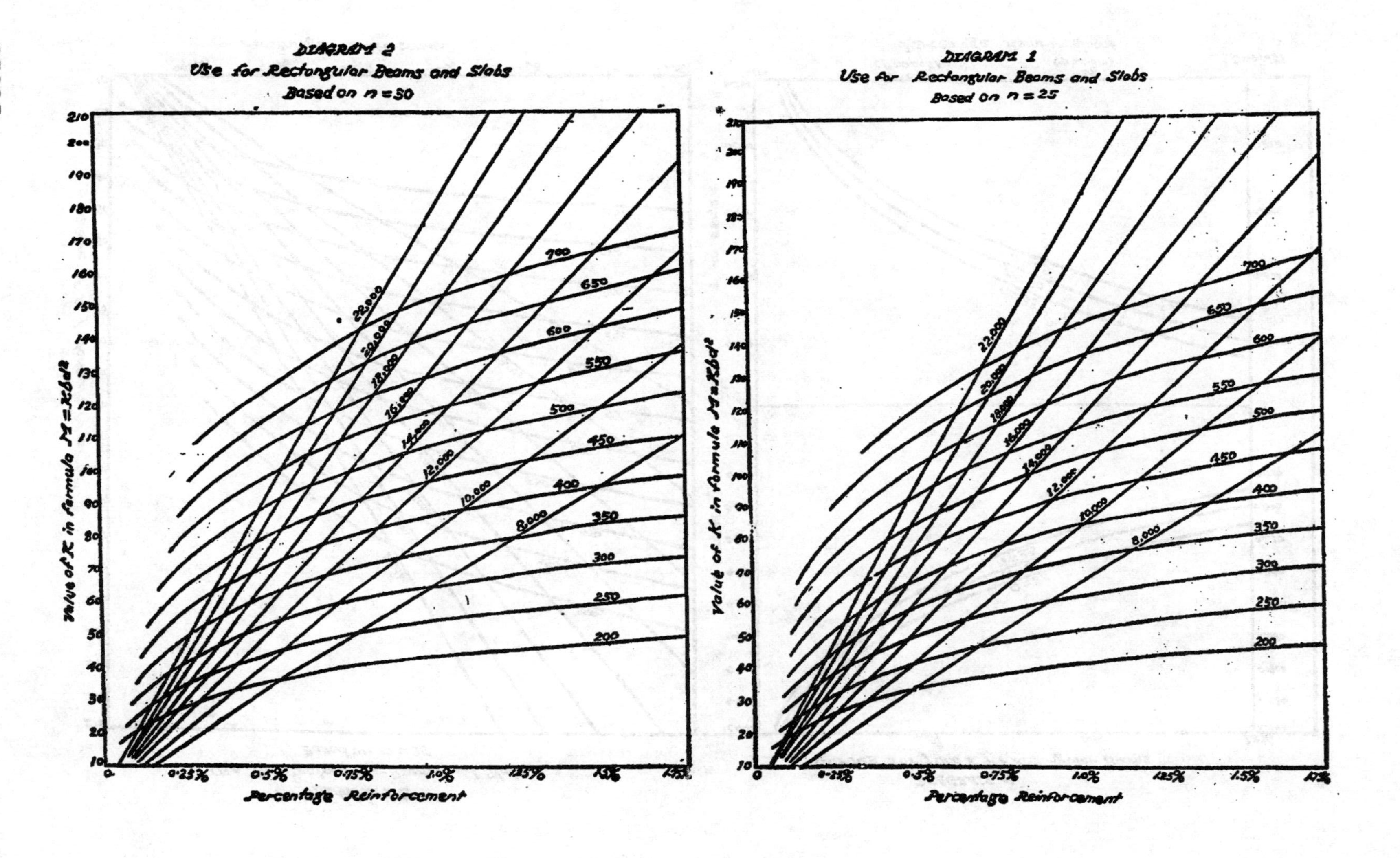
DIAGRAM 2
Use for Rectangular Beams and Slabs
Based on n = 30
Value of K in formula M = Kbd²
Percentage Reinforcement
DIAGRAM 1
Use for Rectangular Beams and Slabs
Based on n = 25
Value of K in formula M = Kbd²
Percentage Reinforcement

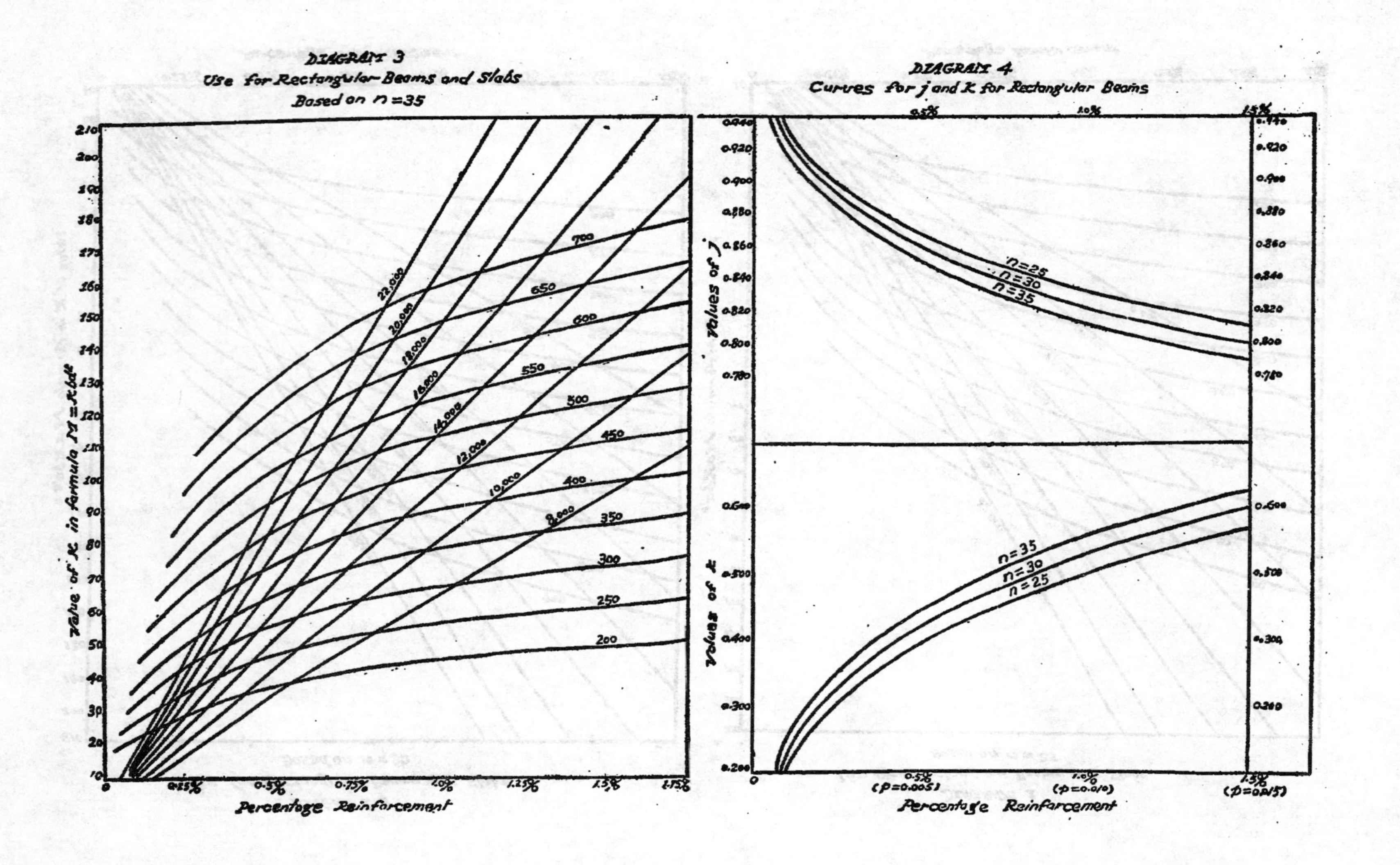
DIAGRAM 3
Use for Rectangular-Beams and Slabs
Based on $n=35$
Value of K in formula $M=Kbd^2$
Percentage Reinforcement
0
0.25%
0.5%
0.75%
1.0%
1.25%
1.5%
1.75%
10
20
30
40
50
60
70
80
90
100
110
120
130
140
150
160
175
180
190
200
210
22,000
20,000
18,000
16,000
14,000
12,000
10,000
8,000
700
650
600
550
500
450
400
350
300
250
200
DIAGRAM 4
Curves for j and k for Rectangular Beams
Values of j
Values of k
$n=25$
$n=30$
$n=35$
0.940
0.920
0.900
0.880
0.860
0.840
0.820
0.800
0.780
0.600
0.500
0.400
0.300
0.200
Percentage Reinforcement
0
0.5%
($p=0.005$)
1.0%
($p=0.010$)
1.5%
($p=0.015$)

A STUDY OF THE CHARACTERISTICS OF THE PEARL RIVER WATER, DATA COVERING A PERIOD FROM APRIL 1933 TO APRIL 1934

By WEI-WEN HUANG

(Separated from
LINGNAN SCIENCE JOURNAL, Vol. 14, No. 1. Jan., 1935)

INTRODUCTION

This is the third paper in the present series. The two previous ones,[1] covering a period of a year each, also appeared in the Journal. Each year new things have come up or some deviation from the past record has appeared here and there, suggesting that surface water, especially that of a river, is a thing capable of many changes. Its changes, however, follow very definitely the changes of the weather. Since the changes of weather are not expected to take place at the same time, to the same degree and under the same conditions every year, it will follow that changes in water with respect to its dissolved constituents will also be variable. However,

---

[1] Huang, Wen-Wei: A Study of the Characteristics of the Pearl River Water, I. Data covering a period from November, 1930, to November, 1931. *Lingnan Science Journal,* 12(1): 1-11, 3 fig., 1 tab. 1933.

Huang, Wen-Wei: A Study of the Characteristics of the Pearl River Water, II. Data covering a period from March, 1932, to March, 1933. *Lingnan Science Journal,* 12(3): 415-421. Hereafter these will be referred to as I and II.

the general trend may be predicted, if a study of the water over a long period of time is made. The Pearl River, because it is close to the sea, offers an interesting source for study. The location where the water samples were taken is at the Lingnan University wharf, and is about 30 miles from the sea. The influence of the sea upon the river is clearly seen throughout the year particularly during the dry season.

The present study is an attempt not only to try to determine the influence of the changing weather and the far reaching effect of the ocean in a short period of time, but will serve also to extend the earlier studies. No study will be of the greatest use, unless repeated results on similar seasons change can be obtained, and this is not in any way possible if an extended time is not covered.

DISCUSSION

Methods used in analysis are those given in the *Standard Methods of Water Analysis*, 7th edition, U. S. Public Health Association (1933). The data of this paper are not reproduced here on occount of their voluminous figures; in their place, and for practical reasons the results have been translated into curves, which will be found at the end of the article.

Turbidity—One of the measurements by which the physical properties of water can be judged is turbidity. It is important because it bears on the appearance of the water. The thing that will first attract people to a water is its sparkling clarity or its perfect transparency. Hence, generally speaking, the better the water, the lower will be its turbidity, though this is not always true. As a rule, almost without exception, the turbidity of surface water is higher than that of ground water. This is due to the fact that surface water, like that of a river, is flowing and thus carries in suspension, matter with it such as sand, clay, and microorganisms. The amounts of these would be uniform at all times if no other factors entered. However, the turbidity of a river water changes almost every day. In the curve here presented representing turbidity, every peak represents a rise of turbidity and this rise is due, in nearly every case to the effect of a heavy rainfall. Since the Pearl River and its tributaries flow through several provinces, if any area along the river receives a rainfall the effect will eventually be seen here (Canton). It is not an uncommon thing therefore, that in a course of several days of clear weather suddenly on one fine morning one finds the river water very muddy. This indicates that somewhere upstream there was a heavy rainfall. Thus we find that not only the rainfall in Canton but also what occurs anywhere upstream will cause a rise in turbidity of the water here. The turbidity of the Pearl River water as a rule is not considered high. For the whole year, from April 1933 to April 1934, with only a few exceptions, the curve is well under the 50 P.P.M. line. The curve starts out with very low turbidity and begins to rise toward the month of May. It then begins to decline and reaches its minimum between September and October, 1933, then goes up slowly and when we come to the end of April, 1934, it shoots up enormously again. This indicates that the rainy season came in toward the end of April and reached its climax in May for both 1933 and 1934. The same thing happened also in 1931 (I), and in 1932 (II).

As far as the amount of the variations is concerned, it depends largely on the amount of rainfall received each time. The heavier the rainfall, the higher will be the curve. Since the amount of rainfall will not be the same each year, it is natural that the curves will not be identical.

Color.—Turbidity differs from color in that the former measures the suspended substances as well as those in a colloidal state, whereas the latter measures only the soluble matter that has color. It is one of the physical properties on which the quality of the water depends a great deal. Water of the first order should be free from any color as well as turbidity. The presence of color is not only due to rainfall but may also be due to any source of contamination that will carry with it infusions of vegetable or animal origin. A rise in turbidity is not necessarily accompanied with the same rise in color, although this may happen. For the first half of the curve, ending in September, the line runs mostly on or above the 20 mark and from then on it runs mostly below this mark. The same change can be found in the curve covering a period from November, 1930, to November, 1931, (I) and another covering a period from March, 1932, to March, 1933. (II) The one which appeared in the second paper, seems to have values varying about the 30 line for the first half of the curve, from March to September. This on the average and for the same period of the year, is higher than those appearing in the first paper and in the present one. For the second half of this year's curve, the line runs well below the 20 mark for the most part. On the whole these three curves on color run pretty much the same course, that is, for the months from April on to September the color is usually higher than that for the months from October on to February or perhaps March. The highest mark of this year was 60 and the lowest was 5.

pH Value.—pH value and hydrogen ion concentrations really express one and the same thing. Mathematically the relationship is given by $pH = \log \frac{1}{(H^+)}$. pH is an expression of the intensity factor of acid or alkaline properties as opposed to the quantity factors, "acidity" and "alkalinity". It is important as far as water purification is concerned. Thus it is well known that water can be best coagulated at a certain pH value and that below or above this value precipitation is not complete. The optimum range has been found to be anywhere between pH 5 to 7, though each water has its own value. Besides, pH value is most important in the control of corrosion. The average pH value for the river water under consideration can be said to be about 7, the point of neutrality on this scale. On the whole, the pH value curve does not fluctuate as much as that which appeared in the second paper. However, the average for both curves will be in the neighborhood of 7. Similar changes can be found here to what occurred before, that is, a rise in free CO_2 usually accompanied a fall in the pH value curve.

Odor.—Odor is usually determined in cold as well as in hot water, for certain odors may not appear at all in the cold but will come out very distinctly when the sample is heated. In general the odor of this water, both cold and hot, is earthy, although there are cases of tarry or oily, and disagreeable odors. The tarry or oily odor was largely derived from petroleum storage houses or boats located upstream and perhaps from the University Water and Power Plant. However, odor, if not excessive, can be easily eliminated after the process of coagulation and filtration.

Hardness (Permanent).—Permanent hardness is defined as that kind of hardness that will not be precipitated by boiling. It consists principally of chlorides, nitrates, and sulfates of calcium and magnesium. In general hardness in water causes serious trouble in laundry, textile, and dye works, print-works and bleach works, in leather and tanning industries, in chemical works, paper mills, distilleries, ice plants, in aerated water or beverage industresi and particularly in steam-raising. The hard and tenacious scale formed by these compounds in the boiler results in decreased efficiency, increase in fuel cost and danger of serious explosions. Under the high pressure and temperature conditions, which obtain in boilers, many hard waters, especially those containing magnesium chloride and nitrate, liberate acids which cause serious corrosion of fittings and plates.[a] The hardness in the Pearl River water, with the exception of the months of February, March and April, is almost equal to zero and in most cases never exceeds 5 P.P.M. For the years 1933 and 1934, the hardness begins to rise at the end of January, reaches its climax in February, goes through March and then declines gradually toward the middle of April. Last year, the highest mark was 47.5 P.P.M. while this year was 180 P.P.M., the highest of all three years. This exceptional case can perhaps be explained by the unusual drought of this year. This is further proved by the unusual rise of

[a] Water Treatment. 23 p. Imperial Chemical Industries, Imperial Chemical House, London, S. W. 1, 1934,

chloride, which occurred at exactly the same tim e time. The Pearl River water, if it is not for this exceptional rise of hardness and chloride each year, may be considered saft and in this respect is good for household use as well asy for any ndustrial application.

Total Alkalinity.—Total alkalinity is the same as temporary hardness in this case (1), meaning that the alkalinity is due to the presence of only the bicarbonates of calcium and magnesium and calculated, as in the case of hardness, in terms of $CaCO_3$. The present curve indicates that for the months from May or June to August, the alkalinity runs mostly below the 50 P.P.M. line and from September to April it runs always above the 50 P.P.M. mark. The unusual rise of hardness during February and March does not seem to show up in the total alkalinity to any marked degree. This means that the increased hardness on account of dry weather and the incoming of sea water during that part of the year, was principally of a permanent nature, though in general there is a very slight increase in temporary hardness of bicarbonate type. The rise and fall of the curve seem to have an opposite movement with the turbidity. For instance, for the month of April, turbidity, runs it lowest course on the average of about 16 P.P.M. whereas the total alkalinity rises high to the 60 P.P.M. line. The next month, May, when turbidity goes up, the total alkalinity, on the whole, turns down. Toward the end of April, 1934, there was a sudden rise of turbidity which took place on account of heavy rainfall. This, curiously enough, however, was also accompanied by a sudden fall of total alkalinity. As a matter of comparison, one may find a fall of alkalinity during the months from May to August or thereabouts in 1931 and 1932, (I and II) similar to that in 1934. As for the rest of the year, the curves begin to rise in September or thereabouts and stay on mostly between the 50 and 60 P.P.M. lines. One would not be far from right were one to say that the average total alkalinity of the river water is in the neighborhood of 50 most of the time.

Free Carbon Dioxide.—Carbon dioxide is the gas derived from organic decomposition and combustion. By a determination of the gas together with the analysis of organic matter, the chemist is able to tell whether the pollution of water is of vegetable or animal nature. High CO_2 usually indicates bad water. It has an opposite relation with pH value, that is, the higher the pH value, the lower will be the CO_2 content. Moreover, a high turbidity usually causes a rise in CO_2. This is due to the fact that rainfall not only raises the turbidity but also brings down CO_2 from the air to the water. On the average, the CO_2 of the water ranges from 5 to 10, which is to be considered low. The highest was 19 P.P.M. and the lowest was 2. As a matter of fact free CO_2 is by no means a bother to us. In the Filtration Plant, situated in the University Campus, a process of aeration is used, which was designed to get rid of the gases and oxidize the iron. But this is not necessary here as the aquatic plants, the Hydrilleae and Vallisnereae types, grow in the raw water reservoir so abundantly! These plants generate much oxygen and take up the free CO_2 that is in the raw water, so that when the water gets to the aerator there is for the most part no CO_2 left to be removed.

Iron.—Iron is considered undesirable in water for household use in that it stains linen and enamels. It also causes color in the water. Water of over 0.1 to 0.2 P.P.M. of iron is undesirable. For river waters the iron is usually much higher than that in well waters and therefore some means must be resorted to in order to get rid of these salts. The usual process is by aeration and coagulation.

Iron in this water is not considered to be high. For the last two years, the average iron content in water was well under the 2 P. P. M. line. There are several jumps on the curves; but these were due to the effects of heavy rains which carried off iron from the soil.

Chloride (*Chlorine*)—Chloride itself has no sanitary significance and is perfectly harmless. The things that go with it are much more important, for kitchen washings carry chloride, as do human and animal discharges and sewage. All these may act as disease carriers and offer an excellent opportunity for bacterial growth. If the normal chlorine of the region is known (I, II), any additional amount will measure the intensity of pollution, though it does not give the time of pollution. The present curve starts out with about 20 P.P.M at the beginning of April and then very slowly but steadily decreases to a few parts per million, through some seven months. Then it begins to rise gradually in November and at the end of December it gets up to about 12 P,P.M. Toward the end of January, 1934, it takes a jump to 150 P,P.M. when the curve shoots up at it has never done before in the course of a number of years.

The highest mark of the year was 1618 P.P.M. This can only be explained by the seasonal change plus that caused by heavy pollution. The water, during the months of February and March, and a great part of April was brackish. In other words, it actually tasted slightly salty. This is unheard of in our records, though there are reports that it happened once before some 6 or 7 years ago.

The present curve in other respects looks pretty much the same as the last one, (II) as was predicted a year ago (II). The only difference between them is that the present curve goes up very much higher than the last one. This is because the sea water, due to tidal movement penetrates far into the region of fresh water. When there is no rainfall to dilute the fresh water, the cumulative effect of sea water can be extreme. Last year, the weather was not so dry as this, and the maximum which took place in February, 1933, was only 156 P.P.M. as compared to 1618 P.P.M. this year. The present curve looks pretty much the same as the hardness curve. The same thing was true for the two curves, namely the chloride and hardness appeared in II. This clearly shows that, part of the chloride goes to make-up of the permanent hardness. Some of the hardness naturally will be $MgCl_2$, which, as it has been mentioned before, under a high pressure and temperature in the boiler will liberate free acid and thus causes corrosion.

The highest peak in the curve, 1618 P.P.M. suggests that some sort of heavy local pollution must have taken place as on the day before, February 15, chloride was recorded as 418 P.P.M. and the day after, February 17, 602. The average chloride for the first half of the month was 134.3 P.P.M. and for the second half was 436.5 P.P.M. Ordinarily an amount of 400 P.P.M. of NaCl will produce taste. Generally chloride thould not exceed 250 P.P.M. in water, for household use particularly.

Consumed Oxygen.—"Consumed oxygen" or oxygen consumed is a measure of the amount or oxygen required to oxidize the organic matter present in water. When a water does not contain unoxidized substances such as nitrites, sulfides, ferrous salts, the carbon of the organic matter takes up oxygen readily. Therefore a determination of this kind in an acid $KMnO_4$ solution for a standard length of time makes it possible to determine approximately the amount of organic matter. It is closely relacted to color. A rise in the color curve usually, if not always, gives a similar rise in the consumed oxygen curve.

For the greater part of the year, the consumed oxygen in this water is well below the 5 P.P.M. line, and this may therefore be considered as water of doubtful character or impure. For the past two or three years, consumed oxygen was about the same and it was generally true for the month of February or thereabouts that the value was somewhat higher than during the rest of the year. Dr. Smart classifies water containing 3.071 P.P.M. of consumed oxygen as of doubtful purity: and according to the French classification, a water having a value from 3 to 4 P.P.M. is suspected and above 4 P.P.M. is impure. However, deeply colored or peaty waters, otherwise pure, will give high results, due to the quantity of organic carbon present as color. Thus for interpretation of results, an inspection of the surroundings together with a measurement of color is necessary and valuable. This test, according to Mason and Buswell,[3] has its greatest usefulness in the study of heavily polluted samples whose source alone condemns them for drinking purpose. Surface water, which usually carries suspended organic matter, always shows a high oxygen consuming capacity.

Dissolved Oxygen.—This is a measure to determine the amount of oxygen dissolved in water. The presence of this gas is important for the existence of life in water and for the oxidation of organic matter. It is usually true that the higher the "oxygen dissolved" the faster the rate of oxidation, and the lower the oxygen dissolved the greater the amount of organic matter present. For further discussion of dissolved oxygen, one may refer to II' The curve of dissolved oxygen for this year is, on the average, about 5 P.P.M. or a little less, which is about the same as that a year ago, II. Oxygen, besides being derived from the air, is generated by the action of micro-organisms or aquatic plants containing chlorophyl. These plants consist principally of Hydrilleae, Vallisnereae and Spirogyra (茜) varieties. The Spirogyra, consisting of long green hair-like plants grows in cold weather and dies off slowly during the month of April. The other two tribes, green and grass-like, grow up to take the place of Spirogyra in April, reach their climax during warm weather and in winter resume the dormant stage and only to grow up again in April. The amount of gas generated by these plants depends largely on the abundance of the plants, the temperature and sunshine. It happens that the raw water reservoir of the University filtration plant is quite full of these plants for the greater part of the year. Cleaning of reservoir has to be resorted to to several times a year in order to get rid of the plants. In the course of analysis some interesting results bearing on this subject were obtained. The table below gives a few of the figures illustrating the effect of these

[3] Mason, Wm. P. and Buswell, A.M. *Examination of Water*. Sixth Ed., revised, 224 p. John Wiley & Sons. New York. 1931.

plants to the quality of the river water.

TABLE 1

Date	CO_2		Dissolved Oxygen				Temp.		Total Alkalinity			
			P. P. M.		% Saturation		C.		River water		reservoir	
	1[1]	2	1	2	1	2	1	2	HCO_3	CO_3	HCO_3	CO_3
1933												
June 30	5	0	4.4	5.9	39.6	53.1	27	27	42.5	—	30	5
July 27	7	0	3.0	6.0	27.9	55.9	29	29	52.5	—	45	2.5
July 29	8	0	2.7	6.5	25.1	60.6	29	29	45	—	37.5	2.5
Aug. 12	10	0	2.6	6.4	24.7	60.9	30	30	42.5	—	50	—
Sept. 23	8 5	0	.24	5.6	39.1	52.2	29	29	52.5	—	57.5	—
Dec. 3	5	0	3.6	7.2	27.9	55.9	19	19	55	—	50	—
1934												
Jan. 15	4.5	0	3.5	6.7	24.5	46.9	14	14	55	—	55	—
Feb. 8	6	2	6.1	6 7	41.8	44.9	13	12	55	—	60	—
Mar. 21	10	4	4.2	6.1	30.6	44.5	16	16	55	—	55	—

[1]Note:— 1 = River water; 2 = Reservoir

The free CO_2 gas, in most cases, was entirely taken up by the plants and in several cases part of the bicarbonate hardness of the river water was destroyed and transformed into carbonate with the loss of CO_2 gas. The dissolved oxygen increased by amounts varying from a few percent (Feb. 8) to as high as 146% (Aug. 12).

A careful study of the curve indicates that for the greater part of the year the dissolved oxygen value was below 5 P.P.M. and only for January and the first half of February (1934) was the curve running above the 5 P.P.M. line. The same change took place in 1933. On the whole the average for all these years on dissolved oxygen was in the neighborhood of 5 P.P.M.

SUMMARY OF RESULTS

Although this is only the third paper of the series, yet certain definite knowledge of the characteristics of the river water has already been established. To summarize what we have learned thus far, I shall attempt to outline the results obtained as follows:

1. When a comparison of curves is made, the close resemblance in every case of the chloride curve to the hardness curve will first strike the attention. They appear about the same and reach their climax at exactly the same period. The hardness in the Pearl River water, with the exception of February, March and April, is almost zero and in most cases does not exceed 5 P.P.M. The unusually high rise in the curve is entirely due to the lack of rainfall. The chloride curve, which usually runs a few parts per million, suddenly shot up in February to such a high value that the water actually tasted brackish. The curves look alike but differ in magnitude. On the whole chloride and hardness are unusually low for the water with the exception of the unusual jump in February, March and part of April.

2. Turbidity is considered to be low for the whole period and is generally higher in April or May, when most of the rainfall occurs, than the rest of the year.

3. Color is not considered to be high and is usually higher for the first part of the curve than the second. This is entirely in agreement with the distribution of rainfall during the year.

4. Odors have never been a problem here. They are mostly earthy and can be got rid of through the processes of aeration, coagulation and filtration.

5. pH value of the river water usually stays very close to 7, which is neutral. It shows an opposite relation to CO_2.

6. Total alkalinity for the first half of a courve, ending in August or September, is usually lower, below 50 P.P.M., than for the period covered is roughly 50 P.P.M. The unusual rise in hardness, during

February, March and April, does not affect very much the shape of the alkalinity curve. It has an opposite relation to turbidity.

7. A rise in free CO_2 usually accompanies a fall in pH. High turbidity often gives rise to high CO_2. The average CO_2 content of the water was in the neighborhood of from 5 to 10 P.P.M. for the period covered. The aquatic plants in the raw water reservior are able to take up the gas present in the river water.

8. The average of iron in the water is usually about 2 P.P.M. which is not considered to be high. It is usually higher during the first half of the curve than the second half.

9. Consumed oxygen is pretty much the same for the last three years. It is closely related to color. The average is well below 5 P.P.M. for all these years.

10. The general average for dissolved oxygen is about 5 P.P.M. The curves reach their maxima on or about the 5 P.P.M. line in January and the first half of February.

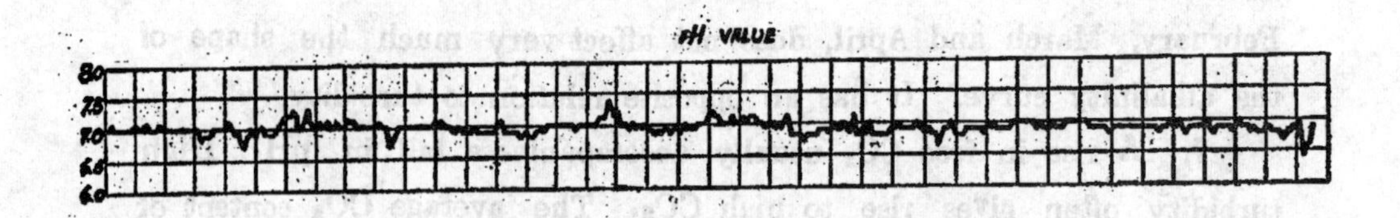
pH VALUE
8.0
7.5
7.0
6.5
6.0

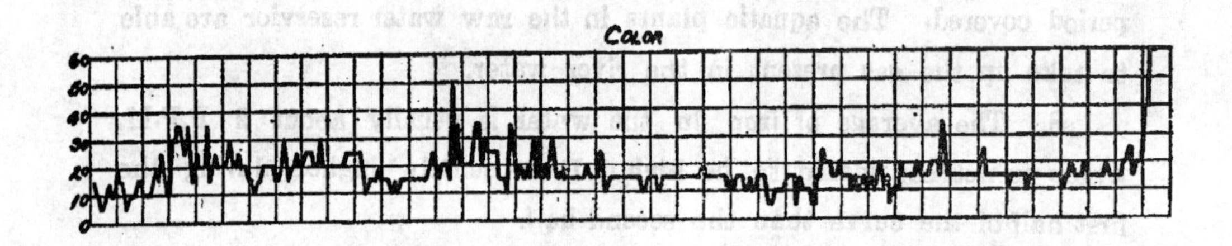
COLOR
60
50
40
30
20
10
0

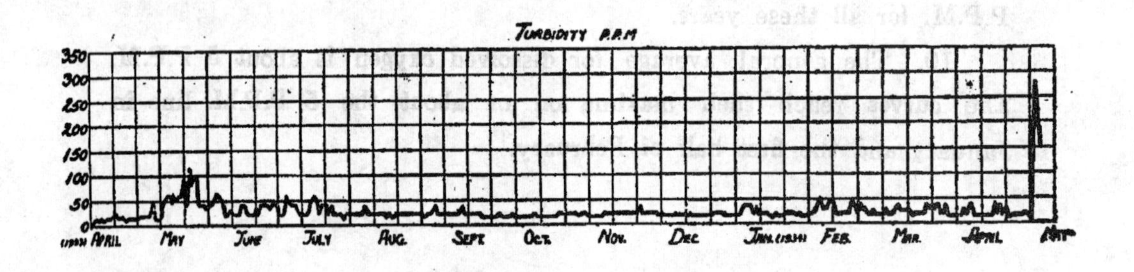
TURBIDITY P.P.M
350
300
250
200
150
100
50
0
APRIL
MAY
JUNE
JULY
AUG.
SEPT
OCT.
NOV.
DEC.
JAN.
FEB.
MAR.
APRIL
MAY

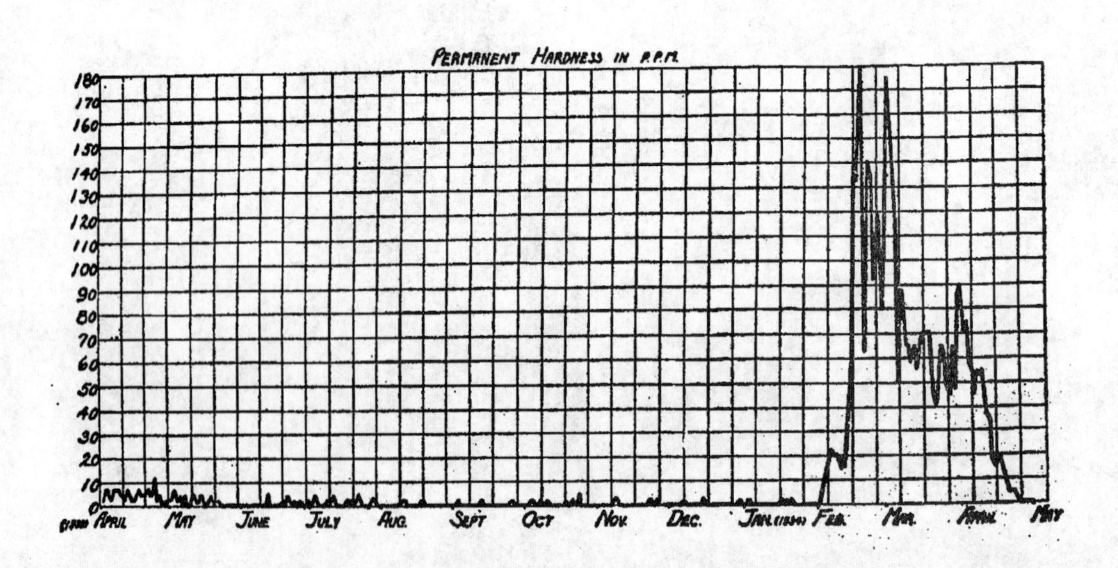
PERMANENT HARDNESS IN P.P.M.
180
170
160
150
140
130
120
110
100
90
80
70
60
50
40
30
20
10
0
APRIL
MAY
JUNE
JULY
AUG.
SEPT
OCT
NOV.
DEC.
JAN.
FEB.
MAR.
APRIL
MAY

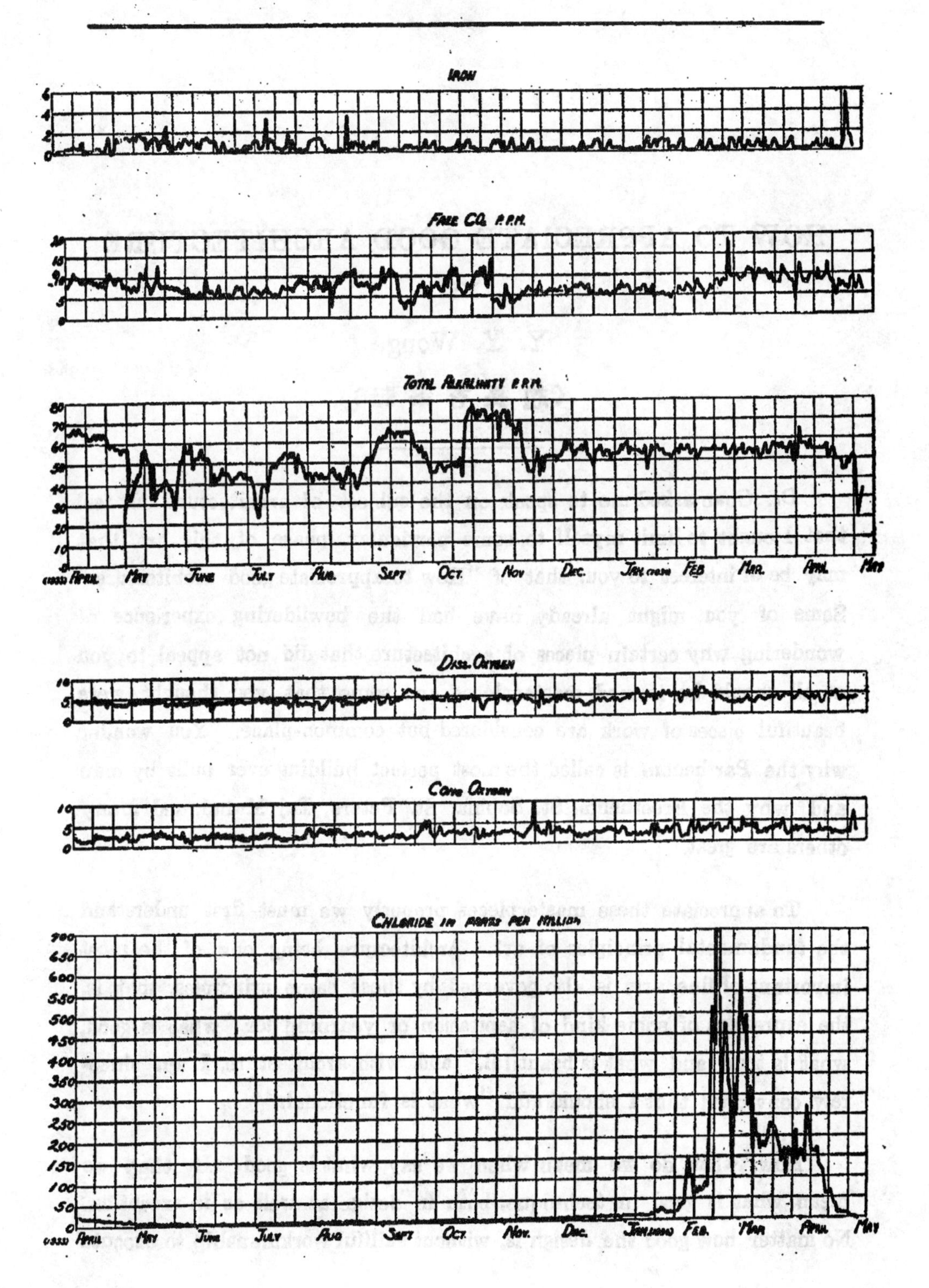
IRON
FREE CO2 P.P.M.
TOTAL ALKALINITY P.P.M.
(1933) APRIL
MAY
JUNE
JULY
AUG.
SEPT
OCT.
NOV.
DEC.
JAN.
FEB
MAR.
APRIL
MAY
DISS. OXYGEN
CONS. OXYGEN
CHLORIDE IN PARTS PER MILLION

HOW TO APPRECIATE GOOD ARCHITECTURE

BY

Y. Y. Wong.

(趙長有筆記)

Dr. Chan asked me to speak on the subject of architecture. I feel that I ought to limit myself to some particular phase of this art, that may be of interest to you, that of "How to appreciate good architecture." Some of you might already have had the bewildering experience of wondering why certain pieces of architecture that did not appeal to you at all should be termed masterpieces and some that you thought were beautiful pieces of work are considered but common-place. You wonder why the Parthenon, is called the most perfect building ever built by man and why the Erectheum, St. Sophia, St. Peters, Taj Mahal, and many others are great.

To appreciate these masterpieces properly we must first understand the fundamental principles of art. Architecture ,being one of the most important of fine arts is also governed by these same principles, that is, the expression of some kind of aspiration or yearning for "what is good, what is true, and what is beautiful," and with architecture, I am sincerely convinced that I should add "what is functional."

Now, what do we mean when we say what is good. By that, we mean what is good in technique, both in design as well as in execution. No matter how good the design is, without skillful workmanship to execute it, it is liable to lose the very expression sought for. In this respect the architect has to depend on others who are not wholly conversant with his thoughts, to execute his work, whereas in the other branches of fine art the artist usually completes his work, by himself. Thus in the very beginning we are limited in what is good only to the extent of what skillful workmen are available, furthermore we are limited by principles of business. When a contract is awarded to a builder. Whether in the Hellenic period, the so-called Golden Age, or in the Renaissance, (the Silver Age of art,) Architects and sculptors usually execute their work on the premises that is, they would have trial pieces put up and change until they are perfectly satisfied with what is there. I believe that most of you can tell good workmanship from bad but to understand whether the execution of the work imparts the feeling intended in the design one would have to understand some theories of design.

To design is to compose the elements together in a natural harmonious order. I have often compared a good design to that of a good government. Whether in plan, or in elevation there must be a dorminating motive, a climax in the design, dominating in interest, important sometimes in mass but unquestionably dominating.

There are of course other motives in the design and they must be subdued, to help and enhance the main motives in the design but never conflicts with it. Now ladies and gentlemen, how true is it with a state? If we should have every district fighting for dominating we would have chaos throughout the state. Sometimes we meet with problems that present two or more motives of equal importance and interest. In that case we would have to tie them up into a harmonious union. In any case if one seeks to express the interest and importance of each motive in its right order one would have a natural good design.

Good Architecture must have good character. Character may be derived from many ways. It may express the feelings, usages, methods of construction, etc. For instance, a church must be religisus in character capable of inspiring what the church tries to teach us. When one stands under the dome of St. Peters one feels that one is really under the very dome of Heaven, and that one is holding communion with God. or what the average man is accustomed to in his active life. Thus stair treads and risers, balconys, stone courses, etc. become the unit of measure in a design. All motives must be consistent in scale. St. Peters Cathedral for instant its original facades had no scale or really out of scale with the steps especially the huge columns and entablature until Bernine later added his famous fore-court colonade which serves as a medium to step down the scale of the facade to a human scale.

Good proportion requires a lot of study in each individual design. However, there are certain geometric proportions which are excellent and complete such as the square, or circle. The two square rectangle or the root-2 or root-3 rectangle are always good proportions by themselves. The Renaissance masters always seem to have an instinct for good proportions and anything they touched were good, thus they were able to take the Classic Orders and break them at will and pile it up it would have good proprotions and good composition. But the cleverness of these masters led the Roccoco architects to a period of folly. The Roccoco style was an imitation of the Renaissance in the breaking up of the Classic Orders but they were not able to pile it back and still retain the good proportions and good composition of the Renaissance masters.

Consequently the Post-Renaissance masters such as Vignola. Palladin wisely made a revival of the Classics and made it a point that the proportions of these orders be the fundamentals of every student of architecture.

St. Sophia, and the Cathedral of Florence are wonderful examples of truthful expressions of their construction. Truthfulness in expression of usage, of the status of the occupants and of his whims and fancy ought to be expressed.

Now with a bank, we ought to express security. a business-like dignity, and sturdiness. With a hospital we express restfulness, sanitation, and a lightness of heart. This had been a very perplexing problem with the skyscraper hospitals in New York. I remember when we were working on the design for the Cornell Extension Hospital in that city and the problem of how to get restfulness in the design came up. A skyscraper is essentially verticle in direction when whereas a man is sick he wishes to rest and nature has acustomed us to a horizontal position in resting. Big business houses on the contrary is ambitious and we are acustomed to look upward towards the sky when we are ambitious. "Hitch your wagon to the star" as the old saying goes. Thus verticality is expressive of ambitious activity.

Most public buildings require a formality in character whereas private buildings is much less rigid in that respect. Symmetry, balance, and reserveness, dignity, usually go to make up formality. Of course there are different degrees ranging from severity of a court-house to the grand foyer of a theatre. Even in your home your different rooms have a different degree of formality—hall—living room—study—dining room etc.

Gaiety and sadness is also expressed in architecture—Study of the human face in action is a good example. Sex can be expressed even. If we take the Taj Mahal for example the whole structure has a feminine character, but go and visit a fort or fortress it certainly is forbidding enough even for a man not to say of a woman. Within recent years we have developed a modern character. What do we mean when we say modern? We mean that concentration of business in a small area forces us up to build skyward and that our working processes are done by machinery etc. straight lines flat surfaces, uses of metals. Speed is a requisite of modern practice and for the lack of better means of expressing speed we just design something that would be of least resistant to wind, the opponent of speed. If we take a fast moving dot it would appear to the eye to blurr into a line, thus the stream-line effect is developed. People hurry so much now that even soda fountains have adopted this streamlining so that they may accommodate their customers in the right spirit.

So you can see why character is so important in architecture.

Of the more difficult techniques to understand are scales and proportion. By scale we mean a standard of measure based on the height of the average man.

What is true in architecture is more easily understood, truthfulness in expression is essential to good architecture. This has been more flagrantly violated than any of the others. Many excuse themselves by saying that a white lie does no harm. I have seen stone lintels to span 20'-0" yet I know stone cannot span that length, methods of construction ought to be truthfully expressed.

What is beautiful is often misinterpreted by the layman. Beauty never become great without the good and the true. That is the reason why an artist of a school of artists if inclined to only one phase can never become great. They are all inter-related.

When we judge beauty we should begin from the structure under the surface. When my critic friend from Shanghai asked me why I didn't disguise the chimney for beauty's sake I felt that it would be a worse folly to do so. In fact I felt so strongly against that kind of practice that I dare compare the structure of our noses on our faces. No doubt if nature had that kind of a sense of beauty she would make our faces all a smooth surface without the protrusion of a nose, the indentation of the eyes or the side hanging of the ears. It is natural and functional where the nose is instead of conducting it to the top of the head. I really think that it is due to these kinds of thories that complicates our lives instead of the simple and direct methods of dealing with the problems.

Thus firstly, beauty must be beautiful in structure and then beautiful in shape and proportions and lastly that of surface ornamentations. We learn what is beautiful from nature and we draw inspirations constantly from the creations of God. He is the greateast artist of all times.

We learn that too much surface embellishment for instant often ruin an otherwise good design, thus one should study how to omit superfluous ornamentations. Use ornamentation only where it is needed and they must be in character. If one has a beautiful face I do not see any advantage to cover that face with cream and powder.

Lastly, I added, what is functional to the requisites of good architecture. I feel justified, at the least in the art of architecture. If a structure is without a function it is useless and have no right to be a structure. Even in pictorial art if a picture has no function at all it is meaningless. That is why I pointed out to students about the use of orders in a hearse it has no function and incidentally that is the only kind of a moter car that employs the orders. The hearse of course leads to the grave.

THE ENGINEER OF THE FUTURE

THE RELATION OF THE STUDENT ENGINEER TO THE STANDING OF THE PROFESSION.

BY

(W. Y. Lam, C. E.)

From time immemorial until our present era, greater and still greater progress has been made in all branches of engineering, until to-day that science is the governing necessity of the age. The complex community life of to-day requires, in ever increasing measure, the services of that small group of men which, through the laws of natural science, bends the forces of nature to the will and service of man.

The civil engineer furnishes mankind with water, highways, bridges, tunnels, sewers, etc.; the mechanical engineer constructs and perfects the countless machines providing heat, means of transportation, the manufacture of foodstuffs and materials, etc.; the electrical engineer gives us light, power, the telephone, radio, etc.; the chemical engineer is instrumental in the manufacture of cement, the various steels, the various by-products of coal tar, etc.; and the mining engineer tears from nature the secrets of the world's formations and enriches the universe with nature's mineral wealth. All these make possible the existence of the community life of to-day.

Strange as it may seem, however, the lay public of our country has had, until very recently, only the slightest conception of the term engineer, while on the other hand law and medicine have been accredited the respect and remuneration which should be accorded a profession. A new bridge, a new highway, all were accepted by the public as a matter of course and the political power of whatever faith, under whose administration the improvements were built, was given the credit for their construction. The chief or consulting engineers and assistants on these projects receives but casual if any mention by the press. Here the engineer himself is to blame for this condition of affairs. Of a mathematical and analytical mind he has been ultra-conservative and retiring. Unlike the lawyer and doctor, he has not let his voice be heard in public affairs,—he has been willing to contribute his all to humanity but has allowed his identity to be lost.

But a new era is dawning for the engineering profession. A number of leaders are making their influence felt in civic life; a number of strong engineering societies through the press and elsewhere are educating slowly but surely the public; and last but not least the engineer himself is awakening to take his place in the sun.

It is nevertheless, the university itself that must assist the embryo engineer in making the influence of his profession felt in the communities, into which, upon graduation it sends him. It should give intensive training in public speaking and should encourage debates upon engineering subjects, it should quicken the imagination as well as the reasoning powers of its engineering students, and it should visualize to them the coming engineering age and what will be required of the graduate.

Thus with the foundations of science thoroughly implanted in his mind and with the ability to express clearly and concisely his thoughts

upon any subject connected with his work, the engineer of the future should command the armies of progress rather than content himself with being an able private in the ranks.

A Brief Survey of the Development of Railway Engineering.

(By Chan Seong Boo.)

Development of the Railway.

It would be hard indeed to say definitely when and where originated the first railway, neither in the modern sense nor in its literal sense. But as long as History recorded, we know that in the days of the Romans, there were the Roman "Stone Roads," which consisted of parallel lines of stone blocks each 2 ft to 6 ft in length, laid 4 ft apart.

These Roman "Stone Blocks" were merely a simple expedient to diminish friction and were the first steps toward a railroad. The railway doubtless originated in the coal districts of the North of England and Wales, where the coal was brought to the boats in wagons, which were hauled along stone laiden roads. Still farther to facilitate the haulage of the wagons, pieces of planking were laid parallel upon wooden sleepers, or imbedded in the oridinary tracks. The use of these wooden rails gradually extended. The wooden rails were farther improved by

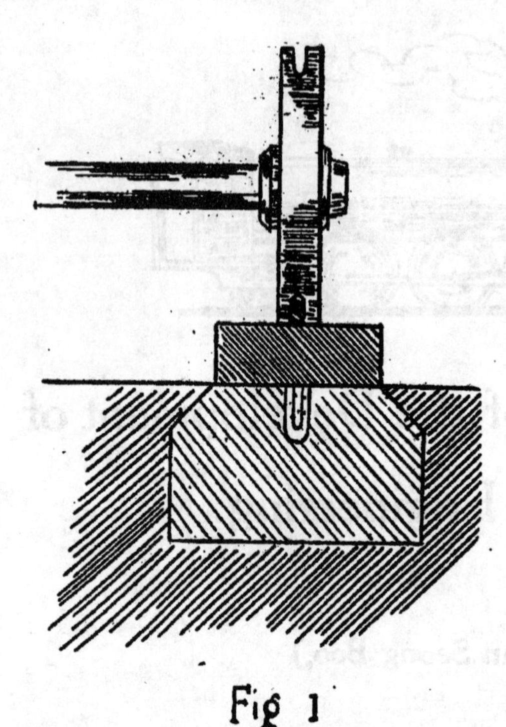

Fig 1

having a rounded surface, like a projecting moulding, and the wagon wheels being of cast iron and hollowed in the manner of a metal pulley readily fitted the rounded surface of the rails (see Fig. 1).

In these rude wooden tracks we find the germ of the modern railroad. Later thin plates of iron were nailed upon their upper surface for

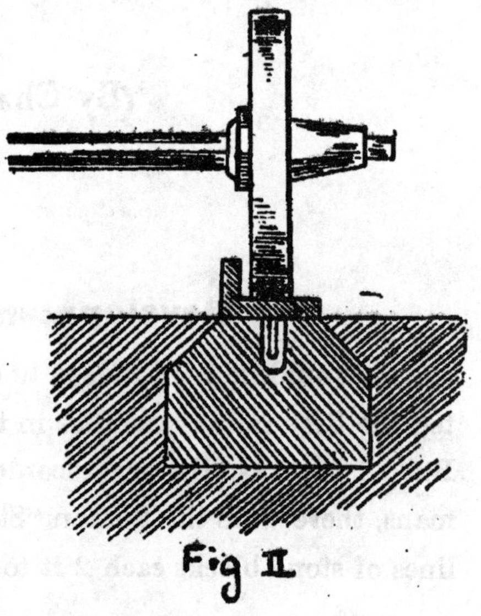

Fig II

the puroose of protecting the parts most exposed to friction. A third stage towards the development of the railroad was the introduction of cast-iron rails about the year 1738. This cast-iron road was denominated a "plate-way", which had a ledge cast on their outer edge to guide the wheels along the road, after the manner shown in the figure. (see Fig. 2)

In 1789, Mr William Iessope introduced the cast-iron edge-rail with flanches cast upon the tire of the wagon wheels to keep them on the track, instead of having the margin or flanch cast upon the rail itself, (see Fig. 3). Later on there were many improvements on the railroad, among which was the famous Half-lap Joint, invented by George Stephenson.

First ExPedient towards the Railway Engine.

With the gradual development of the railroad, men began to think of inventing a machine, which would substitute the horse in hauling the heavy loaded wagons. This would save both time and money. Many expedients were suggested with the view of getting rid of the horse. The power of wind was one of the first expedients proposed. It impelled ships by sea; why should it not be used to impel carriages by land? The first man to take up this idea was Simon Stevins, who invented the first sailing-coach, but after long attempts his results were fruitless. As a practical machine it proved worthless, for the wind could not be depended upon for land locomotion. The coach could not tack as the ship did. Sometimes the wind did not blow at all, while at other times it blew a hurricane. The sailing coach merely remained a toy and a curiosity.

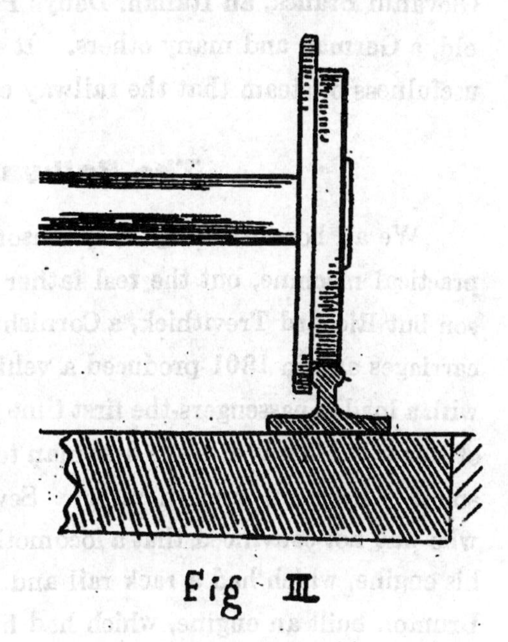

Fig III

Development of the Railway Engine.

The railway engine has a very close relation to the steam engine. The steam engine was the invention of no one man. It was the result of the discoveries and inventions of many centuries, brought to a head by James Watt, who built his first steam engine in 1768. But as far back as 200 B. C. Hero, a mathematician of Alexandria is believed to have made a kind of steam engine. It consisted of a globe with outlet pipes working in a frame over a boiler. When a fire was lighted under the boiler, steam entered the globe through one of the pivots supporting it, filled the globe and as it rushed out of the outlet pipes, rotated the vessel by pressure against the air. Among the many early contributors to the steam engine we have the Marquis of Worcester, an English man, Giovanni Branca, an Italian, Denys Papins, a French man, Jacob Leupold, a German and many others. It was only after the discovery of the usefulness of steam that the railway engine was able to be made.

The Railway Engine.

We all honour George Stephenson for making the railway engine a practical machine, but the real father of the locomotive was not Stephenson but Richard Trevithick, a Cornishman. He made many model steam carriages and in 1801 produced a vehicle which went a trip on the road with a load of passengers-the first time people had ever travelled by means of steam. He was also the first man to make the railway engine run with smoth wheels on smooth rails. Seven years later John Blenkinsop, who was not convinced that a locomotive could run on smooth rails, built his engine, which had a rack rail and a cogwheel driving gear. In 1813 Brunton built an engine, which had hind legs, making it in fect a steam horse. Believing also that no engine with smooth wheels could travel on smooth rails he had his engine pushed by the curious legs, worked by the steam cylinders. The arrangement, however, was a failure. In the same year William Hedley built a locomotive, which had a great puffing noise, and was therefore called "Puffing Billy", but nevertheless it did good work.

Although we must honour Trevithick as the pioneer of the locomotive, and Hedley for proving its commercial practicability, it is to George stephenson more than to any other man that we owe the modern railway engine. Stephenson's first machine was built in 1814 and was driven by means of gears worked by the pistons. Stephenson's contributions to locomotive engineering were many. In addition to establishing direct communication between the cylinders and the wheels, he obtained adhesion of all the wheels by steam blast to increase the combustion.

Present Development of Railway Engineering.

The very early experiments in locomotives had boilers, which were simply vessels of water heated by a furnace in the same way as a kettle is boiled over a fire. But in the present system the fire and hot gases from the furnace are carried through tubes so as to provide a great heating surface and generally the steam is superheated.

Besides this improvement we have the vacuum brake and the wonderful Westinghouse brake. The vacuum brake is operated by allowing air to pass into a vacuum cylinder and to push up the piston which works the cranks that apply the brake blocks to the wheels. The Westinghouse brake, which is very complicated, is the most efficient of all kinds of brakes, and it works automatically. In the olden days, the trains had to waste much time in stopping to take up water for the boilers, but now the water is taken up at speed, without causing the train to stop. This is done by having a trough full of water between the rails for a considerable distance. When the engine approaches the trough, the engine man lowers the scoop, which then dips into the water in the trough, and the water is thus driven on to the tanks by the inertia of the water itself.

Besides this development we have another advance and that is the stream lined railway engine, which is coming into permanent use both in Europe and in America.

New Branches of Railway Engineering.

The utility of the railroad is not limited to level grounds only, but is extended to hill slopes and other high levels. Both in Hong Kong and in Penang we have the hill railways, in which the train is being drawn up the hill slope more than a thousand feet, by mans of strong cables, operted by machines.

We have at present the recent invention of the railplane, which instead of running on the ground is suspended high up in the air, by means of Bridge-like bogies. The railplane looks like a gigantic torpedo, and is driven at the front and back by air screws or propellors, like that of the aeroplane. It looks more like an air ship, but instead of flying in the air it is suspended above the ground on rails. It is going to be the most useful kind of conveyance in the future, especially over uneven grounds, ravines and deep valleys.

Another new branch of railway engineering is the wonderful Robot railway, utilized in England by the Post Office under the London Streets. This consists of a miniature underground railway six and a half miles long, and it is the only kind in the world. The railway links the important post offices with the main office, and the trains that carry the mails travel without guards or drivers. The whole system is a remarkable example of up to date efficiency. One tunnel is for trains going in one direction, and the other for trains going in the opposite direction. There are two sets of rails in each tunnel so that while some trains may stop at one station others may go through without stopping. The line is entirely operated by electricity on a third rail system. The driverless electric engines are automatically controlled, and they draw the trucks at an average speed of 35 miles per hour. The control is exercised by switchmen at a number of control cabins along the route.

The above described railplane has its advantage over very uneven grounds and deep valleys, while the underground Robot railway has its advantage in utilizing the useless space under the busy streets, and thus solving a problem for over packed communication. Both the railplane and the underground railway system are recent contributions to the development of railway engineering and they will be among the most important engineering works in the near future.

DOMESTIC ENGINEERS. LTD.

興利有限公司

本行承接安設及修理凍房雪櫃各種機器工程專做家庭及商店所用電力雪櫃常備各種機件以應修理各式機器

Everything in the Electrical Automatic Refrigeraton and Air Conditioning

辦事處
德輔道中八號二樓
電話 二三五零四

工場
馬頭圍太子道三百二十七號
電話 五九二八一

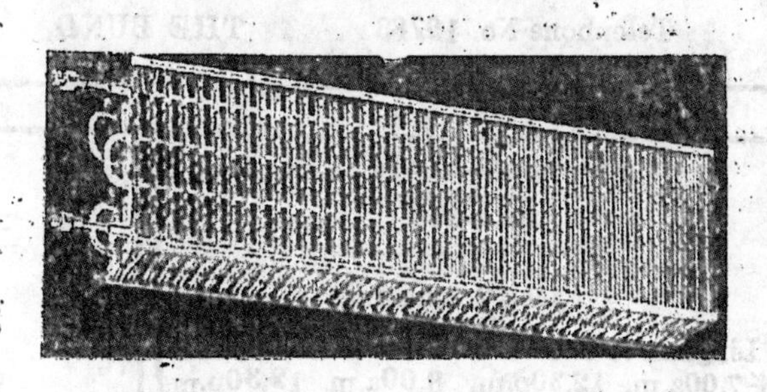

Stock of Parts and Refrigerants on Hand.

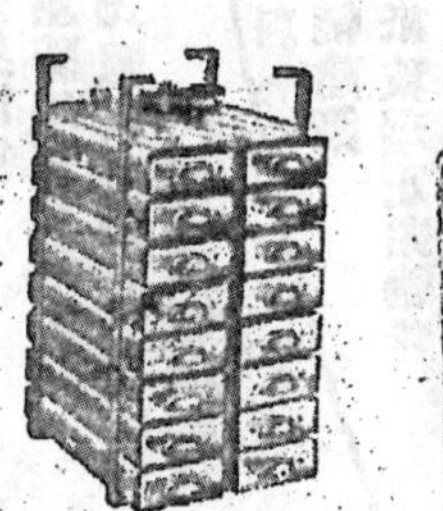

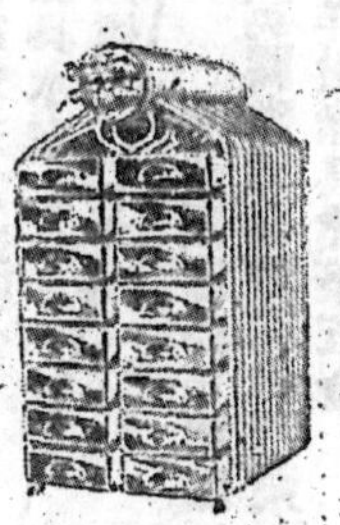

OFFICE:
8 Des Voeux Road Central
1st Floor
Telephone No. 23504

Works:
327 Prince Edward Road
Ma-TauWei, Kowloon
Telephone No. 59281

廣州市長途汽車改善方案

梁卓芹

(壹)緒論

城市日漸發達，人口日漸增加，市區範圍亦隨之擴充；故都市之交通問題亦隨之而發生；且城市居民各執其業，有遠居郊外而執業市中者，有因營業上關係而來往東西或南北者，其對於時間上交通上均爲重要；故一城市之發展，實需迅速便利之交通，而欲再繁榮該城市，則更便利迅速之交通爲最要，本市自通商後商業日漸繁盛，及開闢馬路後，交通則有汽車及手車，比前頗稱利便，各國市城中，手車因遲緩，不能滿足便利迅速之條件，皆已成落伍物，而以電車，及汽車代之，往昔因汽車車價及汽油價格昂貴，故多以電車爲主，迫近汽車事業發達，汽油亦較前爲廉，故多採用公共長途汽車，蓋以其行駛迅速，不藉軌道，任便縱橫全市，即使半途偶有損壞，亦不致如電車之梗塞全路，且電車之車價，及各種建築費，均爲數甚鉅，故城市交通之新設計多捨電車而取公共長途汽車，就本市而論，當亦以採用公共長途汽車較電車爲佳。

本市之長途汽車，始創於民國十年之加拿大公司，及民國十九年已漸發達，惟以路線縱橫錯亂，配置不宜，且多注重繁盛地帶，忽畧待時發展之市區，致形成畸形之城市發展，殊爲憾事。

(二)現有路線及交通現象之批評

(A)路　線

長途汽車之目的，固在便利交通，利便市民來往，及發展市區，依交通原則上，各路線須由城市中心點散佈四郊，換言之亦即將四郊市民帶到市之中心點，然考本市現有路線，均來往於已繁盛之區域，而于急待發展之近郊如河南，白雲路，越秀路，龍津路，及西北區一帶之盤福路，德宜路等，仍付厥如。此實有失交通原意，且現有之路線，皆密集繁盛街道，致有兩路以上長途車幾經同一路線；如倉邊路至黃沙與廣九至黃沙二線幾有四分三之路線相同，交通公司與明華公司之路線，雖異而同，試展本市地圖觀之，則知長途汽車路線皆密集於長堤，一德路，永漢路，惠愛路，十八甫，上下九路一帶，致該處繁盛之路益爲擠擁，致影响各長途車公車營業；試觀不甚繁盛區域則市民幾無車可搭，此不特對於整個城市之發展有碍，交通不能十分暢便，且對於一部份居民亦甚不公平。

(B)總　站

凡屬城市之中心點，大抵商業繁盛，行人擠擁車輛繁多，絕不適宜於設立長途車站，如本市之永漢北路及惠愛路之十字路口，長途汽車起止站多至四處， 全市交通以此處最繁集， 此處車輛來往本已繁多，今再加之各長途車之起止站， 致交通淩亂不堪， 指揮者固屬困難，駛車者，亦不易易，且危險事件之發生，在在堪虞，故各長途車站應移置別處。

(三)改善計劃

本市地勢南隔珠江，北阻粵秀山，雖有珠江鐵橋可貫通南北，惟河南多未築馬路，交通究屬不便，故本市將來之發展，必向東西伸張無疑，欲改善本市長途車各弊，使其功能盡量發展，非依本市未來形勢從新規劃不可，故長途車路線應多以東西爲主：所有總站應分布東西近郊之公路車站，火車站，及住宅區，而路線引入城市之中心，祗經過城市之中心點而不以之爲起止點。

(A)改善票價

本市各路線皆因收費關係不能延長，致多有連接城市之兩中心點，現在票價，雖對於購票及售票者均屬便利，惟搭客如在中途或三分一之路程，搭車則吃虧甚大，且對於半途轉乘別車，均不經濟，爲矯此二弊，須改善票價，將某路分爲若干站，每站票價半毫，此則半途上落者均感經濟，或以影嚮長途汽車公司之營業，此固不難解釋者，蓋此法行後，半途下車者雖多，而半途上車人數亦較前驟增，如此則對於交通經濟雙方均感利便，卽長途車營業方面亦未有若何影响也。

(B)新路線系統

現在之路線多因票價之限制皆止於城市之中心附近不能充份發展延長，致脈絡不能貫澈，今票價改善則雖路線延長，而商方營業亦不致吃虧，玆擬定八大路線及該路線之票價分站如後：

第一路——東山公園至荔枝灣

由東山公園起，經百子路，惠愛路，轉永漢路，經天字碼頭，長堤，西濠口，太平南路，十三行，十七甫，十八甫，第十甫，十五甫多寶路，至荔枝灣。

本市中心點商業區則以永漢路及西濠口長堤一帶爲最盛，此路線連貫二中心點而至東山及荔枝灣，所有東西兩住宅區居民，每日至市中心區內工作及購物者，均可便利，依現有路線，則欲由荔枝灣至西濠口，長堤，極費週折，此路線卽能改善此點。此線可由東山至大東門爲一站，大東門至財廳前爲一站，財廳前至財政局一站，財政局至西濠口爲一站，西濠口至十八甫爲一站，十八甫至荔枝灣一站，共分六站，每站票價半角，全線票價三角，比之現有路線票價並無差別，欲由東山至豐寧路，長壽路，寶華市者，可在大東門轉搭第二路線之車，財廳前站可與第二第三第四第五第六連接，財政局前一站亦與第七路第八路連接，西濠口站可與第四路連接。

第二路——荔枝灣至小北

由荔枝灣起，經多寶路，寶華市，長壽路，豐寧路，惠愛路，經財廳前，大東門，越秀北路，以至小北。

此線將西區市民與本市中心溝通并將市西市內學生直達教育區之大東門及小北。

此線分三站，由荔枝灣至豐寧路一站，豐寧路至財廳前一站財廳前至小北一站，每站票價亦爲半毫。

第三路——黃沙站至廣九站

由黃沙站起，經蓬萊路，恩寧路，第十甫，上下九甫，豐寧路，惠福路，文明路，小東門，越秀南路，至廣九站。

此路線可連接粵漢及廣九兩鐵路總站并將兩鐵路乘客引至本市中

心區。

此線分爲四站，黄沙至第十甫一站，第十甫至惠福西路一站，惠福路至永漢路一站，永漢路至廣九一站，每站票價同上。

第四路——黄沙至小北

由黄沙沿六二三路，經西濠口，入太平南路，經一德路，泰康路，永漢北路，財廳前，惠愛路，倉邊路至小北。

此路線亦是接駁粤漢鐵路搭客入市區，更將小北區市民引入市內。

此路線分四站，由黄沙至西濠口一站，西濠口至泰康路口一站，泰康路至財廳前一站，財廳前至小北一站。

第五路——黄沙至沙河

由黄沙經蓬萊路，龍津路，光復北路，長庚路，盤福路，德宣路，吉祥路，越華路，廣仁路，財廳前，惠愛路直至沙河。

此路線將粤漢鐵路之黄沙站搭客及附近市民引至市內後將沙河之公路車站乘客引至市內。

由黄沙至西門一站，西門至財廳前一站，財廳前至大東門一站，大東門至黄花崗一站，黄花崗至沙河一站，共分五站。

第六路——東山至黄沙

由東山大街，經東華東路，東川路，白雲路，廣九站，越秀南路，小東門，萬福路，維新路，大新路，太平南，拱日路，叢桂路，以達黄沙。

此線連接東山廣九站及黄沙站至市內。

由東山至廣九站一站，廣九站至永漢路一站，永漢路至太平南路一站，太平南至拱日中一站，拱日中至黄沙一站，共分四站。

第七路——鳳凰崗至越秀山

由鳳凰崗，經洪德路，南華路，過海珠橋，維新路，惠愛路，廣大路，越華路，吉祥路，蓮塘路，至越秀山麓。

此路線連河南河北兩岸溝通，亦可將河南及河北市民引入市內。

此路線分二站鳳凰崗至河南橋脚一站，再至越秀山一站。

第八路——小港至西村

由小港橋起，經基立村，南華東路，海珠橋，維新路，中央公園前，惠愛路，長庚路，西華路，至西村。

此路斜過全市，東北及西南區，亦爲河南河北之主要交通線，並爲發展河南近郊，及西村近郊，不可少之路線。

由小港至河南橋脚一站，河南橋脚至公園前一站，公園前至西華路口一站，再至西村一站，共分四站。

結論

依新擬路線交通則四處便利，近郊發展亦可期而待也。蓋路線皆散佈全市而皆經本市之中心，不致如前擠擁一處之弊，而交通管理亦較前易，危險自可減少，且票價改善，各路分站互相連接，搭客可任意轉車數次以達其目的地，需費亦不多，尤較以前便利，是則全市路線均易收聯絡之效矣。

——二十三，十一，十一——

25334

廣州市土杉雜木材之檢討

彭龔原

木材乃工程上最重要之材料，爲吾人所共識。無論宮室，鐵路，橋樑，船艇，燃料，傢私， 以及其他種種工程建築， 非木材不能成功，非木材不足勝其任。

本市現有木材，過半來自外地，如法屬安南山打根等，所謂洋雜木是也，每年金錢外溢之巨，令人聞之咋舌。在提倡國貨，復興農村之今日，一致採用土木材之限制，實有刻不容緩之勢。惟是環顧土產木材，多數木身細小，未能應大任，在工程材料上，素未爲人所重視；此種情形，以土雜木爲尤甚，其原因實有二。

（1）本地植林者，皆爲私人資本，本小力薄，急于謀取蠅頭之利，以爲糊口，當樹稍長，卽伐之沽之，以致因小失大，不能培植巨大之木材，而應需求，古人所謂斧斤不以時入山林也。

（2）各地交通不便，木材之運至本市者，祗賴東西北三江水道，惟是河身狹窄，河水低淺，巨大之木，不能航行，故遇巨大者，亦必在當地開鎅後，方能運到，其最小者，如靑化松木，因航行時不能轉彎之故， 逼將木頭削去一邊， 又如各種稍大圓周之木，必須對邊開鎅，或截成三數尺之圓柱，然後方能運來。其輸運之困難，于此可以概見。

此等情弊，實爲土杉雜木在市塲之致命傷，非謀澈底解決不可。至于救濟方法，第一步必先改良各地交通。如河道之整理，及鐵路之建築等。使輸運利便，運費輕微。第二步當集合當地農民，組織偉大之森林公司。一方面可以救濟農村，一方面可以採用有組織之方法，以培植所需木材。第三步則政府須從速設立木材研究所，任用專門人材，或委托當地對于材料試驗素有成績之大學，將各種木材，作澈底之試驗，以決定某種木材，應爲某項工程之用，及決定各木材在工程計算上，所需要之數目。俾採用者，知所問津，而農民亦知所從事培植。則本地木材，當有頁好之發展矣。

玆將各種土杉雜木，分述于後。

(甲) 土杉木

杉木價平而產額大宗。木質乾潔幼滑，不易腐壞。爲目前工程上重要木材。長由數尺至二大許。圓周由數寸至數尺不等。屋宇之柱樑，樓板間格，橋樑窗牖，小則如傢私什項，箱板，三和土板等皆用之。因價廉物美也。土杉之名稱衆多，各地出產者，各有其名。玆就出產區域而總分之爲三。(1)東江杉，(2)西江杉，(3)北江杉。

(1)東江杉　試驗所得，東江杉之結果，與其他相較，似有遜色，在其外表觀之，因土人上山採木時，常連樹頭掘出，故市面之東江杉，多數樹頭尙存。木身稍柔瘦，其出產區域，包括沿東江流域，其最盛者，如老隆，河源，惠陽，博羅等，石龍乃最重要之樞紐。

每立方英尺重 21.96磅。

(2)西江杉。沿西江兩岸所產之杉，皆屬於此。最盛之區，在梧州以上，有桂江上游之桂林，象江流域之柳州，左江上游之懷西，龍州等。產額甚大，粵桂兩省之杉料，多仰給於此。大木頗多，壽枋之料，亦可取材。在梧州以下，綏江一帶所產之杉，皆聚於四會，然後由四會運來。因此故，又號之爲四會杉。四會較近本市，爲省工作計，常不將杉鑿鼻，而祗用竹篾縛成排，順水流下。木身肥壯，首尾相稱，建築者多樂採此以爲屋宇之樑。每立方英尺重 28.09磅

(3)北江杉。中木與大木多。杉鼻較細。最盛之區，乃仁化，南雄，翁江之翁源，英德，�江口，清遠（在清遠來者，或到清遠然後鑿鼻者，謂之清遠杉），及蘆苞等。湖廣杉亦屬此類。

每立方英尺重 26.79磅。

(乙)土雜木

除杉以外，各種新舊土木，皆謂之土雜木，其所包括之範圍最廣，名目最多，行中人亦不能盡識：近市各地亦有出產，因雜木即各種樹木也。但仍以東西北三江來者爲最大宗，而北江又居三者之首席。其出產之地，與杉木畧同。市面最通行之幾種，大約如下。

(1)松木　　每立方英尺重 42.07 磅

價平而大宗。市面所售之厚薄松板，每井不過四五元。三合土板，比洋松價平。市面價格，每英尺售八九分，而洋松則在一角五分之譜。相差甚遠。用作士敏土桶者，多爲四會松，及其他富有韌力者。湖廣松最易折撓，不適用。其他如箱板，肉枱，砧板，及柴多用松木。但最易生白蟻。不適宜于房舍間格，及傢俬等用。

(2)樟木　　每立方英尺重 36.76 磅

此爲上等雜木之一。味辛辣。能避蟲蟻。樟木衣箱一項，每年輸出甚大宗。其他如船艇及牌片等，多採用之。

(3)梧桐　每立方英尺重 21.93 磅。

質雅潔而密度甚低。色白。木紋直。用作樂器，其音韻最清雅。蚨蝶琴等必需此項木材，牌片亦有用之。

(4)棉木　每立方英尺重 25.28 磅。

質軟而輕。最合作紙床之用。因其不傷刀刃。

(5)柏香　每立方英尺重 45.26磅。

味香。色黃。極油潤。質最幼。歷時久而不變。常用作神主。價頗昂貴。上等之貨，每個總在十元以上。裔枋亦有用此，每副必過千元，現在甚少見。

(6)荷木　每立方英尺重 36.4 磅

色白。紋幼而堅。下等傢私多用之。細者作柴。市面荷柴，每元數十斤。

(7)橡木　每立方英尺重 37.34磅

色白。紋質稍粗而堅韌。轎槓，傢私等用之。

(8)紅桃　每立方英尺重 31.75磅

色橙紅，舊式傢私有用之。類似酸枝坤甸，但體輕而價廉。

(9)紅花森　每立方英尺重 40.00磅

色紅而帶黃。傢私，樂器，槍托等用之。

(10)椆木　每立方英尺重 45.50磅。

色深紅。質堅且韌。農具如犂鋤等，及轎槓用之。

(11)樬木　每立方英尺重 25.53磅

色白而輕。極似梧桐。但全身有黑紋如雲形。一若人工造就，多用爲傢私枱面等。

(32)桔桯　每立方英尺重 45.36磅。

木質最結實。色略紅。用爲農具槍托等。

此外更有較通行者甚多，如桃花森，色深紅。重于紅花森，與木署同。山任，用作砧板及農具，堅而重。木頗大。因對邊開錁後，卽不能用作砧板，故運來時，祇得截成三數尺之圓柱，以便輸運。冷生香，每年用作鞋屐者不鮮。𣞀木產于廣西龍州，地近安南，現已改經安南轉香港而來，遂成洋雜木。鐵木一若鐵抄，其質之堅硬，大有勝于鐵抄。土黃洋木，質最幼而矜貴。廣東黃洋木梳頗有名。仁面，最堅，多用爲大門。其餘如山枝，荔枝，黃皮，秋風，元眼，石班等，現亦祇用作傢私及什項用器具。

以上各木之密度，係在本校試驗室，經多次試驗所得之結果。關於土杉雜木材試驗，現已由黃郁文教授指導下，從事工作中。異日試驗完畢完，獲得良好結果，當再發表。此文不過欲引起社會人士，對於土木材有所認識及關心矣。

鐵路的一切問題

王仁康演講　劉靜懸筆述

一般的講述鐵道，其問題有四，即(一)鐵道史略(二)鐵道行政(三)鐵道業務與(四)鐵道管理，每一問題非有一二句鐘之講述，不能詳其究，今之所述，亦僅道其概要而已。

(一) 鐵 道 史 略

人類文化之演進與傳播其所以臻至今日之地位者，蓋交通發達有以致之也。交通之始，原自陸地，寖至鐵道成立，陸地交通，遂登堂入室。世界鐵道之歷史，迄今已一百二十年，其時雖暫，然其俾益人類，自有史以來未有一物匹其功，鐵道之建設，始自鐵軌及蒸汽機關車之發明。世界最初之一條鐵道爲一八一五年成立之英國「士托頓打令」鐵道，其始專以運煤，嗣且用以載運客貨。此後四十餘年，中國尙未與聞鐵道之建設。

淸同治年間，英人斯雷芬挾其中國全國鐵道大幹線計劃遊說淸室與粵商。翌年，復有英商在北平宣武門外曠地，築約一里之鐵道試駛火車，幷邀淸吏往觀，觀者駭汗驚怪，誣爲妖孽之物，遂迫令英人毀之；同年某英商再倡議築上海與淞間十英里之鐵路，惜因清吏頑迷之反對，遂又事寢。

光緒元年(一八七五年)英商怡和洋行以淞滬間商業繁盛，倡議築鐵路貫之，清廷遂只准築鐵道九里，此實我國鐵道之源始。此路翌年而成，通車後，沿線商業驟盛，鐵路客貨收入，亦利市萬倍，惜因我國當局與英商怡和洋行立約築道之初只准用牛馬拉車，不准用蒸汽機車發動，路成後怡和洋行擅用蒸汽機車頭駕駛，遂動我國公憤，同年且有我國兵士一名，行經該路，因不慎致慘遭輾斃，因是清廷反對甚力，廹令停車，怡和行置之不理，我國願以二十八萬兩銀購斷該路，俟付清路價後，即將該路鐵軌掘起後，遂將同淞滬全路車輛及所有材料，一同運往台灣，盡棄沉於打狗港中。

及後，李鴻章因開採唐山煤礦，運輸不便，乃上奏議築鐵路，清廷不允，謂地近東陵，車聲嗚嗚，有碍靈寢，且烟煤障暗，有傷禾植。不得已遂聲明用騾馬拖駛，然後請准敷築，光緒六年，此路完成，名曰唐胥運煤鐵路。

我國鐵道之興築，其始非由於自動之倡議，實大半出諸列強侵畧政策之要求，蓋列強以為鐵道所及，即其政治經濟侵略勢力之所及，故莫不爭先恐後，奪取路權。清廷積弱，亦唯一任各國之予取予求。膠濟路權為德國索奪後，繼而南滿，平漢，滬寧，津浦，中東，滇越，廣九等均挾侵略政策，但當時廣九訂約性質，畧有不同耳。

粵漢鐵路本為湖南湖北廣東三省巨商奉清廷命自辦，但當時商民建築鐵道全無熱心，無法集欵，故不久即讓與美商合興公司興辦，後美商違約轉賣一部分股份於比利時，我人乃主廢約而收回國營，旋于光緒三十一年八月，由政府向香港借英金壹百壹拾萬磅，向該公司贖回。翌年正月，湘鄂粵三省商民，以該路有切身關係，故請准清廷照價贖歸商辦，所有贖路欵項，概由三省分任償還。嗣湘鄂兩省，因無力籌付贖路欵項，旋由粵省照價承併，是以該路線權，遂歸粵商所有矣。自是招股興築，規定每股五元，分三期繳納，第一期繳一元，第二期繳二元五角，第三期繳一元半，計自光緒三十二年二月至閏四月底止，共招集股份至八百八十一萬七千餘份，應收股本爲四千四百零八萬餘元，嗣因收股不足，并除去利息三百八十萬餘元外，實收得銀二千六百二十三萬餘元，加以當時紙幣低折，損失不下二百餘萬元。至民國五年，已築成廣韶段，嗣以收欵不足，遂停止建築餘段，至民十八年鐵路部計劃分別接築本路未完各段，民十九年廣三亦奉令歸併本段。同年，鐵路部發行公債，抵換商股，收回國辦而直隸鐵路部。

計其始此路之資本除少數歸諸外人外，大半爲國人資金。此外。陵陽，潮汕，新甯，房山，漳厦，洛潼等鐵道全是中國內資之民辦鐵道。

(二) 鐵路之行政

清朝鐵路管理隸屬郵傳部(因當時尚非國營)後屬交通部，其時有路政司，乃稍事整理鐵路事業并釐定關於鐵路行政之法規。

民國十八年我國遂有鐵路部組織；在孫總理建國大綱內，只有交通部而無鐵路部。日本有鐵路省，良以鐵路之建築及管理，事滋體大，勢難隸屬他部，故有鐵路省之設。十餘年來我國交通正待發展，鐵路網實施至繁，爰有鐵部路組織之必要。此部組織共分四部：

(一)總務司。專司人事之任免獎懲，技術人員之訓練教育，及本部經費之預算決算等管理。

（二）業務司。專司鐵道營業運輸之發展暨機車車輛之調設，及規定運價編練鐵路警衛等事項。

（三）財務司。專司鐵路預算決算之編製審核，款項支配及保管與帳目單據之稽核，及鐵路財產之處理等。

（四）工務司。專司鐵路工務之監督管理與擴充改良，建築工程設計及建築材料購置之審核等項。

此外尚有委員會若干，如職工教育委員會，業務改進委員會，技術標準委員會，銓敘委員會等，尤以技術標準委員會之責任最重，幹事亦最繁瑣。

此部組織之後，各路管理局則有局內與局外二部。局內職員在局長副局長（或管理委員會）之下設總務，車務，會計，工務及機務五處。總務處掌理局內一切事務，凡不能歸其他各處辦理者皆屬此處。下設各課，如地畝課或產業課，掌理全路地產之購置，轉讓及租佃。有編譯課，編製全路一切出版物。有庶務課，掌理一切雜務。有警務課，管理全路警務，目前廣韶段用路警千餘人。蓋沿途治安不大平靜也。有材料保管課，專司材料之收發與保管。此外有購料課與購料委員會，購料委員會掌二千元以上之大宗購買，二千元以下之購買則由購料課長辦理，以杜舞弊。此外有公益課，專理全路員工及其眷屬之疾病調護，及其子女之入學等，以解決其子女求學問題，并代員工組織消費合作社，及公餘娛樂體育等事項，務使工作人員安心樂業。

車務處掌理一切招徠客貨，行車運輸，制定運價，及統計電務等事項，爲管理工作之最繁複最重要部分。并設商務調查委員會，調查并諮詢沿線各工廠之運輸事項，務使本路客貨日進繁盛，長此委會者須具遠大眼光。粵漢路現任商務調查委員長是前任車務處長，以示呼應靈活。此外并設有文牘，運轉，計核等課。文牘掌公文及本處局內外人員工作成績等(以爲升黜遷調之根據。)運轉課掌理全路列車之編配，與行動時刻及狀況等。計核課掌理本處之各種統計。

會計處管理及審查全路財政及賬目。工務處掌理路基，橋樑，隧道，號誌，及一切建築物之建築及修養工程事務。機務處掌理機車與車輛之建造，修養及保管事務。

至於沿綫職員則爲數最衆，沿線職員工作之努力與否組織之得當與否，影响於鐵路營業者至深且巨沿線職員之屬於總務處者有警務總段長及分段長，醫院院長，材料廠長，及其助理員，工務處沿線有工務總段長分段長幷設工程司，工務員，及監工以輔助之，幷有段班專修護沿路車軌，其責任至爲重要，尤須熟練久職之工人任之，現每五公里一班，一班增至八人。機務處亦設總段長，分段長，工務及監工，更於適當地點設機廠及機車房。此外，車務處除車務段長之外，又設客票及貨物檢查員，專司稽查客貨運輸，客商與本路員司有無不法情事，對於偸運，揑報，逾重以及運價之核收，車輛之分配，客票有無舞弊情事，皆應隨時隨地考察，犯者依法懲處。

重要行政組織已如上述，至於(三)鐵道之業務與(四)鐵道之管理方面尤屬繁瑣，恐非一夕昕能言盡，倘時乎再來，當與諸君子論之。

測量常識一束

黃道基

一、小平板儀用諾模(Nomograph)

在小平板儀測量當中，對于不甚重要的部份，可以不必用尺直接量度距離。通常是利用照準儀兩直立鈑的劃度和兩覘牌的固定距離來計算各該點與儀器所在地的地平距離和高度差。

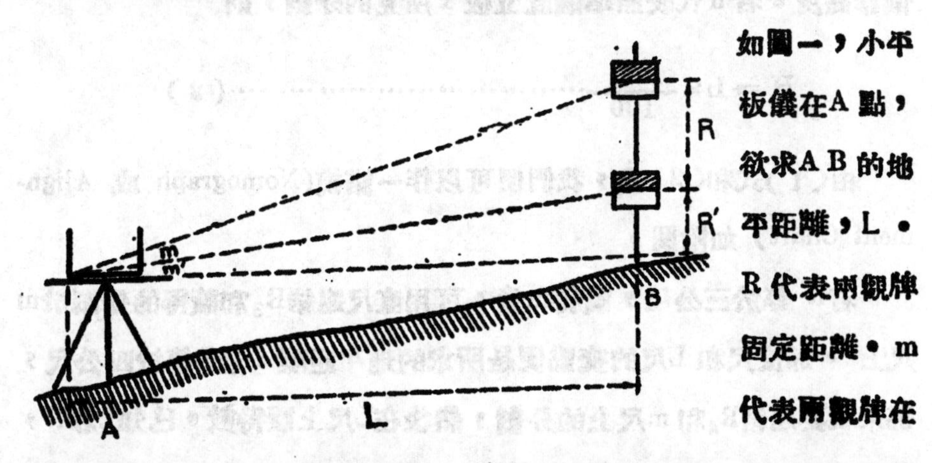

如圖一，小平板儀在A點，欲求AB的地平距離，L。R代表兩覘牌固定距離。m代表兩覘牌在照準儀直立鈑上看見的分劃差。m' 是下覘牌的分劃。因爲直立鈑上一分劃的長度等於兩鈑間格的一百分之一。所以

$$L \times \frac{m+m'}{100} = R',+R',$$

$$L \times \frac{m'}{100} = R';$$

$$\therefore L = \frac{100}{m} \cdots\cdots\cdots\cdots\cdots (1)$$

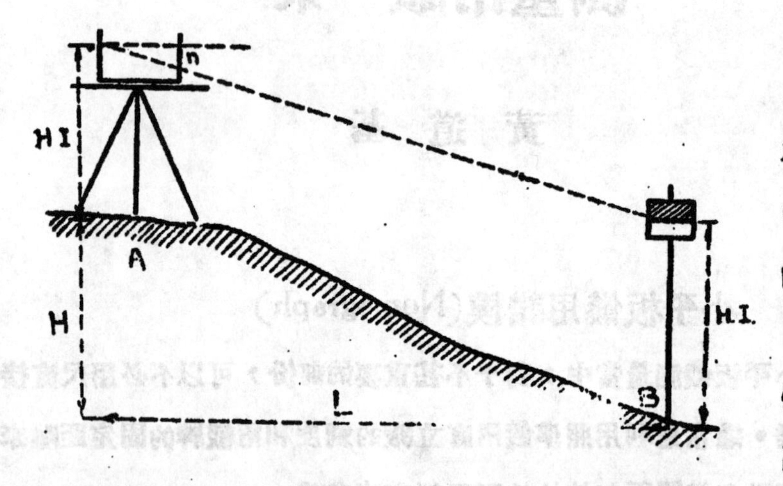

如圖二，小平板儀在A點，欲求AB的高度差，H．先使觀牌的高度等於儀器高度．若n代表照準儀直立板r所見的分劃，則

$$H = L \times \frac{n}{100} \cdots\cdots\cdots\cdots\cdots (2)$$

由(1)式和(2)式，我們便可以作一諾模(Nomograph 或 Alignment Chart) 如附圖．

若R 等於三公尺，讀得m後，可用直尺連結R_3和讀得的分劃於m尺上．那直尺和L尺的交點便是所求的地平距離．若R等於四公尺，那末就要連結R_4和m尺上的分劃，然後在L尺上讀得數．巳知L和N，欲求高度差時，祇須用直尺連結L尺和N尺的數目，該直尺與H尺的交點便是所求的高度差．

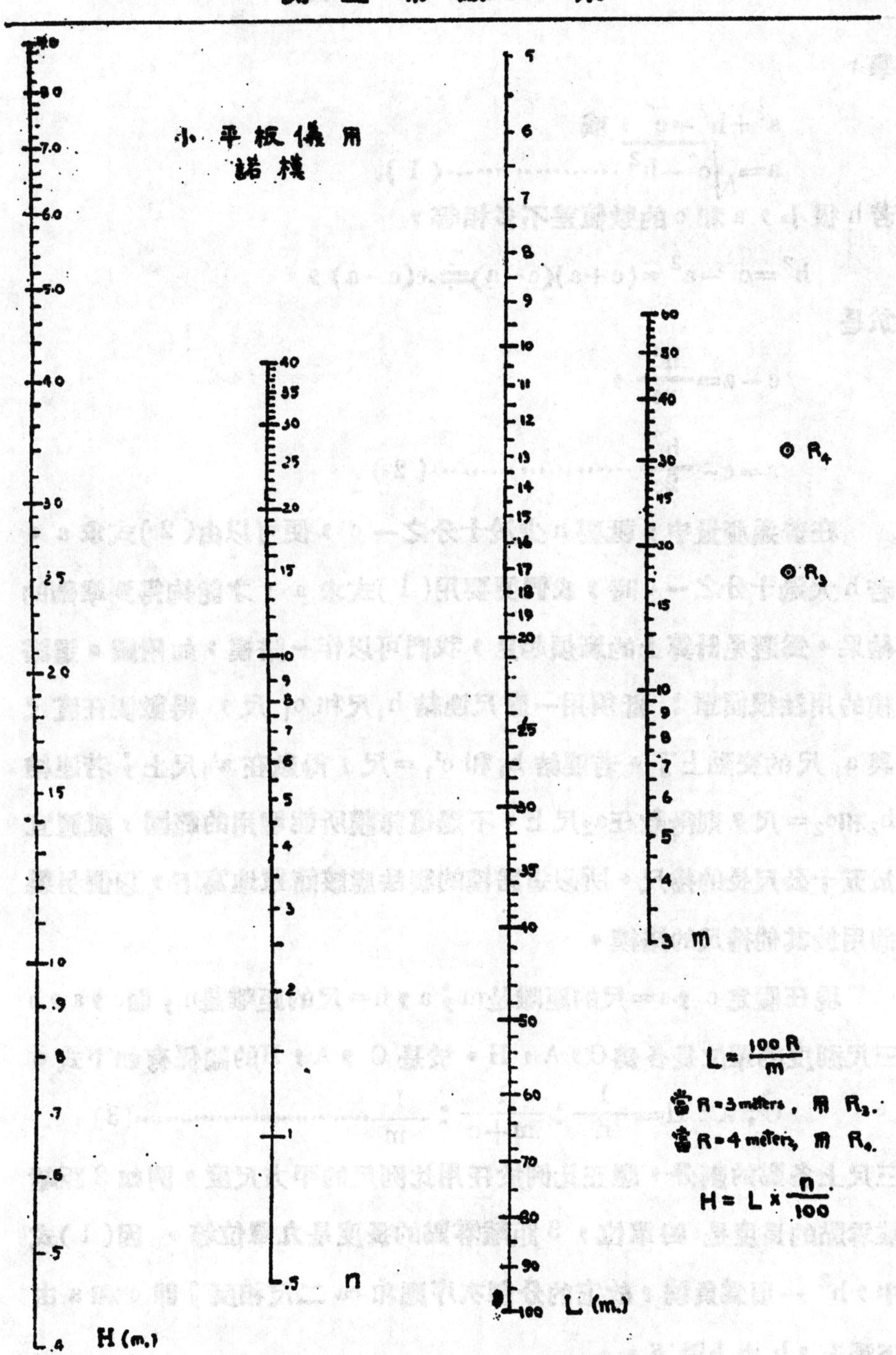

代表勾股弦關係之諾模

在一直角三角形中，若勾爲 h，股爲 c，弦爲 a 時，根據幾何

學，

$$a^2+h^2=c^2，或$$

$$a=\sqrt{c^2-h^2}\cdots\cdots\cdots\cdots\cdots(1)$$

若h很小，a和c的數值差不多相等，

$$h^2=c^2-a^2=(c+a)(c-a)\doteqdot 2c(c-a)，$$

於是

$$c-a=\frac{h^2}{2c}，$$

$$a=c-\frac{h^2}{2c}\cdots\cdots\cdots\cdots\cdots\cdots(2)$$

在普通測量中，祇要h少於十分之一c，便可以由(2)式求a。若h大過十分之一c時，我們便要用(1)式求a，才能夠得到準確的結果。爲避免計算上的麻煩起見，我們可以作一諾模，如附圖。這諾模的用法很簡單：祇須用一直尺連結h_1尺和c_1尺，得數便在直尺與a_1尺的交點上了。若連結h_1和c'_1二尺，得數在a'_1尺上；若連結h_2和c_2二尺，則得數在a_2尺上。不過這諾謨所能應用的範圍，祇適宜於五十公尺長的捲尺。所以這諾模的製法應該簡單地寫下，以便另製適用於其他捲尺的諾模。

現在假定c，a二尺的距離是m；a，h二尺的距離是n；而c，a，h三尺刻度的單位長各爲C，A，H。於是C，A，H的關係有如下式。

$$C:A:H=\frac{1}{n}:\frac{1}{m+n}:\frac{1}{m}\cdots\cdots\cdots\cdots\cdots(3)$$

三尺上各點的劃分，應正比例於任用比例尺的平方尺度。例如2距離零點的長度是四單位，3距離零點的長度是九單位等。因(1)式中，h^2一項爲負號，故它的分劃次序應和ca二尺相反；即c和a由下讀上，h由上讀下。

在製圖時，最好先定c和h二尺的範圍，然後定C和H的比例與及m和n的比例。如c由30至50，h由1至9；則c尺含有

50^2—— 30 即1600單位；而h尺含9^2—1^2即80單位。這樣c尺之單位長與h尺之單位長的比例最好是1：20。代入(3)式我們便知道m等於1，n等於20。決定了c尺和h尺所用的比例尺後，我們便可以先畫出這兩尺。又因當h等於零時，c等於a，所以a尺所用的比例尺不必從(3)式求出；祇須連結h尺的零點和c尺上各點，便可定a尺上各點的位置3.3。

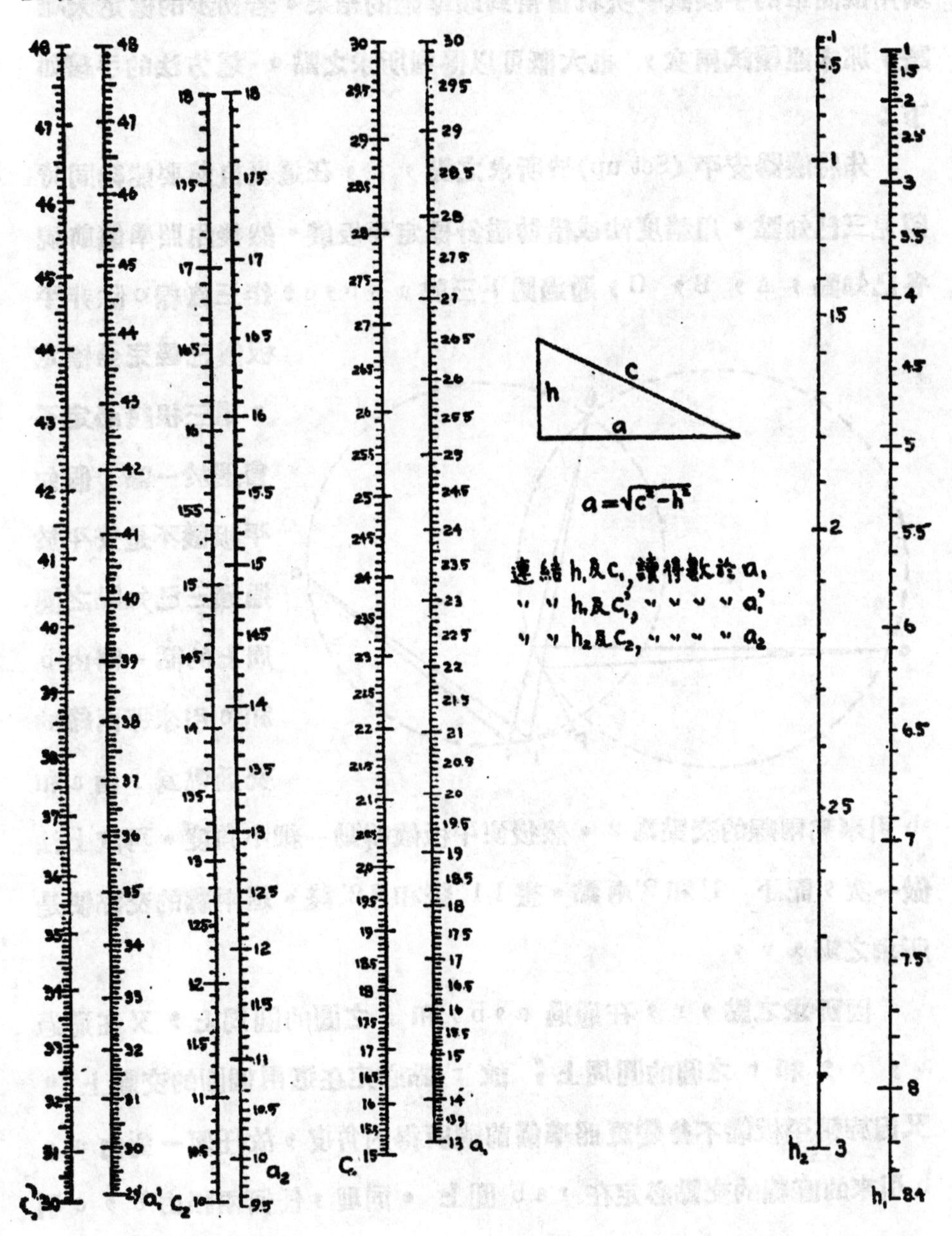

三・三點題別解（譯自 Engineering News Record Jan. 10, 1935

1592年，墨西哥軍隊裏有一個軍曹 Cucunutti 求得一個合用而簡單的三點題解法。雖然這方法所得的結果非絕對精確，因這法以直線代替弧，但在野外工作時，若平板儀已大約標定（Orient），則祗須用很簡單的手續試一次就會得到頗準確的結果。若初步的標定太離譜，那末連續試兩次，也大概可以得到所求之點。這方法的手續如下：

先將儀器安平（Set up）於所求之點，r，在這裏自然要能夠同時望見三已知點。用猜度法或借助磁針標定平板儀。然後用照準儀前視各已知點，A，B，C，通過圖上三點 a，b，c，作三直線。除非平板儀已經完全標定，這三根綫必定不會遇於一點，假如平板儀不是安平於通過三已知點之圓周上的話。稱由 b 和 c 引來那兩綫的交通點爲 1. 由 a 和 b 引來那兩線的交點爲 2。然後使平板儀轉動一細小角度。再依上法做一次，記下 1' 和 2' 兩點。畫 1 1' 綫和 2 2' 綫。這兩線的交點便是所求之點，r。

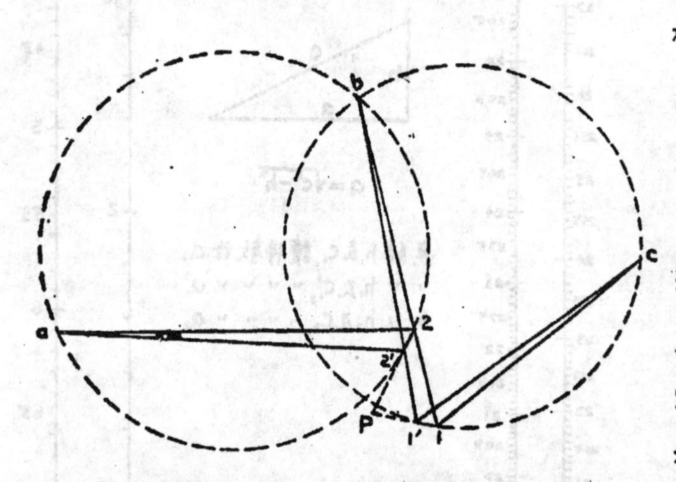

因所求之點，r，在通過 a，b，和 r 之圓的圓周上，又在通過 a，c，和 r 之圓的圓周上；故 r 點必定在這兩個圓的交點上。又因旋轉平板儀不會變更照準儀前視所得的角度，故任何一對由 a，b 引來的直線的交點必定在 rab 圓上。同理，任何兩根由 b，c 引

來的直線的交點必定在 rbc 圓上。若每一圓都可以找出兩個這樣的交點；所求之點，r，自然很易找到。因爲這些點，如1'，1 和-'，2 都很接近；所以我們不妨用直線代替弧來求交點 r。若能夠看出這些圓灣曲的方向，用一塊曲線板便可得到很近似的結果了。

四。作長半徑之弧(譯自 Civl Engineering Vol.4,No.1)

在製圖室中，我們有時須要作一長半徑的弧，但沒有良臂規(Beam Compass)，或製圖桌太小，不能應用長臂規來作弧。這裏有一個頗利便的方法可以解決那問題。

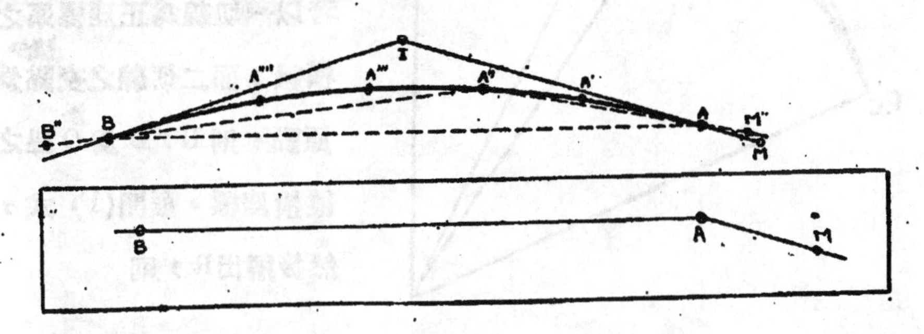

如圖四，已知二切線 IM 和 IN 交於 I 點，而 AI 與 BI 的距離相等。求作一弧，以 A 爲始點以 B 爲終點。

先連結 AB 二點得 AB 弦。在一臘紙或其他透明紙上作 BAM 角。然後移動所作之角，由 A至 B，使A M和AB同時經過A和 B點。當A點繼續變更位置時，用針或其他有尖端的東西從A點插下去，於是在原圖上得 A'，A"，和 A'''…………各點。例如當A點在A"時，則所作之角爲B'' A" M"。

以Railroad Curve或其他曲線板連結A' A", A'''………各點卽得一弧。若該弧很扁平，那末三角板之類的直尺也可以替代曲線板，祗須輕微地使筆尖向出向入移動，便可以得到一根圓滑的弧了。

五，求經過一定點及二切線之曲線方程式

（譯自 Ciyil Engineering Vol, 4, No. 1）

求一曲線經過一定點和二切線的方程式，本來有很多方法。但下面所說的方程式十分簡便，因爲牠祇有一個三角函數，$\tan\frac{\Delta}{2}$

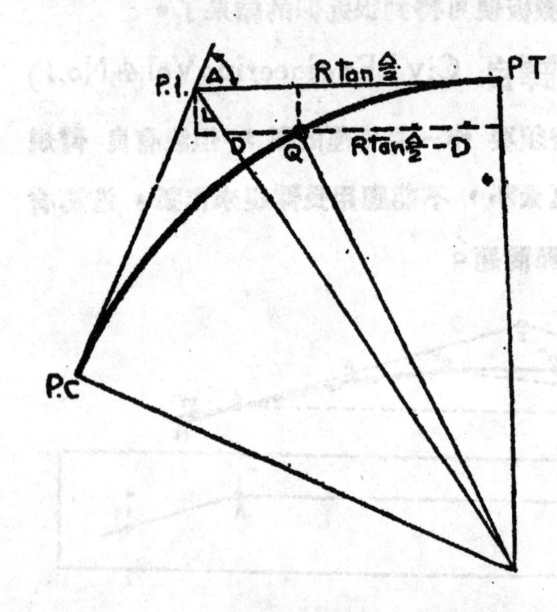

$$R^2=\left(R\tan\frac{\Delta}{2}-D\right)^2+(\Gamma-L)^2 \quad \cdots\cdots(1)$$

若以一切線爲正座標系之橫軸，而二切線之交點爲原點，則 L，D 爲 Q 點之縱橫座標。展開(1)式，然後解出R，則

$$R=\frac{\sqrt{\left(D\times\frac{L}{\tan\frac{\Delta}{2}}\right)^2-\left(D^2+L^2\right)}+\left(D+\frac{L}{\tan\frac{\Delta}{2}}\right)}{\tan\frac{\Delta}{2}} \quad \cdots\cdots(2)$$

然求R時，亦可將各已知數代入(1)式，從那二次方程式解出R。

當 $\Delta=90^\circ$ 時，$\tan\frac{\Delta}{2}=1$. 上列公式又可化爲

$$R=\sqrt{2DL}+D+L \quad \cdots\cdots(3)$$

25352

木樑新設計對於經濟上之影响

(By J. A. Newlin, G. E. Heck, and H. W. March)

陳士敦譯

木樑平衡剪力計算法，近由美國林木試驗所求得一新法。蓋木樑受剪力試驗時，其分裂多緣裂縫底之隣近而起。於是乃得假定木樑上半段所受之力與下半段無關，結果使中立平面之剪力大減。惟此被減剪力乃非近代設計所注意。故舊法盛行，而靡費甚多。間有設計如公路橋樑托樑之設計，或鐵路枕木之設計，常有預定其應力爲最大剪力之二倍或三倍者，然受力部仍未見其失效也。現本文所論者，爲木樑（Checked Beam）受重壓時之彈力狀態。並說明事實與採用舊法預定剪力之差異。此項說明之証明爲數學式之分析與二百餘次實驗之結果。實驗時所用之木樑之面積，乃限於3/4×1½吋與8×16吋之間，根據

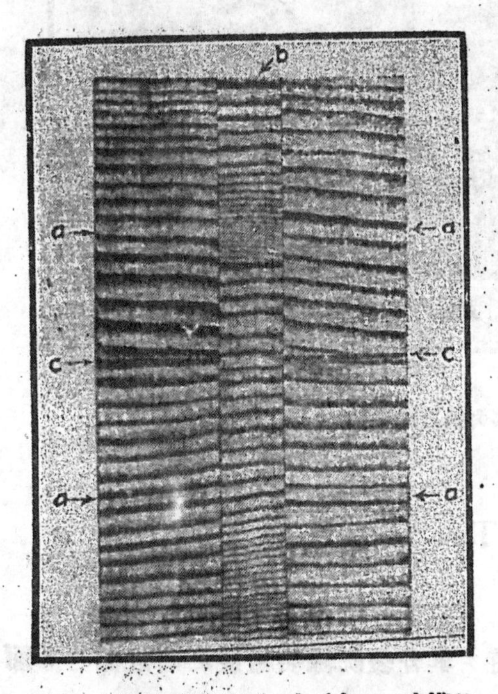

圖一　實驗時特造之 checked beam 之橫段面，a 之四段乃膠黏於中段b. c節乃以臘塗搽使不黏隣近各節之膠且能減少其磨擦力

理論與實驗之結果，吾人遂可精計使剪應力(Shearing Stress)失效之重量。

受試驗之木樑爲人工造成之夾木式，(如圖一)夾木(Checks)乃置於樑之中心位，因木樑受剪力後之失效狀態常發現於此也。凡木料之用以建築者，當受重量施於二點時，(如圖二)其因平衡剪力而失效之狀態，乃甚顯著，至本文所述者，則與上法稍異，因木樑之受力處祇一點而非二點故也。

圖二．——木樑失效之狀態，其失效乃因鄰近中立軸之平衡剪力而起。

夾樑作用 (Two-beam Action)

Checked beam受剪力作用時，其分裂既緣裂縫底之鄰近而起，而其上下二半段既彼此獨立，由是乃得以明瞭 Checked beam 之慣性。剪應力之垂直分配法，根據數理之分析，可由圖三求得之，圖中之剪應力線乃木樑失效時剪應力之中和與樑之全濶度之比率。夾樑作用之存在，依據圖中之曲線遂得以証明。中立平面之剪應力亦因上下二半段之獨立之作用而大減。由數理之分析，更可明証此項作用之存在：

隣近重壓之支點，其反應力 R可以二項數之和代之，

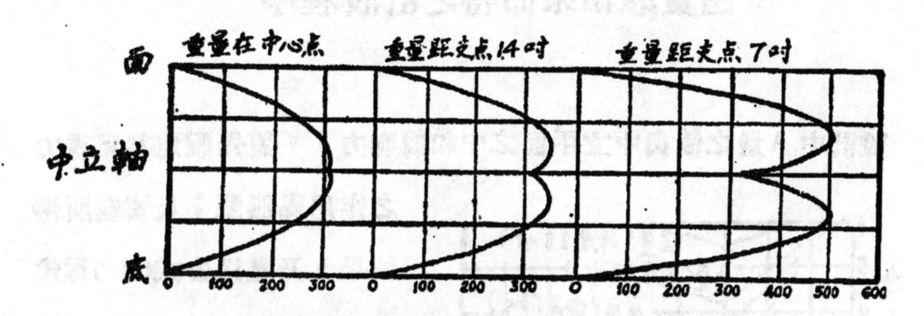

中和翦力與樑之全闊度之比率 lb. per square inch)

圖三·——一集中重置於 Checked beam 各點時所得理論上之平衡翦應力之變率

托樑容積爲 $2\frac{1}{2}$ 乂$4\frac{1}{2}$ 乂45 吋跨度爲四十二吋夾木(Checks)之深度爲一吋

(1) $R = B + A/a^2$

於此公式內，惟 B與中立平面之翦應力有關。

(2) $B = 2/3Jbh$

J 爲中立平面之中和翦力與樑之全闊度之比率。 b 與 h 爲樑之闊與高，B 爲局部反應力，與中立平面之中和翦力 J 相聯絡，爲單樑之局部反應力。(" Single-beam portion " of the reaction)

公式(1)內之a爲重量與隣近支點之距離，A爲一數量，隨樑之容積與楊氏長度率 (Longitudinal Young's Modulus) 而定，且木樑上半段底面之中和縱度排擠，(Longitudinal displocement) 亦爲決定 A 之一要素。反應力之一部 A/a 遂能使樑之上下兩半段之作用獨立，且與中立平面之翦應力脫離關係， 此爲夾樑之局部反應力 ("two-beam portion" of the reaction)。 夾樑及單樑失效時之局部反應力之值，乃根據多次實驗之結果而決定，實驗時之重量乃集中於各不同之點。苟將結果之數值代入公式(1)內，遂可得二公式以求B 與 A。

由實驗結果而得之計設程序

設計中A量之值與中立平面之中和剪應力J，須先假定其與重力之作用點無關；且實驗所得結果，乃盡以公式(1)為代表，A與B恆為常數，故此假定足証為準確。苟於實驗時置一集中重量近於支點，則夾樑之局部反應力大增，而與中立平面剪應力有關之單樑局部反應力則並無影響，祇為反應力之一小分數。

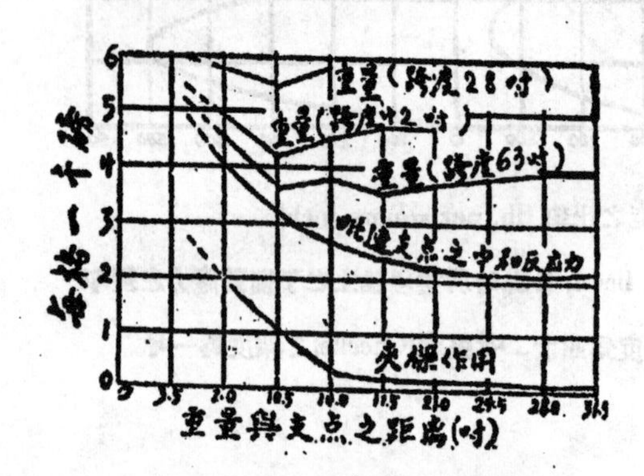

圖四——不同跨度之Checked beams之試驗結果。
注意單樑局部反應力與夾樑局部分應力之關係值。
重量置近支點時夾樑局部反應力漸增。

由是可知使樑受剪應力而失效之最小重量之施用點乃不在支點之內，而在支點相當距離之外。附印之表為實驗二木樑後所得之結果。圖四為其代表線，表中所錄單樑與夾樑之局部反應力，乃合(a=7)時之中和反應力與重壓每點之中和反應力而計算其值，設反應力之結合與上述不同者；其結果與上數表相差不甚遠。

現據多次實驗之結果，照錄其撮要如下：凡(Checked beams)之跨度與高度之比率(Span-depth ratio)為由九至一者，其受剪力而致失效之最輕重量施用點，乃在支點相當距離外，為樑高之三倍。凡稍長之跨度，其臨界點(Critical Point)與支點之距離常較稍短之跨度者為大，惟不論何時，凡重量之施用點與支點之距離為樑高之三倍時，毗近支點之夾樑局部反應力約為六份之一。

25356

特造之 Checked Spruce beems 之實驗結果（長度不同；濶 $2\frac{1}{2}$ 吋；高 $4\frac{1}{2}$ 吋；Checks 深 1 吋）

(a)重量與隣近支點之距程	失效時之重量			隣近支點之中和反應力	單樑作用	夾樑作用	夾樑作用與 a^2 之積
	28吋跨度	42吋跨度	63吋跨度				
吋數	磅數	磅數	磅數	磅數	磅數	磅數	
7.0	5,770	5,050	4,660	4,226	…… …	1,872	91,700
10.5	5,460	4,240	3,560	3,186	2,354	832	91,700
14.0	5,680	4 430	3,780	2,911	2 473	438	85,800
17.5	………	4,800	3,460	2,649	2,349	300	91 900
21.0	………	4,800	3,630	2,410	2,183	227	100,100
24.5	… ……	………	3,780	2,310	2,140	170	102,000
28.0	………	………	4,050	………………因張力(tension)而失效………………			
31.5	………	………	3,970	1,985	1,868	116	115,100

設計內之獻議

1.設重量有可能移動性者，則置于相當位置，使其最大之集中重量與支點之距離爲樑之三倍，此後則可應用下文靜重量與活動重量之獻議。

.2設集中重量距支點僅爲樑之高時，可置不計。

3.設所有集中重量與支點之距離爲樑高之一倍，或二倍三倍時，可假定其距離爲樑高之三倍，而計算其結果反應力。反應力之六份一則可不計，因彼爲夾樑之局部反應力，而與中立平面之剪力無關也。

[注意：凡跨度與高度之比率爲最小(小於六至一)者，可照本文獻議內一與三之法而置重量於跨度之中心點。]

4.假定其他集中重量仍如常。

5.忽計均等重量(Uniform Ioad)百份之十，因彼固屬夾樑作用之支配，

6.根據假設位置之事實而假定毗連平行樑各重量之側面分配。(Lateral distribution)

7.代入樑之總濶度及已知之反應力（此種反應力并不屬夾樑）于公式內，而計算其中立平面之剪應力。

8.安全剪應力(Safe Shearing Stres)之標準，須以美國林木試驗所規定之百份之九十爲準則，因此規定并無影響於夾樑作用之效果也

以上之獻議，苟吾人能依法而行，則材料上大可節省，建築公路橋樑，可爲一例。採用新法與舊法之結果，其差異乃甚顯著，蓋以毗連樑各重量之側面分配之存在也。致公路橋樑新舊二法之之計算法，玆照錄如下：

假定一集中重量 P 緣托樑而移動，跨度爲十六呎，托樑之高 h爲

十六吋，濶度 b 爲五吋，所有容積皆照原度。設重量在跨度之中心點時，則假定位於二托樑間之地板之負荷量爲原重量之1/4。在此情形下，荷重量與支點之距程爲樑高之三倍時，則邊樑所負之重量稍大於重量之百份之四十，中立平面之剪應力 J，其計算值可照上列公式而求之：

$$J\ \ 3/2\times5/6\times\frac{R}{bh}\cdots\cdots\cdots\cdots(4)$$

R ＝重量 P 與支點之距程爲樑高之三倍時之反應力。但依據重量側面分配之假定，則

$$R=6/10\times3/4\times P=9/20P\cdots\cdots\cdots\cdots(5)$$

設安全剪應力每方吋爲一百磅，而照上文所述減去其百份之十，由公式(4)與(5)關係而得

$$90=3/2\times5/6\times9/20\times P/5\times16\cdots\cdots\cdots\cdots(6)$$

P ＝12,800磅，爲剪力之安全重量。

依照舊法之計算，假定重輪脫離跨度時中立平面之剪力爲最大，復用安全剪力之值，(每方吋爲一百磅) 且樑端 并無重量側面分配狀態，則

$$100=3\ 2\times\ P\ 5\times16$$

安全重量 P爲 5,333磅。以與由新法而得之 P值相比較，則前者爲後者之半數。

至單樑之無重量側面分配狀態者，可將公式(5) 與(6) 內之6/10省畧，得

$$R=3\ 4P$$

及 $$90=3/2\times5/6\times3/4\ \times P/5\times16$$

則P＝7,680磅。以與舊法之P值相較前者，則比後者大於百分之四十。

避免滑溜之磚面設計

GEO. F. SCHLESINGER作

沈錫琨譯

以磚鋪路，可免滑溜，故近年來硬度高之鋪砌磚，其使用及製造，均漸次發展；至其發展之途徑有二：

(一)免除瀝青質之填塞物(Filler)附着於路面之方法。

(二)直紋突耳式之鋪砌磚。

於以磚鋪路之工程中，其路面所須要之性質，及能增進磚面抗滑性之填塞物，均爲本篇所討論者。

路面之清除(Surfoce Removal)——以磚鋪路，多用柔韌之填塞物以填塞磚隙，因其旣能抵禦潮濕之內侵，復不受氣候變化之影响；瀝青乃柔韌填塞物中之一，亦卽良好工程之必需品也。車輛之平均速率日增，因滑溜而失事者遂隨之而起，尤以潮濕天時爲甚。磚鋪道每於雨天發生滑溜，而致傷害人命，查其原因，皆由於其路面蓋有瀝青，而瀝青內之礦物復未得適宜之混合。路面瀝青因車輛磨擦而消失之舊磚鋪道，常無滑溜之事發生，是以現下以磚鋪路，莫不極力設法免除填塞物之附着於路面。

一九二九年建築於 Jacksonville,Florida 之橋道之造法書內，曾說明須以手傾倒填塞物於磚隙，以免其附着於路面，惟承辦者則請求准其使用別法，即先以石灰水掃磚面，然後施以瀝青，其遺下路面者，則以剷去之；試驗結果，較用手傾倒法尤佳，承辦者因而得用之法。一九三一年掃灰水之方法復採用於華盛頓街(Washington Street)紀念橋(Memorial Bridge)上之鋪道。普通使用之灰水，掃於磚面即成薄膜，乾後則變為一堅硬之離間層，(Separting layer)若其不受攪動，則瀝青之傾倒可延遲數日。用灰水時，須特別小心，以免其流入磚隙，因其所成之薄膜有持久性故也。

一九三二年愛希愛公路局(Ohio Highway Department)亦利用灰水於磚鋪道，惟其灑掃之方法則不同。其法即將石灰水入於一手動之氣壓桶內，然後利用氣壓使其由一管口噴出，如此則可得一細勻如霧之浪花，且可減少灰水流入磚隙之弊。灰水掃後，未乾即須施以瀝青，因濕膜較硬膜之離間功能為大也；至於遺下地面之瀝青，則於其變冷凝結時以通常之剷雪器清除之。關於可作離間物(Separating medium)之各種溶液，曾經數度之研究，結果以略含澱粉之氯化鈣為最佳，因其能產生少許膠性之溶液，以阻止蒸發或急乾也；現其已成為一最通用之離間物矣。離間物為含水溶液，故用時難免使磚旁潮濕，但無關重要，其於瀝青傾入磚隙時所發出之氣泡，亦不過為離間溶液之蒸發現象耳。至於路面瀝青之清除，則宜用切法(Cutting)，不可用轉法(Rolling)，因恐磚隙內之瀝青亦受牽動也。

有少數工程師常喜用硬膜方法(Hard coating method)，蓋取其能延遲清除瀝青之時間，殊不知其常將磚隙內之瀝青拖出，僅此一端已不能與濕膜方法相比，況採用濕膜方法，尚可避免路面受汚，而瀝青貯藏器又不用常洗，因其沉澱物少故也。於此二法中，瀝青均須乘其熱度高時傾倒，蓋如此始可使其直透磚底，故於其未傾倒之前，切勿將其攤開而使其變冷；至於遺下路面之瀝青，非必須薄，蓋稍厚尚可予路面清除時以便利也。由路面剷回之瀝青，可以復用，但用時須雜以新料一半，而熱量之消耗亦略增。應用此種路面清除之方法可得一面無瀝青，及磚隙填塞完善之磚鋪道，且瀝青可以省用，又無須蓋以粗沙，而致增加費用。

直紋凸耳式磚—鋪砌磚乃由黏土工業中之硬泥方法所製造也。泥板石，或黏土，於研成細末後即與適量之水相混合，一強有力之螺鑽繼而將此硬性之混合物推過一模，使其成爲圓柱狀，最後復將其運至一斷切器(Cutter)，器內有鋼線一排，自動而將其切成磚狀。所謂直紋鋪砌磚，其長度乃由模型所定，而深度即根據斷切器內鋼線之間隔，至於磚旁凸出之耳，則由切線之離心動作(Eccentric motion)所造成。於此種磚中，較滑之一面常用於鋪砌之面，但近年來，路面之抗滑性異常重要，故磚面以略粗爲佳；至於磚隙填塞方面，磚旁之凸耳則爲多數工程師所認爲最重要者。三年前，因爲原料關係，此種直紋凸耳式磚之製造業不甚發達但現已非昔比，而此種磚亦經成爲許多地方中之最普通之鋪砌材料矣。根據一九三三年美國標準局(U. S. Bureau of Standards)對於各輪船所載貨物之調查，直紋凸耳磚竟佔全部貨物百分之四十二又二分一。最普通之直紋凸耳磚，其兩旁各有二槽，每槽之邊附着兩耳，但有時因特別之需求，有僅有耳而無槽者。二者中，其模型均有二棒，此二棒於磚被切後壓下，而造成磚旁之四耳，至於磚旁之槽，需要時亦於模內造成之。

於免除填塞物附着於路面之工作中，工程師常有機會觀察磚旁，及磚端，接合處之填塞是否完善；通常欲用以利便填塞之磚端傾斜，經發覺不能滿足吾人之需求，故最近磚之兩端，亦須製造凸耳。

以「直紋」二字用於線切之鋪砌磚，就描寫方面言之，實屬不合，尤其於免氣製造法進步之後，姑勿論其他，其構造既堅實而無氣泡，當知其無紋也，但此二字沿用日久，且一見即知其所指為何種磚，故至今仍無人將其更改，無氣泡之線切面有時不如氣泡者之粗，但粗滑之程度均可用畸形之切線(Deformed cutting wires)造成之也。

填塞物之研究——史丁臣(Stinson)與羅拔史(Roberts)二人，交與公路研究所之道路調查報告表內，以直紋磚鋪道為各種道路之阻力最大者。彼二人於調查各種道路之後，復於一直紋磚鋪道中；作詳細之考察，發覺其阻力常因日久而減少，蓋瀝青每由磚隙滲出，而將一部份之路面遮蓋也。國立鋪砌磚會(National Paving Brick Association)認識填塞物滲出對於路面抗滑性之影響，特於一九三三年，在愛希愛州立大學(Ohio State University)內發起調查之工作，其目的為欲得一滲出性較小之填塞物以替代瀝青。

工作之總監督為愛希愛州立試驗所教授，保羅博士(Dr. G. A. Fole)，直接管理者則為工程研究師，魯開羅博士(Dr. W. C. Rueckel)。其試驗方法，則為建築一小磚鋪道於一木板上，其接口填以各種不同之填塞物，同時用電施以一定[illegible]熱。其所得之結果如下：

溶解點高之混雜地瀝青，及雜有礦物之瀝青質土敏土，二者之滲出性均小，一種柔軟之硫黃亦頗適用。試驗所得，將以愛希愛公路局所建築於鶴金(Hocking)與化費(Fairfield)兩州間之哥倫布—亞坦司(Columbus Athens)路復驗之；此路共分十四段，每段試以各種不同之填塞物，其結果將可在最近期內得見之。

空氣調劑法導論

John Cushman Fistere

林 家 就 譯

近十年來社會人士對于建築物中之空氣調劑及其設備，悉心考慮鼓勵促進，將有無限之收効。工程界亦曾發表預論——將來一國之發展，工業之興盛，消貨之速率，生產之強弱，皆賴乎空氣完滿之設備，故空氣之重要，由此可見矣。玆將其理論分述如下：

(一)空氣調劑定義

空氣調劑之定義雖淺，惟其涵意甚廣，大底可從兩方面分析之：「一」關於大建築物如戲院，大寫字樓，及公衆場所等，其設備可分爲四類「甲」氣候管理即冷熱配置「乙」空氣中所含濕度若干「丙」潔淨「丁」分配(二)關於住宅，及小建築物等，其所用機件無須如前者之宏大，所採用者僅如下列數類而已。

(甲)熱氣爐，連有送風機及一乾濾清器。

(乙)熱氣爐而有送風機及滌洗機者(或濕度機及乾濾清器)

(丙)可移動濕度機，須用時則插入電氣臼節。

(丁)安置濕度機與熱水或蒸氣管相連

(戊)凍氣機安置房間者，以機器製成之冰爲之。

(己)熱氣爐配有送風機及滌洗機，至夏季則利用水花而使空氣變涼。

(庚)倣用大規模空氣致涼之法而簡爲之使有相當效率。

以上所述各種方式括入此題中，畧覺太泛，惟每一類方式，對於空氣調劑與溫度變態，有切實效能也。

(二)用意

溫度表雖能指示溫度，惟溫度與空氣所含之溫度，及空氣流動互相維繫也。

在某一限度內之空氣，能使大多數人(超過百分之五十)感覺舒暢不冷不熱者，是謂之空氣適中帶。

推此言之，所謂使空氣至適合狀態，即藉機械之力，將室內溫度，變爲空氣適中帶也。

(三)使涼法

何能致涼，空氣與冷面積接觸也，其法有二：曰水花　曰盤旋圈式。若採以水花式法，法無論天然井水、雪溶所得之水，或以機械力致凍之水，皆可適用。雖水喉之水用於水花式甚少，惟倘其溫度在華氏表四十度以下，亦能奏効。至若用盤旋圈式。所用者不外以冰凍水，或別種液體，或凍氣旋轉環繞之，而令熱空氣致涼而已。

四　濕度減輕法

減少空氣中濕度辦法有三，(一)水花式。此法最爲通行，其特殊之點，即先將已成水花之水凍冷，然後輸入水花池，空中水氣因與運入冷水花相觸而凝結，繼而達至浸透點，與空氣滌洗機將空中多量濕氣削除無異也。(二)利用澆圈及水花，同置一室，使室內空氣能與水花及澆圈接觸，就此兩法比較，前者所收効能較高，蓋水花與空中氣體相換熱力較高，且可省購置減低溫度機之費用。(三)與上述完全相異，其材料採用矽石床底，此種矽石，甚堅，且能吸收溫氣，其外表如沙，惟透孔甚多。當溫氣與此種矽石相觸時，密孔即將水氣吸收至達浸透點而止， 矽石既浸透後，可用熱力蒸發之，以待再用。

(五 空氣洗潔與分配

甲、雖然洗潔空氣之法，僅有漏隔或滌洗兩法，惟分配空氣法範圍甚廣，視乎屋宇欵式及大小而定也。

漏隔法有二，即乾濾與膠貼式乾濾也。前者乃以布質隔除塵垢，及別種有碍衛生之物，此種隔帳，宜常轉換，以免穢垢堆積，膠貼式乾濾則利用膠貼質以吸收空中塵埃，至於空氣滌洗，既可將空氣通過水花洗濯，亦可將之通過濕面上而滌洗，或兩者並用，則其効率較大工。照美國程調查所佈告，平常滌洗機効率有百分之七十，而濕度機等，所得効率，則達百分之九十五至九十八焉。

乙。空氣分配計劃，主張各異，例如某一屋宇，每人每分鐘需十立方尺空氣，或在另一範圍限內，則不然，空氣量數或增至三十立方尺不等。風扇摩打，氣管等物，對於空氣所需量數，亦有密切關係，故每凡計劃空氣如何分配，必先有切實查究，各處溫度，方向，然後按圖配備，方可稱善也。分配設計有三：(一)空氣向下流動法，(二)空氣向上流動法，(三)排噴法。無論任何建築物，皆可利用此法。

將種設計，非以樓之層數爲單位，祇依據各房舍位置所有熱度而統算也。至下述屋宇分類時，再詳提論。

(六)屋宇空氣調劑

上述各種空氣冷熱狀態，已稍論及，在此題下祇將使冷法及減少溫度法，畧加說及，近年來屋宇中及寫字樓等，多用繞圈 水花式或此將空氣輸過冰凍法。

用繞式圈法所用之機件多藏於牆中，或另一室中所用配件爲空氣滌洗機，減濕機滌洗機較爲妥善，因其在于極點熱度時，亦不失其効用也。再者，滌洗機之繞圈與冰凍器配件，安置手續簡便，價值相宜，且機下安設小輪，能隨時遷移，如心適從。欲使寬濶地方較凉，可增用此機兩副或三副。房屋中亦間有安設此者機，直透至會客室，日夜可用，晚上可用另一分管通達寢室。

大槪而言，小規模之建築物，用冰凍機者甚少，普通用以調劑空氣者，祇用管束溫度機與空氣旋轉速度器，使空氣達適合溫度帶，則夏日不致酷熱矣。

(七)空氣調劑要素

在未談論如何使大建築物中空氣至適合狀態前，先將設計時對於空氣影响之重要及關於房間熱度率連種種原素，畧說如下：

(1) 室外空氣乾燥極點與濕度極點，必先有的確審定，蓋室內空氣與室外空氣之乾燥點有關係，而室內空氣中含冰份亦因室外濕度高低而定也。

（２）屋內空氣，亦基於屋外間熱氣由墻壁，屋面，或樓板透入之多少與屋內設備傳去熱量之多寡。最繁雜者如窻門寬狹，日之熱力，屋方位向，及各處透氣量數等影响屋內溫度，亦甚重要。

（３）人類所發出熱量，亦不可忽視蓋閒靜之人，每小時發出三百B．T．U熱量，勞動者或跳舞者大約七百B．T．U；他如電燈及機器等亦有熱氣排洩也。

甲．酒樓與大商店

雖然酒樓與商店設計不大相同，然因其同屬營業性質，故可以並題而論．至於兩者重量則顯然相異．酒家則有重量蒸氣機，咖啡爐，電氣焙爐，人客重量不過暫時而巳。大商店則不然，日中人數無大增減，雖空氣調劑與分配法略有不同相差不遠也．在上等酒家及商店內最宜在假天花板內，與橫樑內，配設多量垂配鑲板，使冷氣得而洩出，下列圖樣係現在最普偏者。

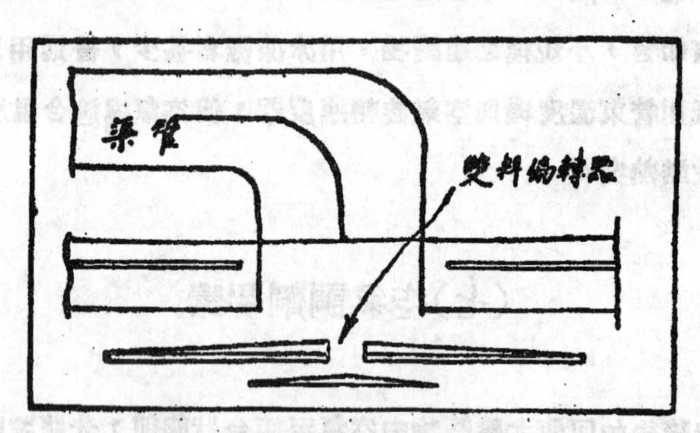

上圖乃一種新垂鑲板式
此種多用於空氣速度被減
少之處與面積廣濶地方

上圖乃明假天花板式新款空氣分配之法

乙·戲院

建設戲院計劃，非有富於調劑空氣經驗之工程師，從中策謀，不能得良善佈置也，在草案策劃中，尤須考慮，例如用機械式冷藏法，壓力凝氣機所應用之水，是否可隨時供給水源，或如何使該機不致有水荒之虞等，至若在樓頂上設一水花池，院內鋼枝亦須預計此項增量也·戲院中空氣分配法，不外三條件，(一)空氣量數應以號位數目而成比例·(二)介於下層與上層者應設流氣管·(三)空氣流動應與觀衆之面相向也·

舊式戲院所用方式與其配備，乃使空氣向上流·此種方式，計供給每人空氣由二十五至三十五 c.f.m.，而輸入之氣與室內溫度相比爲華氏表一度半，空氣向下流之法，氣管則藏於天花板內，傍近牆中或在後牆·臺前空氣則由吊下屋蓋之鐵爬，或孔穴放出，臺後及介於下層與上層者則由垂下鐵鏤板供給·細小屋宇，多用牆後供給空氣之法，多數戲院所供給冷氣由臺前放洩，惟狹窄者則採牆邊施放之法·

現在普通冷氣機多用排噴之法，其所用氣量少且溫度低，冷氣由排噴機發出時速度亦較爲高，據安道馬列著名工程師所說·此法欲得其舒收効最要者，院內屋蓋平坦，勿以垂樑；燈光及柱等與噴出之氣相衝·戲院大小，亦不能超過限數，如高度不逾過二百尺外等，至若高度與長度比例相差太遠，此法亦須審愼考究方可實施，氣之噴出如斯，而濁氣則由牆邊出口與牆後引洩或由地面之出入氣管機與臺底暗管吸收，兩者收効相同也·

丙、辦公室

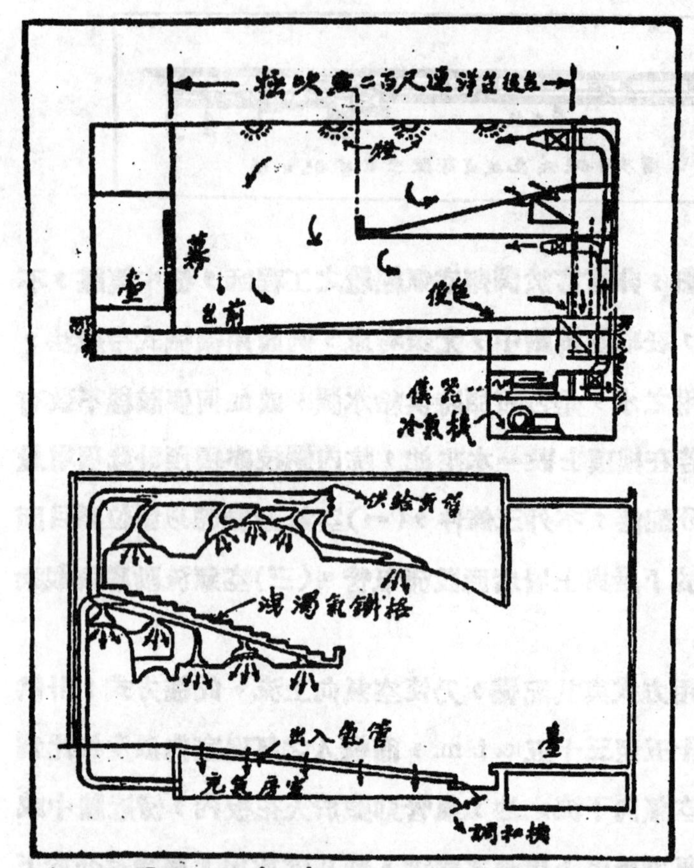

上兩圖乃新建戲院空氣分配方式
上圖為放射法冷氣則由內射口放出
射出時速度甚高下圖乃空氣向下流
法洋臺底暗管與充氣房室用以輸運
濁氣此法較上圖所用空氣倍增

大寫字樓中，所配置氣機繁雜內可分地帶，治理，以單位法管理，樓中策劃最要者莫如選擇機器安設之位置故許多承辦此項工程公司對於配設機件位置多以普通之法為之。如減濕度機安設多與樓面平衡，而每層則加用電器升壓機扇。致涼機則移放於地窖，惟分配着要點則因樓之高度，樓內設備，屋蓋構造等而定。

凝者出口格子多設於屋蓋附近之走廊上，廢氣格子則設於門前或內牆內，排洩外出。走廊亦可作吸廢氣管，近年則安設低壓管嘴於窗下或走廊各部，而將廢氣排之於外，此二法各有所長，用者則視乎屋內策劃適合何種而採用之，大建築之凝結機所需水量甚多，策劃者每多忽略之，故無水花池不足為用也。

(八)循環法

不獨冷氣如何輸入屋中或廣場中爲重要，廢氣之輸出而再使與外來鮮氣渾合以供再用亦同重要，因其於經濟上亦大有帮助也。

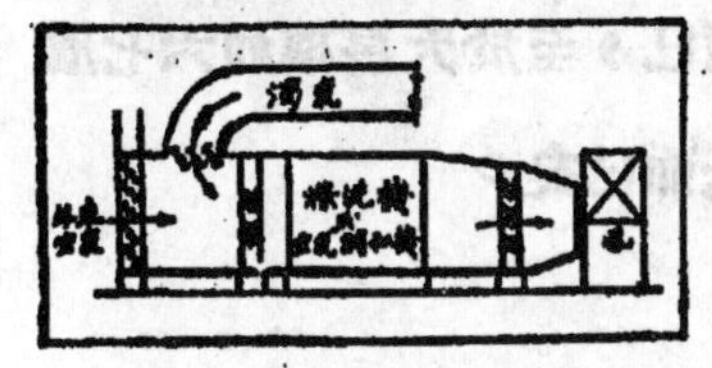

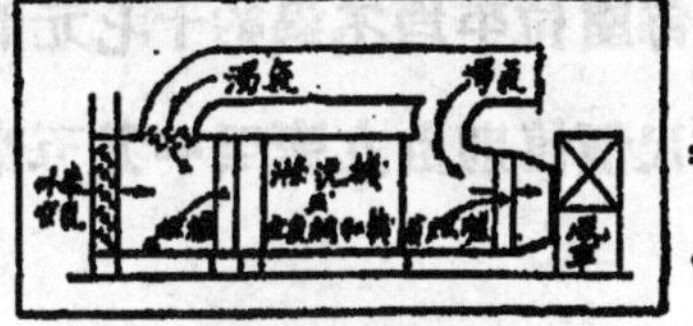

下列二圖雖畧有不同然其理一也。

(九) 隔聲法

劇場之聲，對於觀衆或辦事者殊爲騷擾，爲避免計，應建築暗渠與隔聲幔以阻止聲音外達。下列圖樣詳示隔聲設備之內容。

(十) 價 值

裝置氣機，價目不一，如單位凍機爲住宅及寫字樓等用，其價值則由四百元至八百元，視其容量與欵式定，又如集中調和空氣機除凍氣機配件則由一千元至二千五百元下項數目記示調和機除凍氣附帶機所值價格——

用於房間數目	價值
十	一千二百元
十二	一千五百元
十四	一千八百元
十八	二千元

附帶凍氣機則由二百元至一千元，至於配置酒樓及大商店等之價值，當比較更高，數年前戲院所置凍氣機等價值大爲減跌，因買客日見增多，而消流較廣也。五年前計有凍氣機等之戲院每座位估價由四十元至五十元現每座位平均不過約十七元而已，至於大房樓如六七層者其配置費每方尺價值則由九毫至一元五毫而已矣。

新工藝及其研究一束

("Techical New ands Research"-
The Architecturol Record. January. 1935.)

鄒煥新譯

(I)金屬之塗油外層保護

甲•在芝加哥電力化學會中提出之關於金屬外層塗油保護一文•已盛載於美國各學術雜誌矣。此文大致偏重於英國近年來關於此項問題之試驗，其結果可槪括於下舉四則：

1.金屬之性質。

2.離解塗油與金屬之物質之存在如鐵銹，水，及鹽等。

3.以顏料及油等之性質及其品質而斷定之塗油之性質。

4.空氣，水或油之性質，此種物質與已塗油之金屬接觸者。

塗油於金屬外層之初次油，多屬紅鉛，朱砂，麻子油及混以小量之促乾劑之松節油。此種塗油已風行五十餘年，且收效亦頗著•惟近年來商業部曾發表謂因鐵銹及腐敗之損失，每年達三萬萬元。此項鉅大之損失，誠與建築上所應用之鋼鐵之用途成正比例也。

塗油研究學家經悉心之考驗，皆謂上至銅鋼及電力製造之純鋼，下至含有多量炭及鎂之鋼，皆與塗油發生不同之化學作用，此種作用與濕度，水份，酸素，氣體及鹽實有關係焉。

乙•近年來腐壞之試驗結果

近數年來諸化學師對於此等試驗已不遺餘力。探求著名之紅鉛，何以不能充當防止鐵銹之保護劑。且已作下列數項具體之試驗：

試驗一：取鋼數斤，截之成角，片等形。經沙擦及線擦法磨淨後，以紅鉛，及麻子油與松節油混合溶液，使之曝露於工業中心區域，且近鹽水之建築物之屋頂上。經二載後麻子油已分解殆盡，祗餘一層紅鉛之薄膜及含多量水分之碎鉛矣。蓋紅鉛中之水分成一導電體。經電解作用後該金屬遂告腐壞且生鐵銹。久之此碎鉛更溶解於水而被沖去。此時該金屬已完全曝露於空氣中矣。

試驗二：取氧化鐵和以麻子油與紅鉛混和，此混和之溶液已被公認爲化學上及機械學上用麻子油調開紅鉛之妙品。若以顯微鏡視察此混和溶液，即見紅鉛及氧及鐵之粉末適塡麻子油間之微孔。惟經兩年之考驗，即知水分可透穿此種塗油而使之腐壞也。

試驗三：取氧化鐵和以無孔之塗油媒介劑，用以塗上數金屬片之外層，即起氧化作用而生一層薄鉄銹。此種塗油經此而成一導電體，圍繞該鉄銹之微粒，則可防以後之氧化及生銹矣。經兩載之考驗，此外層塗油已証明爲一種完全之佳品，雖彼常曝露於風雨陽光中亦不雖破裂及水分之存在，故此實驗可稱滿意。可供工業及航海輪船之用。後又証明此塗油更可用於常受磨擦及一切機件焉。

試驗四：取鉻酸鹽和以麻子油，此塗油與紅鉛麻子油之情形無大差異，惟前者較覺耐用也。

試驗五：取純鋅末(百份之九十九之成分)和以麻子油。此種塗油甚不適宜於保護之用以其具有麻子油之微孔故也。

試驗六：將鋅末和桀膠，塗於潔淨之鋼鉄片上，其結果至爲滿意，蓋該塗油媒介劑成一無孔之薄膜，而鋅末又與鋼鉄性質相似。此種塗油曾用數艘渡洋之輪船——未用此塗油之前，先將船身擦淨，並盡去一切以前船身所用之塗油，惟六月以後，此種塗油漸次脫落，若以顯微鏡驗之，卽見此塗油間發生無孔微孔，使此種塗油與金屬之外層脫離焉。

丙·鋅之塗抹試驗

試驗七：用掃帚將鋅塗沫於鋼版上，此工作並非困難，且塗沫之結果亦佳成績較用紅鉛爲美滿。

丁·結論

誠然此種試驗實有助於工藝塗沫之製造及應用，以去除鉄銹及防止腐敗，雖其功效亦有定期，惟最低限度比通常用以初次塗沫各種金屬之紅鉛長久二倍也。

考塗油製造雖能操縱鋼鐵製造事業，惟此種保護塗油事業已成一專門問題，與煉油家之對于汽車噴油事業殊不相伯仲，因此種塗油能適用於大小各項之機件器具也。

從試驗中可知初次塗油之時間，應在用該金屬外層發生氧化作用及除去垢積之後。且此種工作應擧行於製鋼鉄之所，及當鋼鉄甫鑄成時，而於如經建築後之鋼質構造物等情形尤覺重要。

因不良之初次塗油以致鋼鉄容易生銹及腐敗之例，已不勝枚擧，故應用塗油者誠屬要事，使用不良之塗油與浪費金錢有何異哉。

(II)隔熱之研究

甲　邇來隔熱體之應用於建築及各種營造，已成一經濟上之重要問題，其原因有四：

(一)隔熱之物質如蔗渣，乾禾草，廢木及各種含纖維之物，以前皆被認為廢物今已成一大宗應用物矣。此種廢物之應用獲利有二焉：此種廢物之利用其一也，舊材木之應用及由森木間接獲得之新用途其二也。

(二)燃料之新經濟及由此種隔熱物體有在寒暑季候中皆能增加快感之效用，且在一應用隔熱物體完美之房屋，可保持室內溫度之勻調，不致受窘於驟寒或驟熱之苦也。

乙　關於隔熱體之研究，固有根本事實可據。試舉鋼，石，木及棉諸物以作討論之資料無錯。若此四物體充當隔熱體時，則其大概之比率為 1:60:400:1,200 。即謂石比鋼之效用大 60 倍，木比鋼大 400 倍，棉比鋼大1200倍也。

事實尚屬如斯。惟從此等鋼石木棉諸物，更可製造其他隔熱體使其效用能追棉之應用者。如鋼棉與棉之效用比率為 1 與 2；石棉更能與棉並駕齊驅也。

由是觀之，於隔熱價值之立場上而言，物體本身之構造成纖維物體時，吾輩即可利用空氣以作隔熱之物。蓋纖維質能防止熱之傳播及對流作用也一據理論而言，熱之傳播不能在靜止之空氣中發生作用。纖維物體則畧傳熱，此量之大小乃視其物體之本質而定，故上舉諸物其比率懸殊亦此理也。

丙　製造法在隔熱之價值上之影響

纖維質之隔熱價值，乃賴於製造之方法。從同一物質中，能造成兩種不同價值之導熱體。此無非視乎其纖維之排列，凝聚此等纖維質之物(松香，瀝青等)及綑扎之密度等項而定而已耳。作者曾作一成功之實驗，法用細麻皮造成兩種不同之標本，其密度相同。一標本之纖維之排列方向與熱流方向垂直，一標本之纖維之排列方向與熱流方向平行，從試驗所得之結果則知第一標本之隔熱價值較高於第二標本。蓋第二標本之纖維質成一導熱體而第一標本之隔熱價值較高於第二標本者，因第二標本之纖維質成一導熱體而第一標本之纖維質則抵拒熱之傳播也。化學藥品能使纖維質不受潮濕及火患，惟此等藥品足以減低其隔熱之價值，且物體之隔熱能力亦頗賴其質量，大概其質量愈小則其隔熱體愈高，此已爲規律矣。

丁. 空氣之隔熱

試舉磚牆而論，此牆導熱固屬熱之傳導作用，蓋凡固體皆以此作用以導熱也。倘吾等除去此牆中間之磚，則惟取空氣以充滿於間而已，若以木屑代空氣則其隔熱能力必大於磚牆本身，因木屑之導熱能力較低於磚，若取金屬代木屑則當不能增加隔導熱之能力，此空間之物體補充愈疏則隔熱能力愈大，亦一規律也。

吾等試使其牆中之空間絕不補充物體，則此中空之牆之隔熱能力較大于完全牆磚乎？問題則如此，究竟此等牆是否較補充疏鬆之物於空間之牆較爲佳，欲解決此問題則吾等須明瞭空氣之隔熱能力也。空氣具有一特性異於固體及疏鬆之物體者，在乎後者具有之特性爲其物愈厚則導熱能力愈大，空氣則不盡然焉。從極薄層之空氣論起愈厚則隔熱能力愈大惟至四分三吋則其性突變，若至二吋或三，四吋以外，其隔熱能力之增加極微矣。

上舉之空氣之特殊性質，其理由乃如斯：空氣之導熱也作用凡三，卽傳導，對流，及輻射是也。第一作用與普通導熱無異，第二作用則因空氣之流動，於相當限度內其空間愈大則導熱愈多，第一作用與第二作用互處反對狀態，在四分三吋之外則成一同定情形也第三作用與空間之大小絕無關係，此三種作用同時並行。

空氣之特性何故若斯反常之問題茲已可解答矣。今吾等再求何時需以碎屑之物以填充空間之問題之解答。據熱在空間傳播之事實，吾等卽知在狹窄之空間，若補充以物體其傳熱率小於。.6者，則可得隔熱之效果，又若以一傳熱能率高至2.。之物體，填充一寬達三英吋之空間，亦可收隔熱之效， 本來填充之物體之傳熱能率愈低， 則上舉二項之隔熱愈大焉。

戊 熱之反射隔熱作用

邇來利用金屬反射以為隔熱之途劇增。吾人若知在空間有何物發生，則當知一絕好之導熱體能供隔熱之用之玄妙。如上述三種在空間之傳熱作用，其中一屬輻射傳熱。此輻射乃非可見之光，亦非X光，又非無線電波。在其大概之特徵而言，彼實與該三者相似名 infra-red radiation 此種輻射非人眼可能睹， 惟可用 Bolometer（抵抗微熱器）及 Thermopile(熱電堆)試察之。通常實驗中吾人所謂 “Black surface” 者，卽能全吸可睹之光平面也， 若較以 Infra-red 光則此雜誌之紙張已屬 Black surface 矣；卽謂此白及反射可人目能見之光之紙張實為一 Infra-red 光之吸收器。實際上普通供建築之用之石，磚，木，紙若就 Infra-rea radiation 而言，皆超過 90% black 矣，故建築物之牆壁中之空間，乃在良好吸收輻射熱度之物質所圍繞，蓋此輻射之熱侵襲該物質也。

光彩之金屬表面乃良好之熱反射品，彼非但能反射肉眼能見之光線，並能反射 Infra-red 光線。故此種光彩之金屬塗抹，幾成一完全之反射品，而熱之傳播於空間，祇行傳導及對流二作用而已矣。

此塗抹雖防止輻射作用，惟傳導與對流之作用，仍未能免。從實驗之結果中，吾人知熱之傳播於空間，輻射作用佔 50%至80% 故能避免輻射作用。則大體上不成問題顯矣。

(Ⅲ) 建築材料之研究

近來吾等已從人類之食品及衣業之供給之追求，漸移至適應於萬物之化學物品，甚至大自然中最卑賤之有機廢物，或亦有其特別用途者。試將木屑滲透於 Phenolformaldehyde condersation 媒介劑中即成一膠質，能代木質燃料也。邇來經切成細片而有組織之膠質體已爲有異常靭力之片狀物。若將此物夾以經壓成薄片之金屬，則成一具有驚人靭力及耐用之片狀物。將來之汽車及外殼及房屋中室內之打磨材料，將有以賴之。此乃千萬例証之一，由此吾人可明示從無機化學而引用至有機化學之成效也。

木質中之 Lignin 及 cellulose 能作建築用之禦火及抵抗物，此亦一應用有機質原料之例也。通常之葡萄糖及菓子糖亦有相當貢獻，蓋糖乃一絕妙之建築用品，且在最近之將來吾人當能用之以作膠質水管，若據吾輩祖宗而言，以爲糖祇能供食料之用，則大謬矣。

編後語

—編　者—

本刊出版過於忽忙，錯誤自所不免，且未能將全數稿件付梓；如黄謙益先生講述之"城市設計"，葉作霖君譯述之"塌底急流之防禦工程"等。又黄郁文先生之"廣州市空氣調劑設計"，王叔海先生之"意大利之發電機"，彭奠原君之"工程標準名詞"，及容永樂君之"東北江杉木之試驗"均未克脫稿。惟此刊之內容及印刷稍較前刊進步，聊引為快也。

今屆出版部廣告主任梅國超君服務努力，任事精幹，校對主任陳守勤君校對精密，不厭求詳，殊堪敬佩，特誌之以表謝忱。

THE JOURNAL OF THE LINGNAN ENGINEERING ASSOCIATION

VOL. 3 No. 1

Edito-in-chief Chow Woon Sun

	Pang Ting Yuen	Advertising Manager	Chester Moy
Associate Editors	Chan King Hung	Proof-reader	Chan Sau Kan
	Lung Po Yuk	Circulating Manager	Tsang Tak Poo

中華民國二十四年六月十五日出版

南大工程

第三卷 第一期

(每册四角) 郵費··本市二分··國內五分··國外三角

編輯者……………………………………鄒 煥 新

出版者……………………………………南大工程學會

發行者……………………………………南大工程學會

總發售處…………………………………南大書局

分售處……………………………………各大書局

印刷者……………………………………洛陽印務館

25381

廣州嶺南大學工程學會會刊

嶺南工程

【第四卷】　【第一期】

中華民國二十五年六月一日

附鋼筋三合土設計手册

THE JOURNAL OF THE LINGNAN ENGINEERING ASSOCIATON

VOL. 4, NO. 1.　JUNE, 1936.

英商怡和機器有限公司

（即渣甸洋行）

經理各國名廠專家出品

發電採礦航空航海農墾水利紡織救火鐵路印刷各種機械及製冰機冷氣機升降機潔淨具軍用品等諸君惠顧無任歡迎

香港：必打街十四號
廣州：沙面英租界

SIXTY-FIVE OF THE LEADING ENGINEERING MANUFACTURERS OF THE WORLD EACH SPECIALISING IN A PARTICULAR BRANCH OF ENGINEERING ARE REPRESENTED IN CHINA

BY

THE JARDINE ENGINEERING CORPORATION

LIMITED.

(Incorporated under the Companies Ordinances of Hong Kong)

HONG KONG,
14, Pedder St.

CANTON,
Shameen.

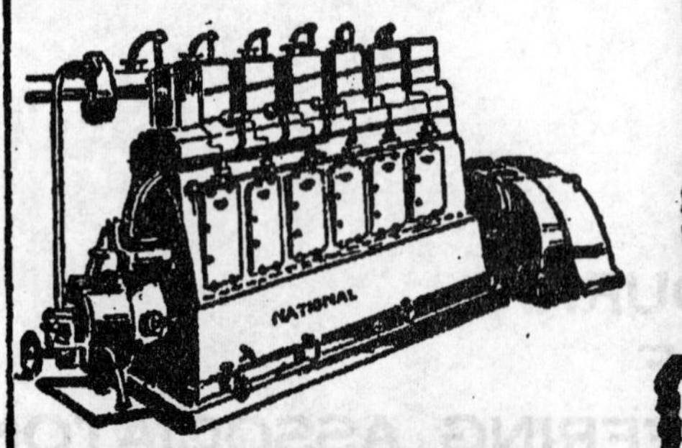

THE NATIONAL GAS ENGINE Co. LTD
ASHTON-UNDER LYNE.

南 大 工 程

廣州嶺南大學工程學會會刊

第四卷第一期目錄

投稿簡章

(一)稿件係屬自撰或翻譯，均以關於工程者爲限。中文或外國文均可。

(二)外國文稿件，暫以英德法三文爲限。

(三)來稿請繕寫清楚，如有附圖附表，尤須繕寫清潔。外文稿件打字寄來尤佳。

(四)如係譯稿，請書明原著書名篇目，出版地點日期。

(五)稿末請注明姓名住址，以便通訊，至揭載時如何署名，聽投稿者自便，作者能將詳細履歷敘名尤佳。

(六)投寄之稿在二千字以上者，如不揭載，編輯部於每期出版後，按址郵寄回。

(七)來稿經揭載後，酌酬本刊是期多本。

(八)來稿編輯部或酌量增删之，但投稿人不願他人增删者，請預先聲明。

(九)來稿請逕寄廣州嶺南大學工程學會出版部收。

連續式橋樑設計

李卓傑

第一節　導言

近世運輸日繁，車輛之載重日增，橋樑之安全及經濟之設計，誠一極重要之問題。我國現有之鐵路橋樑，多因設計時之年代過久，不獨無一定之規範標準，即橋身之載重能力，亦不能與現定之標準載重相符。其中有跨度較大者，設計時因利計算簡易故，多用數座單式橋聯合而成之。然此種設計，從結構上而言之，橋身當然不甚剛結，而車行其上，必需將速率減底，以減少衝擊應力，使橋安全。此種橋樑，不獨乘客及貨物之全安發生問題，而路局本身，對于車輛之載重減少，及速率之減底，亦爲一重大之損失。是故如何能使車輛之載重及速率增加，如何能使貨客安全，實乃路局當前急切之任務。然欲解決此問題，橋樑之加固，及改建剛結之橋身，即其要也。

較大跨度之橋樑，近世所採用者有多種，如懸臂式橋樑，連續式橋，懸橋等，此上各種，各有其特殊優點，但何者爲最適宜，則須視該處地勢情形而定。本文所述，只限于連續式橋樑而已。連續式橋樑較諸單式及懸臂式橋之優點如下：(一)可以節省材料；(二)可以用懸臂式方法安架；(三)建造可以免去一切膺架等材料；(四)可得剛結之結構；(五)免除鉸鏈端設計之麻煩。以上諸端尤以第二點較爲重要。普通連續式橋樑所需要之條件，除良好之地基外，更需橋墩較高，橋空大約相等，及跨度較大。而跨度愈大，則連續式橋樑較諸他式愈爲經濟。

第二節　橋之形式

普通鐵橋之形式分有多種：

（一）上行橋，(Deck Bridge)。上行橋又可分爲：工字梁(I-Beam Bridge)，鐵板橋 (Steel Plate Ginder Bridge)，鐵桁橋(Steel Truss Bridge)

（二）下行橋 (Through Bridge)，下行橋又可分爲：鐵板橋，鐵桁橋數種。然用於連續式橋樑者，普通以鐵板橋及鐵桁橋爲多。

鐵桁橋之中又可分爲白式(Pratt Truss)，波式(Battimore)，必式(Petit)，及王式(Warreinr Truss)等式。但白式，波式等式用在單式橋梁，則甚經濟。至於連續式橋，則以王式(Warren Truss)爲佳。蓋其腰支撑(System of Web Bracing)甚適應於連續式橋受力情形故也

第三節　連續式鐵橋路之衝擊力

連續式鐵路橋之衝擊力，在各規範書中尚未有規定。但在已成之連續橋樑設計所用之衝擊應力公式，乃根據設計者之經驗而定。例如美國奧亥奧省橫跨奧亥奧河之(Sciotoville Bridge)橋，乃一兩空鏈軌連桁橋，每跨度之長等於七百七十五尺。其所用之衝擊力公式：

$$I = \frac{L}{D+L} \times \frac{1200+\frac{a}{n}}{600+4a}$$

D ＝ 靜重應力（磅）

L ＝ 活動應力（磅）

a ＝ 橋受最大應力時在車頭後之車長（尺）

n ＝ 受最大應力時，荷重軌道之數

橫跨(Nelson River)之三孔連桁橋，兩傍跨度爲300尺，中間跨度爲400尺，依古柏民活重E-50單軌設計，其所用之衝擊力應力公式：

$$I = \frac{(L+0.4L^{1})^{2}}{L+D}$$

L ＝ 最大活重應力

L ＝ 最小活重應力

D ＝ 靜重應力

美國 奧亥奧省，Cincinati，之三孔連桁鐵路橋所用之衝擊應力公式：

受相同性質之應力者（Stresses of Single Character）

$$I = \frac{300}{300+L} \times \frac{\text{活重應力}}{\text{活重應力}+\text{靜重應力}}$$

L ＝ 受最大活重應力時之荷重長度。

受二種與反應力者（Reveral of Stresses）

$$I = \frac{300}{300+L} \times \left(1 + \frac{m}{2M}\right)$$

M ＝ 同性質之活重應力及靜重應力之代數和。

m ＝ 相反性質之活重應力及靜重應力之代數和。

但若相反性質之活重應力小於靜重應力時，則上式不能應用

第四節　跨度之經濟比率與橋孔之數

橋樑之經濟與剛硬度，乃與連續橋孔之數成正比。橋孔愈多，則此橋愈爲經濟及硬剛，然此理祇限至三個或四個橋孔而已，因橋空太多，則橋樑之伸縮設備甚難，且建橋時因橋托多之故，而橋托高度需用較準（Jacking Operation）亦多。如二空橋，祇有一橋托高度需用較準（Jacking Adjustment），而六空橋，則有五個橋托高度需用較準因此之故，橋空宜多至三個或四個爲限。

兩空之連續橋，從經濟及美觀方面而言，當以兩空相等爲佳。

三空之連續橋，其跨度長之經濟比率，大約是 7:8:7 ，倘此比率需要變更

時，亦當以不影響其經濟比率爲條件。

四空之連續橋，其跨度長之經濟比率，大約是 3:4:4:3。但此比率可以變更少許而不致於影響橋樑之經濟問題。

但有時跨度之佈置，常依地勢之情形，及建築時之便利而定之。

第五節 應力之計算

連續濡梁應力分析之第一步驟，乃先繪任一未知反應力 (Unkown Reaction)之感應線(Influence Line)。此感應線乃將來結構各部之應力計算之基礎。感應線，可以用在受力點施一單位之變位 (Unit Displacement) 時此生之撓度圖 (Deflection Diagram) 之方法繪之。因爲此撓度圖，或此曲線與結構之真正荷重無關，祇乃結構內之彈性關係而已。是故此感應線又可稱之爲彈性曲線 (Elastic Curve)

連續橋樑設計之初步，乃先假定該橋爲一不變慣性力率(Constant Moment of Inertia)之結構，繪就彈性曲線，計算應力，定該結構各部之剖面，然後依此剖面，從新計算該結構之各剖面之慣性力率，再繪一變動慣性力率彈性曲線 (Elastic Curve for Varable Moment of Inertia)，用以前之方法設計之。

一•不變慣性力率之彈性曲線 (Elastic Curve for Constant Moment of Inertia)——假設一不變慣性力之二孔連續樑A B C ，承托於三支點之上，如第一圖。

第 一 圖

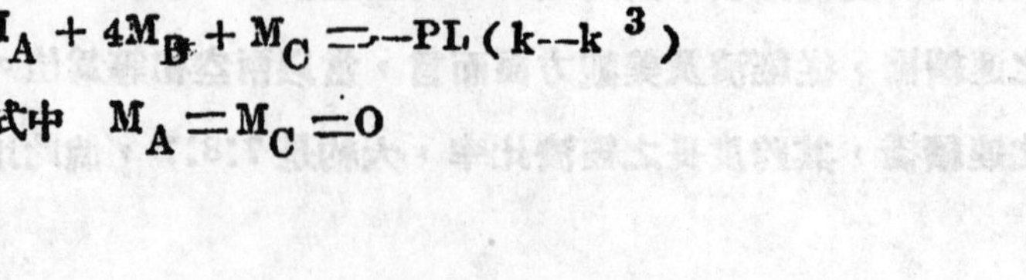

$\leftarrow Kl \rightarrow$ P

A B C

L L

R_A R_B R_C

在任一跨度上如 A B ，施一集中荷重 P 。根據三力率之定理：

$$M_A + 4M_B + M_C = -PL(k-k^3)$$

於上式中 $M_A = M_C = 0$

得 $M_B = -\frac{PL}{4}(k-k^3)$ …………………………(1)

由公式(1)得A點之應力：—

$$R_A \times L - P(l-kL) = M_B = -\frac{PL}{4}(k-k^3)$$

$$R_A \times L = \frac{PL}{4}(4-5k+k^3)$$

$$R_A = \frac{P}{4}(4-5k+k^3) \text{ ……………………(2)}$$

C 點之應力：—

$$R_C = -\frac{P}{4}(k-k^3) \text{ ……………………(3)}$$

B點之應力：—

$$R_B = P - R_A - R_C$$

$$= P - \frac{P}{4}(4-5k+k^3) + \frac{P}{4}(k-k^3)$$

$$= \frac{P}{2}(3k-k^3) \text{ ……………………(4)}$$

今假定P 爲一移動單位集中荷重，則 A 點應力之感應線或彈性曲線，依公式(2)計算其結果如第二圖將示。

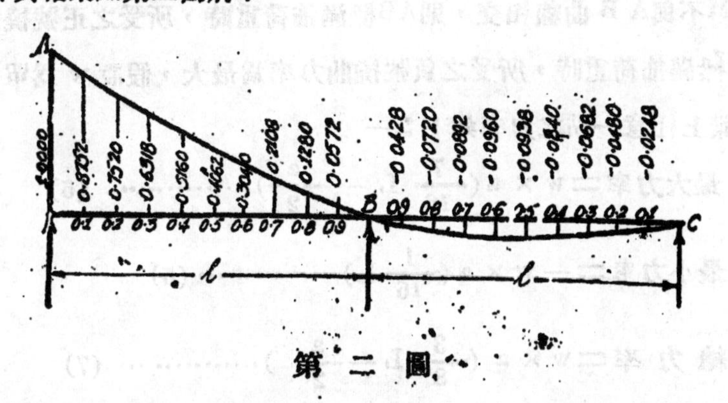

第二圖

感應線與 AB弦所成之面積，等於$+\frac{7}{16}$ l；感應線與 BC 絃所成之面積等於$-\frac{1}{16}$l。其總面積則等於$\frac{7}{16}L-\frac{1}{16}L=\frac{3}{8}L$。若AB樑受滿佈荷重 w

時，則A點之應力爲 $\frac{7}{16}$ wL；若BC樑滿佈重荷w時，A點之應力等於 $-\frac{7}{16}$ wL；若全樑 ABC 皆滿佈荷重 w，則 A 點之應力，當爲 $\frac{3}{8}$ wL

面積在感應線之下者，乃使A 點生正號之應力，面積在感應線之上，乃使A 點生負號之應力。

二・撓曲力率之感應圖——在 A B 樑上任何一點m之感應圖，可繪--直線聯A及m兩點而成之・m點之撓曲力率：則等於應圖之面積乘以A至m之距

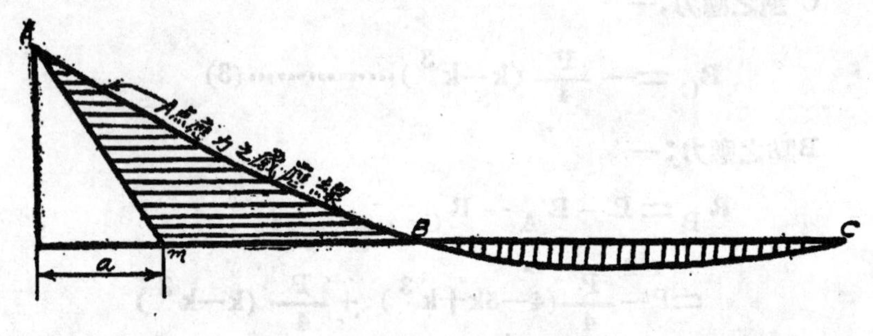

第 三 圖

離a・此數a乃該感應圖之常數也。在感應線以下之面積，乃代表正號撓曲力率，在感應線以上之面積，乃代表負號之撓曲力率。

若AM不與A B 曲線相交，則AB樑滿佈荷重時，所受之正號撓曲力率爲最大；B C樑滿佈荷重時，所受之負號撓曲力率爲最大，假設 w 爲單位長之荷重，則AB樑上任意一點之力率如下：—

最大力率 $= w \times a(\frac{7}{16}L - \frac{a}{2})$ ……………… (5)

最小力率 $= -w \times a(\frac{1}{16}L)$ ……………… (5)

總力率 $= w \times a(\frac{3}{8}L - \frac{a}{2})$ ……………… (7)

若全樑皆滿佈荷重，則樑在 $a = \frac{3}{4}L$ 之處，有一反撓曲點(Countraflexure)此點之撓曲力率 $= 0$。而樑之最大撓曲力率等于 $\frac{9}{128}Wl^2$ 乃在 $a = \frac{3}{8}L$ 之

處。若祗AB滿佈荷重，則樑之反撓曲點(M＝0)在$a=\frac{7}{8}L$。而此時樑之最大撓曲力率乃在$a=\frac{7}{16}L$之處，其量等於$\frac{49}{512}wL^2=0.957wL^2$

若a超過0.8L時，AM與AB線相交使成分段荷重之感應圖。故此臨界點(Critical Point)，a＝0.8L，稱爲固定點(Fixed Point)。

兩等空之連續樑各點之最大，最小，及總力率，依公式計算，其結果列于第一表中。

點(L)	最大力率(wL^2)	最小力率(wL^2)	總力率(wL^2)
a＝0	0	0	0
a＝0.1	+0.0388	−0.0062	+0.0325
a＝0.2	+0.0675	−0.0125	+0.0550
a＝0.3	+0.0862	−0.0188	+0.0675
a＝0.4	+0.0950	−0.0250	+0.0700
a＝0.5	+0.0938	−0.0312	+0.0625
a＝0.6	+0.0825	−0.0375	+0.0450
a＝0.7	+0.0612	−0.0438	+0.0175
a＝0.8	+0.0300	−0.0500	−0.0200
a＝0.9	+0.0061	−0.0736	−0.0675
a＝1.0	0	−0.1250	−0.1250

活重用　　靜重用

三•剪力之感應圖 — AB樑上任一點S之剪力感應圖，乃由S點劃一單位高度之線與ASB相垂直，然後再繪一直線AS聯A點如第四圖所示，其面積至感應之下者，代表正號剪力，其面積正感應之上者，代表負號剪力。若SB佈滿荷重時，則S點受最大之正號剪力，AS及BC滿佈荷重時，則S點所受之負號剪力爲最大。

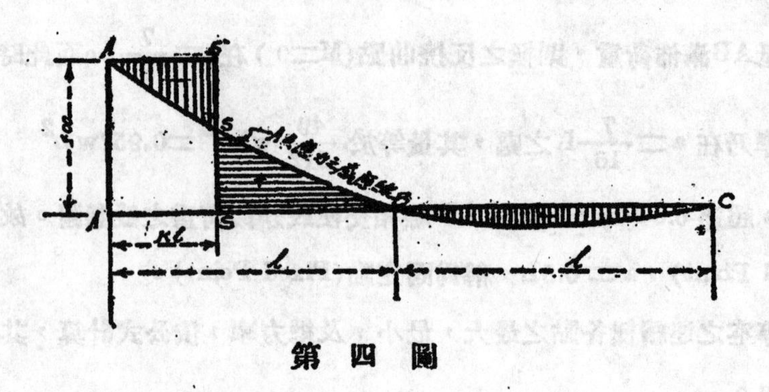

第四圖

由公式（2）：$R_A = \frac{P}{4}(4-5k+k^3)$

設y 爲此弾性曲線之縱座標

則 $y = \frac{P}{4}(4-5k+k^3)$

$$= P\left(1-\frac{5}{4}k+\frac{k^3}{4}\right)K$$

AA'SS' 之面積

$$A = \int_0^k y\,dk$$

$$= P\int_0^k \left(1-\frac{5}{4}k+\frac{k^3}{4}\right)dK$$

$$= P\left(k-\frac{5}{8}k^2+\frac{k^4}{16}\right) \quad \cdots\cdots(8)$$

設 w 爲滿佈單位長之荷重

則離A點kl之剖面S所受最大正號剪力

$$= wL\left[\frac{7}{16}-\left(k-\frac{5}{8}k^2+\frac{k^4}{16}\right)\right] \quad \cdots\cdots(9)$$

離A點kl之剖面S所受最大負號剪力

$$= -\frac{1\,wL}{16}-wL\left[k-\left(k-\frac{5}{6}k^2+\frac{k^4}{16}\right)\right]$$

$$= -wL\left(\frac{1}{16}+\frac{5}{8}k^2-\frac{k^4}{16}\right)$$

若全樑均滿佈荷重W時，S剖面之總剪力

$$=wL\left(\frac{3}{8}-k\right)\cdots\cdots\cdots\cdots(10)$$

根據以上公式（8），（9），（10），計算兩等空連續樑之最大，最小及總剪力，其結果如下表所示：——

點（L）	最大剪力(wL)	最小剪力(wL)	總剪力(wL)
a=0	+0.4375	—0.0625	+0.3750
a=0.1	+0.3437	—0.0687	+0.2750
a=0.2	+0.2624	—0.0874	+0.1750
a=0.3	+0.1932	—0.1182	+0.0750
a=0.4	+0.1359	—0.1609	—0.0250
a=0.5	+0.0898	—0.2148	—0.1250
a=0.6	+0.0544	—0.2794	—0.2250
a=0.7	+0.0287	—0.3537	—0.3250
a=0.8	+0.0119	—0.4369	—0.4250
a=0.9	+0.0027	—0.5277	—0.5250
a=1.0	0	—0.6250	—0.6250

（最大剪力、最小剪力）活重用　（總剪力）靜重用

四·不等橋孔之連續橋樑—— 不等橋空之連續樑，亦可用上文所述之方法及手續設計之。假設一兩不等橋空之連續樑ABC，其跨度長爲 l 及 l_1（l_1=nl）。如第五圖，在樑AB上施一集中荷重P：——

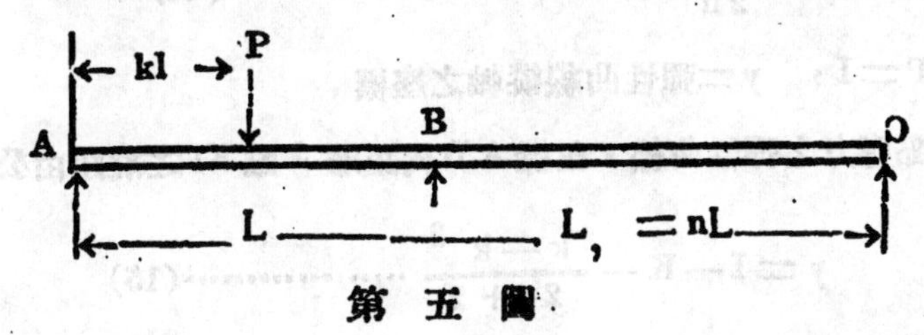

第五圖

25395

根據三力率之原理

$$M_A L+2M_B(L+nL)+M_C nL=-PL^2(k-k^3)$$

$$M_a+2M_B(1+n)+M_C n=-PL(k-k^3)$$

於上式中 $M_A=M_C=0$

則 $M_B=\dfrac{-PL}{2(1+n)}(k-k^3)$ ……………………(11)

從公式(B)得A點之應力

$$R_A\times L-P(L-kL)=-\frac{PL}{2(n+1)}(k-k^3)$$

$$R_A-P(1-K)=\frac{-P}{2(n+1)}(k-k^3)$$

$$R_A=P\left(1-k-\frac{k-k^3}{2n+2}\right)\text{……………………(12)}$$

C 點之應力

$$RC=\frac{M_B}{nL}=\frac{-PL}{nL(2n+2)}(k-k^3)$$

$$=-\frac{P}{2n(1+n)}(k-k^3)\text{…………(13)}$$

B 點之應力

$$R_B=P-R_A-R_B$$

$$=P-P\left(1-k-\frac{k-k^3}{2(n+1)}\right)+\frac{P}{2n(1+n)}(k-k^3)$$

$$=P\left(k+\frac{k-k^3}{2n}\right)$$

$$=\frac{Pk}{2n}(1+2n-k^2)\text{……………(14)}$$

若令 $P=I$， y=彈性曲線縱軸之座標

則 A點應力之彈性曲線，在樑AB內距離A點Kl之縱距由公式(9)

$$y=I-K-\frac{k-k^3}{2n+2}\text{………………(15)}$$

25396

B點應力彈性曲線，在樑AB內距離A點Kl之縱距

$$y = \frac{R}{2n}(1+2n-K^2) \cdots\cdots\cdots\cdots (16)$$

假設在樑BC，距C點KL，處施一單位之集中荷重，依同一之方法，可得： A點應力彈性曲線在樑BC內距C點KL，處之縱距

$$y = -\frac{n^2}{2n+2}(k-k^3) \cdots\cdots\cdots\cdots\cdots\cdots (17)$$

B點應力之彈性曲綫，在梁BC內距C點Kl，處之縱距

$$y = \frac{kn}{2}\left(1+\frac{2}{n}-k^2\right) \cdots\cdots\cdots (18)$$

不等橋空之連續橋梁，其感應綫，撓曲力率之感應圖，剪力之感應圖，及應力之計算等之方法及手續，均與等橋空之連續橋者無甚差異，故從畧。

五•變動慣性力率之彈性曲線（Elastic Curve for Variable Moment of Inertia）—— 連續式橋樑設計之初步，可用前篇所述之方法設計之。但最後之設計，仍需再依照初步設計時所定橋身各處剖面之慣性力率，繪一變動慣性力率之彈性曲線，然後計算橋樑之各部剖面。繪變動慣性力率之彈性曲線之方法：—

第一•假設一樑ABC，承托於B及C二點之上，如第六圖（a），然後在A點施單位之重量，計算樑ABC各點之撓曲力率。

第二•將各點之撓曲力率M，除以該點剖面之慣性力率I，則得各點之$\frac{M}{I}$。此$\frac{M}{I}$之值，又可稱之曰彈性荷重（Elastic Weight）。見第六圖（B）。

第三•假定該ABC樑，A點固定，B及C兩點乃鉸着（Hinged）如第六圖（C），樑上各點則負該點之彈性荷重。依此荷重情形計算撓曲力率之曲線，其所得撓曲力率曲線，即所求A點之彈性曲線也。此彈性曲線，可用圖解法，繪彈性荷重$\left(\frac{M}{I}\right)$之索多角形（Funicular polygon）以求之，或用分析方法，由各點剪力總和（Summation of Shears）以求之。但二法之中，後者較為簡易。

今若假定樑ABC負着彈性荷重，及平支在A及C點之上，則所得之撓曲力率曲線變為B應力之彈性曲線。

上述彈性曲線所需之值，乃一相率之比率，並非絕對之值。故在所求之應力點，使之等於一。其後者，依次比例而得之。

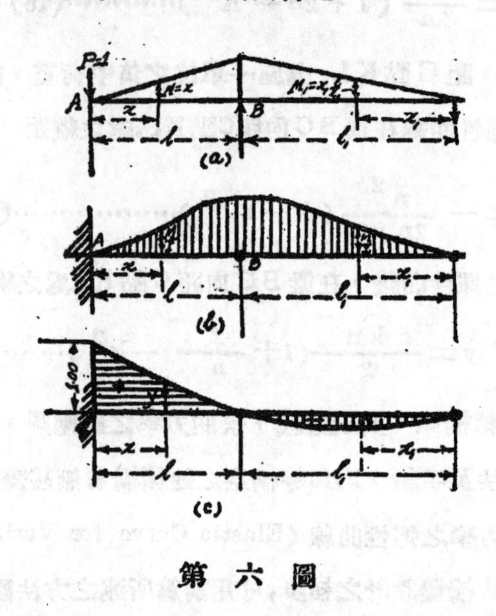

第六圖

六•連續式鈑桁橋之彈性曲綫

連續式鐵桁橋之初步設計，亦先假定該橋為不變慣力率之結構，用不變慣性力率之感應綫，或用第一表及第二表之撓曲力率，與剪力之值以設計之如上文所述。至於其最後正式設計，仍須用較準確之變動慣性力率之彈性曲綫以計算各支杆(Members)之應力也。

此彈性曲綫之求法，乃假設一單位荷重在 A 點，如第七圖，以求出各(Panel Point)之撓度。其方法可先用示力多角形(Force Polygon)求出各支杆之應力，然後用韋樂氏變位圖(Williot Displacemert Diagram)以求各幅點之撓度，將所得之撓度曲線之縱座標，在A點等於一，其後者依次比例。使得A點應力之感應線。

求彈性曲線，除上述之圖解方法外。亦可用分拆方法(Analytical Method)以求之•此法乃求在各幅點施以彈性荷重W時之力率圖，其理如下：

假設$\triangle C$為絃支杆之變長，r為力率距U為施一單位荷重在所求撓度之點時，絃支杆所受之應力，則該點之撓度$=\triangle C.U.$結構各點之撓度，依照此法求之，可成一撓度曲線]，然此撓度曲線，實乃等於支杆之應力感應線乘以△C而得之者。

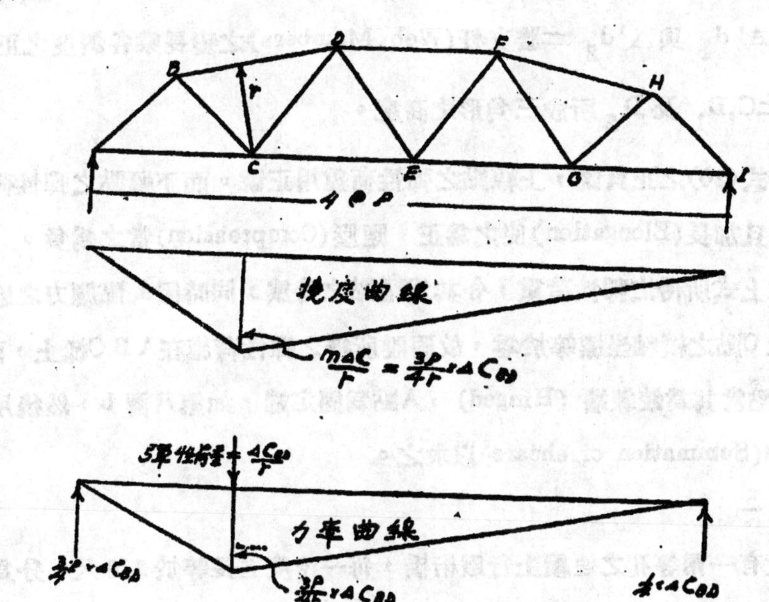

第七圖

第七圖絃BD之感應線在C點之縱距＝mp÷r，則在C點所生之撓度＝$\frac{mp}{r}\times\triangle C_{BD}$。此種情形等於在C點施一荷重$\left(\frac{\triangle C_{BD}}{r}\right)$時所生之力率曲線無異，然此荷重並非一絕對之數值，不過在分析中假定其為一假設力而已。

若假設腰支杆之應力所生之影響不計，則在各幅點之彈性荷重w，如公式(19)

$$W=\frac{\triangle C}{r}\cdots\cdots\cdots\cdots\cdots(19)$$

若為精確起見，在最後設計中，腰支杆應力所生之影響應計算在內，則王式桁鉄橋(Warren Touas)可用下式(20)：--

$$W=\pm\frac{\triangle'C-\triangle'D_1\ \triangle'D_2}{h}\cdots\cdots\cdots\cdots(20)$$

在上式中：—

$\triangle'c$＝絃支杆之變長乘其斜度之正割(Secant)

$\triangle A'd_1$ 與 $\triangle'd_2$ ＝腰支杆(Web Members)之變長乘各斜度之正割

h＝C,D_1 及 D_2 所成三角形之高度。

公式(20)之正負號，上幅點之彈性荷重用正號。而下幅點之彈性荷重則用負號，且加長(Elongation)使之爲正，應壓(Compression)當之爲負。

從上式所得之彈性荷重，令其爲幅點之荷重，同時因A 點應力之感應線在B 點及C點之縱軸坐標等於零，故假設所得之彈性荷施在A B C樑上，而樑之B點及C點當其爲鉸鍊端 (Hinged) ，A點爲固定端，如第八圖 b，然後用各點剪力總和(Summation of shears)以求之。

實例：—

設有一兩等孔之連續上行鐵桁橋，每一跨度之長等於240尺，分爲九幅，如第八圖。此橋乃依古柏氏E—50荷重設計，其在初步設計時所定各支桿之剖面面積，見第三表中之第五柱。

爲簡便及清楚起見，將各支桿之長度變爲比率，而以一幅長之長爲單位，同時令彈性系數 E＝1. 其所得之結果及手續均詳列於第三表及第四表中。

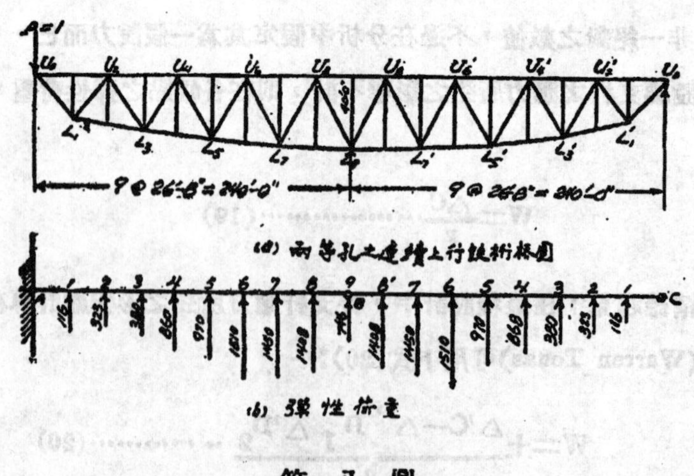

第八圖

第三表——彈性荷重之計算

	支杆	(1) 撓曲力率 M	(2) 支杆應力之係數	(3) 支杆之應力 S 1×2	(4) 支杆之長度 (L)	(5) 剖面之全面積	(6) 支杆受力後之變長 10,000△L (3×4÷5)	(7) 支杆斜度之正割	(8) △'L (6×7)	(9) 高度 h	(10) 彈性荷重 w (8÷9)
L_1	$U_0\ L_1$	1.0	1.41	--1.41	1.424	60.92	300	1.424	— 423		
	$U_0\ U_2$	1.0	1.00	+1.00	2.00	45.92	440	1.00	+ 440	1.010	-- 116
	$L_1\ U_2$	7.88	0.18	+1.422	1.424	38.42	527	1.424	+ 750		
U_2	$L_1\ L_3$	2.0	0.893	--1.785	2.04	60.92	593	1.02	— 600	1.125	-- 323
L_3	$L_3\ U_2$	9.33	0.114	--1.063	1.580	26.88	625	1.58	-- 336		
	$U_2\ U_4$	3.0	0.815	+2.445	2.00	73.67	664	1.00	+ 664	1.225	-- 380
	$L_3\ U_4$	11.65	0.0825	+0.970	1.580	20.46	750	1.53	+1185		
U_4	$L_3\ L_5$	4.0	0.766	—3.070	2.00	67.75	906	1.00	— 906	1.306	-- 868
L_5	$L_5\ U_4$	15.00	0.065	—0.974	1.70	29.28	565	1.70	— 960		
	$U_4\ U_6$	5.00	0.725	+3.620	2.00	57.29	1262	1.00	+1262	1.376	— 970
	$L_5\ U_6$	20.88	0.046	+0.960	1.70	31.32	522	1.700	+ 887		
U_6	$L_5\ L_7$	6.0	0.70	--4.200	2.00	45.25	1860	1.00	—1860	1.430	—1510
L_7	$L_7\ U_6$	32.167	0.0317	--1.020	1.773	54.67	332	1.776	— 589		
	$U_6\ U_8$	7.00	0.682	+4.770	2.00	45.92	2080	1.00	+2080	1.466	—1450
	$L_7\ U_8$	54.80	0.0193	+1.055	1.776	60.92	303	1.776	+ 546		
U_8	$L_7\ L_9$	8.00	0.672	—5.375	2.00	54.92	1960	1.00	—1960	1.490	—1408
L_9	$L_9\ U_8$	191.0	0.0061	—1.145	1.800	91.98	224	1.80	— 404	1.50	-- 746
	$U_8\ U_9$	9.0	0.666	+6.00	1.0	84.17	713	1.00	+ 713		

註：--上表之結果，乃一相對之比率，並非絕對之數值也。

上表中之第一項，乃在A點施一單位集中荷重，如第八圖，求出對各支杆之力率中心（Moment Center）所生之撓曲力率。第二項，是支杆應力之係數；例如支杆 U_2U_4，其力率中心是 L_3。在A點施一單位集中荷重時，則 L_1 點之撓力率＝1×3＝3，而 U_2U_4 之應力＝$M\times\frac{1}{r}=3\times\frac{1}{1.125}$＝3×(0.816)＝2.445。因此該數0.816便是 U_2U_4 之應力係數。腰支杆之力率中心，可將其相關之上下絃引長至相交處而得之。例如 L_1U_2 之力率中心，是絃 U_1U_2 及 L_1L_2 引長綫之相交點。若設每一幅長＝1，則其相交點與A點之距離＝7.88，故在A點施一單位荷重時，對於 L_1U_2 之力率中心所生撓曲力率＝7.88。而 L_1U_2 之應力＝$\frac{7.88}{(7.88+2)}\times\frac{1.410}{1.01}=7.88\times(0.18)=+1.422$。則該數(0.18)是便 L_1U_2 應力之係數。至於其餘各項，在表中均詳明。故無需解釋。

C點之剪力：

$$M_B=0$$

$$R_C=\frac{1}{9}(116\times8+323\times7+380\times6+868\times5+970\times4+1510\times3+1450\times2+1408\times1)$$

$$=2503$$

第四表　A點應力之感應線

	彈性荷重 (W)	剪　力	力　率	感應線之縱軸坐標
C_0	0		0	0
		--2503		
1	-- 116		—2503	0.0284
		--2387		
2	-- 323		—4893	0.0555
		--2064		
3	-- 380		--6957	0.0790
		—1684		
4	— 868		--8641	0.0980
		— 816		
5	-- 970		--9457	0.1075
		154		
6	--1510		—9303	0,1055
		1664		
7	--1450		--7639	9.0867
		3114		
B 8	—1408		—4522	0.0513
		4522		
9	— 746		0	0
		5268		
8	--1408		5268	0.0598
		6676		
7	--1450		11944	0.1360
		8126		
6	—1510		20070	0. 228
		9636		
5	— 970		29706	0. 337
		10606		
4	-- 868		40312	0. 458
		11474		
3	-- 380		51786	0. 588
		11854		
2	— 323		63634	0. 722
		12177		
1	— 116		75811	0. 860
		12293		
A.0	0		88104	1. 000

上表中之第一項，是彈性荷重施各幅點上如第八圖 (a)，第二項是各幅當之剪力，首先求出C點之應力，然後向左便加起，便得各點之剪力，第四項乃各點之力率，從各點剪力之面積求之，因爲假設各幅長等於一，故此各點之力

率便等於該點以上之剪力總和。第五項，是令力率在A點等於一。以下各各點照其比率求出，乃得A點應力感應線之縱軸坐標。

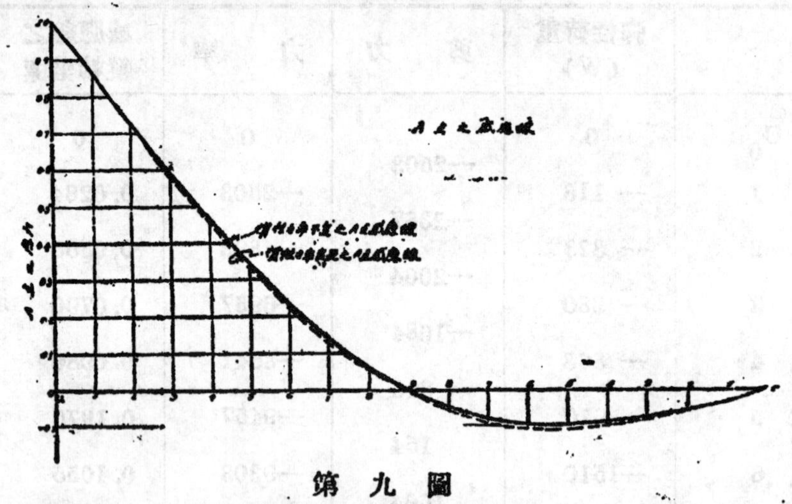

第九圖

七。連續式鐵桁橋之撓曲力率及剪力之感應圖：一 連續式桁橋之撓曲力率及剪力之感應圖與前篇所述無甚差異，其不同者，僅腰支桿之感應圖而已，玆爲清楚及簡便起見，將各支桿之感應圖繪列如下：一。

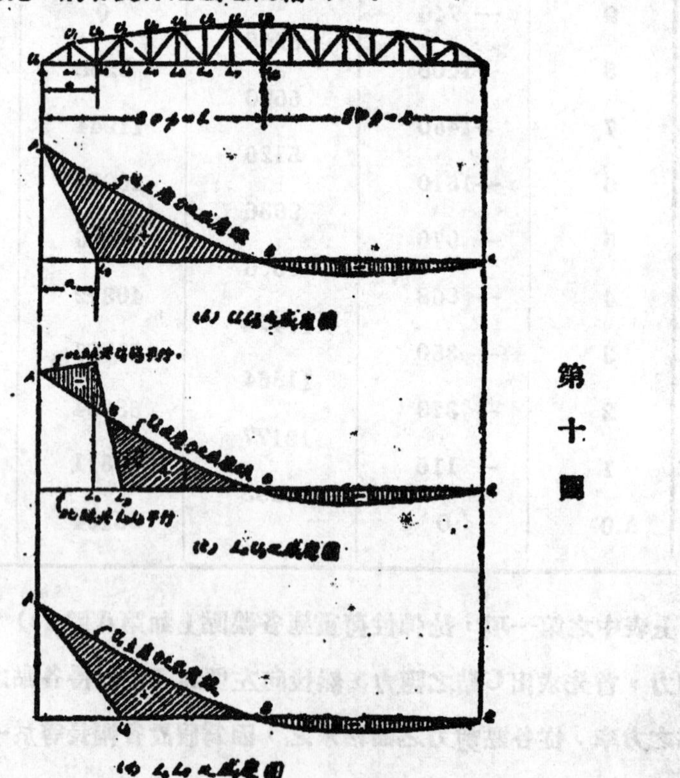

第十圖

八·多孔之連續濤　多孔之連續橋梁，亦可用上文所述之方法設計，但若孔數過多，用彈性曲線以計算應力，其手續以甚麻煩。不若直接求出樑上各點之感應線以計算應力，較為便利，連續式橋樑之感應線，在許多鋼筋凝土書中，均有詳細之討論，尤以 Gray："Reinforced Concrete Water Towers, Bunkers, Silos, and Grantries" 一書中所用之方法，較為簡易。玆將其要點分述於后：

倘若一多孔連續樑之任一跨度荷重，則其餘未負荷重之跨度所生之撓曲力率，乃以直綫代表之如第十一圖，而此直線不論負荷重跨度所受任何重量，必過各跨度之定點(Fixed point) F_1, F_2, F_3, ……等，並且由A點等於零起，漸次向荷重之跨度增大，增至受荷重跨度之端時，則撓曲力率為最大。

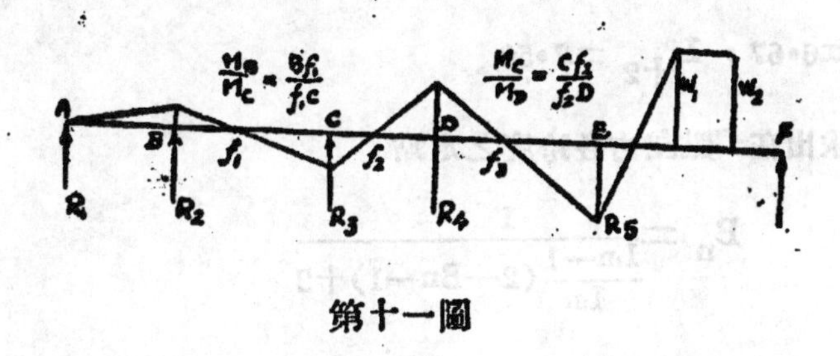

第十一圖

由上圖之關係，假設任一端之撓曲力率已知，其餘各跨度之撓曲力率可由比例而得之

$$\frac{MB}{Mc}=\frac{Bf_1}{f_1C}$$

$$\frac{M_B}{M_D}=\frac{Cf_2}{f_2D}$$

因此求感應綫之手續，先求出跨度之定點(Fixed point)，及任一承托 (Support)之撓曲力率感應綫，然後基此以計算各點之力率，求任一承托之撓曲力率感應綫之方法如下例所示：—

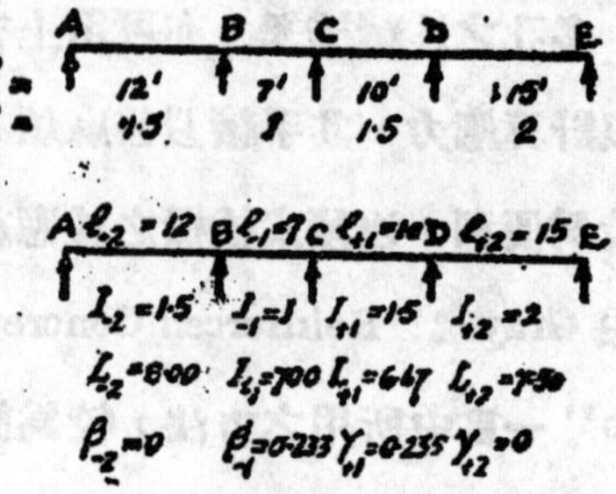

A　B　C　D　E
l = 12'　7'　10'　15'
I = 1.5　1　1.5　2

A $l_{-2}=12$ B $l_{-1}=7$ C $l_{+1}=10$ D $l_{+2}=15$ E
$I_{-2}=1.5$　$I_{-1}=1$　$I_{+1}=1.5$　$I_{+2}=2$
$L_{-2}=8.00$　$L_{-1}=7.00$　$L_{+1}=6.67$　$L_{+2}=7.50$
$\beta_{-2}=0$　$\beta_{-1}=0.233$　$\gamma_{+1}=0.235$　$\gamma_{+2}=0$

第十二圖

(1)任意選擇一承托C爲基點，凡在C點左方之跨度，以L_{-1}，L_{-2}………代表之，凡在C點右方之跨度，以L_{+1}，L_{+2}………代表之

(2)計算每跨度之$L=\frac{1}{I}$，在上例中，$L_{-2}=8.00$，$L_{-1}=7.00$，$L_{+1}=6.67$，$L_{+2}=7.50$

(3)用下式求出在C點左方各跨度之定點

$$B_n=\frac{1}{\frac{In-1}{In}(2-Bn-1)+2}$$

上式中之n代表跨度離C點之次第

由上式：　因$B_{-2}=0$

$$B_{-1}=\frac{1}{\frac{8}{7}(2-0)+2}=0.233$$

(4)用下式求出在C點右方各跨度之定點

$$y_n=\frac{1}{\frac{I_{n+1}}{In}(2-y_{n+1})+2}$$

由上式：　因$y_{+2}=0$

$$y_{+1}=\frac{1}{\frac{7.50}{6.67}(2-0)+27}=0.235$$

(5)用下公式求A

$$A=(2-B_{-1})L_{-1}+(2-y_{+1})L_{+1}$$
$$=(1\cdot767\times7)+(1\cdot765\times6\cdot67)=24\cdot14$$

(6)從已知A,1,1,B及y之值,求 k_{-2}, k_{-1} ……等。$k_{-1}=-\frac{L_{-1}^2}{AI_{-1}}$

見第六表。命 $t=\frac{X}{l}=0$,·1,·2…………,代表感應綫之橫軸坐標

(7)用B,y,及 $\frac{X}{t}$ 之值在第十一圖中,求出 l_{-2},l_{-1},l_{+1} 及 l_{+2} 各跨度之 Y_{-2},Y_{-1},Y_{+1} 及 Y_{+2}。 例如在跨度 l_{-2},因 $B_{-2}=0$,所以在 $t=\frac{X}{l}=0\cdot1$ 時 $Y_{-2}=0\cdot099$;在 $t=0\cdot2$ 時,$Y_{-2}=0\cdot192$。 倘若t之值=0時,Y之值等零。

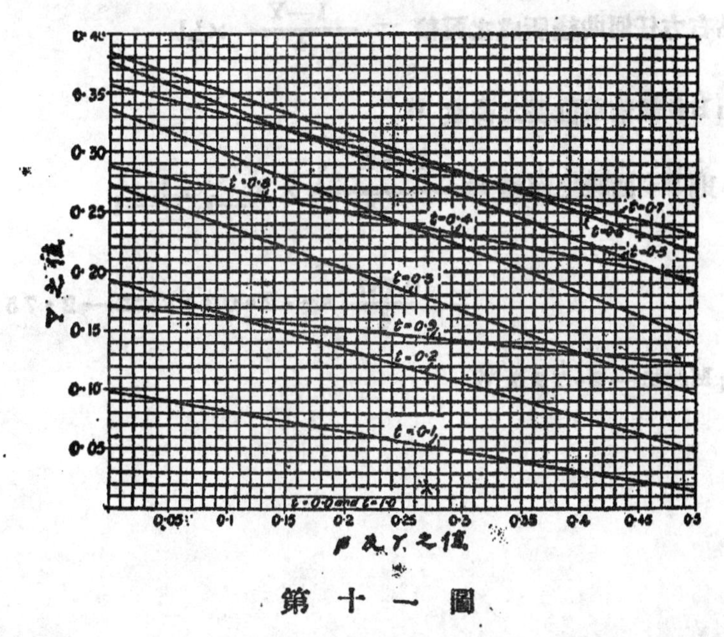

第 十 一 圖

(8)在六表中,i代表C點力率感應綫之縱軸坐標,其值等於Y與K之乘積,如 $L_{-2}=k_{-2}\times Y_{-2}$ ………

繪感應綫時,凡負號跨度(如 L_{-1})之原點(Origin)應從該跨度之左

25407

端起，凡正號跨度（如L_{+1}，L_{+2}……）之原點，應從該跨度之右端起。

（9）倘若C點右方之跨度多過二時，則 k_{+3} 之值如下

$$k_{+3}=-\frac{y_{+3}\ y_{+2}\ L_{+3}^{2}}{AI_{+3}}$$

依同理 C點左方第三跨度K之值

$$k_{-3}=-\frac{B_{-1}\ B_{-2}\ l_{-3}^{2}}{AI_{-3}}$$

（10）在C點左方任何曲綫所成之面積 $=\frac{1-B}{4}\times kL$

在C點右方任何曲綫所成之面積 $=\frac{1-Y}{4}\times kL$

例如BC滿佈單位長之荷重 W

則BC曲綫之面積 $=\frac{1-B_{-1}}{4}\times k_{-1}\times L_{-1}$

$$=\frac{0.767}{4}\times-2\cdot03\times7=-2\cdot755$$

因此 $Mc=-2\cdot725$ W

第六表

跨度	K 之值	$f=\frac{x}{l}$	0	0.1	0.2	0.3	0.4	0.5	0.6	0.7	0.8	0.9	1.0
		Y 及 C 之值											
L_{-2}	$k_{-2}=+\frac{B+L_{-2}^2}{Al_{-2}}=+0.93$	Y_{-2}	0	0.099	0.192	0.273	0.336	0.375	0.384	0.357	0.288	0.171	0
		$i_{-2}=k_{-2}Y_{-2}$	0	0.092	0.178	0.254	0.322	0.349	0.356	0.332	0.268	0.159	0
L_{-1}	$k_{-1}=+\frac{L_{-1}^2}{Al_{-1}}=-2.03$	Y_{-1}	0	0.059	0.124	0.190	0.247	0.288	0.307	0.293	0.243	0.148	0
		$i_{-1}=k_{+1}Y_{-1}$	0	-0.120	-0.252	-0.386	-0.502	-0.585	-0.623	-0.595	-0.493	-0.301	0
L_{+1}	$k_{+1}=-\frac{L_{+1}^2}{Al_{+1}}=-2.76$	Y_{+1}	0	0.058	0.124	0.189	0.246	0.287	0.306	0.292	0.242	0.147	0
		$i_{+1}=k_{+1}Y_{+1}$	0	-0.160	-0.342	0.521	-0.678	-0.791	-0.845	-0.805	-0.668	-0.301	0
L_{+2}	$k_{+2}=\frac{y_{+1}L_{+2}^2}{Al_{+2}}=+1.003$	Y_{+2}	0	0.099	0.192	0.273	0.336	0.375	0.384	0.357	0.288	0.171	0
		$i_{+2}=k_{+2}Y_{+2}$	0	0.108	0.210	0.299	0.369	0.411	0.421	0.393	0.317	0.188	0

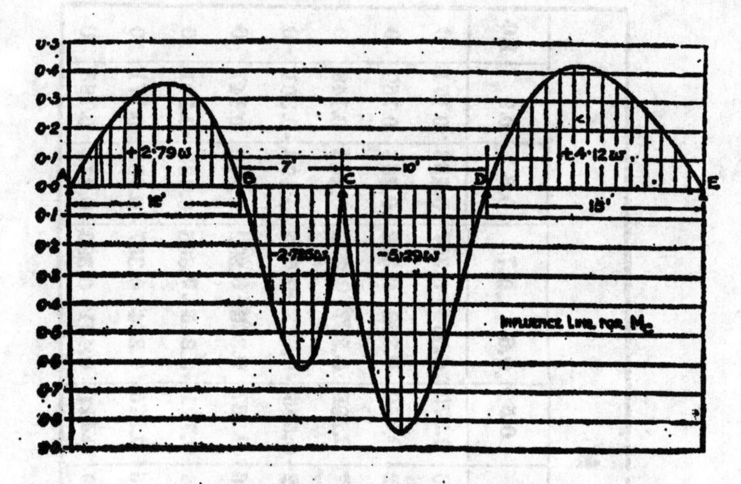

第十三圖 C點之力率感應線

General Considerations in Earth Dam Dasign

Y. L. Chen

Introduction

In irrigation and hydro-electric projects water is impounded either by embanking a depression, which receives the run off of the rainfall from a catchment area, or by holding up a certain river by a dam, weir, or earthen bank. Among the above devices for pondage or storage of water for various purposes, the earthen embankment is usually cheaper in comparison with masonry dam of heights up to 60 feet, while both can be made equally stable and lasting. In places where labor is cheap, and materials for other kinds of construction are rare, earthen dams are much more economical especially in irrigation works, which are a great necessity in the interior parts of our country.

Earthen embankments for the purpose of water storage have been built to a height of 330 feet by the Oigawa Dam Project in Japan, which is as high as the highest masonry arched dams. Under usual circumstances the limiting height is considered to be around 100 feet. For higher earthen embankments, special attention should be given to their economy and stability.

The record of failure in earthen dams is much higher than in any other kind and almost all of them caused complete destruction of the project. The following table, which is given by J. D. Justin in his book "Earth Dam project," shows the causes of the recorded failures and their percentages, which will draw our attention to the important things to be considered in design in order to have a stable structure.

Cause of failure	Percentage of total number of failures from given cause as stated in report	Percentage of total number of failures from given cause as deduced by author from the same date
Overtopping	34.5	39
Leakage along conduit	16.5	19
Other leakages	26.4	29
Slides	15.5	5
Miscellaneous	3.6	4
Not stated	3.6	4

From the above table, it is to be seen that the most common cause is lack of sufficient spillway capacity to take care of floods, and next is leakage which results in undermining and wash-out of the structure.

Criteria for the Design ot Earthen Dams.

The section of embankment depends upon the angles of repose of the soil of which it is furmed, when dry and when saturbted by the water of the reservoir or by rainfall, and on the nature of its material, and of the foundation. The height to which the work has to be raised, and the importance of the work are also governing factors in the determination of the section of the embankment.

Generally the practical criteria for the design of earth dems may be stated briefly as follows:

(1) The spillway capacity is so great that there is no danger of overtopping.

(2) The line of saturation is well within the downstream toe.

(3) The upstream and downstrem slopes must be such that, with the materials used in the construction, they will be stable under all conditions.

(4) There must be no oportunity for the free passage of water from the upstream to the downstream face.

(5) Water which passes through and under the dam must, when it rises to the surface, have a velocity so small that it is incapable of moving any of the material of which the dam or its fonndation is composed.

(6) The freeboerd must be such that there is no danger of overtopping by wave action.

Griterion 1

Earth dams should be designed with sufficient spillway capacity that it will never be overtopped. As shown in the above table, the hiehest percentage failure is caused by overtopping which, almost always, means complete destruction. Createst flood to be expected in a long term of years should therefore be determined, either by actual observation or by calculation with empirical formula, and the spillway designed accordingly with a fair factor of safety.

Criterion 2

The line of saturation should be well within the downstream toe to reduce the velocity of percolating water, end to prevent piping. The Formula for the point where the line of saturation intersects the base is derived from the Slichter formula

$$q=K(ps/h)$$

for flow of water through soil. In the formula

q = discharge, in cubic feet per minute ;

p = difference in pressure loss of head, in feet ;

h = the length of column of soil, in feet ;

s = the area of cross section of the colum, in square feet ;

K = the " Transmission Constant" (Value can be obtained in the table at the end of this article).

The formula derived is as in figure 1.

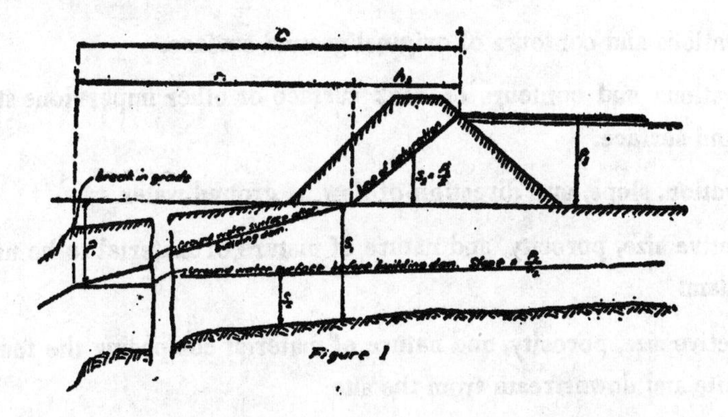

K_1 = Slichter's transmission constant for the soil forming the foundation of the dam and the soil below the dam site ;

K_2 = Slichter's transmission constant for the soil forming the dam proper.

(The derivation of the formula is fully treated by J. D. Justin in his book "Earth Dam Projects"

The cross.section of a dam shown in finure 2 shows several percolation lines (lines of saturation). A B is the line of saturation for a dam of extreme perous material, A C is the saturation line for dams construeted of material which is more compact but still somewhat porous (dams badly constructed with earth).

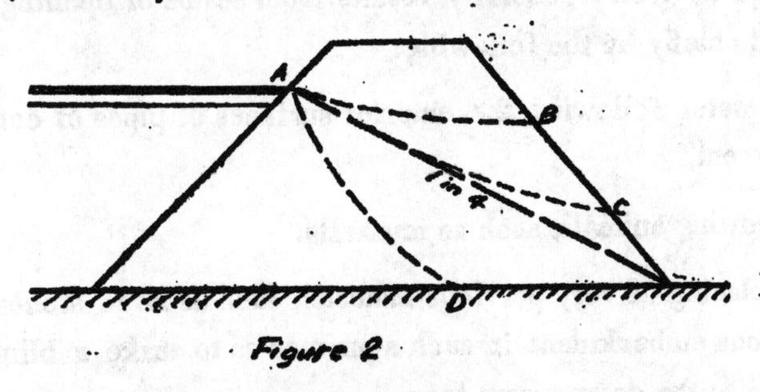

Figure 2

$$h=\frac{K_2 P_2 S_2 C}{K_1 P_1 S_1 + q_2 + K_2 P_2 S_2}$$

AD is the line of saturation for dams made of thoronghly consolidated watertight. materials, more especially for those having their downstream portion formed of self-draining material, and is underlain by base drains so as te secure a dry add stable downstream toe. The line 1 in 4 is the line of saturation obtained from actual observation on some dams in Bombay. In good dams, the line is somewhat steeper.

When the above formula is to be used, the following essential features should be determined in the field beforehand:—

1. Elevations and contours of original ground surface.
2. Elevations and contours of rock surface or other impervious strata below the ground surface.
3. Elevation, slope, and direction of flow of ground water.
4. Effective size, porosity, and nature of mature of material to be used in building the dam.
5. Effective size, porosity, and nature of material composing the foundation soil at the site and downstream from the site,

The possession of these data also gives thorough subsurface information.

Criterion 3

Stable slopes required. This requirement is governed by the angle of repose of the material under water for the upstream side. The slope of the downstream face, however, should be flatter than the angle of repose of the material of that part of the dam even if the position of the line of saturation would permit the use of a steeper slope, and should become flatter as depth below the top of the dam increases.

Criterion 4

Free passage of water generally results from seams or openings through the dam which are caused chiefly by the following:

1. By By water following the exterior surfaces of pipes or conduits through the embankment.
2. By burrowing animals, such as muskrats.
3. By the placing of very pervious material, such as large stones, in an otherwise impervioas embankment, in such a manner as to make a blind drain from the upstream to the downstream face.
4. By failure to bond and compact the succeeding layers of the embankment properly.

5. By the use of a layer of extremely pervious material over a layer of impervious material.
6. By failure to bond the lower layers of an earth dam properly to the foundation.
7. By water following the smooth surfaces of concrete abutments or other concrete structures.

In order to fulfil the requirement that "there must be no opportunity for the free passage of water from the upstream to the downstream face" special attention should be given to the things mentioned above in both design and construction.

Criterion 5

The velocity of percolating water must be too low to cause piping. The theoretical wel- ocity required to move particles of various sizes by jet action is given by the formula

$$C = \sqrt{\frac{W_1}{2F}} \quad \text{approx.}$$

in which

C = the velocity of the water as its leaves the ground;
W_1 = the effective weight of a particle of the material; and
F = the area of the particle exposed to jet action.

In the table below, values of C are given with the corresponding diameter of the particles.

Dia. of particle in mm.	Vel., in ft. per min., reqd. to move by jet action	Dia. of particle in mm.	Vel., in ft. per min., reqd. to move by jet action
5.0	43.4	0.1	6.0
3.0	33.7	0.08	5.5
1.0	19.4	0.05	4.3
0.8	17.4	0.03	3.4
0.5	13.8	0.01	1.94
0.3	9.6	......	......

It is assumed in the table that W, the weight of the material, is 159.5 lb. per. cubic foot, and that W_1, the effective weight, is 97 lb. per cubic foot.

In using the above table, a factor of safety of 4 is generally used. Thus a permissible velocity of 1.94÷4 or 0.5 ft. per minute is recommended for general use.

The formula for necessary thickness of dam to prevent piping can be also be derived from Slichter's formula

$$Q = K(p/h) \text{ for 1 foot run of the dam.}$$

In order to obtain an expression for velocity, it is merely necessary to divide K by P, the porosity of the material. Therefore,

$$V = Kp/Ph$$

where V is the permissible velocity which is equal to 0.5 feet per minute as mentioned above.

Therefore, $h = Kp/PV$

in which h is the necessary thickness of embankment or base at the given head, p, to prevent piping. In using this formula, it should be borne in mind that, with very fine material, h might be very small, in which case the width of the base and the side slopes will be determined by other criteria. With some coarse materials, h will require a great width of base, and this criterion may determine the dimensions.

Criterion 6

To prevent an earth dam from overtopping, a sufficient free board should be provided so that water raised by wave action or flood will surmount the crest of thedam. The height to which waves will rise on the upstream slope can be obtained by the following formulas:

a. Stevenson's

$$h = 1.5\sqrt{f}$$

b. Stevenson's

$$h = 1.5\sqrt{f} + (2.5 - \sqrt{f})$$

c. Hawksley's

$$h = 0.025\sqrt{L}$$

in which

f = the fetch, in miles, and

L = the fetch, in feet.

In comparison with actual observations, given in Justin's "Earth Dam Projects," formula (a) is more suitable for fetches from 10 miles up, and formulas (b) and (c) for fetches from 10 miles down.

Depth of Cutoff for Pervious Foundation

In cases where the foundations are so pervious that the percolating velocity through them mnst be reduced, sheet piling is generally used. The depth of cutoff or sheet piling should be determined by the following method.

Case I. When the sheet piling or rutoff is located at or near the upstream toe of the dam. as in figures 3 a and b:

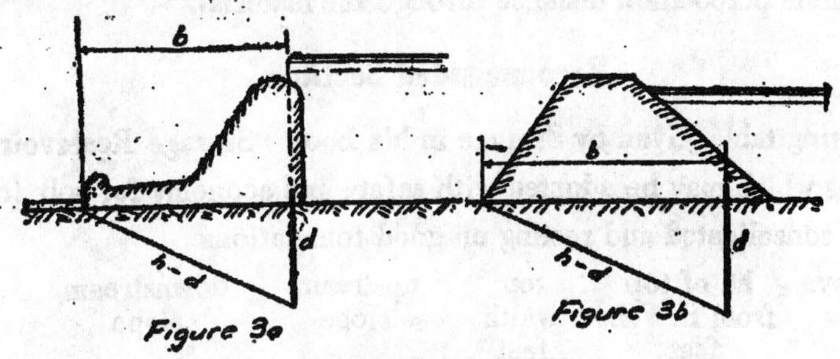

Figure 3a Figure 3b

$$h^2 - 2dh + d^2 = d^2 + b^2$$

Substituting for h its value Kp/VP, and transposing, we have

$$d = Kp/2VB - b^2VP/2Kp$$

Case II. Wheu the sheet piling or cutoff is located at or near the center line of the dam, as in figure 4:

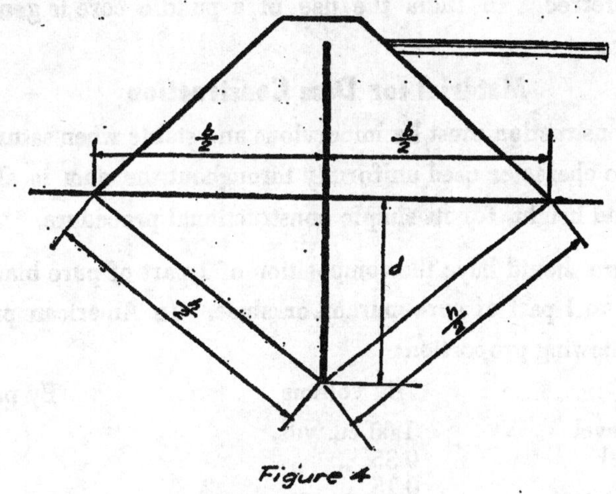

Figure 4

$$h^2/4 = d^2 + b^2/4$$

Substituting for h its value Kp/VP, and transposing, we have

$$d = \sqrt{(Kp/VP)^2 - b^2/2}$$

In the above equations:

K = transmission constant of the material composing the foundation (see table for value)

p = the difference in elevation between the two given points, in feet;

P = porosity of the material expressed as a decimal;

V = permissible velocity of flow through the material, usually taken as 0.2 to 0.5 ft. per minute; V is selected by judgment between these limits, giving consideration to importance of structure, value of water lost, probable water-tightness of cutoff, etc.;

b = effective width of base of the dam; may include in some cases apron or blanket; and

h = minimum percolation distance through the material.

Recommended Sections

The following table, given by Strange in his book "Storage Reservoirs," gives the general sections which may be adopted with safety and economy for soil (ordinary good soils) properly consolidated and resting on good foundations.

Ht. of Dem above ground level feet	ht. of top from H.W.L. feet	top width feet	upstream slope	downstream slope	width at H.W.L. feet
15' and under	4 - 5	6	2 - 1	1 1/2 - 1	20 - 23 1/2
15 - 25	5 - 6	6	2 1/2 - 1	1 3/4 - 1	27 1/4 - 31 1/2
25 - 50	6	8	3 - 1	2 - 1	38
50 - 75	7	10	3 - 1	2 - 1	45

The above sections are for monolithic dams. Where good soil is available, monolithic construction is proferred. In India the use of a puddle core is generally limited to very large reservoirs.

Material for Dam Construction

Soil for dam construction must be impervious and stable when saturated with water. Material of the same cheracter used uniformly throughout the dam is always preferred, for dams of moderatd height, for its simple constructional procedure.

Artificial mixture should have the composition of 1 part of pure black "cotton soil" or other clayey soil to 1 part of pure muram, or shale. In American practice, Fanning recommends the following proportion:

	By volnme	By percent
Coarse gravel	1.00 cu. yds.	59
Fine gravel	0.35 „ „	20
Sand	0.15 „ „	9
Clay	0.20 „ „	12
Total when loose	1.70 cu. yds.	100
Total when compacted	1.25 „ „	

This mixture, when properly consolidated, would be free from voids, but the small amount of clay used would apparently not make the mass sufficiently impervious, and would prevent it from possessing mnch cohesive stability. A gravel that would puddle, or "binding gravel" will answer this requirement.

To test the stability of such a gravel, the sample is mixed with water in a pail to the consistancy of moist earth as generally used in dams. If on turning the pail upside down the gravel remained in the pail, it would be of the right character for use; but if it dropped out, it would by too gritty for employment, and should be rejected.

Transmission Constant from which the Velocity of Water in Sands of Various Effective Sizes of Grain can be Obtained.

(Table computed for temperature of 60° F.; Results for other temperatures can be found by the use of succeeding table)

Diam. of soil grains, in mm.	Porosity						Kind of soil
	30%	32%	34%	36%	38%	40%	
.01	.000033	.000040	.000650	.009060	.000072	.000085	Silt
.02	.000131	.000162	.000198	.000239	.000286	.000339	
.03	.000296	.000364	.000446	.000538	.000645	.000763	
.04	.000527	.000648	.000394	.000958	.001145	.001355	
.05	.000822	.001012	.001240	.001495	.001790	.002120	Very fine sand
.06	.001182	.001458	.001784	.002150	.002580	.003050	
.07	.001610	.001983	.002430	.002930	.003510	.004155	
.08	.002105	.002590	.003175	.003825	.004585	.005425	
.09	.002660	.003280	.004018	.004845	.005800	.006860	
.10	.003282	.004050	.003960	.005980	.007170	.008480	Fine sand
.12	.004725	.005830	.007130	.008620	0.1032	.0122	
.14	.006430	.007940	.009720	.01172	.01404	.01662	
.15	.007390	.00912	.01115	.01345	.01611	.01910	
.16	.008410	.01036	.01268	.01531	0.1835	.0217	
.18	.01064	.01311	.1605	.01940	.02320	.02745	
.20	.01315	.0162	.01983	.0239	.02865	.0339	
.25	.0205	.0253	.031	.0374	.0448	.053	Medium sand
.30	.0296	.0364	.0446	.0538	.0645	.0763	
.35	.04025	.0496	.06075	.0733	.0879	.1039	
.40	.0527	.0648	.0794	.09575	.1145	.1355	
.45	.0665	.082	.1005	.1211	.145	.1718	
.50	.0822	.1012	.124	.1495	.178	.212	
.55	.0994	.1225	.15	.181	.2165	.2566	
.60	.1182	.1458	.1784	.215	.258	.306	
.65	.139	.171	.2095	.253	.303	.358	
.70	.161	.1983	.243	.293	.351	.4155	
.75	.185	.2278	.2785	.3365	.403	.477	Coarse sand
.80	.2105	.259	.3175	.3825	.4585	.5425	
.85	.2375	.2925	.358	.4325	.5175	.6115	
.90	.266	.328	.4018	.4845	.58	.686	
.96	.2965	.365	.447	.54	.646	.765	
1.00	.3282	.405	.496	.598	.717	.848	
2.00	1.315	1.62	1.983	2.39	2.865	3.39	
3.00	2.96	3.64	.446	.538	6.45	7.63	
4.00	5.27	6.48	7.94	.9575	11.46	13.55	
5.00	8.22	10.12	12.4	14.95	17.9	21.2	

The constant is for the Sliehter's Formula.

q=K (ps/h)

q = discharge, in cubic feet per minute;
K = transmission constant;
p = difference in pressure loss of head, in feet
h = length of column of soil column. in square feet;
s = the area of cross section of the soil column, in square feet.

Variation, with the temperature, of the flow of water at various temperatures through sand, 60°F. being taken as the standard temperature

Temperatnre, in degrees F	Percentage of relative flow	Temperature, in degrees F	Percentage of relative flow
32	.64	70	1.15
35	.67	75	1.23
40	.73	80	1.30
45	.80	85	1.39
50	.86	90	1.47
55	.93	95	1.55
60	1.00	100	1.61
65	1.08	...	...

"Relative flow" means flow at given temperature compared with flow at 60°F. It is expressed as a percentage.

References

"Earth Dam Projects" by J. D. Justin.

"Indian Reservoir and Earth Dams" by W. L. Strange.

"Engineering for Masonry Dams" by W. P. Creaker.

"Irrigation Practice and Engineering Vol. III" by Etcheverry.

"The Practical Design of Irrigation Works" by Blich.

真空管電壓調節之研究

馮秉銓

(一) 引言

吾人日常所用交流電源，其電壓每因外輸担負之隨時變更，以及引擎速度之不常，而發生漲落現象(Fluctustion)。以本校情形而論，原有一一〇伏脫電源，最低時竟降至七十五伏脫。普通日間上午八時至下午五時，恆爲九十五伏左右。五時以後，用電戶逐漸增加，電流担負過大，電壓輒降至八十至九十伏間。入夜十二時後，復增至一百乃至一百十伏脫。此種漲落影響所及，每使用電戶感覺極大不便。蓋電壓過低，則所用電燈不明，電爐不暖，電扇太慢，無線電收音機無聲；爲求電壓復常起見，勢必加用小型可變變壓器，隨時調節。而電源或時而突然增加，用戶偶爾不愼，所用電具，即有損壞之虞，而以此種電源，作比較精確之科學實驗者（如物理學中X光，無線電，以及光學種種實驗，莫不需要極常之電壓），雖終日小心翼翼，仍不免時時受此種危險威脅。蓋實驗進行已至半途，如電源驟變，每使已得結果全盤改變致前功盡棄也。

職是之故，電壓調節器之設計，近年以來，乃成爲一般電工界人士趣味中心之一。市上所售電壓調節器，不外二種：一爲 Tirril 式，一爲反電勢式（註一）。前者爲 Tirril 氏所發明，流行市上已有多年。後者爲奇異電器公司最近出品。二者製造原理，雖微有不同，而施用方法，皆係利用電機續器，使振動接觸器隨時矯正電壓之高低。二者相較，後者畧佳，然其弊端，均有如下述：(1)電壓之矯正，全恃繼續器接觸點之振動，接觸處畧不合式，即失其作用。(2)振動器用久易壞，須時常更換。(3)所得調節係斷續的而非繼續的，振動器雖較正至極靈敏處，亦不易隨電壓之漲落而移動，使其間無時間之落後。

(Time Lag)(4)構造複雜，使用不易。

近年以來，無線電學發展一日千里，眞空管之應用亦幾無境不入。一九三六年，乃有 verman 氏(註二)首創利用三極眞空管之特性，以矯正交流電機電源之漲落。繼其說者多人(註三)其理論皆大同小異。至本年(一九二五年)二月復有 D.H. Craig 氏(註四)著說行世，利用外柵三極眞空管，施用於遞能線間。法簡効宏，尤堪稱述焉。

本校物理學系，年來致力於此項問題者頗不乏人。以下所述種種實驗，均在進展之中。原理所宗，與上述諸家，均大同小異。主其事者，實爲本系教授許湞陽先生及研究生高兆蘭女士。(誌此以示不敢掠美矣。)

(二) Thy ratron 三極眞空管電壓調節器，施用於交流電機。

Thyratron 構造原理，與普通三極眞空管大致相同。惟管內含有多量氣體，當放電作用(Discharge)發生時，其板極電路中可負頗大之電流，是其特點。放電作用之發生，視燈絲熱度，板柵二極電壓之高底而定，在普通應用電路中，眞空管燈絲熱度，每爲常數。故放電作用之能否發生全視板柵二極電壓之互相關係而定。板壓愈高，柵壓愈低(假定柵極電壓爲負極)，則放電作用之發生亦愈易。反之亦可類推。

柵壓愈高則放電發生亦愈難。其在放電可能界中最高之値，曰臨界柵電壓(Criticial grid potential)。柵電壓高過此値時，則眞空管不能放電。臨界柵壓之値視板極電壓之高低而定。板壓愈高，則此値亦愈大。

今設有交流電壓施於此種眞空管，其板柵各極電壓之相位關係，有如圖一所示，則可見全週中，板極電路，均無電流通過，蓋柵壓旣高於臨界之値，而板柵二極相位之『相差』(Phase difference)又適爲一百八十度，板極負時當然無電，正時又因柵極電壓之過高，放電作用，自不能發生。

第一圖 

虛線示臨界柵壓

今設板柵二極之電壓不變而設法移其相位，使當板極爲正壓時，柵極有時低於其臨界之值，則一週之中，必有一部份有板極電流通過。此板極電流。顯然呈脈動之值。其振幅之大小，視二壓相位之關係而變。若以合式之電容器濾淸其波，此板極電流可使成紋波甚低之直流。圖二示板柵二電壓之相位成正交時之情形，全週四分之一有板極電流通過。

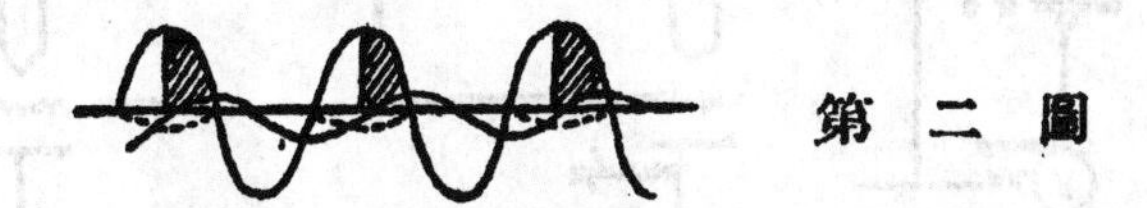
第二圖

移相之法可利用電阻電容降勢正交之理。如圖三所示：R 極大時，板柵二極之電壓『相差』可至一百八十度，此時板極電流幾等於零(如圖一)。今若減小 R 之值，使板柵二電壓之『相差』減至九十度，則板極電流漸增。(如圖二)R 極小時，二壓漸歸同位，全週中均有板極電流通過，其值亦至大。

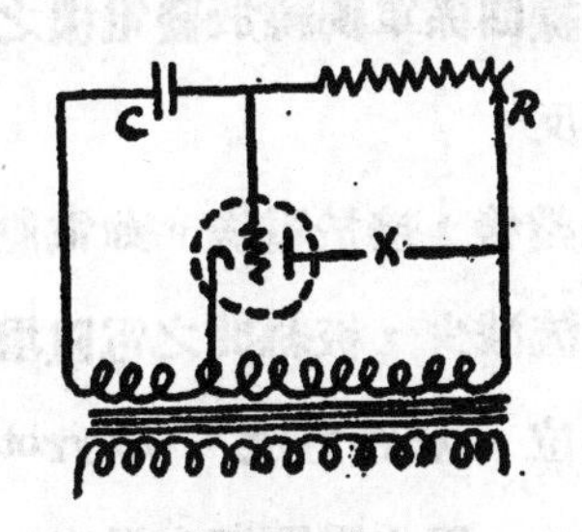

第三圖

是以板極電流之強弱，實視板柵二極相位之變遷，而板柵相位之變遷，又決於R 爲值之大小。今若能設計一綫路，使電源電壓之漲落，能影響及此電阻 R 之値，因而變動眞空管板極電流之強弱，而卽以此變化之板流，反施於電源失常之發電機之磁場綫圈中，如各常數配設合宜，自可得相當之調節。

吾人又知普通三極眞空管之板絲電阻，實爲板極電流之函數。設板柵電壓不變而燈絲電流驟然加減，則板絲間電阻必因之而漲落。此種條件，適爲上述及電阻所需要者。是以吾人所用之初步綫路乃有如圖四所示。

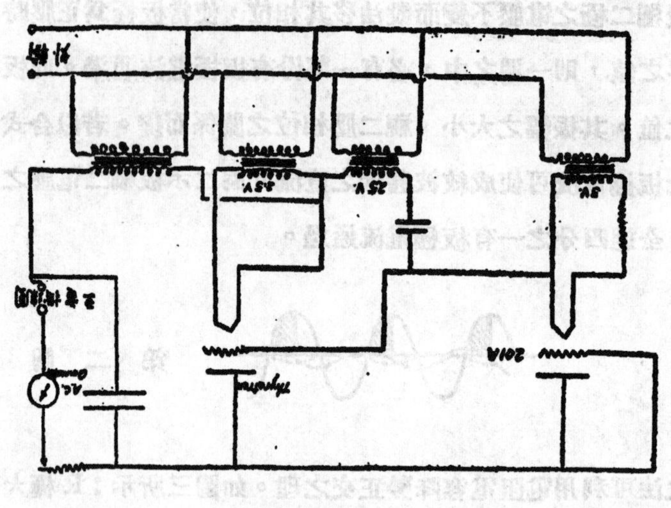

第　四　圖

圖示以一普通 201--A 管代電阻 R 。此管之絲極電流，由小變壓器 T 所供給，T 之正圖，接所欲矯正之電源。Thyrotron 板極電流經濾波器後，通過發電機之新磁場線圈。此線圈係單獨繞於發電機之磁場鐵心者，與原有線圈不連，其方向與原有磁場相反。

今設調節器之各部經調整後，接於電機。如電壓忽降 ，則 201— A 燈絲熱度隨之降低，結果板極電流減少，板絲間之電阻增加（卽 R 增加）。R 增加之結果，使柵板二壓之相位，相距愈遠，Thyrotron 板極電流因而減少，其反原向之磁力綫亦隨之減少，因之電壓漸次增高。

反之可類推。

(三)外柵式三極眞空管 kathetron 電壓調節器——施用於遞能綫。

上述綫路，缺點厥有二端：(1)只能施用於發電機本身。用電戶方面之無法自動調節。(2)所用 Thyrotron 柵板二路中均有多量電流 ，眞空管易於損壞。

Craig 氏之說 ，係用普通二極汞氣整流管（如 RCA866,872 等皆是）外面加以網形柵極，利用柵極靜電場之變化，板極電流之強弱 ， 可因之隨意變更。

今設有線路如圖五。S 示所欲調節之不定電源。T 示一變壓器，此處用以作反射電抗作用者。如圖示，電燈 L 兩端之電壓，自較電源爲低。今若關閉電鍵 K，則變壓器之副圈短路，反射於正圈之電抗，因之減少而電燈兩端之電壓亦因而增加。二極眞空管整流之作用，實與此電鍵K殊途同歸。蓋此種眞空管，當板極電路無電流通過時。

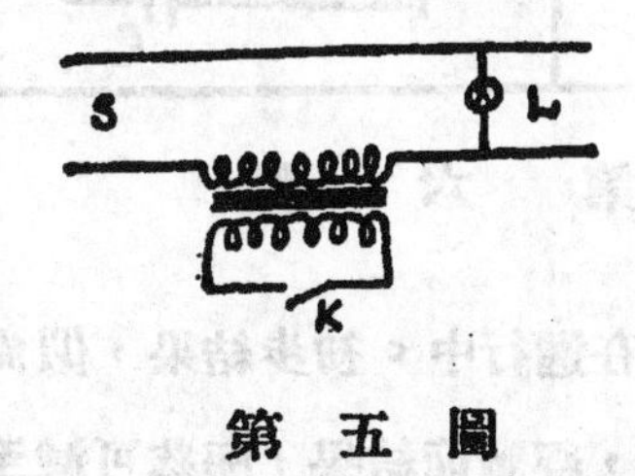

第五圖

其板絲二極間之電抗極高，反之當板極有電流通過時，電抗則降至極低之値。今若將此眞空管板絲二極，接於變壓器之副圈，以代電鍵K，則當板極無電流時，電抗極高，其作用有類斷路。反之當電流通過時，眞空管電抗驟減，其作用又有似短路。此種短路作用之久暫，全視通過電流之強弱，而電流之強弱又依外柵電位之高低。今若即以漲落不定之電源，經過相當變壓而以之供給外柵之電壓，則外柵電壓高低，可影響板極電流之強弱，因而隨時變更短路作用之久暫，使反射於正綫路之電抗，隨時增減，設綫路配合得法，電壓之調節良非難事。

圖六示可能綫路之一種。外柵電壓，由所欲調節之電源，經過變壓器T，電阻R，間接供給。阻力 r 及電容C 係用以矯正柵板二極電壓之相位者，爲綫路中之常數，一經規定，無需隨時變更。今設調節器已接於電源。電壓若驟然降落，則外柵電隨之低減，板極電流增加，因而增加所謂『短路作用』而使反射於正路之電抗減低，結果外輸兩端之電加壓必見增，因得平衡，反之可類推。

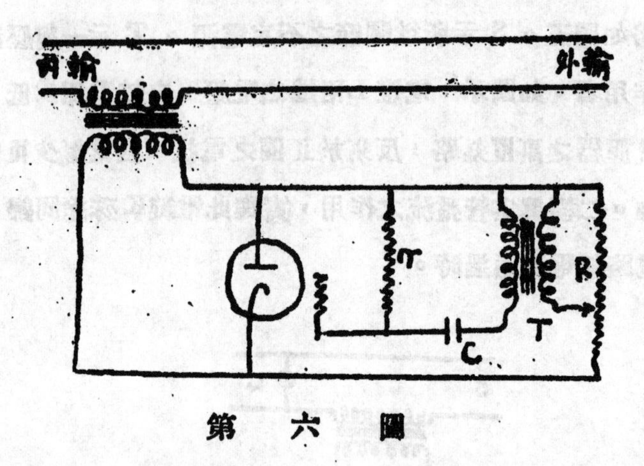

第 六 圖

上述兩種方法，實驗均在進行中。初步結果，似尚差強人意：交流電源從一一〇突降至一〇〇伏脫時，經調節結果，漲落可減至二伏脫左右。線路改良後，結果或可較佳。詳細結果，容後刋佈之。

註一： Drwls: Electrical Engineering Vol. I. p.p. 361--363

註二： R.S.I. 1930 p.p. 581—590

註三： R.S.I. pp. 479,1933. Electronics pp132—133，1932.
pp,164—165. 1932 R.S.I. pp 9—19, 1932

註四： Electronics p.p.70—72, 1933

三等分任何角

彭 萸 原

用圓規及尺，三等分任何角，乃幾何學之不能解决難題。自古代希臘以至于今，致力于此問題者甚衆，惜竟無成功。

德國數學專家 Felix Rlein在"Association for the Advancement of the teaching of Mathematics and the Natural Science,曾發表論文，証明三等分任何角永遠不能成爲事實。

此篇非基于幾何學而三等分任何角，不過欲以最單簡之方法，求得較準確之答案，而便實用于工程或其他問題而耳。

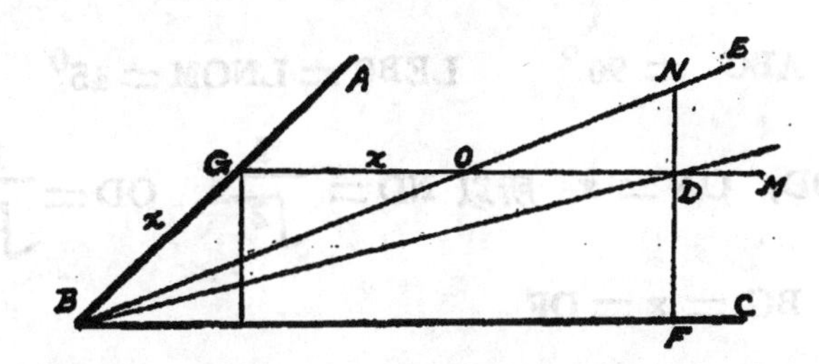

1. 設ABC爲任何角。X爲任意長度。由AB截出BG=X GM與BC平行。以BE等分 L ABC，令BE與GM相交于O。ON=X，由N作直垂線至BC，交BC于F。NF交GM于D，聯BD並使直伸。

$$L\ DBC = \frac{1}{3} L\ ABC\ (約)$$

証：設 L $ABC=60^{\circ}$， L $EBC=30^{\circ}$ =L NOM. 在△NOD，ON=X，

所以 $ND=\frac{X}{2}$； $OD=\frac{X}{2}\sqrt{3}$　由G直垂GP至BC，接于P。

25427

因 $\angle GBP = 60^{\circ}$ 所以 $BP = \frac{X}{2}$，$GP = \frac{X}{2}\sqrt{3} = DF$

$$NF = ND + DE = \frac{x}{2} + \frac{x}{2}\sqrt{3} = \frac{x}{2}(1+\sqrt{3})$$

$$BF = BP + PF = BP + GO + OD = \frac{x}{2} + x + \frac{x}{2}\sqrt{3}$$

$$= \frac{x}{2} + (1+2+\sqrt{3}) = \frac{x}{2}(3+\sqrt{3})$$

$$\tan \angle DBF = \frac{DF}{BF} = \frac{x}{2}\sqrt{3} \div \frac{x}{2}(3+\sqrt{3}) = \frac{1.7321}{4.7321}$$

$\angle DBF = 20^{0}\,6'\,15''$　　$\frac{1}{3}\angle ABC = \frac{60}{3} = 20^{0}$

相差約　$6'\,15''$

設 $\angle ABC = 90^{0}$　　$\angle EBC = \angle NOM = 45^{0}$

在 $\triangle NOD$，$ON = x$ 所以 $ND = \frac{1}{\sqrt{2}}x$，$OD = \frac{1}{\sqrt{2}}x$

$OG = x$　　$BG = x = DF$

$$\tan \angle DBF = x \div \left(\frac{x}{\sqrt{2}} + x\right) = 1 \div \left(\frac{1}{\sqrt{2}} + 1\right)$$

$\angle DBF = 30^{0}\,21'\,37''$　　$\frac{1}{3}\angle ABC = \frac{90}{3} = 30^{0}$

相差約　$21'\,37''$

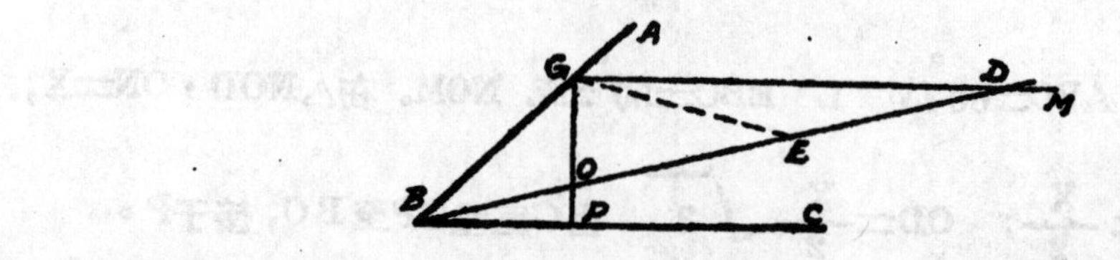

2. 設 ABC 爲任何角。以任意 x 長爲單位。

由 BA 截出 BG = x　　GM與 BC 平行。

由 G 直垂 GP 于 BC，相接于 P。

以一尺放于 B 之上，擇取直線 BOD (O爲 GP上之一點，D 則在 GM)。

使OD=2x　　即得 $\angle DBC = \frac{1}{3}\angle ABC$。

証：聯GE　　$\angle DBC = \angle DGE = \angle GDE = \frac{1}{2}\angle GEB$.

因 DE = GE = BG，所以 $\angle GBE = \angle GEB$.

即是 $\angle GBD = 2\angle DBC$，$\angle DBC = \frac{1}{3}\angle ABC$。

第一法所得之答案，非絕對準確。以六十度及九十度角之答案相較，則可見其角愈大，則三等分所得之錯誤亦愈大。第二法較爲準確。作圖時倘能將直綫 BOD之OD 段確等于2BG(DO=2BG)，則三等分何任三角亦不成問題矣。

25429

窗牖與室内光綫之關係及其計算

龍 實 銓

吾人設計屋宇，恆顧慮及光綫問題，而光綫之供給，宜爲窗牖是賴。在歐美各國及通都大邑中，常人恒以爲窗牖與光綫無若何關係，窗牖祇便空氣流通而已，至光綫則電燈足以供給之，與窗牖無關也。殊不知電燈耗力，費用綦昂，遠不若窗牖之取諸自然，無禁無竭也。更就本國情形而論，市鎮亦非盡有電力，縱使有之，亦恒熄滅光暗靡常，若荒言村及消夏等地，當無供給矣。似此，則窗牖之於光綫問題，固仍不失其重要性也。

光有綫強弱之分，光度 (Light Intensity) 因時因地而異。日中強於日午，夏強於冬，烈曬強於陰晦，在赤道之北者，南方強於北方，在赤道之南者則反是。工程界丈量光綫之單位爲呎燭 (Foot Candle)。一呎燭者卽一標準燭枝(Standard Candle) 放射於距離一呎之一方呎面積上，其所發生之強度是。標準燭爲鯨魚油 (Sperm oil) 所製，每燭六枝共重一英磅；每燭一枝，每一小時燃去重 120 英釐。(每英釐爲一磅之七千分之一) 光綫強度，通常在工作平面測量之。工作平面者，卽做工作之平面，(Workirg plan) 此工作平面之高度，約與檯椅之高度相埒，卽在離地面高二呎半至三呎之間。例如月光在工作平面之光度爲$\frac{1}{20}$呎燭，尋常燈光在工作平面爲10至20呎燭上。欲辨別顏色約需二十至二百呎燭，實驗室，珠寶商或其他顯微鏡下工作約需二十至一百呎燭。

規定所需光綫之最低限度，各時代人之眼光不同，惟其數目則漸趨增加，茲將十五年來規定需要呎燭數目列下，以資比較。

規定需要光綫最低限度

	十五年前	八年前	現在
通廊，升降機，儲物室	0·5—1	3—6	5十
教堂，大堂	1—3	2—4	3—8
商店	3—7	7—10	10+
粗工業	2—4	3—6	6+
書寫	3—8	5—12	10+
精緻工業	4—8	6—12	10+
繪圖工作	5—10	10—15	15+

據現在一般學者規定，在工廠或辦公室工作，所需最少光綫限度在工作平面上爲十呎燭。茲將 1931 年威斯丁侯斯電燈公司 (Westing house lamp company) 之佈告 E--108 號所載各種工業所需要光綫標準，撮要列下：

	美滿的	最少的
樓梯，石工，鋼廠	3	2
儲物室	12	8
汽車工廠，織布，樹膠	15	10
鞋廠，製餅場，礶頭，糖業，乳業，洗衣，油漆	12	8
書籍，釘裝，製手套，製帽，木匠，印刷	15	10
普通化學工業	6	4
建築材料製造，各秤機房，梘皂	8	5
皮革，包裹，製紙	10	6
珠寶，鐘錶	75	25
繪圖工作	25	15
糖廠，煙草	25	10

光度愈佳，則工作之効率亦愈增，日光照耀地面上，夏天六月天朗氣清之正午，室外強度幾有至一千呎燭，但此強度仍未能傷人類之視神經，良以人類之眼球能使大小以就之也，迹至現在，仍未有光線最高限度之規定，故愈光則

愈妙。求其能小均分配各處，不至乍強乍弱，則不患其過強也。

設計光綫之多寡，須以最劣之情境而估計之，例如：

上午八時——在工作時間中，冬天上午八時爲最黑。十二月——一年內最弱光綫之月上蓋全蔽——無通天面積，不見太陽直射。

在此情況內，室內光綫有十呎燭者，斯可以言足。在單層之建築最簡便適當之設計，須窗牖總面積約等於地面積百分之三十，如是則縱在十二月侵晨之候，想仍有十呎燭光度之供給也。

屋宇甫在落成之時，光綫常感甚佳，迨相當期間後，光綫漸趨於弱，蓋窗戶經過時間常失其一部份之效用。最足以阻礙光綫者，厥爲窗牖上所積之塵垢。尤其燭製造物品之工廠內，其窗牖常汚穢不堪。通常工廠或商店之窗戶，每年洗濯最多不過二次，在中國則每年僅一次，即在每年歲杪而已。甚有因其爲工廠，不求美觀而通年亦不施洗濯一次者，其不潔情況更毋待言。最足稱異者，則窗牖循相當之時間，其汚垢泰半積于窗內。據大約估計，積於窗內者四分三，積于窗外者四分一。此或窗外常受風雨侵蝕，故較爲清潔，尤其在甚斜或高徹之窗戶，其內面不易洗濯，有以使其然也。

建築物之窗牖，多爲垂直者，然亦有傾斜至相當角度者，亦有與地面平行者。在中國屋宇，恆比鱗而立，故尤恆用天窗。然則何者較爲適用？何者損失光綫較多？根據美國地賽鋼鐵公司（Detroit Steel Products Company）與米西根大學(Univeritys of Michiqan)聯合驗之佈告其結果爲：

(甲)直窗牖六閱月之積垢後，失其透入光綫效率百分之五十。

(乙)傾斜三十度角之窗，在六月之積垢後，失其效率百分之七十五。

(丙)傾斜六十度角者，失其效率百分之八十三。

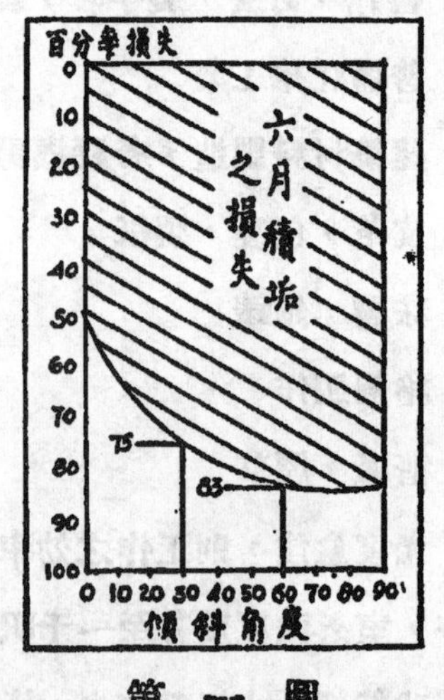

第一圖

所得結果，亦在第一圖表出，由此可知斜窗損失光綫較直窗爲多。六月之積垢實爲普通情形，在中國，窗牖積垢所受之損失，想尤不宜此也，在中國普通習慣，每屆廢曆年杪爲洗濯窗牖規定時間，在歐美各國亦多在冬季舉行之，此實大悖其理。窗一不宜于冬季洗濯，其理由有二。其一，光綫最暗淡之月爲冬季，尤其爲廢曆十二月國曆十一月，故洗濯應在此時之先，使黑暗月份不感太甚。其二，冬季天寒，工人必不欲沾水濡手，故濯必草率從事，使窗牖不净。由是觀之，洗濯似舉行于秋季爲宜，雖則度舊曆年關，有除舊換新之義，然亦距此時非遠也。

吾人苟由一暗淡之室，驟進一光亮之室，則必覺視物反不大清楚，眼睛反不大舒服。平均分配之光綫爲吾人所需要者。苟在一室內，此方之光綫甚強，彼方之光綫極弱， 共結果實影响於工作。 然而事實上，室內光綫之有差異，爲斷不可避免之事實，求其差異不至太甚而已。室內最光與最暗之處應持一定之比例限度。通常此比例限度爲三，即最光之處不強于最暗之處之三倍。二與一之比尤爲適意可人，但此甚不易得。前此曾言，窗牖面積應爲樓面面積百分之三十，惟窗牖最要平均分配於各處，不不宜一方過多，一方不足，致感覺不舒服。

建築物通常所用爲牆間窗亦當爲直窗，即垂直於地平面。直窗在經驗上，可測光綫所測遠近其定律爲：光線充裕之地方，其長度爲窗高之二倍。

光綫充裕云者，即足敷十呎燭之謂如圖二甲所示，光綫由一方射入，室闊25呎，窗高爲5呎2吋，在下繪光綫曲綫圖所示

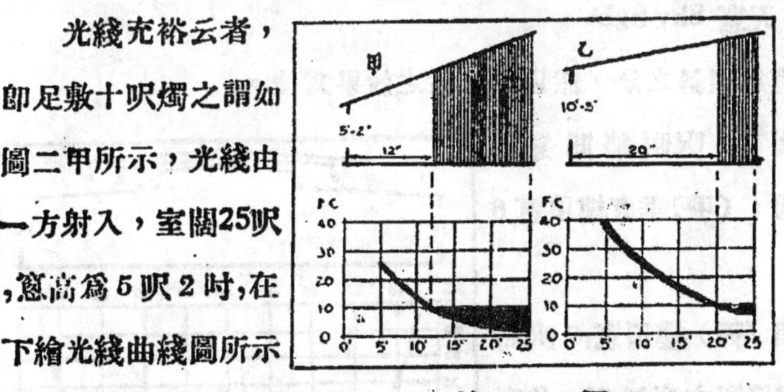

第 二 圖

，最亮處爲25呎燭，位於距窗5呎之點。由距離12呎起，則不敷10呎燭之數。可知祗得約半室充足光綫，其餘半室爲黑暗不便工作。光綫充裕之長度(12呎)

約爲窻高(5呎2吋)之二倍。最亮處與標準光度比爲2½倍。二圖乙所示同一室廣，光綫亦由一方射入，窻之高度增爲10呎3吋，光綫曲綫圖所示，最光處爲40呎燭。距窻20呎之處爲10呎燭，過此則在10呎燭以下。如是卽表示室內五分四地方爲光綫美滿。20呎長度仍約爲10呎3吋之二倍，最亮處爲標準光度之四倍(太大)。

設光線由室之雙方射入，由經驗上亦可測光線所及遠近。卽光線充裕之地方，其長度爲窻高之三倍。第三之圖屋，其長度爲第二圖之2倍(50呎)。每房建窻高8呎6 3/4時。光線曲線圖所示，每窻射入光線最高爲35呎燭。

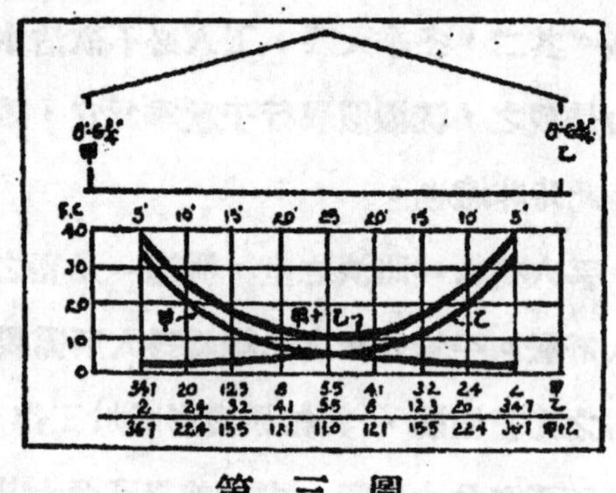
第　三　圖

17呎6吋以後卽不足10呎燭，但光線爲二窻共同射入者，故室內光線爲二窻之和。室之中間每窻射入者，如是則每窻射入室內中點之光綫爲5.5呎燭，二窻共射入11呎燭)最暗處(11呎燭)爲最亮處(35呎燭)之三分一，室內光線充裕處(50呎)約爲窻總高(17呎)之三倍。

通常如一室甚廣闊，牆邊窻不足以供給光線，故恆在屋頂開窻，以增光線之透入。屋頂窻牖可分爲三類(甲)樓頂窻 Monitor windows (乙)鋸牙式窻Saw-teeth windows (丙)天窻 Sky light

樓頂窻亦有垂直與傾斜之分，然以垂直者之效果爲佳。

圖四示一建築闊100呎兩牆開窻高12呎，屋頂開窻如圖，(甲)垂直樓頂窻6呎高(乙)6呎樓頂

窻傾斜30°角者(丙)樓頂窻6呎高傾斜60°角者。(樓頂窻之傾斜60°角者亦可稱之爲天窻。)以上三種窻之面積均相等。

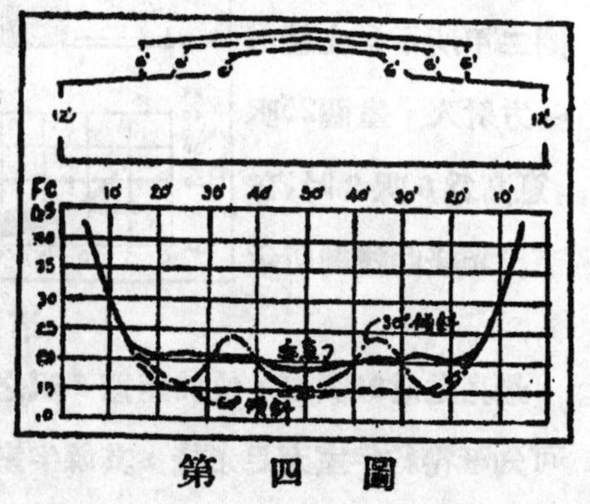
第　四　圖

光線曲線所示爲三種窗，各射于工作平面之光線，其單位爲呎燭。垂直窗不特在室中供給最高光線(約25呎燭)，且能使之甚均匀。30° 角者能供給15至24呎燭，但不均匀。30°角者所給最少，僅得1之5數而已。

天窗旣透光効率低劣，而前又已知其易於藏垢，則天窗之不宜用，其理甚明。我國房室常感光線不佳，是則或與多用天窗不無關係也。天窗之所以易於藏垢之理，亦可忖測得之。60° 角及平過之玻璃片，與地面之所成角度小甚，在窗外最適宜於煤屑塵穢之停留，不能受地吸力而滑走下在窗內，平面投影面積亦大，最足以使烏煙及熱空氣所帶上之塵垢所汚玷。固有人謂可常洗濯之，惟此良非易易也。天窗高徹屋頂，汚垢之百分七十五又集於窗內，危立斜梯以洗濯之，殊有生命危險，欲求安全，宜設永久平梯，如是又費用浩繁。洗濯時汚垢下墮，適足以防碍工作及毀壞儀器，天窗之不宜于用，固無疑義矣。

垂直樓頂窗旣屬適于實用，然宜使其廣闊抑短狹？窗徹抑低下？此亦爲吾人所宜留意者。

垂直樓頂窗之闊度，最簡便適當之方法，莫如使之等於室闊之一半，不宜少於此數。

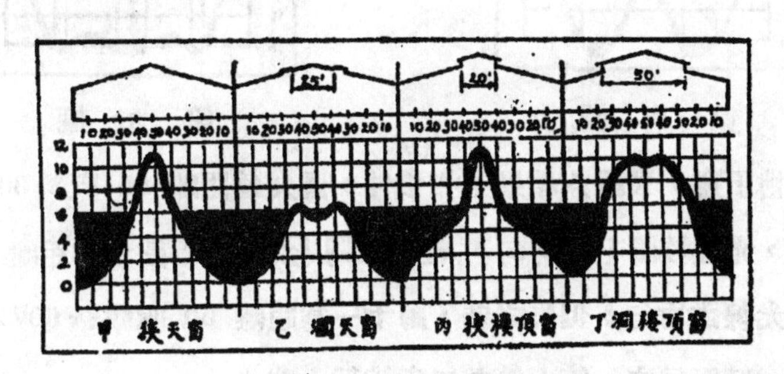

第　五　圖

圖五所示爲天窗與樓頂窗之光線曲線之比較。圖內牆邊窗之光線未有算在內，故取 6 呎燭爲規定光線，若連牆窗而計算之，足有10呎燭也。圖甲卽爲相連天窗，室中之間光線甚強，惟兩旁黑暗之處太多，注意曲線圖之突高。每邊約有35呎，地面爲不足 6 呎燭。圖乙爲遠闊之天窗，距離25呎，光線較均匀，但嫌其弱。使天窗距離較遠則增加室中間之黑暗，距離較近，則促曲線縮高如圖

甲。圖丙爲樓頂窗之距離20呎者，光線失之均匀，暗處亦多，圖丁爲樓頂窗之距離50呎者，光線既均匀，黑暗面積亦少。由圖內曲線表比較之可知後者最爲適宜。

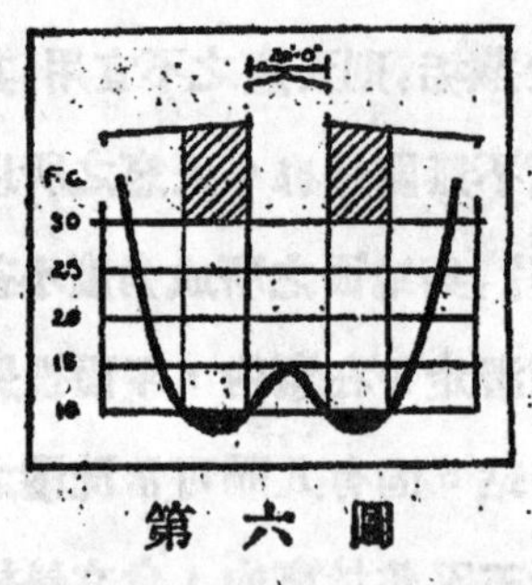

第 六 圖

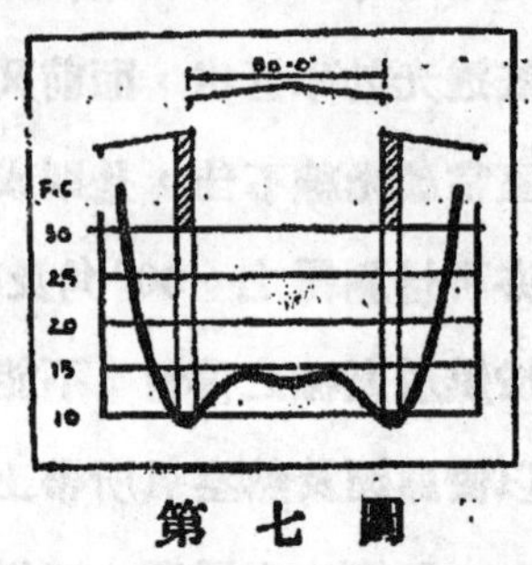

第 七 圖

圖六至圖十一所示，爲建築物100呎闊之直窗，30°與 60° 各種窗之比較。圖六距離20呎之爲直窗，在光線曲線圖內有二處爲不敷規定之10呎燭者。圖七爲距離50呎之直窗幾乎全室充足光線，給予頗滿意之結果。圖八爲距離20呎

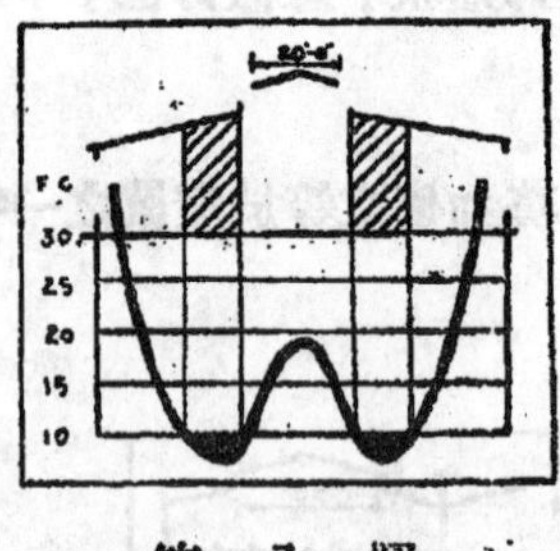

第 八 圖

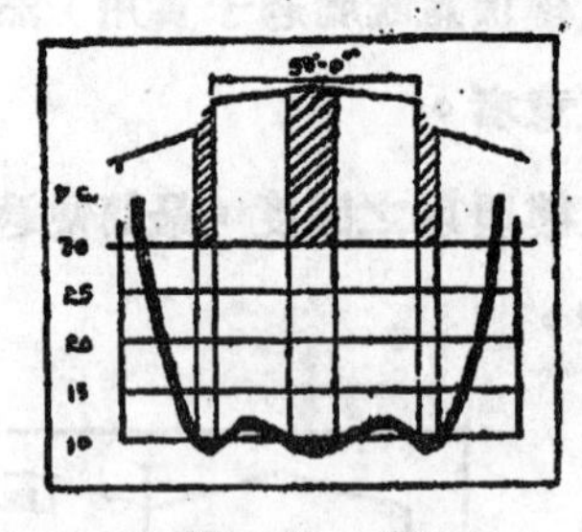

第 九 圖

傾斜 30° 之樓頂窗，與圖六結果不相伯仲，圖九爲距離 50 呎斜 30° 者，其結果亦頗滿意，光線較圖七爲均匀，惟不如圖七之強。圖十爲距離 20 呎傾斜 60° 之天窗光線雖均匀，但失之弱，圖十一爲距離 50 呎傾斜 60° 者，室之中間光綫甚暗。統而言之，第七第九二者較稱人意也。

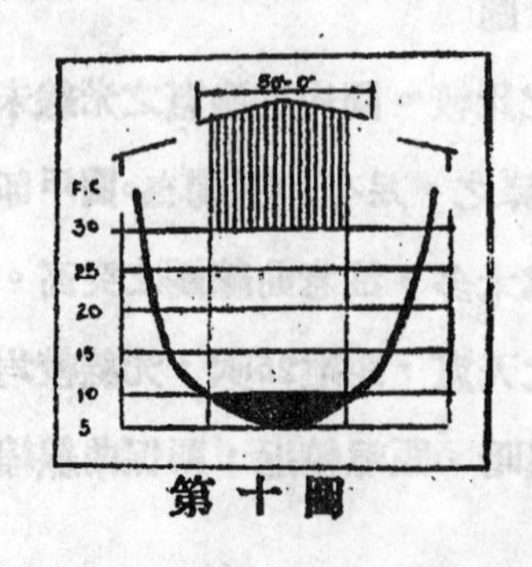

第 十 圖

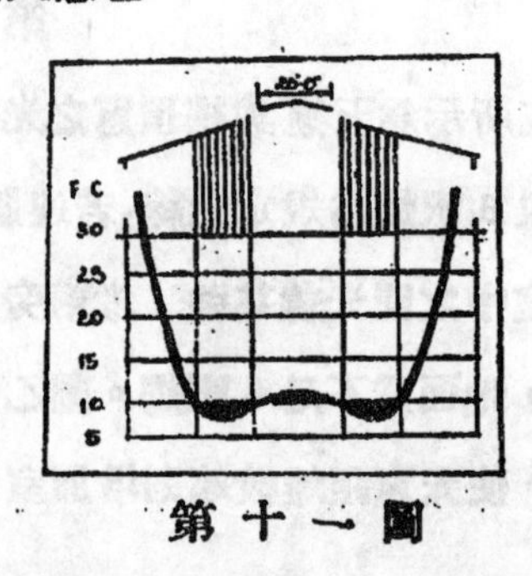

第 十 一 圖

在可能範圍內，窗牖宜高不宜低，高窗牖之利益爲能使最暗處增加光綫較亮處所增加者其速率爲快。惟通常窗之高或底，以房屋之高度及美觀爲標準，故此關係尙屬渺少也。

鋸牙式窗常用于室宇過廣闊，單獨牆窗不足以供給光之者。鋸牙式窗最大利益之點在能使光綫均勻，但此種窗似嫌其不美觀。故通常多在工廠內用之。此種泰半爲不設玻璃片，通常均爲向北，以避免日光射入之塵垢，惟此點未必全臻乎理；苟使設置得宜，使鋸牙式窗向南，則室較明朗，蓋南方之光線常強於北方者也。設計鋸牙式窗，亦適用以前所說關于直窗之條例，以求光綫之充裕美滿。宜加以注意者，則爲使窗之跨度 Span 小些，能使最少光處增長快于多光綫處。此點可由圖十二表明之。圖十二左方爲40呎之鋸牙式窗，右方爲20呎之鋸牙式窗，窗之高度未有改變，由左右二圖比較，最大光綫增加甚微，而最少光綫則增加較多矣。故亦使光綫較趨均勻。

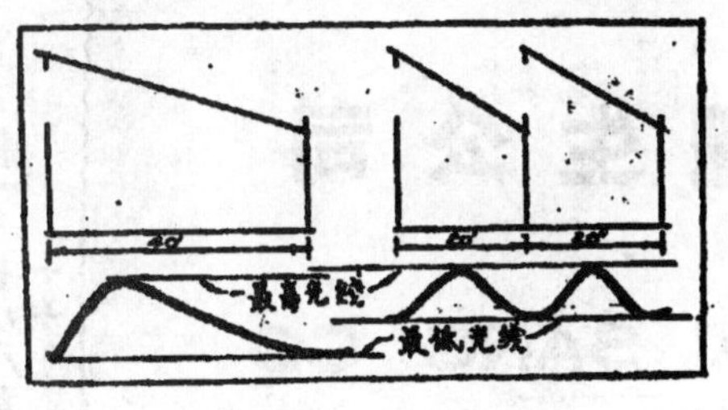

第十二圖

此外於窗牖應注意之微點亦多。例如窗內油漆宜爲淡色。普通白色與黑色二者之比較，能增加百分之十至二十五，視窗之高，及建築情形而定。玻璃片之形狀與傳光亦有關係。璃片之形式甚繁，有光滑，稜形，磨沙，刻蝕，凸紋，透明與半透明等。全白色璃片當其全潔時，假算無光綫反折 Refraction 及擴散Diffusion，其傳光律爲百分之八十二，直紋及橫紋者約爲百分之七十二，但經六月之積垢，其傳光律亦幾相等矣。結晶之璃片，不大傳光，因其反光故也。璃片之有紋面宜向外，因其滑面，向內易於洗濯也。

至于設計窗牖，計算室內光綫之法，以前學者曾作許多實驗，吾輩可取之爲計算根據。表一所列爲十三種高度直牆邊窗（5'--2'，　6'--10 3/8"，

8'--6 3/4" 等)，由窗至室中不同距離(5'，10'，15' 等) 給予之光綫數目，其單位爲呎燭在工作平面上。 表二爲各種高度之樓頂窗（3'—6'，5'—2" 等，）在四種離工作平面之高度（15'，25'，35'及45'），由窗至室中不同距離（5'，10'，15'等）所給予之光綫數目。表三爲四種高度之30°傾斜窗(3'，6'，9'，及12'）在四種離工作平面之高度（15'，25'等）由窗至室不同距離(5'，10'）給予之光綫數目。 此數表均爲專家之有系統之實驗所得之結果， 而其實驗之窗牖型模，均曾蒙六閱月之積垢者，其單位均爲呎燭。（下期待續）

來往廣州私立嶺南大學

南樂電船開行時間

國歷二十五年印行

康樂碼頭開	省城碼頭開
（上午）	（上午）
÷七點	÷八點
÷八點	÷八點半
八點半	九點
九點	九點半
九點半	十點
十點	十點半
十點半	十一點
十一點	十一點半
十一點半	十二點
十二點	（下午）
（下午）	十二點半
十二點半	一點
一點	一點半
一點半	二點
二點	二點半
二點半	三點
三點	三點半
三點半	四點
四點	四點半
四點半	五點
五點	五點半
五點半	六點
六點	六點半
六點半	七點

÷此符號指明每日有船開

惟星期日停開

廣州嶺南大學工程學會會務要聞

本屆職員

趙宗武（主席） 梁鼎燊（秘書） 彭眞原（出版）

林家就（財政） 黃惠光（學術） 梁健卿（總務）

崔世泰（體育） 鄒煥新（交際）

本會會員名册

李權亨院長 廣州嶺南大學

梁綽餘教授 廣西遷江合山煤鑛公司

羅石麟教授 香港九龍深水埔福華街廿七號二樓現址廣州東山新河浦十三號三樓

王叔海教授 廣州市法政路二號現址嶺南大學西南區十一號

黃郁文教授 香港深水埔長沙灣道二八七號四樓

黃玉瑜教授 廣州太平路廣東信托公司

金肇組教授 廣州越秀北路安樂道二號韶園

馮兆端教授 廣州東山美華北十六號三樓

藥福安教授 九龍深水埔福榮街一一八號三樓

李文邦教授 廣州白雲路廣東治河委員會

李鈞雄教授 廣州惠吉東路一號三樓

以上本校工學院教授

劉栽和 職 湖南衡陽粤漢鐵路株韶段第六總段第一分段

彭震東 職 浙江省𣄃江鐵路局

黃錦裳 職 第一集團軍總司令部技士

崔兆鼎　　留美米西根大學土木工程

梁綽芹　職　廣東治河會

李廷芬　職　武昌縣金口鎮國營金水農場

陳銘珊　職　嶺南大學附僑任數學主任

黎均霖　職　嶺南大學附中數學及機械畫教員

陳國柱　　美國 M. I. T. 習土木工程

李文遴　職　山中縣翠亨鄉中山紀念中學

文宗堯　　廣州都府街五號二樓

容永樂　職　粵漢鐵路株韶段工程局第三工程總處

梁寶琨　職　廣東建設廳現派任龍川大江橋監工

陳書馨　職　廣西省南寧集團軍總司令建置委員會

莫佐基　職　廣東建設廳

鄒漢新

盧志勤　　留美米西根大學習營造學

梁卓華　　留美米西根大學習電機工程

彭延匡　　留美米西根大學習土木工程專修水利學

鄧樹翼　　留美

余錦煥　　留美

雷攀桀　　留美

梁　綮　　清華大學工學院

關祖舜　　留美米西根大學

簡石農　　復旦大學

潘祖芬　　留美 Ohio 大學

以上離校會員

王宏猷　廣州市河南後樂園街二十八號

李卓傑　廣州市河南同慶四街四十二號

陳元力　廣州市東山竹絲崗勝園

趙宗武　廣州市長堤永安人壽保險公司轉
黃惠光　廣州市長堤華盛頓餐室
黃樹邦　廣州市河南嶺南大學怡樂村或台山城縣前馬路豐華銀號
崔世泰　廣州市第十甫馬路平盛里三號
陳守勤　廣州市中華北百靈街一百三十二號新丰
夏進興　香港中環永和街和興昌
陳秉淮　香港灣仔厚豐里八號
龍寶遜　廣州市第十甫曾二巷十二號
趙長有　中山斗門李邊水
曾德甫　廣州西關十二甫西廿三號
彭奐原　廣州河南南華中路一百一十八號
林家献　廣州廣九路廿三號轉
沈錫琨　廣西鐵象街福亭號
陳士敦　廣州市西關昌華新街廿八號三樓
鄒煥新　廣州市德宣東路九功坊十二號
陳振泰　南洋檳城新街頭廿九號成裕公司
陳尙武　南洋比能埠新街頭十一號 F
陳棣濂　香港大道中四十九號二樓
陳樹彬　香港大道東二百五十七號二樓
陳自仁　福州嶺下竹梅山館
李梁材　本校
黎辛才　香港尖沙咀山林道五號
吳潤蕃　廣州市中大第二醫院吳潤芬醫生轉
梁健卿　廣州市高第街一百六十九號三樓
梁鼎燊　南洋檳城羅郎士拿勿街二號 E
關廣培　香港雲咸街廿九號
林文贊　廣州市沙基東中約華益建築公司

李卓平 香港深水埗長沙灣道 289 號四樓

黄崐昌 佛山叠滘汁黄有山醫所

黄文海 香港永樂街一廣九號

温兆明 廣州市西關和息里二十四號

王鋭鈞 廣州市河南後樂園街三十號

楊天用 本校

余廣才 本校

莫華泿 香港九龍窩打老道 73 號

李祺煥 廣州市天成路怡和

余槐欽 廣東台山荻海中和路廣美昌

夏傑榮 廣東新會城蘭桂里八號

余有庸 廣州市小東門三角市德祥號

伍瑞明 香港堅尼地道一二四號華義書院

梅冷堅 廣東南海九江閘邊市區立三小學校

盧炳煌 廣州市上芳村鎮東廣安里九號

陳簾良 武昌曇華林倫敦會陳宅

陳寧畿 廣西貴縣萬壽街四十一號

黎廣杰 廣州市多寶路一六三號

楊啓驤 廣州市西關逢源北街第九號

馮葆鎏 順德大良莘村蘇巷一巷六號

區錫齡 廣州市小市街西廣元金舖

馮佑增 順德大良莘村蘇巷一巷二號

廣州私立嶺南大學
工程學會

鋼筋三合土設計手冊

南大工程學會編　　主編者羅石麟

附言：　鋼筋三合土之設計，每多慊其計算繁雜。爲補救起見，各國工程界均事先預備各種圖表，以爲設計時之需。化繁爲簡，工程師視爲必需之利器。

本校羅石麟教授，業將鋼筋三合土設計圖表編就，以便同學應用，而補課本之不足。本會深信此種圖表，不特爲功課中之重要參考物，抑亦設計工程師所樂用者，故特請羅教授將各圖表編成手册，附刊于南大工程，俾讀者有所採用焉。

此手册之圖表，除少數由別處（如Turneaure，Maurer，Singleton，Thomas，Nicholson，Morsch，Manning，Waddell，Hool and Johnson 等各書本中）借用外，餘槪爲羅教授與本會會員計算所得。我國規定萬國權度通制（Metric System）爲標準，而習慣相沿，又多喜用英美權度制（Foot Pound System）。爲利便起見，本手册兩種制度均備。其屬于前者，則各圖表上有（公制）二字，其屬于後者，則有（英制）二字。若二者（公英制）皆可通用，則亦如之，以免混亂。各圖表之准許單位應力，乃參照國內各大城市所准許者爲標準。廣州及上海各種設計之准許應力圖表，公制或英制俱皆齊備。

外國工程名詞之繙譯，尙未有劃一之規定，圖表所用之名詞，乃由主編者選擇較適當者用之。因時間關係，本期祗能將英制之圖表刊出。公制之圖表，下期繼續發表。尙望中外工程師惠然指導，俾得隨時改善，以竟未完之工，則幸甚矣。

彭奠原

南大工程學會出版部

出版部啓事

本期印刷時間短促，

手民錯誤，誠恐不免，

尚祈讀者指正。

第一表（英制）

每英尺塊面之鋼筋面積及圓周

中距	鋼筋	1/4圓	3/8圓	3/8方	1/2圓	1/2方	5/8圓	3/4圓	3/4方	7/8圓	1圓	1方	1 1/8方	1 1/4方
距	重量	.167	.376	.478	.668	.850	1.04	1.50	1.91	2.04	2.67	3.40	4.30	5.31
3	面積	0.20	0.44	0.56	0.79	1.00	1.23	1.77	2.25	2.40	3.14	4.00	5.06	6.25
	圓周	2.14	4.71	6.00	6.28	8.00	7.85	9.43	12.0	11.0	12.6	16.0	18.0	20.0
3.5	面積	0.17	0.38	0.48	0.67	1.[illegible]6	1.05	1.51	1.93	2.06	2.69	3.43	4.34	5.36
	圓周	2.69	4.04	5.14	5.38	6.86	6.73	8.08	10.3	9.42	10.8	13.7	15.4	17.1
4	面積	0.15	0.33	0.42	0.59	0.75	0.92	1.32	1.69	1.80	2.36	3.00	3.80	4.69
	圓周	2.36	3.53	4.50	4.71	6.00	5.89	7.06	9.00	8.24	9.42	12.0	13.5	15.0
4.5	面積	0.13	0.29	0.37	0.52	0.67	0.82	1.18	1.50	1.60	2.09	2.67	3.37	4.17
	圓周	2.09	3.14	4.00	4.19	5.33	5.23	6.28	8.00	7.32	8.38	10.7	12.0	13.3
5	面積	0.12	0.27	0.34	0.47	0.60	0.74	1.06	1.35	1.44	1.99	2.40	3.04	3.75
	圓周	1.88	2.83	3.60	3.77	4.80	4.71	5.65	7.20	6.59	7.54	9.60	10.8	12.0
5.5	面積	0.11	0.24	0.31	0.43	0.55	0.67	0.96	1.23	1.31	1.71	2.18	2.76	3.41
	圓周	1.71	2.53	3.27	3.43	4.36	4.28	5.14	6.55	6.00	6.85	8.73	9.82	10.9
6	面積	0.10	0.22	0.28	0.39	0.50	0.61	0.88	1.13	1.20	1.57	2.00	2.53	3.12
	圓周	1.57	2.36	3.00	3.14	4.00	3.92	4.71	6.00	5.50	6.28	8.00	9.00	10.0
6.5	面積	0.09	0.20	0.26	0.36	0.46	0.57	0.82	1.03	1.10	1.45	1.85	2.34	2.88
	圓周	1.45	2.17	2.77	2.90	3.69	3.62	4.35	5.54	5.07	5.80	7.39	8.31	9.24
7	面積	0.08	0.19	0.24	0.34	0.43	0.53	0.76	0.97	1.03	1.35	1.71	2.17	2.68
	圓周	1.35	2.02	2.57	2.69	3.43	3.36	4.04	5.15	4.71	5.38	6.86	7.72	8.57
7.5	面積	0.08	0.18	0.22	0.31	0.40	0.49	0.71	0.90	0.96	1.26	1.60	2.02	2.50
	圓周	1.26	1.86	2.40	2.51	3.20	3.14	3.77	4.80	4.40	5.03	6.40	7.20	8.00
8	面積	0.07	0.17	0.21	0.29	0.38	0.46	0.66	0.84	0.90	1.18	1.50	1.89	2.34
	圓周	1.18	1.77	2.25	2.36	3.00	2.94	3.53	4.50	4.12	4.71	6.00	6.75	7.50
8.5	面積	0.07	0.16	0.20	0.28	0.35	0.44	0.62	0.79	0.85	1.10	1.41	1.79	2.21
	圓周	1.11	1.66	2.12	2.22	2.82	2.77	3.33	4.24	3.88	4.43	5.65	6.36	7.06
9	面積	0.07	0.15	0.19	0.26	0.33	0.41	0.59	0.75	0.80	1.05	1.33	1.69	2.08
	圓周	1.05	1.57	2.00	2.09	2.67	2.62	3.14	4.00	3.66	4.19	5.33	6.00	6.67
9.5	面積	0.06	0.14	0.18	0.25	0.32	0.39	0.56	0.71	0.76	0.98	1.26	1.60	1.97
	圓周	0.99	1.49	1.89	1.98	2.53	2.48	2.98	3.79	3.47	3.97	5.05	5.68	6.32
10	面積	0.06	0.17	0.17	0.24	0.30	0.37	0.53	0.67	0.72	0.94	1.20	1.52	1.87
	圓周	0.94	1.41	1.80	1.88	2.40	2.35	2.83	3.60	3.30	3.77	4.80	5.40	6.00
10.5	面積	0.06	0.13	0.16	0.22	0.29	0.36	0.50	0.64	0.69	0.89	1.14	1.44	1.79
	圓周	0.90	1.35	1.71	1.79	2.29	2.24	2.69	3.53	3.17	3.59	4.57	5.14	5.72
11	面積	0.05	0.12	0.15	0.21	0.27	0.34	0.48	0.61	0.66	0.86	1.09	1.38	1.70
	圓周	0.86	1.28	1.54	1.71	2.18	2.14	2.57	3.27	3.00	3.43	4.37	4.91	5.46
11.5	面積	0.05	0.12	0.15	0.21	0.26	0.32	0.46	0.59	0.63	0.81	1.04	1.32	1.63
	圓周	0.82	1.23	1.57	1.80	2.09	2.05	2.46	3.13	2.87	3.28	4.17	4.70	5.22
12	面積	0.05	0.11	0.14	0.20	0.25	0.31	0.44	0.56	0.60	0.79	1.00	1.27	1.56
	圓周	0.79	1.18	1.50	1.57	2.00	1.96	2.39	3.00	2.75	3.14	4.00	4.50	6.00
15	面積	——	0.09	0.11	0.16	0.20	0.25	0.35	0.45	0.48	0.63	0.80	1.02	1.25
	圓周	——	0.94	1.20	1.26	1.60	1.57	1.89	2.40	2.20	2.51	3.20	3.60	4.00
18	面積	——	0.07	0.09	0.13	0.17	0.21	0.29	0.37	0.40	0.53	0.67	0.85	1.04
	圓周	——	0.79	1.00	1.05	1.33	1.31	1.57	2.00	1.83	2.09	2.56	3.00	3.33

第二表A.（英制）

樑之最小寬度

（英吋）

直徑或邊長	鋼筋形式	每層鋼筋之根數（鋼筋端末無鈎者） 1	2	3	4	5	6	7	8	9	10	11	12	每加一根鋼筋加寬
1/4	圓	2 1/4	3 1/2	4 3/4										1 1/4
5/16	圓	2 5/16	3 5/8	4 15/16										1 5/16
3/8	圓	2 3/8	3 3/4	5 1/8	6 1/2									1 3/8
3/8	方	2 3/8	3 3/4	5 1/8	6 1/2	7 7/8								1 3/8
1/2	圓	2 1/2	4	5 1/2	7	8 1/2	10							1 1/2
1/2	方	2 1/2	4 1/4	6	7 3/4	9 1/2	11 1/4							1 3/4
5/8	圓	2 5/8	4 1/4	5 7/8	7 1/2	9 1/8	10 3/4	12 3/8						1 5/8
5/8	方	2 5/8	4 1/2	6 3/8	8 1/4	10 1/8	12	13 7/8	15 3/4					1 7/8
3/4	圓		4 5/8	6 1/2	8 3/8	10 1/4	12 1/8	14	15 7/8	17 3/4				1 7/8
3/4	方		5	7 1/4	9 1/2	11 3/4	14	16 1/4	18 1/2	20 3/4				2 1/4
7/8	圓		5 1/16	7 1/4	9 7/16	11 5/8	13 13/16	16	18 3/16	20 3/8	22 9/16			2 3/16
7/8	方		5 1/2	8 1/8	10 3/4	13 3/8	16	18 5/8	21 1/4	23 7/8	26 1/2	29 1/8		2 5/8
1	圓		5 1/2	8	10 1/2	13	15 1/2	18	20 1/2	23	25 1/2	28	30 1/2	2 1/2
1	方		6	9	12	15	18	21	24	27	30	33	36	3
1 1/8	方		6 1/2	9 7/8	13 1/4	16 5/8	20	23 3/8	26 3/4	30 1/8	33 1/2	36 7/8	40 1/4	3 3/8
1 1/4	方		7	10 3/4	14 1/2	18 1/4	22	25 3/4	29 1/2	33 1/4	38	41 3/4	45 1/2	3 3/4

注意：表中尺度尚未計及鋼絡佔有之位置

第二表B.（英制）

樑之最小寬度

（英吋）

直徑或邊長	鋼筋形式	每層鋼筋之根數 1	2	3	4	5	6	7	8	9	10	11	12	每加一根鋼筋加寬
		鋼筋端末有鈎者												
1/4	圓	2 1/4	3 1/2	4 3/4										1 1/4
/16	圓	2 5/16	3 5/8	4 15/16										1 5/16
3/8	圓	2 3/8	3 3/4	5 1/8	6 1/2									1 3/8
3/8	方	2 3/8	3 3/4	5 1/8	6 1/2	7 7/8								1 3/8
1/2	圓	2 1/2	4	5 1/2	7	8 1/2	9	10 1/2						1 1/2
1/2	方	2 1/2	4 1/4	6	7 3/4	9 1/2	10 1/4	12						1 3/4
5/8	圓	2 5/8	4 1/4	5 7/8	7 1/2	9 1/8	10 3/4	12 3/8	14					1 5/8
5/8	方	2 5/8	4 1/4	5 7/8	7 1/2	9 1/8	10 3/4	12 3/8	14					1 5/8
3/4	圓		4 1/2	6 1/4	8	9 3/4	11 1/2	13 1/4	15	16 3/4				1 3/4
3/4	方		4 5/8	6 1/2	8 3/8	10 1/4	12 1/8	14	15 7/8	17 3/4				1 7/8
7/8	圓		4 3/4	6 5/8	8 1/2	10 3/8	12 1/4	14 1/8	16	17 7/8	19 3/4			1 7/8
7/8	方		5 1/16	7 1/4	9 7/16	11 5/8	13 13/16	16	18 3/16	20 3/8	22 9/16	24 3/4		2 3/16
1	圓		5	7	9	11	13	15	17	19	21	23	25	2
1	方		5 1/2	8	10 1/2	13	15 1/2	18	20 1/2	23	25 1/2	28	30 1/2	2 1/2
1 1/8	方		5 15/16	8 7/8	11 9/16	14 3/8	17 3/16	20	22 13/16	25 5/8	28 7/16	31 1/4	34 1/16	2 13/16
1 1/4	方		6 3/8	9 1/2	12 5/8	15 3/4	18 7/8	22	25 1/8	28 1/4	31 3/8	34 1/2	37 5/8	3 1/8

注意：表中尺度尚未計及鋼絡佔有之位置

英4.

第三表（公英制）

曲鋼筋之長度

a	45°		30°	
	a	兩曲之加長	a	兩曲之加長
1.5	2	1	3	1
2	3	2	4	1
2.5	3.5	2	5	1.5
3	4	2	6	1.5
3.5	5	3	7	2
4	5.5	3	8	2
4.5	6.5	4	9	2
5	7	4	10	2.5
5.5	8	5	11	3
6	8.5	5	12	3
6.5	9	5	13	3.5
7	10	6	14	3.5
7.5	10	6	15	4
8	11.5	7	16	4
8.5	12	7	17	4.5
9	13	7	18	5
9.5	13	8	19	5
10	14	8	20	5
11	16	9	22	6
12	17	10	24	6
13	18	11	26	7
14	20	12	28	7
15	21	13	30	8
16	23	14	32	8
17	24	15	34	9
18	26	16	36	10
19	27	16	38	10
20	28	17	40	11
22	31	18	44	12
24	34	20	48	13
26	37	22	52	14
28	40	24	56	15
30	42	25	60	16
32	45	26	64	17
34	48	28	68	18
36	51	30	72	19
38	54	32	76	20
40	57	31	80	21
42	59	35	84	23
44	62	36	88	24
46	65	38	92	25
48	68	40	96	26
50	71	42	100	27
52	74	44	104	28
54	76	45	108	29
56	79	46	112	30
58	82	48	116	31
60	85	50	120	32
62	88	51	124	33
64	90	53	128	34
66	93	55	132	35
68	96	56	136	36
70	99	58	140	38

第四表 （英制）

數根鋼筋合成之面積，圓周，及重量

直徑或邊長	鋼筋形式	項目	鋼筋之根數 1	2	3	4	5	6	7	8	9	10	12	14
1/4	圓	面積	0.05	0.10	0.15	0.20	0.25	0.29	0.34	0.39	0.44	0.49	0.59	0.69
		圓周	0.79	1.57	2.36	3.14	3.93	4.71	5.50	6.28	7.07	7.85	9.42	11.0
		重量	.167	.334	.501	.668	.835	1.00	1.17	1.34	1.50	1.67	2.00	2.34
5/16	圓	面積	0.08	0.15	0.23	0.31	0.38	0.46	0.54	0.61	0.69	0.77	0.92	1.07
		圓周	0.98	1.96	2.94	3.93	4.91	5.89	6.87	7.85	8.84	9.82	11.8	13.7
		重量	.261	.522	.783	1.04	1.30	1.56	1.82	2.08	2.34	2.61	3.12	3.64
3/8	圓	面積	0.11	0.22	0.33	0.44	0.55	0.66	0.77	0.88	0.99	1.10	1.32	1.55
		圓周	1.18	2.36	3.53	4.71	5.89	7.07	8.25	9.42	10.6	11.8	14.2	16.5
		重量	.375	.750	1.12	1.50	1.88	2.25	2.63	3.00	3.38	3.75	4.50	5.25
3/8	方	面積	0.14	0.28	0.42	0.56	0.70	0.84	0.98	1.13	1.27	1.41	1.69	1.97
		圓周	1.50	3.00	4.50	6.00	7.50	9.00	10.5	12.0	13.5	15.0	18.0	21.0
		重量	.478	.956	1.43	1.91	2.39	2.87	3.34	3.82	4.30	4.78	5.74	6.70
1/2	圓	面積	0.20	0.39	0.59	0.78	0.98	1.18	1.37	1.57	1.77	1.96	2.36	2.75
		圓周	1.57	3.14	4.71	6.28	7.85	9.42	11.0	12.6	14.1	15.7	18.9	22.0
		重量	.668	1.34	2.00	2.67	3.34	4.01	4.68	5.35	6.02	6.68	8.02	9.35
1/2	方	面積	0.25	0.50	0.75	1.00	1.25	1.50	1.75	2.00	2.25	2.50	3.00	3.50
		圓周	2.00	4.00	6.00	8.00	10.0	12.0	14.0	16.0	18.0	20.0	24.0	28.0
		重量	.850	1.70	2.55	2.40	4.25	5.10	5.95	6.80	7.65	8.50	10.2	11.9
5/8	圓	面積	0.31	0.61	0.92	1.23	1.54	1.84	2.15	2.46	2.76	3.07	3.68	4.30
		圓周	1.96	3.93	5.89	7.85	9.82	11.8	13.7	15.7	17.7	19.6	23.6	27.5
		重量	1.04	2.09	2.13	4.17	5.20	6.24	7.28	8.32	9.36	10.4	12.5	14.6
5/8	方	面積	0.39	0.78	1.17	1.56	1.95	2.34	2.73	3.12	2.52	2.91	4.69	5.47
		圓周	2.50	5.00	7.50	10.0	12.5	15.0	17.5	20.0	22.5	25.0	30.0	35.0
		重量	1.33	2.66	3.98	5.31	6.64	7.97	9.29	10.6	12.0	13.3	15.9	18.6
3/4	圓	面積	0.44	0.88	1.33	1.77	2.21	2.65	3.09	3.53	3.98	4.42	5.30	6.19
		圓周	2.36	4.71	7.07	9.42	11.8	14.1	16.5	18.8	21.2	23.6	28.3	33.0
		重量	1.50	3.00	4.51	6.01	7.51	9.01	10.5	12.0	13.5	15.0	18.0	21.0
3/4	方	面積	0.56	1.13	1.69	2.25	2.81	3.38	3.94	4.50	5.06	5.62	6.75	7.87
		圓周	3.00	6.00	9.00	12.0	15.0	18.0	21.0	24.0	27.0	30.0	36.0	42.0
		重量	1.91	3.83	5.74	7.65	9.57	11.5	13.4	15.3	17.2	19.1	23.0	26.8
7/8	圓	面積	0.60	1.20	1.80	2.41	3.01	3.61	4.21	4.81	5.41	6.01	7.22	8.42
		圓周	2.75	5.50	8.25	11.0	13.7	16.5	19.2	22.0	24.7	27.5	33.0	38.5
		重量	2.04	4.09	6.13	8.18	10.2	12.3	14.3	16.4	18.4	20.4	24.6	28.6
7/8	方	面積	0.77	1.53	2.30	3.06	3.83	4.59	5.36	6.12	6.89	7.66	9.19	10.7
		圓周	3.50	7.00	10.5	14.0	27.5	21.0	24.5	28.0	31.5	35.0	42.0	49.0
		重量	2.60	5.21	7.81	10.4	13.0	15.6	18.2	20.8	23.4	26.0	31.2	36.4
1	圓	面積	0.79	1.57	2.36	3.14	2.93	4.71	5.50	6.28	7.07	7.85	9.42	11.0
		圓周	3.14	6.28	9.42	12.6	15.7	18.9	22.0	25.1	28.3	31.4	37.7	44.0
		重量	2.67	5.34	8.01	10.7	13.3	16.0	18.7	21.4	24.0	26.7	32.0	37.4
1	方	面積	1.00	2.00	3.00	4.00	5.00	6.00	7.00	8.00	9.00	10.0	12.0	14.0
		圓周	4.00	8.00	12.0	16.0	20.0	24.0	28.0	32.0	36.0	40.0	48.0	56.0
		重量	3.40	6.80	10.2	13.6	17.0	20.4	23.8	27.2	30.6	34.0	40.8	47.6
1 1/8	圓	面積	1.27	2.53	3.80	5.06	6.33	7.59	8.86	10.1	11.4	12.7	15.2	17.7
		圓周	4.50	9.00	13.5	18.0	22.5	27.0	31.5	36.0	40.5	45.0	54.0	63.0
		重量	4.30	8.61	42.9	17.2	21.5	25.8	30.1	34.4	38.7	43.0	51.7	60.3
1 1/4	方	面積	1.56	3.13	4.69	6.25	7.81	9.38	10.9	12.5	14.1	15.6	18.8	21.9
		圓周	5.00	10.0	15.0	20.0	25.0	30.0	35.0	40.0	45.0	50.0	60.0	70.0
		重量	5.31	10.6	15.9	21.3	26.6	31.7	37.2	42.5	47.8	53.1	63.7	74.4

第五表　（英制）

各種截面之面積，重量，及惰性率

深度 d（英寸）	正方形 面積 平方英寸	正方形 磅/英尺	正方形 I_{1-1} (英寸)^4	圓形 面積 平方英寸	圓形 磅/英尺	圓形 I_{1-1} (英寸)^4	八邊形 面積 平方英寸	八邊形 磅/英尺	八邊形 I_{1-1} (英寸)^4	長方形 面積 平方英寸	長方形 磅/英尺	長方形 I_{1-1} (英寸)^4
6	36	38	108	28	29	64	30	31	71	6.0	6.3	18
7	49	51	203	38	40	118	41	43	132	7.0	7.3	29
8	64	67	344	50	52	201	53	55	224	8.0	8.3	43
9	81	84	549	64	67	322	67	70	360	9.0	9.4	61
10	100	104	830	79	82	491	83	87	548	10.0	10.1	83
11	121	126	1221	95	99	719	100	104	802	11.0	11.5	111
12	144	150	1728	113	118	1018	119	124	1136	12.0	12.5	144
13	169	176	2379	133	138	1402	140	146	1565	13.0	13.5	183
14	296	204	3206	154	160	1886	162	169	2105	14.0	14.6	229
15	225	235	4215	177	184	2485	186	194	2775	15.0	15.6	281
16	256	267	5456	201	210	3216	212	221	3591	16.0	16.7	341
17	289	301	6953	227	237	4100	239	249	2577	17.0	17.7	409
18	324	338	8748	255	265	5153	268	280	5753	18.0	18.8	486
19	361	376	10868	284	295	6397	399	312	7142	19.0	19.8	572
20	400	417	13333	314	327	7854	331	345	8708	20.0	20.8	667
21	441	459	16212	346	361	9547	365	381	10658	21.0	21.9	772
22	484	504	19514	380	396	11499	401	418	12837	22.0	22.9	887
23	529	551	23322	416	433	13737	438	457	15335	23.0	24.0	1014
24	576	600	27648	452	471	16286	477	497	18181	24.0	25.0	1152
25	625	651	32550	491	511	19175	518	539	21406	25.0	26.1	1302
26	676	704	38090	531	553	22432	560	583	25042	26.0	27.1	1465
27	729	759	44280	573	597	26087	604	629	29123	27.0	28.1	1640
28	784	817	51212	616	642	30172	650	677	33683	28.0	29.2	1829
29	841	876	58928	661	688	34719	697	726	38759	29.0	30.2	2032
30	900	937	67500	707	736	39761	746	777	44388	30.0	31.2	2250
31	961	1001	76973	755	786	45333	799	829	50609	31.0	32.3	2483
32	1024	1067	87392	840	838	51472	848	884	57462	32.0	33.3	2731
33	1085	1134	98835	855	891	59214	902	940	64988	33.0	34.4	2995
34	1159	1204	111350	908	946	65597	958	998	73231	34.0	35.4	3275
35	1225	1276	125055	962	1002	73662	1015	1057	82234	35.0	36.5	3573
36	1296	1350	139680	1018	1060	82448	1074	1118	92043	36.0	37.5	3880
37	1369	1426	156177	1075	1120	91998	1134	1181	102704	37.0	38.5	4221
38	1444	1504	173774	1134	1181	102354	1196	1246	114265	38.0	39.6	4574
39	1521	1585	192777	1195	1244	113561	1260	1313	126777	39.0	40.6	5943
40	1600	1667	213320	1257	1309	124664	1326	1381	140288	40.0	41.7	5333
41	1681	1750	235463	1310	1375	138708	1393	1451	154852	41.0	42.7	5743
42	1764	1838	259308	1385	1443	152744	1462	1523	170521	42.0	43.8	6174
43	1849	1925	284918	1452	1512	167819	1533	1598	187850	43.0	44.8	6626
44	1936	2015	312356	1521	1585	183983	1604	1670	205396	44.0	45.8	7099
45	2025	2110	341730	1590	1657	201287	1678	1748	224714	45.0	46.9	7594
46	2116	2202	373106	1662	1731	219785	1754	1826	245365	46.0	47.9	8111
47	2209	2300	406344	1735	1808	239529	1831	1908	267407	47.0	49.0	8652
48	2304	2400	442368	1810	1885	260574	1909	1988	290901	48.0	50.0	9216
49	2401	2500	480396	1886	1965	282977	1990	2074	315911	49.0	51.1	9804
50	2500	2605	520650	1964	2046	306794	2072	2159	342500	50.0	52.1	10417

第六表 （英制） .7英

鋼筋惰性率 （英寸）[4]

力矩 英寸	$\frac{1}{2}$寸 圓	$\frac{1}{2}$寸 方	$\frac{5}{8}$寸 圓	$\frac{5}{8}$寸 方	$\frac{3}{4}$寸 圓	$\frac{3}{4}$寸 方	$\frac{1}{8}$寸 圓	$\frac{1}{8}$寸 方	1寸 圓	1寸 方	$1\frac{1}{8}$寸 圓	$1\frac{1}{4}$寸 方
1										1	1	2
1•5					1	1	1	2	2	2	3	4
2	1	1	1	2	2	2	2	3	3	4	5	6
2•5	1	2	2	2	3	3	4	5	5	6	9	10
3	2	2	3	4	4	5	5	7	7	9	26	14
3•5	2	3	4	5	5	7	7	9	10	12	16	19
4	3	4	5	6	7	9	10	12	13	16	20	25
4•5	4	5	6	8	9	11	12	15	16	20	26	32
5	5	6	8	10	11	14	15	19	20	25	32	39
5•5	6	8	9	12	13	17	18	23	24	30	38	47
6	7	9	11	14	16	20	22	28	28	36	46	56
6•5	8	11	13	17	19	24	25	22	33	42	54	66
7	10	12	15	19	22	28	29	38	39	49	62	77
7•5	11	14	17	22	25	32	34	43	44	56	71	88
8	13	16	20	25	28	36	39	49	50	64	81	100
8•5	14	18	22	28	32	41	43	55	57	72	92	113
9	16	20	25	32	26	49	49	62	64	81	103	127
9•5	18	23	28	35	40	51	54	69	71	90	114	141
10	20	25	31	39	44	56	60	77	79	100	127	156
10•5	22	28	34	43	49	62	66	85	87	110	149	172
11	24	30	37	47	53	68	73	93	95	121	153	189
12•5	26	33	41	52	58	75	80	101	104	132	168	207
12	28	36	44	56	64	81	87	110	113	144	182	225
12•5	31	39	48	61	69	88	94	120	123	156	198	244
13	33	42	52	66	75	95	102	129	183	169	214	264
13•5	36	46	56	71	80	103	109	140	144	182	232	284
14	38	89	60	77	87	110	118	150	157	196	248	306
14•5	42	53	65	82	93	119	126	161	166	210	267	328
15	44	56	69	88	99	127	135	172	177	225	284	352
15•5	48	60	75	94	106	135	144	186	190	240	305	375
16	50	64	79	100	113	144	154	196	201	256	324	400
17	57	72	89	113	128	162	174	221	227	239	366	452
18	64	81	99	127	143	182	195	248	255	324	410	504
19	71	90	111	141	160	203	217	276	284	361	457	564
20	79	100	128	156	177	225	241	306	314	400	506	625
21	87	110	135	172	195	248	265	338	346	441	558	689
22	95	121	148	189	214	272	291	370	380	484	613	756
23	104	132	162	206	234	298	318	405	416	529	670	827
24	113	144	177	225	254	324	346	441	452	576	729	900
25	123	156	192	244	276	354	376	478	491	625	791	977
26	133	169	207	264	299	380	407	518	531	676	856	1056
27	146	182	226	284	321	410	438	558	576	729	926	1137
28	157	196	243	306	345	441	471	600	619	784	996	1223
29	168	210	261	328	370	473	504	644	664	841	1068	1311
30	180	225	279	351	396	506	540	689	711	900	1142	1403
31	192	240	298	376	423	541	477	736	760	961	1220	1500
32	205	256	318	400	451	576	615	785	809	1024	1300	1597
33	218	272	388	425	479	612	654	833	860	1089	1383	1698
34	231	289	359	451	509	650	694	885	914	1156	1468	1804
35	245	206	380	479	539	689	735	939	968	1225	1556	1911

英 8.

第七圖 （英制）

矩樑設計圖

n=15

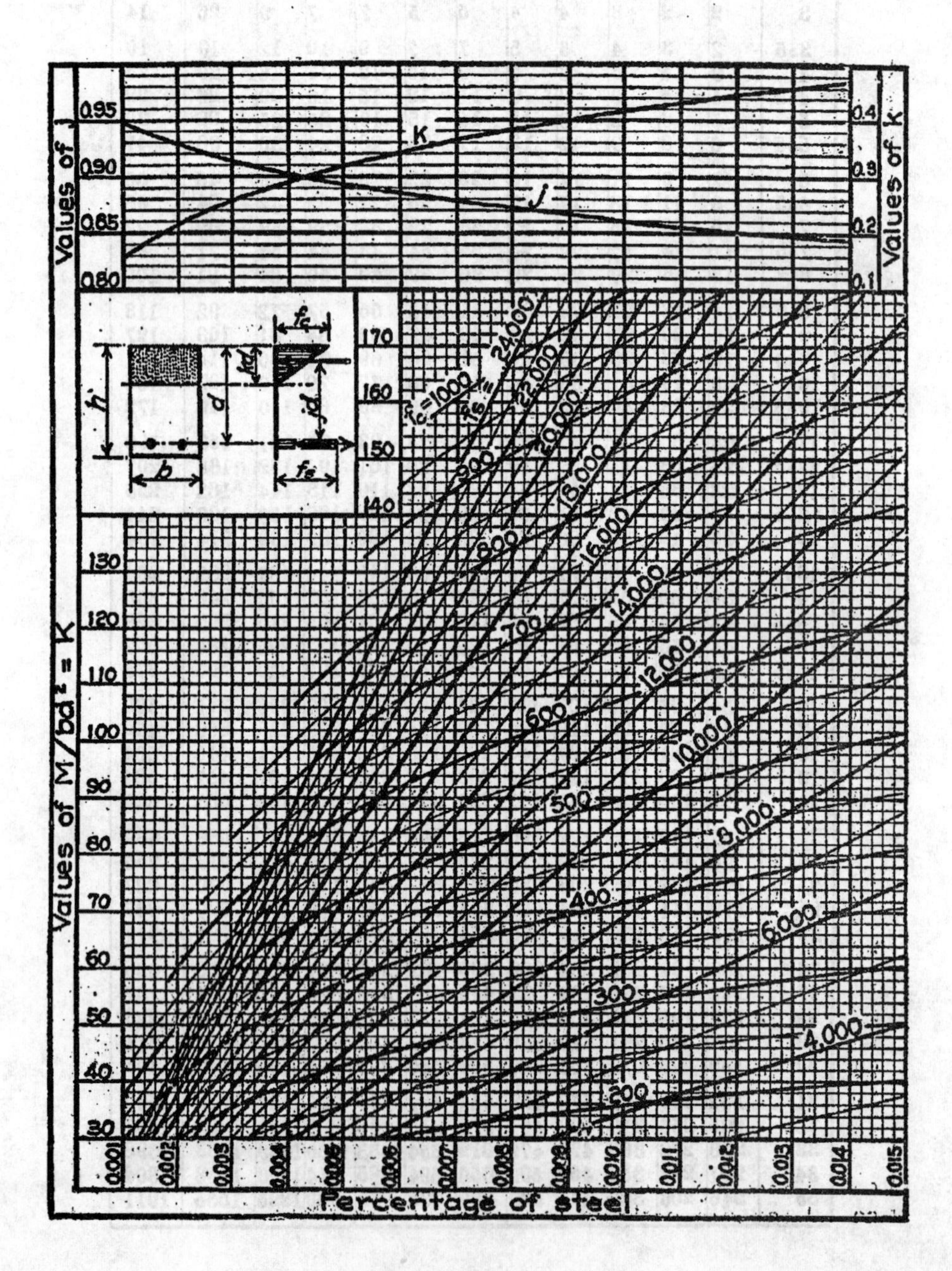

第八表(英制) 矩樑設計表 (樑寬爲一英尺)

fs＝18000 fc＝650 n＝15 R＝100.8 P＝0.0063

高度	矩式樑桁		壓力鋼筋			
			d'＝2"		d'＝3"	
d 寸	M	As	Mc	At	Mc	At
3	0.907	0.23	動率	單位	爲1000	呎磅
1/4	1.065	0.25	M＝	平衡	鋼筋動	率
1/2	1.235	0.27				
3/4	1.417	0.29	Mc＝	每方	吋壓力	鋼筋動
4	1.61	0.30		率		
1/4	1.82	0.32				
1/2	2.04	0.34	At＝	每方	吋壓力	鋼筋所
3/4	2.27	0.36		加之	引力鋼	筋面積
5	2.52	0.38		方吋		
1/4	2.78	0.40				
1/2	3.05	0.42				
3/4	3.33	0.44				
6	3.63	0.46	.16	.03		
1/4	3.94	0.48	.29	.04		
1/2	4.26	0.49	.42	.06		
3/4	[illegible].59	0.51	.57	.08		
7	4.94	0.53	.71	.09		
1/4	5.30	0.55	.86	.11		
1/2	5.37	0.57	1.01	.12		
3/4	6.05	0.39	1.15	.13		
8	6.45	0.61	1.31	.15		
1/4	6.86	0.63	1.47	.16		
1/2	7.28	0.65	1.63	.17		
3/4	7.72	0.67	1.77	.18		
9	8.16	0.68	1.95	.19	0.230	.026
1/4	8.62	0.70	2.11	.19	0.365	.093
1/2	9.10	0.72	2.29	.20	0.503	.052
3/4	9.58	0.74	2.15	.21	0.611	.0[illegible]3
10	10.08	0.76	2.61	.22	0.776	.074
1/4	10.59	0.78	2.78	.22	0.917	.081
1/2	11.11	0.80	2.96	.23	1.07	.095
3/4	11.65	0.82	3.13	.21	1.22	.105
11	12.20	0.84	3.29	.24	1.36	.113
1/4	12.76	0.86	3.45	.25	1.51	.122
1/2	13.33	0.87	2.30	.25	1.71	.134
3/4	13.92	0.89	2.81	.26	1.82	.139
12	14.52	0.91	3.99	.27	1.97	.148
1/2	15.70	0.95	4.35	.28	2.28	.160
13	17.00	0.99	4.69	.28	2.60	.173
1/2	18.40	1.03	5.05	.29	2.94	.187
14	19.80	1.07	5.40	.30	3.25	.197
1/2	21.20	1.10	5.76	.31	3.60	.208
15	22.70	1.14	6.12	.31	3.92	.218
1/2	24.20	1.18	6.48	.32	4.23	.225
16	25.8	1.22	6.84	.33	4.90	.24
1/2	27.4	1.26	7.20	.33	4.05	.25
17	29.1	1.29	7.57	.34	5.28	.25
1/2	30.9	1.33	7.91	.34	5.63	.26
18	32.7	1.37	8.30	.35	5.98	.266
1/2	34.5	1.41	8.65	.35	6.32	.272
19	33.4	1.45	9.03	.36	6.38	.278
1/2	38.3	1.48	9.35	.36	7.02	.284

高度	矩樑桁		壓力鋼筋			
			d'＝2"		d'＝3"	
d	M	As	Mc	At	Mc	At
20	40.3	1.52	9.76	.36	7.39	.290
1/2	42.4	1.56	10.14	.37	7.75	.295
21	44.5	1.60	10.50	.37	8.10	.300
1/2	46.6	1.64	10.88	.37	8.47	.305
22	48.8	1.67	11.24	.38	8.82	.31
1/2	51.0	1.71	11.61	.38	9.[illegible]7	.31
23	53.3	1.75	11.98	.38	9.54	.32
1/2	55.7	1.79	12.38	.38	9.90	.32
24	58.1	1.83	12.73	.39	10.26	.33
25	63.0	1.90	13.47	.39	10.98	.33
26	68.1	1.98	14.21	.39	11.71	.[illegible]4
27	73.5	2.05	45.00	.40	12.44	.[illegible]5
28	78.0	2.13	15.70	.40	13.18	.35
29	84.3	2.21	16.50	.41	13.91	.36
30	90.7	2.28	17.20	.41	14.65	.36
31	96.9	2.36	18.00	.41	15.4	.37
32	103	2.43	18.70	.42	16.1	.37
33	110	2.51	19.50	.42	16.9	.37
34	117	2.59	20.20	.42	17.6	.38
35	123	2.56	20.[illegible]0	.42	18.3	.38
36	131	2.74	21.70	.43	19.1	.39
37	138	2.81	22.50	.43	19.8	.39
38	146	2.89	23.20	.43	20.6	.39
39	153	2.97	24.00	.43	21.3	.39
40	161	3.04	24.70	.43	22.1	.40
41	169	3.12	25.50	.44	22.8	.40
42	178	3.20	26.20	.44	23.6	.40
43	186	3.27	27.00	.44	24.3	.41
44	195	3.35	27.70	.44	25.1	.41
45	204	3.42	28.50	.44	25.8	.41
46	213	3.50	29.20	.44	26.6	.41
47	223	3.58	30.00	.45	27.3	.41
48	232	3.65	30.80	.45	28.1	.42
49	242	3.73	31.15	.45	28.8	.42
50	252	3.80	31.50	.45	29.6	.42
51	262	3.88	32.65	.45	30.3	.42
52	273	3.96	33.80	.45	31.0	.42
53	283	4.03	34.50	.45	31.9	.42
54	294	4.11	35.30	.45	32.5	.43
55	305	4.18	36.10	.45	33.4	.43
56	316	4.26	36.80	.46	34.1	.43
57	32	4.34	37.50	.46	34.3	.43
58	330	4.41	38.40	.46	35.5	.43
59	351	4.49	39.00	.46	36.3	.43
60	363	4.56	39.84	.46	37.0	.43
61	375	4.64	40.60	.46	37.8	.44
62	387	4.72	41.3	.46	38.6	.44
63	400	4.79	42.10	.46	39.1	.44
64	413	4.87	42.90	.46	40.[illegible]	.44
65	426	4.95	43.[illegible]0	.46	4[illegible].[illegible]	.44
66	439	5.02	44.40	.46	41.5	.44
67	452	5.10	45.10	.46	42.4	.44

英10　第九表（英制）矩樑設計表（樑寬爲一英尺）

fs＝1,000 fc＝700 n＝15 R＝11.31 p＝0.00716

高度	矩式樑桁		壓力鋼筋			
			d'＝2"		d'＝3"	
d 寸	M	As	Mc	At	Mc	At
3	1.02	0.26	勵	率單	位爲	一千
1/4	1.20	0.28	呎磅			
1/2	1.39	0.30				
3/4	1.59	0.32	M＝	平衡	鋼筋	勵率
4	1.81	0.34	Mc＝	每方	吋壓	力鋼
1/4	2.04	0.37				
1/2	2.29	0.39		筋勵	率	
3/4	2.55	0.41	At＝	每方	吋壓	力鋼
5	2.83	0.43		筋所	加之	引力
1/4	2.04	0.47				
1/2	3.42	0.47		鋼筋	面積	方吋
3/4	3.74	0.49				
6	4.07	0.52	.31	.05		
1/4	4.42	0.54	.46	.07		
1/2	4.78	0.56	.62	.09		
3/4	5.15	0.58	.76	.11		
7	5.54	0.60	.92	.12		
1/4	5.94	0.62	1.07	.14		
1/2	6.36	0.64	1.2	.15		
3/4	6.79	0.67	1.41	.16		
8	7.24	0.69	1.57	.18		
1/4	7.70	0.71	1.74	.19	0.05	.01
1/2	8.17	0.73	2.0	.20	0.18	.02
3/4	8.66	0.75	2.08	.21	0.33	.04
9	9.16	0.77	2.27	.22	0.47	.05
1/4	9.68	0.79	2.44	.23	0.61	.07
1/2	10.21	0.82	2.7	.24	0.76	.08
3/4	10.75	0.84	2.80	.24	0.91	.09
10	11.31	0.86	2.99	.25	1.06	.09
1/4	11.88	0.88	3.17	.26	1.22	.11
1/2	12.47	0.90	3.4	.26	1.38	.12
3/4	13.07	0.92	3.53	.27	1.53	.13
11	13.69	0.95	3.72	.28	1.70	.14
1/4	14.31	0.97	3.90	.28	1.85	.15
1/2	14.96	0.99	4.1	.29	2.02	.16
3/4	15.6	1.01	4.28	.29	2.18	.17
12	16.3	1.03	4.47	.30	2.36	.17
1/2	17.7	1.07	4.9	.31	2.70	.19
13	19.1	1.12	5.23	.32	3.05	.20
1/2	20.6	1.16	5.6	.32	3.39	.22
14	22.2	1.20	6.00	.33	3.76	.23
1/2	23.8	1.25	6.4	.34	4.12	.24
15	25.4	1.29	6.77	.35	4.48	.24
1/2	27.2	1.33	7.2	.35	4.84	.26
16	29.0	1.37	7.55	.36	5.21	.27
1/2	30.8	1.42	8.0	.36	5.58	.28
17	32.7	1.46	8.34	.37	5.96	.28
1/2	34.6	1.50	8.74	.38	6.34	.29
18	36.6	1.55	9.12	.38	6.71	.30
1/2	38.7	1.56	9.51	.39	7.09	.31
19	40.8	1.63	9.92	.39	7.47	.31
1/2	43.0	1.98	10.3	.39	7.85	.32

高度	矩式樑桁		壓力鋼筋			
			d'＝2"		d'＝3"	
d	M	As	Mc	At	Mc	At
20	45.2	1.72	10.71	.40	8.23	.32
1/2	47.5	1.76	11.10	.40	8.60	.33
21	49.9	1.80	11.50	.40	9.00	.33
1/2	52.3	1.85	11.90	.41	9.37	.34
22	54.7	1.89	12.3	.44	9.77	.34
1/2	57.3	1.93	12.7	.41	10.20	.35
23	59.8	1.98	13.1	.42	10.55	.35
1/2	62.5	2.02	13.5	.42	10.90	.36
24	65.1	2.06	13.9	.42	11.33	.36
25	70.7	2.15	14.7	.43	12.11	.37
26	76.5	2.23	15.5	.43	12.90	.37
27	82.4	2.32	16.3	.44	13.69	.38
28	88.7	2.41	17.1	.44	14.47	.38
29	95.1	2.49	17.9	.44	15.3	.39
30	102.0	2.58	18.7	.45	16.1	.39
31	109.0	2.66	19.5	.45	16.9	.40
32	116.0	2.75	20.3	.45	17.7	.40
33	123.0	2.84	21.1	.45	18.5	.41
34	131.0	2.92	22.0	.46	19.3	.41
35	139.0	3.01	22.8	.46	20.1	.42
36	147.0	3.09	23.6	.46	20.9	.42
37	155.0	3.18	24.4	.46	21.7	.43
38	163.0	3.26	25.2	.47	22.5	.43
39	172.0	3.35	26.0	.47	23.3	.43
40	181.0	3.44	26.8	.47	24.1	.43
41	190.0	3.52	27.6	.47	24.9	.44
42	199.5	3.61	28.5	.47	25.7	.44
43	208.0	3.69	29.3	.48	26.5	.44
44	219.0	3.78	30.1	.48	27.3	.44
45	229.0	3.87	30.9	.48	28.1	.45
46	239.0	3.95	31.7	.48	28.9	.45
47	250.0	4.04	32.5	.48	29.7	.45
48	261.0	4.12	33.3	.48	30.5	.45
49	272.0	4.21	34.1	.48	31.3	.45
50	283.0	4.30	34.9	.49	32.1	.46
51	294.0	4.38	35.7	.49	32.9	.46
52	306.0	4.47	36.6	.49	33.8	.46
53	318.0	4.55	37.4	.49	34.6	.46
54	330.0	4.64	38.2	.49	35.4	.46
55	342.0	4.73	39.0	.49	36.2	.47
56	355.0	4.81	39.8	.49	36.9	.47
57	367.0	4.90	40.6	.49	37.8	.47
58	380.0	4.98	41.5	.49	38.6	.47
59	394.0	5.07	42.2	.49	39.5	.48
60	407.0	5.16	43.1	.50	40.2	.47
61	421.0	5.24	43.3	.50	41.0	.47
62	435.0	5.33	44.7	.50	42.0	.47
63	449.0	5.41	45.5	.50	43.6	.47
64	463.0	5.50	46.3	.50	44.4	.48
65	478.0	5.58	47.1	.50	45.2	.48
66	493.0	5.67	47.9	.50	46.2	.48
67	508.0	5.76	48.8	.50	46.6	.48

第十表(英制) 矩樑設計表 (樑寬爲一英尺)

fs=18000 fc=800 n=15 k=0.4

高度	矩式樑桁		壓力鋼筋				高度	矩式樑桁		壓力鋼筋			
			d'=2"		d'=3"					d'=2"		d'=3"	
d 寸	M	As	Mc	At	Mc	At	d	M	As	Mc	At	Mc	At
3	1.248	0.32	動率單	位爲	1000	呎磅	20	55.5	2.14	12.60	.17	9.92	.38
1/4	1.465	0.35	M ＝	平衡	鋼筋	動率	1/2	58.3	2.19	13.05	.47	10.36	.38
1/2	1.70	0.37	Mc ＝	每方	吋壓	力鋼	21	61.2	2.24	13.51	.47	10.8	.39
3/4	1.95	0.40					1/2	64.1	2.30	13.97	.48	11.25	.40
4	2.22	0.43		筋動	率		22	67.1	2.35	14.42	.48	11.69	.41
1/4	2.51	0.45	At ＝	每方	吋壓	力鋼	1/2	70.2	2.40	14.86	.49	11.13	.41
1/2	2.81	0.48		筋所	加之	引力	23	73.4	2.46	15.3	.49	12.58	.42
4/4	3.13	0.51		鋼筋	面積	方吋	1/2	76.6	2.51	15.8	.49	13.02	.43
5	3.47	0.53					24	79.9	2.56	16.3	.49	13.47	.43
1/4	3.82	0.56					25	86.7	2.67	17.2	.50	14.37	.43
1/2	4.20	0.59					26	93.8	2.78	18.1	.50	15.3	.44
3/4	4.59	0.61					27	101.0	2.88	19.0	.505	16.2	.45
6	4.99	0.64	.62	.10			28	109	2.99	19.9	.51	17.1	.46
1/4	5.42	0.67	.79	.12			29	117	3.10	20.9	.52	18.0	.46
1/2	5.86	0.69	.83	.14			30	125	3.20	21.8	.52	18.9	.47
3/4	6.32	0.72	1.15	.16			31	133	3.31	22.7	.53	19.8	.47
7	6.80	0.75	1.3	.18			32	142	3.42	23.6	.53	20.7	.48
1/4	7.29	0.77	1.52	.195			33	151	3.52	24.5	.53	21.6	.48
1/2	7.80	0.80	1.7	.21			34	160	3.63	25.5	.53	22.5	.49
3/4	8.33	0.83	1.91	.22			35	170	3.74	26.4	.54	23.5	.49
8	8.88	0.85	2.10	.23			36	180	3.84	27.3	.54	24.4	.49
1/4	9.44	0.88	2.3	.245			37	190	3.95	28.3	.54	25.3	.49
1/2	10.02	0.91	2.5	.26			38	200	4.06	29.2	.54	26.2	-49
3/4	10.62	0.93	2.69	.27			39	211	4.17	30.1	.54	27.1	.51
9	11.23	0.96	2.90	.28			40	222	4.27	32.0	.54	28.1	.51
1/4	11.87	0.99	3.12	.29			41	233	4.38	32.0	.55	29.0	.51
1/2	12.52	1.01	3.31	.30			42	245	4.49	32.9	.55	29.9	.51
3/4	13.19	1.04	3.52	.305			43	256	4.59	33.8	.55	30.8	.51
10	13.87	1.07	3.73	.31	1.63	.16	44	269	4.70	34.8	.55	31.7	.52
1/4	14.57	1.09	3.94	.318	1.81	.17	45	281	4.81	35.7	.55	32.7	.52
1/2	15.30	1.12	4.15	.325	2.0	.18	46	293	4.91	36.6	.55	33.6	.52
3/4	16.0	1.15	4.36	.332	2.19	.19	47	306	5.02	37.5	.56	34.5	.52
11	16.8	1.17	4.58	.34	2.38	.20	48	320	5.13	38.5	.56	35.4	.52
1/4	17.6	1.20	4.79	.345	2.57	.21	49	333	5.23	39.4	.56	36.4	.53
1/2	18.3	1.23	5.0	.35	2.76	.22	50	347	5.34	40.3	.56	37.3	.53
3/4	19.1	1.25	5.22	.355	2.95	.23	51	361	5.45	41.2	.56	38.2	.53
12	20.0	1.28	5.44	.36	3.15	.23	52	375	5.55	42.2	.56	39.1	.53
1/2	21.7	1.33	5.9	.37	3.55	.25	53	390	5.66	43.1	.57	40.0	.53
13	23.4	1.39	6.32	.38	3.95	.27	54	404	5.77	44.0	.57	41.0	.54
1/2	25.3	1.44	6.8	.39	4.36	.28	55	420	5.87	45.9	.57	42.0	.54
14	27.2	1.50	7.20	.40	4.77	.29	56	435	5.98	45.9	.57	43.0	.54
1/2	29.2	1.55	7.7	.40	5.18	.30	57	451	6.09	46.8	.57	43.9	.54
15	31.2	1.60	8.09	.41	5.60	.31	58	467.5	6.19	47.8	.57	44.8	.54
1/2	33.3	1.66	8.9	.42	6.22	.32	59	483	6.30	48.7	.57	45.7	.54
16	35.5	1.71	8.98	.43	6.45	.33	60	499.2	6.45	49.6	.57	46.6	.54
1/2	37.8	1.76	9.5	.43	6.87	.34	61	516	6.51	50.5	.57	47.5	.55
17	40.1	1.82	9.88	.44	7.30	.35	62	533.0	6.62	51.5	.57	48.4	.55
1/2	42.5	1.87	9.33	.445	7.73	.35	63	550	6.73	52.4	.57	49.3	.55
18	44.9	1.92	10.78	.45	8.17	.36	64	568.0	6.84	53.4	.57	50.2	.55
1/2	47.5	1.98	10.23	.455	8.61	.37	65	586	6.94	54.3	.57	51.2	.55
19	50.1	2.03	11.69	.46	9.04	.37	66	604	7.05	55.2	.57	52.2	.55
1/2	52.7	2.08	12.15	.465	9.48	.38	67	623	7.16	56.2	.57	53.1	.55

第十一表 塊面安全荷重表（塊面固重不在內） 荷重以每平方尺磅計算

fs=1.8000 fc=650 n=15

塊面厚度未超過5吋者，禦火層用3/2吋超過五吋者用1吋，數目在粗線之右者，其單位剪力未過42•7 設有連續樓面每方呎承190磅。跨度爲11呎 $(M=\frac{1}{2}wL^2)$.由表得樓面厚度爲6吋每呎需鋼筋•5吋方。若用7吋所需鋼筋爲 $As=\frac{190+88}{272+88}\times•584=451$ 方吋

				塊面跨度（英呎）																							
				3•5	4	45	5	5•5	6	6•5	7	7•5	8	8•5	9	9•5	10	10•5	11	11•5	12	13	14	15	16	17	18
3	38	•250	1/8	296	217	164	125	97	75	59	45	35	26	19	12												
			1/10	379	281	217	166	131	104	83	66	53	42	33	25	19	13										
			1/12	462	344	264	207	164	132	107	87	71	58	47	38	30	23										
$3\frac{1}{2}$	44	•292	1/8	452	336	256	199	157	125	100	80	64	51	40	31	23	17	11									
			1/10	576	431	331	260	207	167	136	111	91	75	61	50	40	32	25	19								
			1/12	701	526	407	321	258	209	172	142	118	99	82	69	57	47	39	31	25							
4	50	•333	1/8	644	482	371	290	232	186	252	124	102	83	68	55	44	35	27	20	14							
			1/10	818	561	475	375	302	246	202	167	139	116	97	81	68	57	47	38	31	24	13					
			1/12	990	747	610	460	370	304	270	210	177	149	127	108	92	78	66	56	47	39	26	15				
$4\frac{1}{2}$	56	•375	1/8		561	503	397	318	259	212	175	146	121	101	84	70	57	47	38	30	23	17					
			1/10		829	644	510	412	338	280	233	196	165	140	119	101	86	73	61	51	42	28	16				
			1/12		1006	794	624	505	416	347	291	246	210	179	144	132	114	98	84	72	62	44	31	10			
5	63	•417	1/8		835	646	508	403	330	270	222	184	153	127	105	87	71	57	46	35	26	11					
			1/10		1064	814	654	521	430	355	297	249	209	177	149	127	107	90	76	63	51	33	18				
			1/12		1290	990	800	642	532	442	371	313	267	227	195	167	143	123	106	90	77	54	37	22	10		
$5\frac{1}{2}$	69	•458	1/8		945	721	578	466	379	312	258	215	180	151	127	106	88	73	60	49	39	22					
			1/10		1200	920	741	595	402	408	342	288	244	208	177	151	129	110	94	79	67	46	29	16			
			1/12		1455	1120	903	727	605	505	424	360	307	263	227	196	166	147	127	110	95	70	50	34	21	10	
6	75	•500	1/8		1185	920	731	590	485	402	336	284	240	204	174	148	127	103	92	77	65	44	28	15			
			1/10		1499	1176	934	753	625	521	440	373	319	274	236	204	177	154	134	116	100	74	54	37	24	124	
			1/12		1925	1460	1205	975	815	685	550	495	425	369	321	280	223	215	190	167	147	1145	89	68	50	30	24
$6\frac{1}{2}$	81	•541	1/8				895	725	597	497	418	353	301	257	221	189	163	130	121	103	88	63	43	27	15		
			1/10				1139	929	766	641	541	461	395	341	295	257	224	195	171	149	131	99	75	55	38	25	13
			1/12				1369	1120	930	785	565	570	491	425	370	324	285	253	223	197	175	137	106	81	62	46	32
7	88	•584	1/8					870	718	599	505	248	366	314	270	234	203	170	152	132	114	84	60	41	13		
			1/10					1112	922	772	652	551	478	414	360	314	275	241	212	186	164	127	97	74	54	33	24
			1/12					1352	1122	942	851	687	592	515	449	394	347	307	272	241	215	178	134	106	82	63	47

表13　第12表(英制)　塊面安全荷重表(塊面固重不在內)荷重以每平方英尺磅計算

$f_s=18000$　$f_c=700$　$n=15$

塊面厚度未超過五吋者，其禦火層為¾吋。若超過五吋，則禦火層為1吋。數目在粗曲線之右者，其單位剪力不過40，粗線之左則由四十至六十

設有連續樓面每方呎上承200磅，其跨度為10尺($M=\frac{1}{12}WL^2$)由表得樓面厚度為5½吋，其安全載重為206磅。每呎需$\cdot 39\times\frac{200+69}{206+69}=\cdot 381$方吋。

若用7吋，所需鋼筋為$\cdot 52\times\frac{200+88}{400+88}=\cdot 307$方吋

塊面厚度(英寸)	每方尺重量(英磅)	鋼筋平方英寸面積	動因率數	跨度(英尺) 3½	4	4½	5	5½	6	6½	7	7½	8	8½	9	9½	10	10½	11	11½	12	13	14	15	16	17	18
3	38	•17	1/8	257	188	141	106	82	62	48	36	26	18														
			1/10	331	244	185	143	111	87	69	54	42	33	25	18												
			1/12	406	301	230	179	141	112	90	73	59	47	37	29												
3½	44	•21	1/8	418	310	236	182	143	113	90	71	57	44	34	26	19											
			1/10	533	398	306	239	190	153	124	100	82	66	54	43	35	27	20									
			1/12	650	486	376	295	237	192	157	129	107	88	74	61	50	41	33	26								
4	50	•26	1/8	615	460	352	275	219	176	143	116	95	77	63	50	40	31	24	17								
			1/10	780	586	452	357	286	233	191	158	131	109	91	75	63	52	42	34	27	21						
			1/12	947	713	553	438	253	290	239	199	167	141	119	100	85	72	61	51	42	35						
4½	56	•30	1/8		637	492	387	310	252	206	170	141	117	97	81	67	55	44	36	28	21						
			1/10		810	629	499	410	329	272	227	190	160	136	115	97	82	70	58	49	40	26	15				
			1/12		984	784	609	694	406	338	283	240	204	174	149	128	110	95	81	70	59	43	29				
5	63	•34	1/8		842	652	517	415	340	280	233	195	163	137	116	98	82	68	57	47	37	23					
			1/10		1067	831	611	535	440	365	306	259	220	187	160	137	118	101	87	74	63	44	29	18			
			1/12		1297	1007	805	655	540	451	380	323	277	243	205	178	154	134	117	101	88	66	48	34	22		
5½	69	•39	1/8				665	536	441	365	306	257	217	185	158	134	114	97	82	70	58	39	25				
			1/10				849	688	568	473	398	338	289	248	214	185	160	139	120	104	90	47	48	33	21		
			1/12				1031	840	655	582	492	420	361	312	271	236	206	180	158	139	122	93	71	53	38	26	
6	75	•43	1/8				830	674	555	459	387	328	279	239	205	176	151	130	112	96	82	59	40	26	14		
			1/10				1055	860	711	595	502	428	367	317	275	239	208	181	159	139	121	92	69	51	36	23	15
			1/12				1280	1047	869	729	617	529	455	395	345	301	265	233	206	182	161	125	98	76	58	42	30
6½	81	•47	1/8					823	679	566	476	405	347	297	257	222	193	167	145	126	109	81	58	41	26	14	
			1/10					1049	865	729	617	527	451	392	341	298	261	221	201	177	156	121	93	71	53	37	25
			1/12					1276	1059	891	757	649	560	488	425	374	329	291	258	229	204	461	129	101	79	61	46
7	88	•52	1/8						815	684	577	492	422	303	316	274	238	207	181	158	138	105	78	57	39	25	13
			1/10						1042	877	743	657	548	476	414	363	319	281	248	220	195	153	120	93	71	53	38
			1/12						1272	1078	912	782	677	589	516	454	400	355	317	282	251	201	162	129	103	81	63

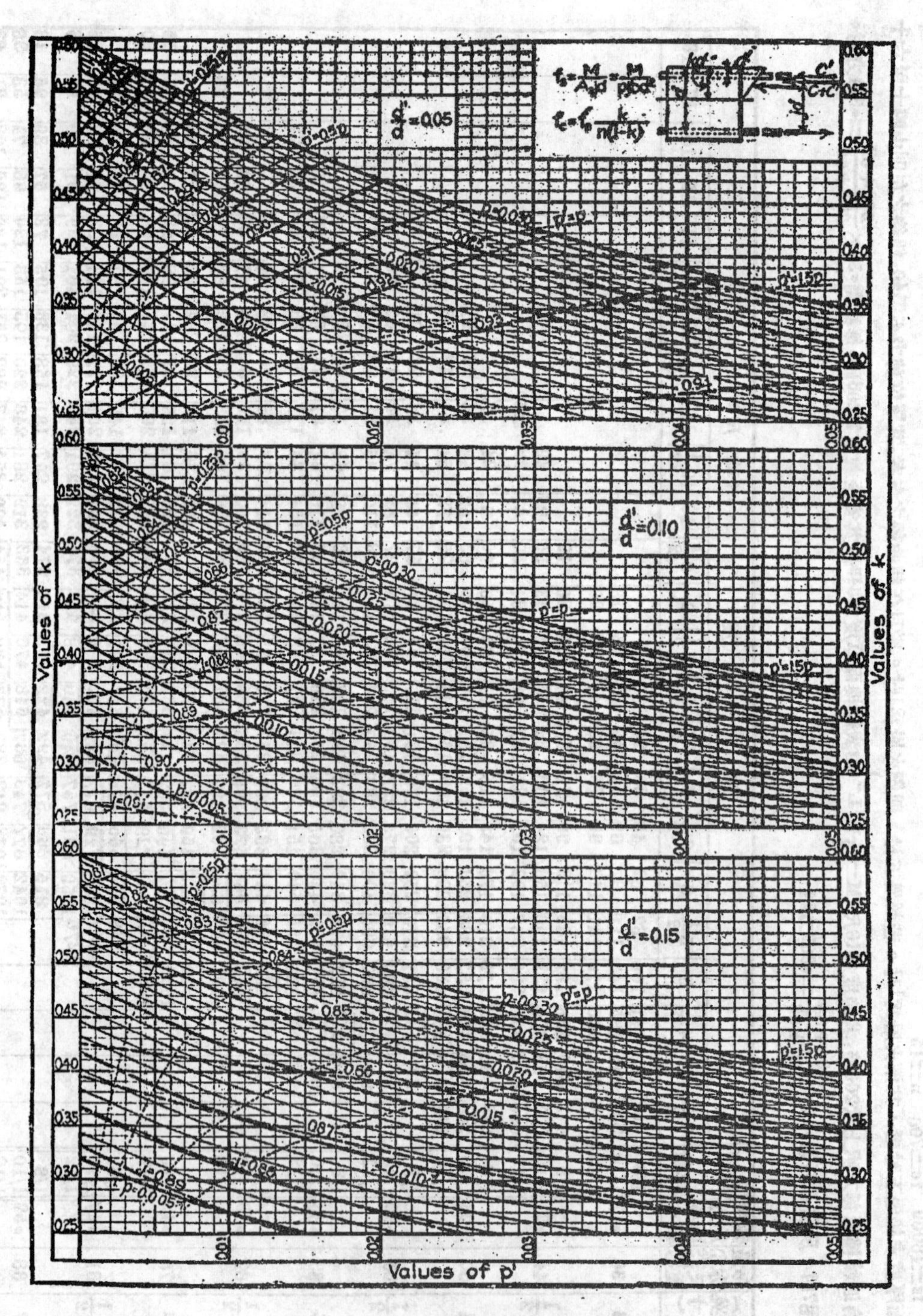

第十三圖　複筋矩梁計力圖

n＝15

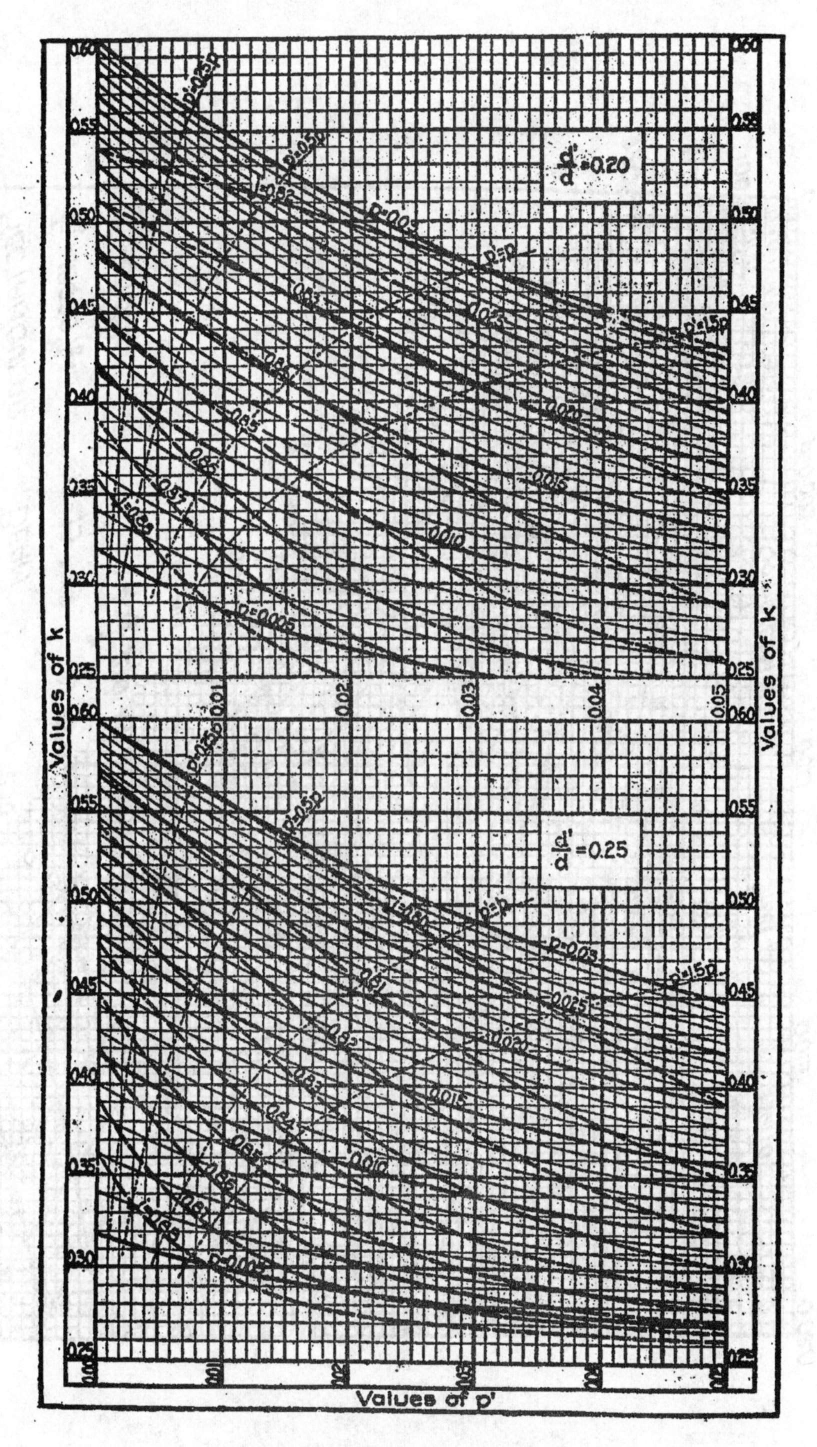

第十四圖　複筋矩梁計力圖

n＝15

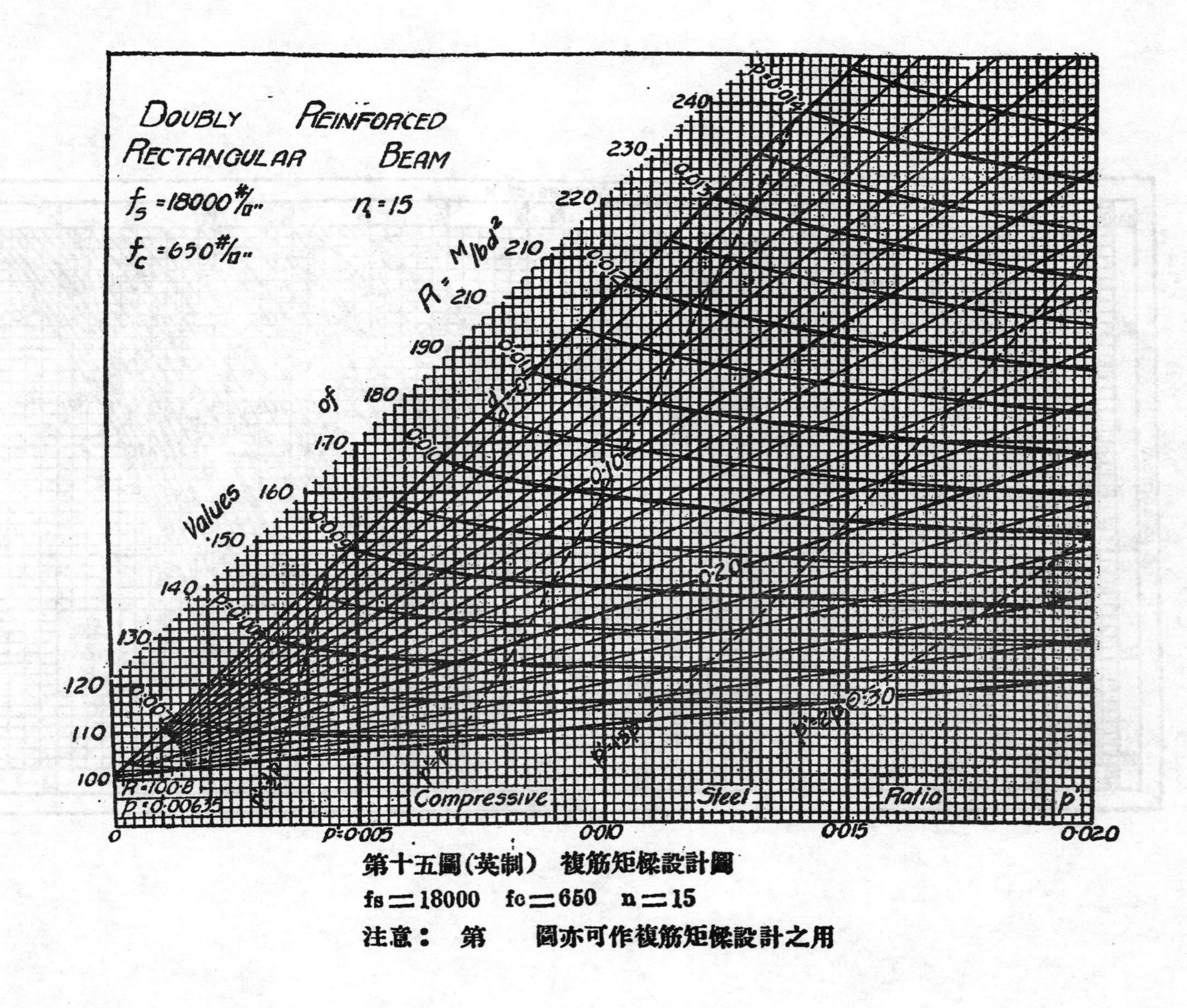

第十五圖(英制) 複筋矩樑設計圖

fs＝18000 fc＝650 n＝15

注意： 第　　圖亦可作複筋矩樑設計之用

Moments of Resistance for Double Reinforced Beams

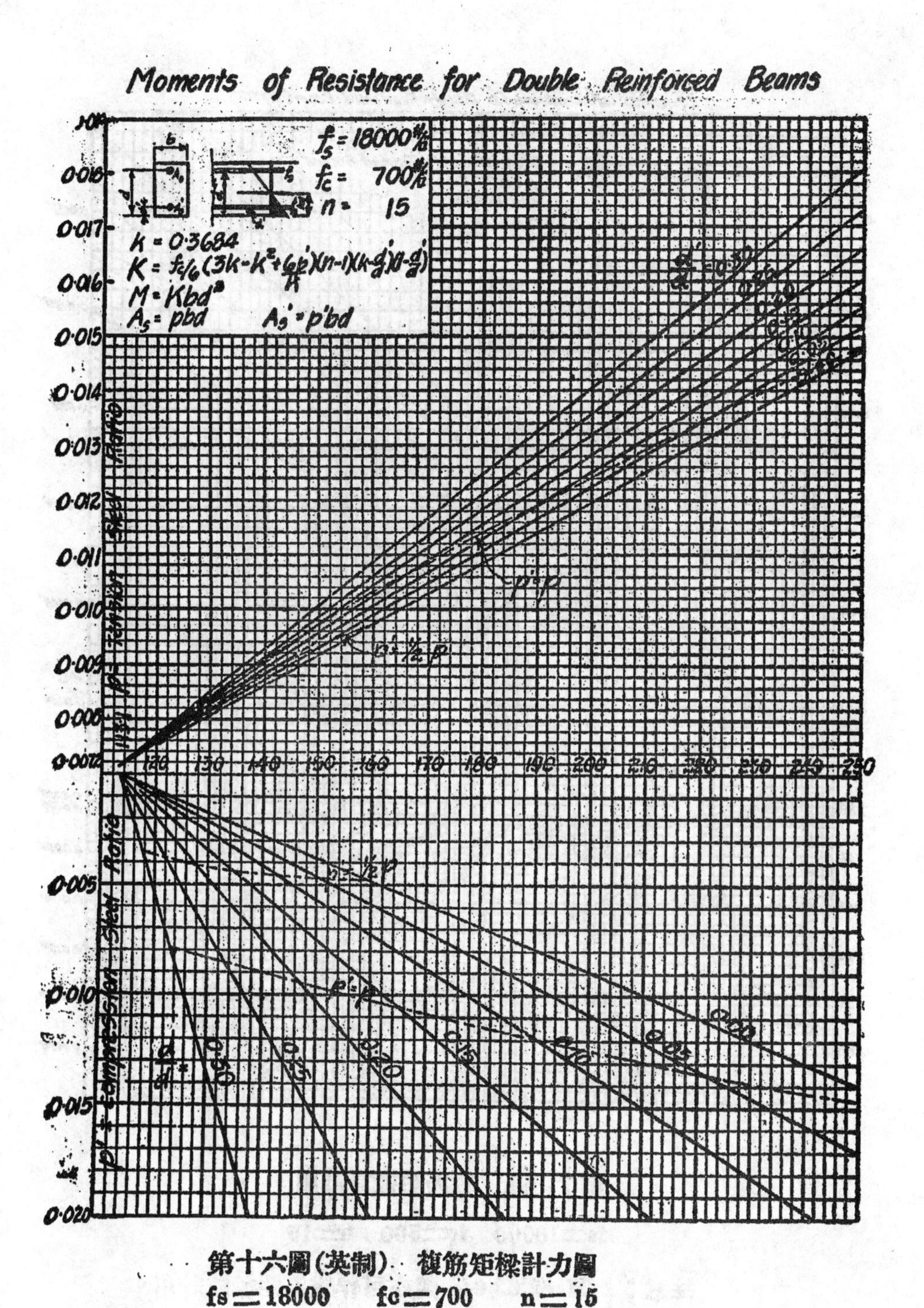

第十六圖(英制) 複筋矩樑計力圖

fs＝18000　fc＝700　n＝15

注意： 第 62,63,64· 圖亦可作複筋矩樑設計之用

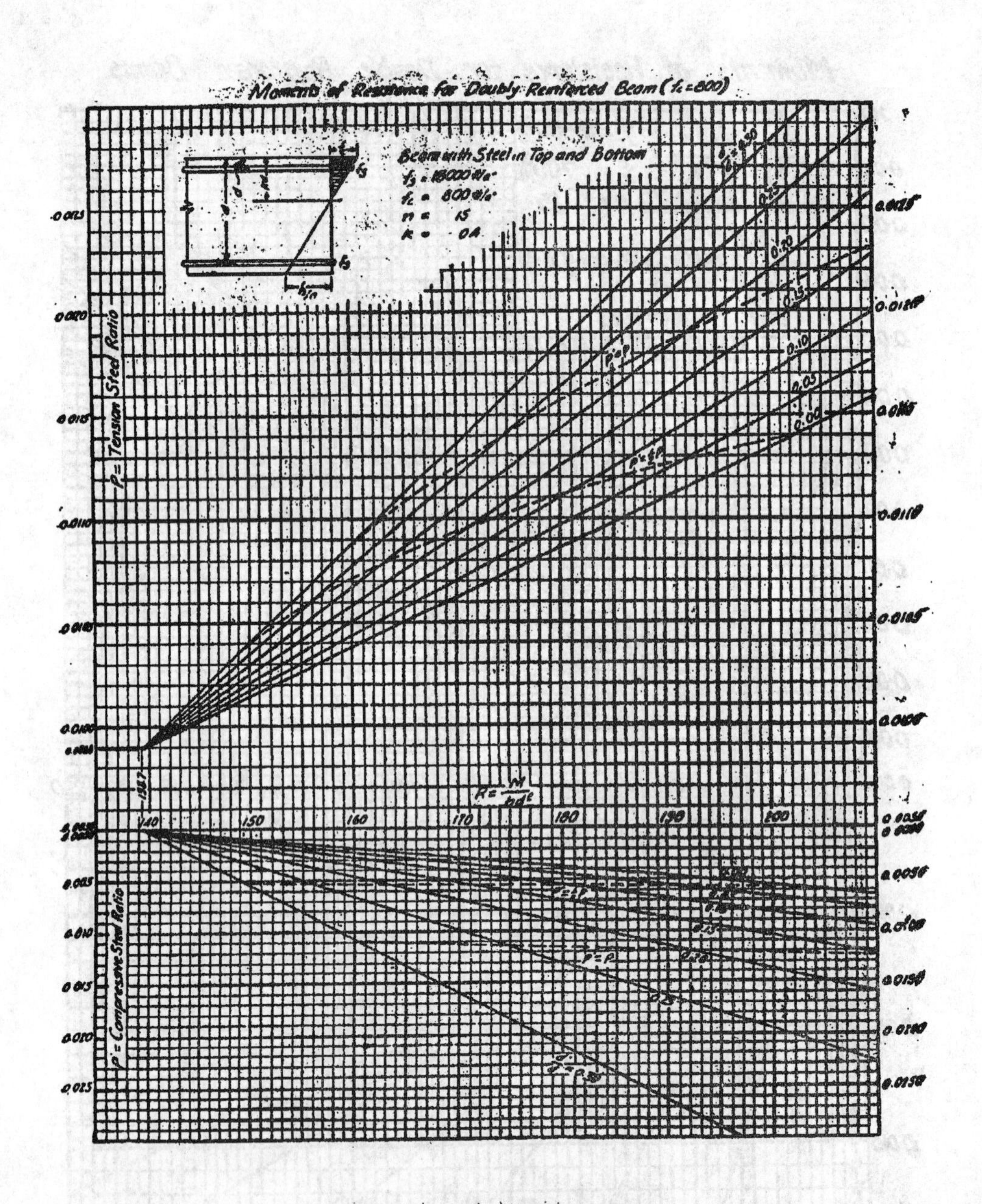

第十七圖　複筋矩樑設計圖

fs＝18000　fc＝800，n＝15

注意：　第 62,63,64. 圖亦可作複筋矩樑設計用。

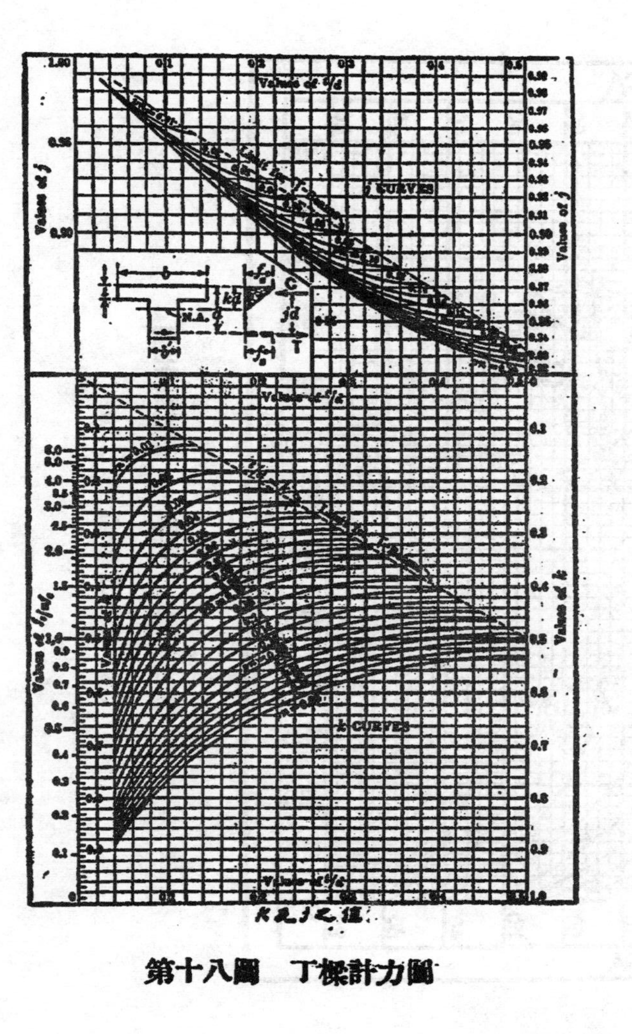

第十八圖　丁樑計力圖

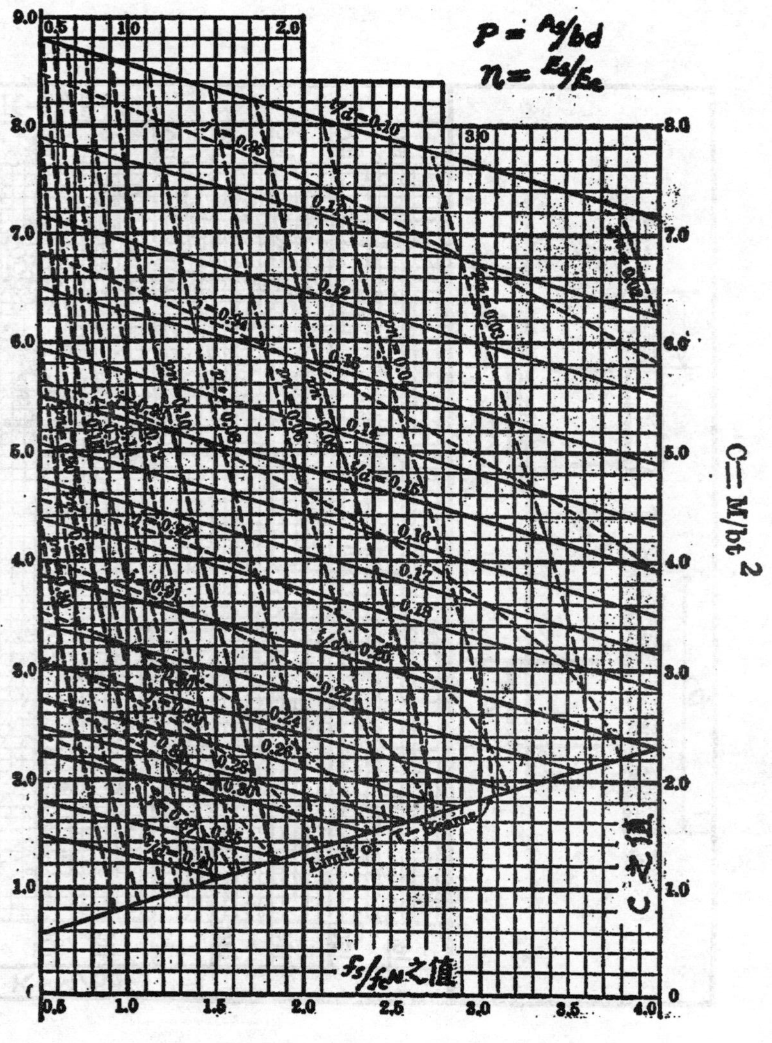

第十九圖　丁樑設計圖

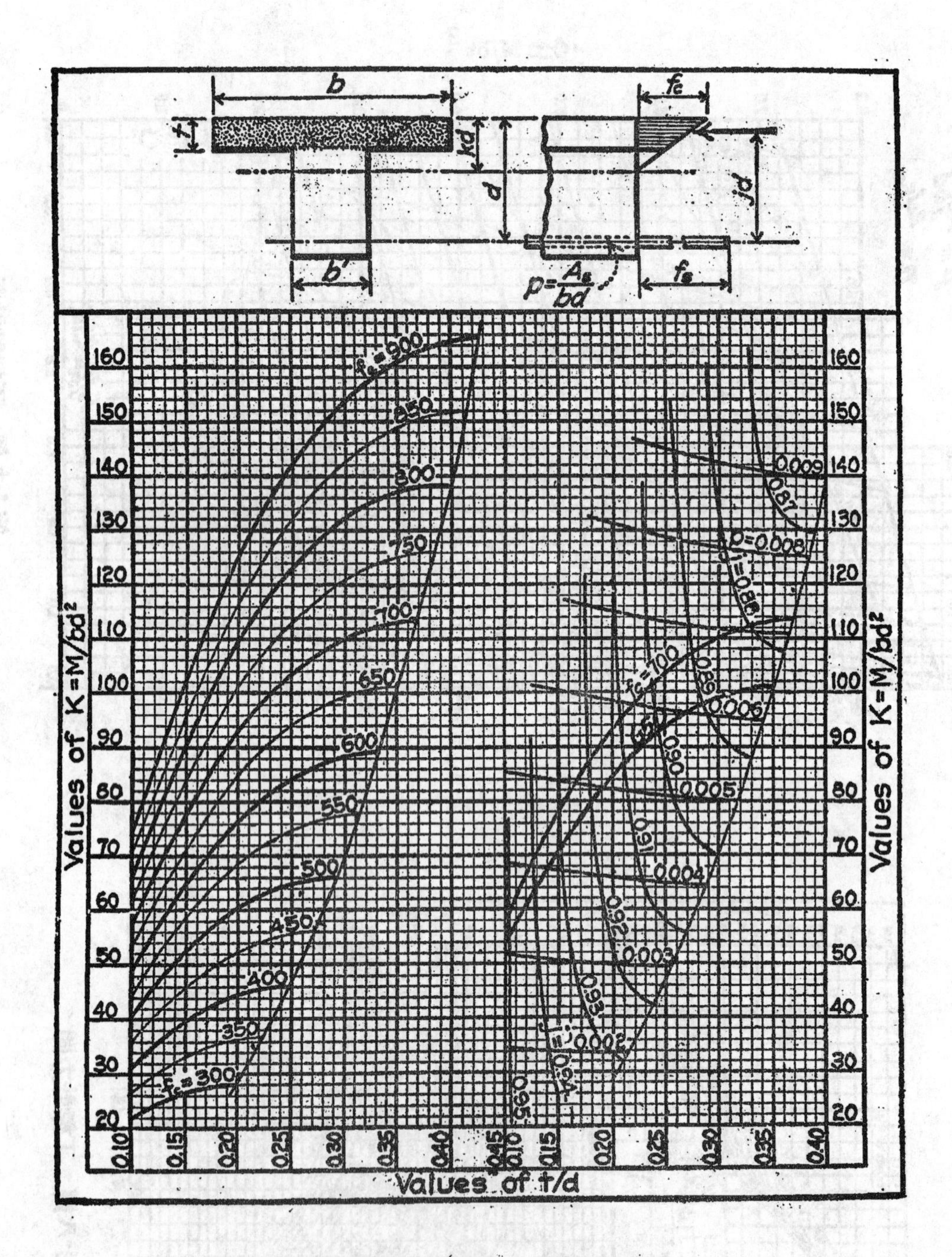

第二十圖（英制）丁樑設計圖

fs=18000 n=15

第二十一表(英制) 丁樑設計表 (樑寬爲一英尺)

勤率以1000呎磅爲單位　　　　每呎闊所需鋼筋面積以方吋計

fs=18,000　　fc=650　　n=15

深度 d	2 吋 樓面		2½ 吋 樓面		3 吋 樓面		3½ 吋 樓面		4 吋 樓面		4½ 吋 樓面		5 吋 樓面	
	勤率	鋼筋	勤率	鋼筋	勤率	鋼筋	勤率	鋼筋	勤率	鋼筋	鋼筋	勤率	勤率	鋼筋
6	3.62	0.46												
7	4.80	0.51												
8	6.02	0.56	6.39	0.60										
9	7.25	0.59	7.88	0.66	8.15	0.68								
10	8.49	0.62	9.40	0.70	9.91	0.74	10.08	0.76						
11	9.75	0.64	10.94	0.73	11.72	0.80	12.12	0.83						
12	11.01	0.66	12.49	0.76	13.54	0.84	14.19	0.89	14.49	0.91				
13	12.28	0.68	14.05	0.79	15.4	0.87	16.3	0.94	16.9	0.97	17.0	0.99		
14	13.55	0.69	15.6	0.81	17.2	0.90	18.4	0.98	19.2	1.03	19.7	1.06		
15	14.83	0.70	17.2	0.83	19.1	0.93	20.6	1.01	21.6	1.08	22.3	1.12	22.6	1.14
16	16.1	0.71	18.8	0.84	21.0	0.95	22.7	1.04	24.1	1.12	25.0	1.17	25.6	1.20
17	17.4	0.72	20.4	0.86	22.9	0.97	24.9	1.07	26.5	1.15	27.7	1.22	28.5	1.26
18	18.7	0.73	22.0	0.87	24.8	0.99	27.1	1.10	29.0	1.19	30.5	1.26	31.5	1.31
19	20.0	0.74	23.6	0.88	26.7	1.01	29.3	1.12	31.5	1.21	33.2	1.29	34.6	1.36
20	21.2	0.74	25.2	0.89	28.6	1.02	31.5	1.14	34.0	1.24	36.0	1.33	37.7	1.40
21	22.5	0.75	26.8	0.90	30.5	1.04	33.7	1.16	36.5	1.26	38.8	1.36	40.7	1.43
22	23.8	0.75	28.4	0.91	32.4	1.05	35.9	1.17	39.0	1.28	41.6	1.38	43.8	1.47
23	25.1	0.76	30.0	0.92	34.3	1.06	38.2	1.19	41.5	1.30	44.4	1.41	46.9	1.50
24	26.4	0.76	31.7	0.92	36.2	1.07	40.4	1.20	44.0	1.32	47.2	1.43	50.0	1.52
25	—	—	—	—	38.2	1.08	42.7	1.21	46.6	1.34	50.1	1.45	53.1	1.55
26	—	—	—	—	40.1	1.09	44.9	1.23	49.1	1.35	52.9	1.47	56.2	1.57
27	—	—	—	—	42.0	1.09	47.1	1.24	51.6	1.37	55.7	1.47	59.4	1.60
28	—	—	—	—	43.9	1.10	49.3	1.25	54.2	1.38	58.6	1.50	62.5	1.62
29	—	—	—	—	45.8	1.11	51.6	1.26	56.8	1.39	61.5	1.52	65.7	1.63
30	—	—	—	—	47.8	1.11	53.8	1.27	59.3	1.40	64.3	1.53	68.8	1.65
31	—	—	—	—	49.7	1.12	56.1	1.27	61.0	1.41	67.2	1.55	72.0	1.67
32	—	—	—	—	51.7	1.12	58.3	1.28	64.4	1.42	70.0	1.56	75.2	1.68
33	—	—	—	—	53.6	1.13	60.6	1.29	67.0	1.43	72.9	1.57	78.3	1.70
34	—	—	—	—	55.5	1.14	62.8	1.29	69.5	1.44	75.8	1.58	81.5	1.71
35	—	—	—	—	57.5	1.14	65.1	1.30	72.1	1.45	78.7	1.59	84.7	1.73
36	—	—	—	—	59.5	1.15	67.3	1.31	74.7	1.16	81.5	1.60	87.9	1.74
37	—	—	—	—	—	—	—	—	77.3	1.47	84.4	1.61	91.1	1.75
38	—	—	—	—	—	—	—	—	79.9	1.47	87.3	1.62	94.3	1.76
39	—	—	—	—	—	—	—	—	82.4	1.48	90.2	1.63	97.5	1.77
40	—	—	—	—	—	—	—	—	85.0	1.49	93.1	1.64	101.0	1.78
41	—	—	—	—	—	—	—	—	87.6	1.49	96.0	1.65	104.0	1.79
42	—	—	—	—	—	—	—	—	90.1	1.50	98.8	1.65	107.0	1.80
43	—	—	—	—	—	—	—	—	—	—	—	—	110.0	1.81
44	—	—	—	—	—	—	—	—	—	—	—	—	113.0	1.82
45	—	—	—	—	—	—	—	—	—	—	—	—	117.0	1.82
46	—	—	—	—	—	—	—	—	—	—	—	—	120.0	1.83
47	—	—	—	—	—	—	—	—	—	—	—	—	123.0	1.84
48	—	—	—	—	—	—	—	—	—	—	—	—	127.0	1.85

第二十二表(英制) 丁樑設計表 (樑寬爲一英尺)

勳率單位 1000 呎磅　　　　每呎闊所需鋼筋面積以方吋計

fs＝18,000　　　fc＝650　　　n＝15

深度 d	5.5吋樓面		6吋樓面		6.5吋樓面		7吋樓面		7.5吋樓面		8吋樓面		9吋樓面	
	勳率	鋼筋	勳率	鋼筋	勳率	鋼筋	勳率	鋼筋	勳率	鋼筋	勳率	鋼筋	勳率	鋼筋
16	25.8	1.22												
17	29.0	1.29												
18	32.2	1.35	32.6	1.37										
19	35.5	1.40	36.1	1.43	36.4	1.45								
20	38.8	1.45	39.7	1.49	40.2	1.51	40.3	1.52						
21	42.1	1.50	43.2	1.54	44.0	1.57	44.1	1.59						
22	45.5	1.54	46.9	1.59	47.8	1.63	48.5	1.66	48.8	1.67				
23	48.9	1.57	50.5	1.63	51.7	1.68	52.6	1.72	53.1	1.74	53.3	1.75		
24	52.3	1.61	54.2	1.67	55.6	1.73	56.8	1.77	57.5	1.84	58.0	1.82		
25	55.7	1.64	57.8	1.71	59.6	1.77	61.0	1.82	62.0	1.86	62.6	1.89		
26	59.1	1.67	61.8	1.75	63.6	1.81	65.2	1.87	66.4	1.92	67.3	1.95	68.1	1.98
27	62.5	1.69	65.2	1.78	67.5	1.85	69.4	1.91	70.9	1.97	72.1	2.00	73.3	2.05
28	65.9	1.72	69.0	1.81	71.5	1.89	73.7	1.95	75.5	2.01	76.9	2.06	78.6	2.12
29	69.4	1.74	72.7	1.82	75.5	1.92	78.0	1.99	80.0	2.05	81.7	2.11	83.9	2.18
30	72.8	1.76	76.4	1.86	79.6	1.95	82.3	2.03	84.6	2.09	86.5	2.15	89.2	2.23
31	76.3	1.78	81.2	1.88	83.6	1.98	86.6	2.06	89.2	2.13	91.4	2.19	94.6	2.29
32	79.8	1.80	84.0	1.91	87.7	2.00	91.0	2.09	93.8	2.17	96.3	2.23	100.0	2.34
33	83.3	1.82	87.7	1.93	91.7	2.03	95.3	2.12	98.4	2.20	101.0	2.27	105.0	2.39
34	86.8	1.84	91.5	1.95	95.8	2.05	99.7	2.14	103.0	2.23	106.0	2.31	111.0	2.43
35	90.2	1.85	95.3	1.97	99.9	2.07	104.0	2.17	108.0	2.26	111.0	2.34	116.0	2.47
36	93.7	1.87	99.1	1.98	104.0	2.09	108.0	2.19	112.0	2.29	116.0	2.37	122.0	2.51
37	97.3	1.88	103.0	2.00	108.0	2.11	113.0	2.22	117.0	2.31	121.0	2.40	127.0	2.55
38	101.0	1.89	107.0	2.02	112.0	2.13	117.0	2.24	122.0	2.34	126.0	2.43	133.0	2.59
39	104.0	1.90	110.0	2.03	116.0	2.15	122.0	2.26	126.0	2.36	131.0	2.46	138.0	2.62
40	108.0	1.92	114.0	2.04	120.0	2.17	126.0	2.28	131.0	2.38	136.0	2.48	144.0	2.65
41	111.0	1.93	118.0	2.06	124.0	2.18	130.0	2.30	136.0	2.40	141.0	2.50	150.0	2.68
42	115.0	1.94	122.0	2.07	129.0	2.20	135.0	2.31	141.0	2.42	146.0	2.53	155.0	2.71
43	118.0	1.95	126.0	2.08	133.0	2.21	139.0	2.33	145.0	2.44	151.0	2.55	161.0	2.74
44	122.0	1.96	130.0	2.10	137.0	2.22	144.0	2.35	150.0	2.46	156.0	2.57	166.0	2.76
45	125.0	1.97	133.0	2.11	141.0	2.24	148.0	2.36	155.0	2.48	161.0	2.59	172.0	2.79
46	129.0	1.98	137.0	2.12	145.0	2.25	153.0	2.38	160.0	2.50	166.0	2.61	178.0	2.81
47	132.0	1.99	141.0	2.13	149.0	2.26	157.0	2.39	164.0	2.51	171.0	2.63	183.0	2.84
48	136.0	1.99	145.0	2.14	154.0	2.27	162.0	2.40	169.0	2.53	176.0	2.64	189.0	2.86
49			149.0	2.15	158.0	2.27	166.0	2.42	174.0	2.54	181.0	2.66	195.0	2.88
50			153.0	2.16	162.0	2.30	171.0	2.43	179.0	2.56	186.0	2.68	200.0	2.90
51			156.0	2.16	166.0	2.31	175.0	2.44	183.0	2.57	191.0	2.70	206.0	2.92
52			160.0	2.17	170.0	2.32	179.0	2.45	188.0	2.58	196.0	2.71	212.0	2.94
53			164.0	2.18	174.0	2.33	184.0	2.46	193.0	2.60	202.0	2.72	217.0	2.96
54			168.0	2.19	179.0	2.34	188.0	2.47	198.0	2.61	207.0	2.74	223.0	2.98
55							193.0	2.48	203.0	2.62	212.6	2.75	229.0	2.99
56							197.0	2.49	207.0	2.63	217.0	2.76	234.0	3.01
57							202.0	2.50	212.0	2.64	222.0	2.77	240.0	3.02
58							206.0	2.51	217.0	2.65	227.0	2.79	246.0	3.04
59							211.0	2.52	222.0	2.66	232.0	2.80	251.0	3.05
60							215.0	2.53	227.0	2.67	237.0	2.81	257.0	3.06

第二十三表(英制) 丁樑設計表 樑寬爲一英尺

勵率單位1000呎磅， 每呎闊所需鋼筋面積以方吋計 fs=18,000 fc=700 n=15

深度 d	2"吋塊面 勵率	2"吋塊面 鋼筋面積	2½"吋塊面 勵率	2½"吋塊面 鋼筋面積	3"吋塊面 勵率	3"吋塊面 鋼筋面積	3½"吋塊面 勵率	3½"吋塊面 鋼筋面積	4"吋塊面 勵率	4"吋塊面 鋼筋面積	4½"吋塊面 勵率	4½"吋塊面 鋼筋面積	5"吋塊面 勵率	5"吋塊面 鋼筋面積
6	4.1	.51												
1/2	4.8	.54												
7	5.3	.57	5.5	.60										
1/2	6.0	.60	6.4	.64										
8	6.6	.62	7.1	.67										
1/2	7.4	.64	7.9	.70	8.1	.74								
9	8.0	.65	8.7	.73	9.1	.77								
1/2	8.7	.67	9.6	.75	10.0	.80	10.2	.82						
10	9.3	.68	10.4	.77	11.0	.83	11.3	.86						
1/2	10.0	.69	11.0	.79	12.0	.85	12.4	.90						
11	10.6	.70	12.0	.81	13.0	.88	13.5	.93	13.7	.95				
1/2	11.4	.71	12.9	.83	14.0	.90	14.6	.96	15.0	.98				
12	12.0	.72	13.7	.84	14.9	.92	15.7	.99	16.2	1.02				
1/2	12.8	.73	14.6	.85	15.9	.94	16.9	1.02	17.5	1.06	17.7	1.08		
13	13.4	.74	15.4	.86	19.9	.96	18.0	1.04	18.7	1.09	19.1	1.11		
1/2	14.1	.74	16.3	.87	17.9	.98	19.2	1.06	20.0	1.11	20.5	1.15		
14	14.8	.75	17.1	.88	18.9	.99	20.3	1.08	21.3	1.14	21.9	1.18	22.1	1.20
1/2	15.5	.76	18.0	.89	19.9	1.00	21.5	1.10	22.6	1.17	23.3	1.21	23.8	1.24
15	16.1	.77	18.8	.90	20.9	1.02	22.6	1.12	23.9	1.19	24.7	1.24	25.3	1.28
1/2	16.8	.77	19.7	.91	21.9	1.03	23.8	1.14	25.2	1.21	26.2	1.27	27.0	1.31
16	17.5	.78	20.5	.92	23.0	1.04	25.0	1.16	26.5	1.23	27.7	1.30	28.5	1.34
1/2	18.3	.78	21.4	.93	24.0	1.05	26.2	1.17	27.9	1.25	29.2	1.32	30.2	1.37
17	18.9	.78	22.2	.94	25.0	1.06	27.3	1.18	29.2	1.27	30.6	1.35	31.6	1.40
18	20.3	.79	23.9	.95	27.0	1.08	29.7	1.20	31.8	1.30	33.6	1.39	34.9	1.45
19	21.7	.80	25.6	.96	29.1	1.10	32.1	1.23	34.5	1.33	36.6	1.42	38.2	1.50
20	23.0	.81	27.4	.97	31.1	1.11	34.4	1.25	37.2	1.36	39.6	1.46	41.5	1.54
21	24.4	.81	29.1	.98	33.2	1.13	36.8	1.26	39.9	1.38	42.6	1.49	44.8	1.58
22	25.8	.82	30.8	.99	35.3	1.14	39.2	1.28	42.6	1.41	45.6	1.52	48.1	1.61
23	27.2	.82	32.6	.99	37.3	1.15	41.6	1.30	45.4	1.43	48.6	1.54	51.4	1.64
24	28.6	.83	34.3	1.00	39.4	1.16	44.0	1.31	48.1	1.44	51.7	1.57	54.8	1.67
25	—	—	—	—	—	—	46.4	1.32	50.8	1.46	54.7	1.59	58.2	1.70
26	—	—	37.7	1.02	43.6	1.18	48.8	1.33	53.6	1.48	57.8	1.61	61.6	1.72
27	—	—	—	—	—	—	51.2	1.35	56.3	1.49	60.9	1.63	64.9	1.75
28	—	—	41.2	1.03	47.7	1.20	53.6	1.36	59.1	1.50	63.9	1.64	68.3	1.77
29	—	—	—	—	—	—	56.1	1.37	61.8	1.52	67.0	1.66	71.7	1.79
30	—	—	44.7	1.04	51.9	1.21	58.5	1.38	64.5	1.53	70.1	1.67	75.1	1.81
31	—	—	—	—	—	—	60.9	1.38	67.3	1.54	73.2	1.69	78.6	1.82
32	—	—	—	—	56.1	1.22	63.3	1.39	70.1	1.55	76.3	1.70	82.0	1.84
33	—	—	—	—	—	—	65.8	1.40	72.8	1.56	79.4	1.71	85.4	1.85
34	—	—	—	—	60.2	1.23	68.2	1.40	75.6	1.57	82.5	1.72	88.8	1.87
35	—	—	—	—	—	—	70.6	1.41	78.4	1.58	85.6	1.73	92.3	1.88
36	—	—	—	—	64.3	1.24	73.0	1.42	81.1	1.59	88.7	1.74	95.7	1.89
37	—	—	—	—	—	—	—	—	83.9	1.59	91.8	1.75	99.1	1.91
38	—	—	—	—	—	—	—	—	86.7	1.60	94.9	1.76	102.5	1.92
39	—	—	—	—	—	—	—	—	89.5	1.61	97.0	1.77	106.0	1.93
40	—	—	—	—	—	—	—	—	92.2	1.61	101.2	1.78	109.6	1.94
41	—	—	—	—	—	—	—	—	95.0	1.62	104.0	1.79	113.0	1.95
42	—	—	—	—	—	—	—	—	97.8	1.63	107.3	1.79	116.3	1.96
43	—	—	—	—	—	—	—	—	—	—	—	—	120.0	1.97
44	—	—	—	—	—	—	—	—	—	—	—	—	123.4	1.97
45	—	—	—	—	—	—	—	—	—	—	—	—	127.0	1.98
46	—	—	—	—	—	—	—	—	—	—	—	—	130.1	1.99
47	—	—	—	—	—	—	—	—	—	—	—	—	134.0	2.00
48	—	—	—	—	—	—	—	—	—	—	—	—	137.2	2.00

Ⅹ24 第二十四表(英制) 丁式樑桁設計表 (樑寬爲一英尺)

勤率單位1000呎磅　　每呎闊所需鋼筋面積以方时計

fs＝18,000　　fc＝700　　n＝15

深度 d	5½"时塊面		6"时塊面		6½"时塊面		7"时塊面		7½"时塊面		8"时塊面		9"时塊面	
	勤率	鋼筋面積	勤率	鋼筋面積	勤率	鋼筋面積	勤率	鋼筋面積	勤率	鋼筋面積	勤率	鋼筋面積	勤率	鋼筋面積
15	25.4	1.29												
1/2	27.2	1.33												
16	28.9	1.37												
1/2	30.6	1.40												
17	32.3	1.44	32.6	1.46										
18	35.8	1.50	36.4	1.53	36.6	1.55								
19	39.4	1.56	40.2	1.60	40.7	1.62	40.8	1.63						
20	42.9	1.61	44.0	1.66	44.8	1.70	45.2	1.72						
21	46.5	1.65	47.9	1.71	48.9	1.76	49.5	1.79	49.8	1.80				
22	50.2	1.69	51.7	1.76	53.1	1.82	54.0	1.86	54.5	1.88	54.7	1.89		
23	53.8	1.73	55.7	1.81	57.3	1.87	58.5	1.82	59.2	1.95	59.7	1.97		
24	57.5	1.77	59.7	1.85	61.5	1.92	63.0	1.97	64.0	2.02	64.8	2.04		
25	61.1	1.80	63.7	1.89	65.8	1.96	67.5	2.02	68.8	2.08	69.8	2.11	70.7	2.15
26	64.8	1.83	67.7	1.92	70.1	2.00	72.1	2.07	73.6	2.13	74.9	2.17	76.3	2.23
27	68.5	1.86	71.7	1.96	74.4	2.04	76.6	2.12	78.5	2.18	80.0	2.23	81.9	2.30
28	72.2	1.88	75.7	1.99	78.7	2.08	81.3	2.16	83.4	2.23	85.2	2.29	87.6	2.37
29	76.0	1.91	79.7	2.01	83.0	2.11	85.9	2.20	88.3	2.27	90.4	2.34	93.3	2.43
30	79.7	1.93	83.8	2.04	87.4	2.14	90.5	2.23	93.3	2.31	95.6	2.38	99.1	2.49
31	83.4	1.95	87.8	2.06	91.7	2.17	95.2	2.27	98.2	2.35	101.0	2.43	105.0	2.55
32	87.2	1.97	91.9	2.09	96.1	2.20	99.9	2.30	103.0	2.39	106.0	2.47	111.0	2.60
33	90.9	1.99	95.9	2.11	100.0	2.22	105.0	2.33	108.0	2.42	111.0	2.51	116.0	2.65
34	94.7	2.00	100.0	2.13	105.0	2.25	109.0	2.35	113.0	2.45	117.0	2.54	122.0	2.69
35	98.4	2.02	104.0	2.15	109.0	2.27	114.0	2.38	118.0	2.48	122.0	2.57	128.0	2.73
36	102.2	2.03	108.3	2.17	114.0	2.29	119.0	2.40	123.0	2.51	127.0	2.61	134.0	2.77
37	106.0	2.05	112.0	2.18	118.0	2.31	123.0	2.43	128.0	2.54	133.0	2.64	140.0	2.81
38	109.7	2.06	116.3	2.20	123.0	2.33	128.0	2.45	133.0	2.56	138.0	2.67	146.0	2.85
39	114.0	2.08	121.0	2.22	127.0	2.35	133.0	2.47	138.0	2.59	143.0	2.69	152.0	2.88
40	117.3	2.09	124.5	2.23	131.0	2.36	138.0	2.49	144.0	2.61	149.0	2.72	158.0	2.92
41	121.0	2.10	129.0	2.24	136.0	2.38	142.0	2.51	149.0	2.63	154.0	2.74	164.0	2.95
42	125.0	2.11	132.9	2.26	140.0	2.40	147.0	2.53	154.0	2.65	160.0	2.77	170.0	2.98
43	129.0	2.12	137.0	2.27	145.0	2.41	152.0	2.54	159.0	2.67	165.0	2.79	176.0	3.01
44	132.6	2.13	141.1	2.28	149.0	2.43	157.0	2.56	164.0	2.69	171.0	2.81	182.0	3.03
45	136.0	2.14	145.0	2.29	154.0	2.44	162.0	2.58	169.0	2.71	176.0	2.83	188.0	3.06
46	140.2	2.15	149.2	2.30	158.0	2.45	166.0	2.59	174.0	2.73	181.0	2.85	195.0	3.09
47	144.0	2.16	154.0	2.32	163.0	2.46	171.0	2.61	179.0	2.74	187.0	2.87	201.0	3.11
48	147.6	2.17	157.8	2.33	167.0	2.48	176.0	2.62	184.0	2.76	192.0	2.89	207.0	3.13
49	——	——	162.0	2.34	172.0	2.49	181.0	2.63	190.0	2.77	198.0	2.91	213.0	3.15
50	——	——	166.1	2.34	176.0	2.50	186.0	2.65	195.0	2.79	203.0	2.92	219.0	3.17
51	——	——	170.0	2.35	181.0	2.51	191.0	2.66	200.0	2.80	209.0	2.94	225.0	3.19
52	——	——	174.0	2.36	185.0	2.52	195.0	2.67	205.0	2.81	214.0	2.95	231.0	3.21
53	——	——	178.0	2.37	189.0	2.53	200.0	2.68	210.0	2.43	220.0	2.97	237.0	3.23
54	——	——	182.6	2.38	194.0	2.54	205.0	2.69	215.0	2.84	225.0	2.98	243.0	3.25
55	——	——	——	——	——	——	210.0	2.70	221.0	2.85	231.0	2.99	250.0	3.27
56	——	——	——	——	——	——	215.0	2.71	226.0	2.86	236.0	3.01	256.0	3.28
57	——	——	——	——	——	——	219.0	2.72	231.0	2.88	242.0	3.02	262.0	3.30
58	——	——	——	——	——	——	224.0	2.73	236.0	2.89	247.0	3.03	268.0	3.32
59	——	——	——	——	——	——	229.0	2.74	241.0	2.90	253.0	3.05	274.0	3.33
60	——	——	——	——	——	——	234.0	2.75	246.0	2.91	253.0	3.06	280.0	3.35

第二十五表（英制）T樑設計表（樑寬爲一英尺）

勁率單位1000呎磅　每呎闊所需鋼筋面積　fs=18,000　fc=800　n=15

深度 d	2吋樓面		2½吋樓面		3吋樓面		3½吋樓面		4吋樓面		4½吋樓面		5吋樓面	
	勁率	鋼筋	勁率	鋼筋	勁率	鋼筋	勁率	鋼筋	勁率	鋼筋	勁率	鋼筋	勁率	鋼筋
5	3.47	0.53												
6	4.59	0.62												
6.5	5.7	0.66												
7	6.36	0.69	6.74	0.74										
7.5	7.2	0.71	7.7	0.78										
8	7.87	0.73	8.55	0.81	8.85	0.85								
8.5	8.7	0.75	9.5	0.84	10.0	0.89								
9	9.39	0.77	10.41	0.87	11.0	0.93	11.23	0.96						
9.5	10.2	0.79	11.4	0.90	12.1	0.97	12.5	1.01						
10	10.95	0.80	12.29	0.92	13.18	1.00	13.71	1.05	13.87	1.07				
10.5	11.7	0.81	13.3	0.94	14.3	1.03	15.0	1.09	15.3	1.12				
11	12.48	0.82	14.2	0.95	15.4	1.05	16.2	1.12	16.7	1.16	16.8	1.17		
11.5	13.3	0.83	15.2	0.97	16.6	1.08	17.6	1.16	18.2	1.20	18.4	1.22		
12	14.05	0.84	16.1	0.99	17.7	1.10	18.8	1.19	19.6	1.24	19.9	1.27		
12.5	14.8	0.85	17.1	1.00	18.9	1.12	20.2	1.22	21.1	1.28	21.5	1.32	21.6	1.33
13	15.6	0.86	18.1	1.01	20.0	1.14	21.4	1.24	22.5	1.31	23.1	1.36	23.4	1.39
13.5	16.4	0.87	19.0	1.03	21.2	1.16	22.8	1.26	24.0	1.34	24.9	1.40	25.2	1.44
14	17.2	0.88	20.0	1.04	22.3	1.17	24.1	1.28	25.4	1.37	26.3	1.44	27.0	1.48
14.5	18.0	0.88	21.0	1.05	23.6	1.20	25.4	1.30	26.9	1.40	28.2	1.47	28.8	1.52
15	18.8	0.89	21.9	1.06	24.9	1.22	26.7	1.32	28.4	1.42	29.7	1.50	30.5	1.56
15.5	19.6	0.89	22.9	1.06	25.9	1.24	28.1	1.34	30.0	1.45	31.4	1.53	32.4	1.59
16	20.3	0.90	23.9	1.07	26.9	1.25	29.4	1.36	31.5	1.47	33.1	1.56	34.2	1.62
16.5	21.1	0.90	24.9	1.08	28.1	1.26	30.8	1.38	33.0	1.49	34.8	1.59	36.1	1.66
17	21.9	0.91	25.9	1.09	29.3	1.27	32.1	1.39	34.5	1.51	36.4	1.61	37.9	1.69
18	23.5	0.92	27.8	1.10	31.6	1.28	34.8	1.41	37.6	1.54	39.3	1.65	41.6	1.74
19	25.1	0.93	29.8	1.11	33.9	1.30	37.5	1.44	40.6	1.57	43.2	1.69	45.4	1.79
20	26.7	0.93	31.8	1.12	36.3	1.31	40.3	1.46	43.3	1.60	46.7	1.73	49.2	1.83
21	28.3	0.94	33.7	1.13	38.7	1.33	43.0	1.48	46.8	1.63	50.1	1.76	53.0	1.87
22	29.8	0.95	35.7	1.14	41.0	1.34	45.7	1.49	49.9	1.65	53.6	1.76	56.3	1.91
23	31.4	0.95	37.7	1.15	43.4	1.35	48.5	1.51	53.1	1.67	57.1	1.81	60.7	1.94
24	33.0	0.96	39.7	1.16	45.7	1.36	51.2	1.53	56.1	1.69	60.6	1.84	64.5	1.97
25	—	—	—	—	48.1	1.36	54.0	1.54	59.3	1.71	64.1	1.86	68.3	2.00
26	—	—	43.6	1.18	50.5	1.37	56.8	1.55	62.4	1.72	67.6	1.88	72.2	2.03
27	—	—	—	—	52.9	1.38	59.5	1.56	65.6	1.74	71.1	1.90	76.1	2.05
28	—	—	47.6	1.19	55.2	1.38	62.3	1.58	68.8	1.75	74.3	1.92	80.0	2.07
29	—	—	—	—	57.6	1.39	65.0	1.58	71.0	1.77	78.2	1.93	83.9	2.09
30	—	—	51.6	1.20	60.0	1.40	67.8	1.59	75.0	1.78	81.7	1.95	87.7	2.11
31	—	—	—	—	62.4	1.41	70.6	1.60	78.2	1.79	85.2	1.96	91.7	2.13
32	—	—	—	—	64.8	1.41	73.4	1.61	81.3	1.80	88.7	1.98	95.6	2.15
33	—	—	—	—	67.1	1.42	76.1	1.62	84.5	1.81	92.3	1.99	99.5	2.16
34	—	—	—	—	69.5	1.42	78.9	1.63	87.6	1.82	95.9	2.00	103.4	2.18
35	—	—	—	—	71.9	1.43	81.7	1.63	90.8	1.83	99.4	2.01	107.0	2.19
36	—	—	—	—	74.3	1.43	84.4	1.64	94.0	1.84	102.0	2.02	111.3	2.20
37	—	—	—	—	—	—	—	—	97.1	1.84	106.0	2.03	115.0	2.22
38	—	—	—	—	—	—	—	—	100.3	1.85	110.2	2.04	119.3	2.23
39	—	—	—	—	—	—	—	—	104.0	1.86	114.0	2.05	123.0	2.24
40	—	—	—	—	—	—	—	—	106.7	1.87	117.3	2.06	127.2	2.25
41	—	—	—	—	—	—	—	—	110.0	1.87	121.0	2.07	131.0	2.26
42	—	—	—	—	—	—	—	—	113.1	1.88	124.2	2.08	134.9	2.27
43	—	—	—	—	—	—	—	—	—	—	—	—	139.0	2.28
44	—	—	—	—	—	—	—	—	—	—	—	—	142.8	2.29
45	—	—	—	—	—	—	—	—	—	—	—	—	147.0	2.30
46	—	—	—	—	—	—	—	—	—	—	—	—	150.3	2.30
47	—	—	—	—	—	—	—	—	—	—	—	—	155.0	2.31
48	—	—	—	—	—	—	—	—	—	—	—	—	158.9	2.32

第二十六表(英制) 丁樑設計表(樑寬爲一英尺)

勵率單位1000呎磅　　　每呎闊所需鋼筋面積

fs=18,000　　fc=800　　n=15

深度 d	5½吋樓面 勵率	5½吋樓面 鋼筋	6吋樓面 勵率	6吋樓面 鋼筋	6½吋樓面 勵率	6½吋樓面 鋼筋	7吋樓面 勵率	7吋樓面 鋼筋	7½吋樓面 勵率	7½吋樓面 鋼筋	8吋樓面 勵率	8吋樓面 鋼筋	9吋樓面 勵率	9吋樓面 鋼筋
14	27•2	1•49												
15	31•0	1•59	31•2	1•60										
16	35•0	1•67	35•4	1•70										
17	39•0	1•75	39•7	1•79	40•0	1•81								
18	43•0	1•31	44•0	1•37	44•8	1•90	44•9	1•92						
19	47•1	1•87	48•1	1•94	49•3	1•98	49•3	2•01	50•1	2•03				
20	51•2	1•93	52•7	2•00	54•0	2•06	54•3	2•10	55•3	2•12	55•5	2•13		
21	55•3	1•97	57•2	2•06	58•3	2•13	59•0	2•18	60•3	2•21	61•1	2•23		
22	59•5	2•02	61•7	2•11	63•6	2•19	65•0	2•25	66•0	2•30	66•7	2•33		
23	63•7	2•06	66•3	2•16	68•4	2•24	70•1	2•31	71•5	2•37	72•4	2•41	73•3	2•45
24	67•0	2•09	70•8	2•20	73•3	2•29	75•3	2•37	77•0	2•44	78•2	2•49	78•3	2•55
25	72•1	2•13	75•4	2•24	78•2	2•34	80•5	2•47	82•5	2•50	84•0	2•56	86•0	2•64
26	76•3	2•16	79•0	2•28	83•1	2•38	85•3	2•48	88•1	2•56	89•9	2•63	92•5	2•72
27	80•6	2•19	84•5	2•31	88•0	2•42	91•1	2•52	93•7	2•31	95•3	2•69	99•0	2•80
28	84•3	2•21	89•1	2•34	93•0	2•46	96•4	2•57	99•3	2•36	102	2•74	106	2•37
29	89•1	2•24	93•8	2•37	98•0	2•49	102	2•61	105	2•71	108	2•79	112	2•94
30	93•3	2•26	98•4	2•40	103	2•53	107	2•64	111	2•75	114	2•84	119	3•00
31	97•5	2•28	103	2•43	108	2•56	112	2•68	116	2•79	120	2•39	125	3•06
32	102	2•30	108	2•45	113	2•59	118	2•71	122	2•33	126	2•93	132	3•11
33	106	2•32	112	2•47	118	2•61	123	2•74	128	2•36	132	2•97	139	3•16
34	111	2•34	117	2•79	123	2•64	129	2•77	134	2•90	138	3•01	146	3•21
35	115	2•36	122	2•51	128	2•66	134	2•30	139	2•93	144	3•05	152	3•26
36	119	2•37	126	2•53	133	2•68	139	2•33	145	2•96	150	3•08	159	3•30
37	123	2•39	131	2•55	138	2•72	145	2•35	151	2•99	156	3•11	166	3•34
38	128	2•40	136	2•57	143	2•73	150	2•87	157	3•01	163	3•14	173	3•38
39	132	2•42	141	2•58	148	2•74	156	2•90	162	3•04	169	3•17	180	3•42
40	136	2•43	145	2•60	153	2•76	161	2•92	168	3•06	175	3•20	187	3•45
41	141	2•44	150	2•61	158	2•78	167	2•94	174	3•08	181	3•23	194	3•48
42	145	2•45	155	2•33	164	2•80	172	2•96	180	3•11	187	3•25	201	3•51
43	149	2•46	159	2•64	169	2•81	178	2•97	186	3•13	194	3•27	207	3•54
44	154	2•47	164	2•66	174	2•82	183	2•99	192	3•15	200	3•30	214	3•57
45	158	2•48	169	2•67	179	2•34	188	3•01	198	3•17	203	3•32	221	3•60
46	162	2•49	173	2•38	184	2•85	194	3•02	203	3•18	212	3•34	228	3•63
47	167	2•50	178	2•39	189	2•87	199	3•04	209	3•20	218	3•36	235	3•65
48	171	2•51	183	2•70	194	2•38	205	3•05	215	3•22	225	3•38	242	3•37
49	—	—	188	2•71	199	2•89	211	3•07	221	3•23	231	3•40	249	3•70
50	—	—	192	2•72	205	2•90	216	3•08	222	3•25	237	3•41	256	3•72
51	—	—	197	2•73	210	2•91	222	3•09	233	3•27	244	3•43	263	3•74
52	—	—	202	2•74	215	2•92	227	3•11	239	3•28	250	3•45	270	3•76
53	—	—	207	2•75	220	2•93	233	3•12	245	3•29	256	3•46	277	3•78
54	—	—	211	2•76	225	2•95	238	3•13	251	3•31	262	3•48	284	3•80
55	—	—	—	—	—	—	244	3•14	257	3•32	269	3•49	291	3•82
56	—	—	—	—	—	—	249	3•15	263	3•33	275	3•51	298	3•84
57	—	—	—	—	—	—	255	3•16	268	3•34	281	3•52	306	3•85
58	—	—	—	—	—	—	260	3•17	274	3•35	287	3•53	313	3•87
59	—	—	—	—	—	—	266	3•18	280	3•36	294	3•54	320	3•88
60	—	—	—	—	—	—	171	3•19	286	3•37	300	3•55	327	3•90

表27 25471

第二十七表　丁樑設計表　（英制）

尺度及重量			塊	樑	寬	圓	鋼	筋		樑之跨度（呎）											
樑之總深	腰寬	腰重 b'(h—t)	面厚度	fc＝650	fc＝700	直		曲		10	11	12	13	14	15	16	17	18	19	20	21
h 吋	b' 吋	磅/呎	t 吋	b 吋	b 吋	數目	尺度	數目	尺度	均佈荷重（磅/呎）同重在內											
12	6	50	4	19	17	1	1	1	3/4	1298	1073	901	768	662	577	507					
	8	67	4	28	25	1	1 1/8	1	1	1872	1547	1300	1108	955	831	730					
	10	83	4	34	30	2	1	1	7/8	2281	1885	1584	1350	1164	1014	891					
14		83	4	26	23																
	8	79	4½	26	23	1	1 1/8	1	1 1/8	2508	2073	1741	1484	1279	1115	979	868	774	695	627	
		75	5	26	23																
		104	4	34	30																
	10	99	4½	34	30	2	1	1	1 1/8	3233	2672	2245	1913	1649	1437	1263	1118	998	895	808	
		94	5	34	30																
		125	4	41	37																
	12	119	4½	41	37	2	1	2	1	3962	3275	2753	2347	2023	1762	1548	1371	1223	1099	991	
		112	5	41	37																
16		100	4	29	26																
	8	92	5	28	25	1	1 1/4	1	1 1/4	3573	2953	2481	2114	1822	1588	1395	1236	1102	990	894	811
		83	6	28	25																
		125	4	34	31																
	10	115	5	33	30	2	1 1/8	1	1 1/8	4374	3615	3038	2588	2232	1944	1709	1513	1350	1212	1093	992
		104	6	33	30																
		150	4	43	39																
	12	138	5	41	37	2	1 1/4	1	1 1/4	5374	4444	3734	3182	2743	2390	2100	1861	1659	1489	1344	1219
		125	6	41	37																

$M=\frac{WL^2}{8}$，　$f_s=18000$，　$f_c=650$ 或 700　$n=15$　鋼筋之禦火層不少于1.5吋　數目在粗線之左者其剪力大于120。

第二十八表　丁樑設計表（英制）

尺度及重量			塊面厚度	樑寬		圓鋼筋				樑之跨度（呎）															
樑之總深	腰寬	腰重 (b'h—t)		fc＝650	fc＝700	直		曲		12	13	14	15	16	17	18	19	20	21	22	23	24	25	26	27
h 吋	b' 吋	磅／呎	t 吋	b 吋	b 吋	數目	尺度	數目	尺度	均佈荷重（磅呎）固重在內															
18	8	117	4	28	25	2	1	1	1 1/8	2849	2429	2093	1823	1602	1419	1266	1136	1025	930	847	775	712			
		108	5	27	24																				
		100	6	27	24																				
		92	7	27	24																				
	10	146	4	35	32	2	1 1/8	1	1 1/4	3555	3029	2613	2275	2000	1771	1580	1419	1280	1161	1058	968	880			
		135	5	34	30																				
		125	6	34	30																				
		115	7	34	30																				
	12	175	4	43	39	3	1	2	1	4363	3717	320[illegible]	2792	2454	2174	1939	1740	1570	1424	1298	1187	1090			
		163	5	40	36																				
		150	6	40	36																				
		138	7	40	36																				
20	8	133	4	30	27	2	1 1/8	1	1 1/8	375[illegible]	3197	2756	2401	2110	1869	1667	1496	1350	1225	1116	1021	938	864	799	741
		125	5	28	25																				
		117	6	27	24																				
		108	7	27	24																				
	10	167	4	38	34	2	1 1/4	1	1 1/4	4605	3924	3383	2947	2590	2295	2047	1837	1658	1503	1370	1253	1151	1061	981	90[illegible]
		156	5	34	31																				
		146	6	34	30																				
		135	7	34	30																				
	12	200	4	44	40	3	1	2	1 1/8	5410	4610	3975	3462	3043	2696	2404	2158	194[illegible]	1766	1609	1472	1352	12[illegible]	115[illegible]	10[illegible]
		188	5	41	37																				
		175	6	40	36																				
		162	7	39	35																				

$M = \frac{1}{8}WL^2$　$f_s = 18000$，　$f_c = 650$ 或 700，　$n = 15$　鋼筋下之保火層不少于1.5吋　數目在粗線之左者其剪力大于120

第二十九表　丁 樑 設 計 表（英制）

尺度及重量			塊面厚度	樑寬		圓鋼筋				樑之跨度（呎）															
樑之總深	腰寬	腰重 b'(h—t)		fc=650	fc=700	直		曲		19	20	21	22	23	24	25	26	27	28	29	30	31	32	33	34
h 吋	b' 吋	磅╲呎	t 吋	b 吋	b 吋	數目	尺度	數目	尺度	均佈荷重（磅/呎）固重在內															
22	8	150 133 117	4 6 8	31 27 27	28 24 24	2	1 1/8	1	1 1/4	1794	1619	1470	1389	1225	1125	1035	957	887	825	769	719				
	10	188 167 146	4 6 8	38 34 34	35 30 30	2	1 1/4	2	1	2239	2021	1833	1670	1528	1403	1293	1196	1109	1031	961	898				
	12	225 200 175	4 6 8	47 41 40	43 37 36	2	1 1/4	2	1 1/4	2679	2418	2193	1998	1828	1679	1547	1431	1327	1234	1150	1075				
24	10	208 188 167	4 6 8	41 34 34	37 31 30	2	1 1/4	2	1 1/8	2705	2442	2215	2019	1848	1697	1564	1446	1341	1247	1162	1086	1017	954		
	12	250 225 200 175	4 6 8 10	50 42 40 40	45 38 36 36	3	1 1/8	2	1 1/4	3319	2995	2716	2475	2265	2080	1917	1772	1643	1528	1424	1331	1246	1170		
26	10	229 208 188 167	4 6 8 10	42 35 34 34	39 32 30 30	2	1 1/4	2	1 1/4	3253	2936	2662	2426	2219	2038	1879	1737	1611	1498	1396	1305	1222	1147	1078	1016
	12	275 250 225 200	4 6 8 10	51 43 41 40	48 39 37 36	3	1 1/8	3	1 1/8	3974	3587	3253	2964	2712	2490	2295	2122	1968	1830	1706	1594	1493	1401	1317	1241

$M=\frac{1}{8}WL^2$　fs=18000　fc=650 或 700　n=15　包鋼筋之禦火層為 15 吋　數目在粗線之左者其剪力大于 120

第三十表 丁樑設計表 (英制)

尺度及重量			塊面厚度	樑寬		圓鋼筋				樑之跨度 (呎)															
樑之總深	腰寬	b'(h—t) 腰重		fc=650	fc=700	直		曲		25	26	27	28	29	30	31	32	33	34	35	36	37	38	39	40
h 吋	b' 吋	磅/呎	t 吋	b 吋	b 吋	數目	尺度	數目	尺度	均佈荷重 (磅/呎) 固重在內															
28	10	240	4	45	41	3	1 1/8	3	1	2251	2080	1929	1794	1672	1563	1463	1373	1291	1215	1148	1085	1027	973	924	
		229	6	36	33																				
		208	8	34	30																				
		188	10	34	30																				
	12	300	4	54	49	4	1	4	1	2649	2449	2271	2112	1969	1840	1723	1617	1520	1432	1352	1278	1209	1147	1088	
		275	6	42	39																				
		250	8	40	36																				
		225	10	39	35																				
	14	350	4	66	60	4	1 1/8	3	1 1/4	3210	2966	2750	2558	2382	2226	2085	1956	1840	1733	1635	1546	1463	1387	1317	
		321	6	52	48																				
		292	8	48	43																				
		262	10	48	43																				
30	10	271	4	49	45	3	1 1/8	3	1 1/8	2694	2491	2310	2148	2002	1871	1752	1644	1546	1457	1375	1299	1230	1166	1107	1052
		250	6	39	35																				
		229	8	36	32																				
		208	10	35	31																				
		188	12	35	31																				
	12	325	4	57	52	3	1 1/4	1	1 1/4	3120	2884	2675	2487	2318	2166	2029	1904	1790	1687	1592	1504	1424	1350	1282	1218
		300	6	46	41			2	1 1/8																
		275	8	41	37																				
		250	10	40	36																				
		225	12	40	36																				
	14	380	4	66	60	4	1 1/4	4	1	3665	3387	3141	2921	272[illegible]	2544	2383	2237	2103	1980	1870	1767	1672	1586	1505	1432
		350	6	54	48																				
		321	8	48	43																				
		292	10	47	42																				
		263	12	47	42																				

$M=\frac{1}{8}WL^2$　fs=18000　fc=650 或 700　n=15　包鋼筋之禦火層爲 15 吋　數目在粗線之左者其剪力大于 120。

第三十一表　丁樑設計表（英制）

樑之總深 h 吋	腰寬 b' 吋	樑重 b'h 磅／呎	圓鋼筋 直 數目—尺度	圓鋼筋 曲 數目—尺度	10	11	12	13	14	15	16	17	18	19	20	21	22	23	24	25
					樑之跨度（呎）：均佈荷重（磅／呎）固重在內															
12	6	75	1—5/8	1—5/8	787	651	547	466	402	350	307									
	8	100	1—3/4	1—3/4	1111	917	772	657	567	494	434									
	10	125	2—7/8	1—7/8	1442	1192	1002	853	736	641	563									
	12	150	5—5/8	2—5/8	1575	1301	1092	932	803	700	615									
14	6	88	1—3/4	1—3/4	1293	1066	895	763	657	573	504	447	398	357						
	8	117	1—7/8	1—7/8	1738	1437	1207	1029	887	772	679	602	537	482						
	10	146	1—3/4+1—5/8	1—3/4+1—5/8	2154	1780	1496	1275	1099	957	842	745	665	597						
	12	175	2—3/4	2—3/4	2579	2132	1792	1527	1316	1147	1008	892	796	714						
16	6	100	1—3/4	1—3/4	1605	1327	1115	950	819	713	627	555	495	445	402	364				
	8	133	1—7/8	1—7/8	2187	1807	1519	1294	1116	972	854	757	675	606	547	496				
	10	167	1—1	1—1	2852	2356	1981	1687	1456	1267	1115	987	881	790	712	647				
	12	200	1—7/8+1—3/4	1—7/8+1—3/4	3613	2986	2509	2138	1843	1606	1412	1250	1115	1001	903	819				
18	8	150	1—1	1—1	3173	2622	2204	1877	1619	1411	1240	1098	979	879	793	719	656	600	551	
	10	188	1—1 1/8	1—1 1/8	3984	3293	2767	2357	2032	1771	1557	1378	1230	1103	996	903	822	753	692	
	12	225	2—7/8	2—7/8	4802	3968	3334	2843	2450	2135	1876	1662	1482	1330	1200	1088	992	907	833	
	14	262	1—1+1—7/8	1—1+1—7/8	5566	4600	3865	3293	2840	2473	2174	1926	1717	1542	1392	1262	1150	1052	966	
20	8	167	2—3/4	2—3/4			2813	2397	2067	1801	1532	1402	1250	1122	1012	918	837	766	703	648
	10	208	1—7/8+1—3/4	1—7/8+1—3/4			3408	2905	2505	2182	1918	1698	1516	1359	1227	1113	1015	927	852	786
	12	250	2—7/8	2—7/8			3928	3347	2887	2514	2210	1958	1746	1567	1413	1284	1169	1068	982	905
	14	292	2—1	2—1			4987	4249	2664	3192	2805	2485	2217	1989	1795	1628	1483	1357	1247	1149

$d = h - 2.5''$　　$M = \frac{1}{10} WL^2$　　$f_s = 18000$　　$f_c = 800$（在承托點）　　$n = 15$　　包鋼筋之禦火層爲1.5吋

數目在粗線之左者其剪力大于120　　如用作 $f_c = 700$，表中之數應乘于0.805　　如用作 $f_c = 650$，表中之數應乘于0.713

注意：以下四表雖爲丁樑，實屬複筋矩樑，其荷重爲該處三合土應力所限制，此數表之計算亦以此爲依據

第三十二表　丁樑設計表（英制）

尺度及重量			圓鋼筋		樑之跨度															
樑之總深	腰寬	樑重 b'h	直	曲	16	17	18	19	20	21	22	23	24	25	26	27	28	29	30	31
h 吋	b' 吋	磅／呎	數目—尺度	數目—尺度	均佈荷重（磅／呎）固重在內															
22	8	183	1—1 1/8	1—1 1/8	1996	1767	1577	1415	1277	1158	1056	967	887	817	756	701	652	607		
	10	229	2—7/8	2—7/8	2452	2172	1937	1739	1570	1423	1297	1187	1090	1004	929	861	801	747		
	12	275	1—1+1—7/8	1—1+1—7/8	2842	2517	2243	2015	1817	1649	1502	1375	1262	1164	1076	997	927	864		
	14	321	1—1 1/8+1—1	1—1 1/8+1—1	3522	3120	2782	2497	2254	2044	1862	1704	1565	1442	1333	1237	1150	1071		
24	10	250	1—1+1—7/8	1—1+1—7/8	3057	2707	2415	2167	1956	1777	1617	1479	1358	1252	1157	1073	998	930	869	814
	12	300	2—1	2—1	3573	3167	2823	2535	228[illegible]	2075	1889	1729	1588	1464	1352	1254	1167	1088	1017	952
	14	350	2—1 1/8	2—1 1/8	4332	3837	3423	3072	2772	2515	2292	2097	1926	1774	1641	1522	1415	1318	1232	1154
	16	400	1—1 1/4+1—1 1/8	1—1 1/4+1—1 1/8	4900	4338	3871	3475	3135	2844	2592	2373	2177	2006	1854	1719	1600	1491	1392	1305
26	10	271	2—1	2—1	3742	3315	2957	2652	2395	2172	1979	1812	1663	1533	1417	1314	1222	1139	1065	997
	12	325	1—1 1/8+1—1	4—1 1/8+1—1	4355	3858	3442	3087	2788	2527	2302	2107	1935	1783	1649	1529	1422	1325	1239	1159
	14	379	2—1 1/8	1—1 1/8	4914	4352	3882	3485	3145	2852	2599	2378	2184	2012	1861	1726	1604	1496	1397	1309
	16	433	2—1 1/4	2—1 1/4	5916	5241	4675	4196	3787	3433	3129	2862	2631	2424	2242	2077	1932	1801	1083	1576
28	10	292	2—1	2—1			3548	3005	2712	2461	2242	2050	1883	1736	1605	1487	1383	1290	1206	1129
	12	350	2—1 1/8	2—1 1/8			4132	3708	3347	3036	2766	2531	2324	2142	1980	1836	1707	1592	1487	1393
	14	408	1—1 1/4+1—1 1/8	1—1 1/4+1—1 1/8			4712	4225	3812	3462	3154	2885	2650	2442	2258	2093	1946	1815	1696	1587
	16	467	2—1 1/4	2—1 1/4			5207	4674	4218	3826	3486	3189	2929	2700	2496	2314	2152	2006	1875	1756
30	12	375	2—7/8+2—3/4	2—7/8+2—3/4					3850	3492	3782	2911	2673	2464	2278	2112	1964	1831	1711	1602
	14	437	2—1 1/4	2—1 1/4					4508	4092	3729	3412	3133	2887	2667	2475	2300	2146	2004	1877
	16	500	2—1+2—7/8	2—1+2—7/8					5127	4654	4237	3877	3561	3282	8034	2813	2616	2438	2278	2134
	18	563	4—1	4—1					5787	5250	4783	4375	4018	3[illegible]04	3423	3175	2950	2750	2571	2408

$d = h - 2.5''$　$M = \frac{1}{10}WL^2$　$f_s = 18000$　$f_c = 800$（在承托點）　$n = 15$　包鋼筋之禦火層爲1.5吋　數目在粗線之左者其剪力大于120

如$f_c = 700$，表中之數應乘0.805　如$f_c = 650$，表中之數應乘于0.713

第三十三表(英制) 丁樑設計表

尺度及重量			圓鋼筋		樑之跨度															
樑之總深	腰寬	樑重 b'h	直	曲	10	11	12	13	14	15	16	17	18	19	20	21	22	23	24	25
h 吋	b' 吋	磅/呎	數目尺度	數目尺度	均佈荷重(磅/呎)固重在內															
12	6	75	1—5/8	1—5/8	945	781	656	559	482	420	369									
	8	100	1—3/4	1—3/4	1333	1101	926	789	681	593	521									
	10	125	1—7/8	1—7/8	1731	1430	1202	1024	883	769	676									
	12	150	2—5/8	2—5/8	1890	1561	1311	1118	964	840	738									
14	6	88	1—3/4	1—3/4	1548	1279	1074	916	789	688	605	536	478	429						
	8	117	1—7/8	1—7/8	20[illegible]6	1724	1449	1235	1065	927	815	722	644	578						
	10	146	1—3/4+1—5/8	1—3/4+1—5/8	258[illegible]	2136	1795	1530	1319	1149	1010	894	798	716						
	12	175	2—3/4	2—3/4	3095	2558	2150	1832	1579	1376	1210	1071	955	857						
16	6	100	1—3/4	1—3/4	1926	1592	1338	1140	983	856	752	666	594	534	482	437				
	8	133	1—7/8	1—7/8	2625	2169	1823	1553	1339	1168	1025	908	810	727	656	595				
	10	167	1—1	1—1	3422	2827	2377	2025	1747	1521	1338	1185	1057	948	855	776				
	12	200	1—7/8+1—3/4	1—7/8+1—3/4	4336	3583	3011	2566	2212	1927	1694	1500	1338	1201	1084	983				
18	8	150	1—1	1—1				2253	1943	1693	1488	1318	1175	1055	952	863	787	720	661	
	10	188	1—1 1/8	1—1 1/8				2829	2439	2125	1868	1654	1476	1324	1195	1084	987	904	830	
	12	225	2—7/8	2—7/8				3412	2940	2562	2251	1994	1778	1596	1440	1306	1190	1089	1000	
	14	262	1—1+1—7/8	1—1+1—7/8				3952	3408	2968	2609	2311	2061	1850	1670	1514	1380	1262	1159	
20	8	167	2—3/4	2—3/4					2480	2161	1899	1682	1500	1347	1215	1102	1004	919	844	778
	10	208	1—7/8+1—3/4	1—7/8+1—3/4					3006	2618	2302	2038	1819	1631	1472	1336	1218	1113	1022	943
	12	250	2—7/8	2—7/8					3465	3017	2652	2350	2095	1880	1696	1540	1403	1282	1179	1086
	14	292	2—1	2—1					4397	3830	3366	2982	2660	2387	2154	1954	1780	1629	1496	1379

$d = h - 2.5''$　$M = \frac{1}{12} WL^2$　$f_s = 18000$　$f_c = 800$（在承托點）　$n = 15$　包鋼筋禦火層爲1.5吋

數目在粗線之左者其剪力大于120　如 $f_c = 700$ 表中之數應乘0.805　如 $f_c = 650$ 表中之數應乘于0.713

第三十四表（英制）丁樑設計表

尺度及重量			圓鋼筋		樑之跨度															
樑之總深	腰寬	樑重 b'h	直	曲	16	17	18	19	20	21	22	23	24	25	26	27	28	29	30	31
h 吋	b' 吋	磅／呎	數目—尺度	數目—尺度	均佈荷重（磅／呎）固重在內															
22	8	183	1—1 1/8	1—1 1/8	2395	2121	1802	1698	1533	1390	1267	1160	1064	981	907	841	782	729		
	10	229	2—7/8	2—7/8	2943	2607	2325	2087	1884	1708	1557	1424	1308	1205	1115	1033	961	896		
	12	275	1—1+1—7/8	1—1+1—7/8	3419	3020	2692	2418	2180	1979	1803	1650	1515	1397	1291	1197	1112	1037		
	14	321	1—1 1/8+1—1	1—1 1/8+1—1	4226	3744	3330	2997	2705	2453	2235	2045	1878	1731	1600	1484	1380	1286		
24	10	250	1—1+1—7/8	1—1+1—7/8			2898	2601	2347	2129	1940	1775	1630	1502	1389	1288	1198	1116	1043	977
	12	300	2—1	2—1			3388	3042	2744	2490	2267	2075	1906	1757	1623	1505	1400	1306	1220	1142
	14	350	2—1 1/8	2—1 1/8			4108	3687	3327	3018	2750	2516	2311	2129	1969	1826	1698	1582	1479	1385
	16	400	1—1 1/4+1—1 1/8	1—1 1/4+1—1 1/8			4645	4170	3762	3413	3110	2848	2612	2407	2225	2062	1920	1789	1671	1566
26	10	271	2—1	2—1					2874	2606	2375	2174	1995	1840	1700	1577	1466	1367	1278	1197
	12	325	1—1 1/8+1—1	4—1 1/8+1—1					3346	3033	2762	2528	2322	2140	1979	1835	1706	1590	1487	1391
	14	379	2—1 1/8	2—1 1/8					3774	3423	3119	2854	2621	2415	2233	2071	1925	1795	1677	1571
	16	433	2—1 1/4	2—1 1/4					4545	4120	3755	3435	3157	2909	2690	2492	2318	2161	2020	1891
28	10	292	2—1	2—1							2690	2460	2260	2083	1926	1785	1660	1548	1447	1355
	12	350	2—1 1/8	2—1 1/8							3319	3037	2789	2570	2376	2203	2049	1910	1785	1672
	14	408	1—1 1/4+1—1 1/8	1—1 1/4+1—1 1/8							3785	3462	3180	2930	2710	2512	2335	2178	2035	1905
	16	467	2—1 1/4	2—1 1/4							4183	3827	3515	3240	2995	2777	2582	2407	2250	2107
30	12	375	2—7/8+2—3/4	2—7/8+2—3/4									3208	2957	2734	2535	2357	2197	2053	1923
	14	437	2—1 1/4	2—1 1/4									3760	3465	3200	2970	2760	2575	2405	2252
	16	500	2—1+2—7/8	2—1+2—7/8									4273	3938	3641	3376	3139	2926	2734	2561
	18	563	4—1	4—1									4822	4445	4108	3810	3540	3300	3085	2890

$d = h - 2.5''$　$M = \frac{1}{12}WL^2$　$f_s = 18000$　$f_c = 800$（在承托點）　$n = 15$　包鋼筋之禦火層爲1.5吋　數目在粗線之左者其剪力大于120

如$f_c = 700$，表中之數應乘0.805　如$f_c = 650$，表中之數應乘于0.713

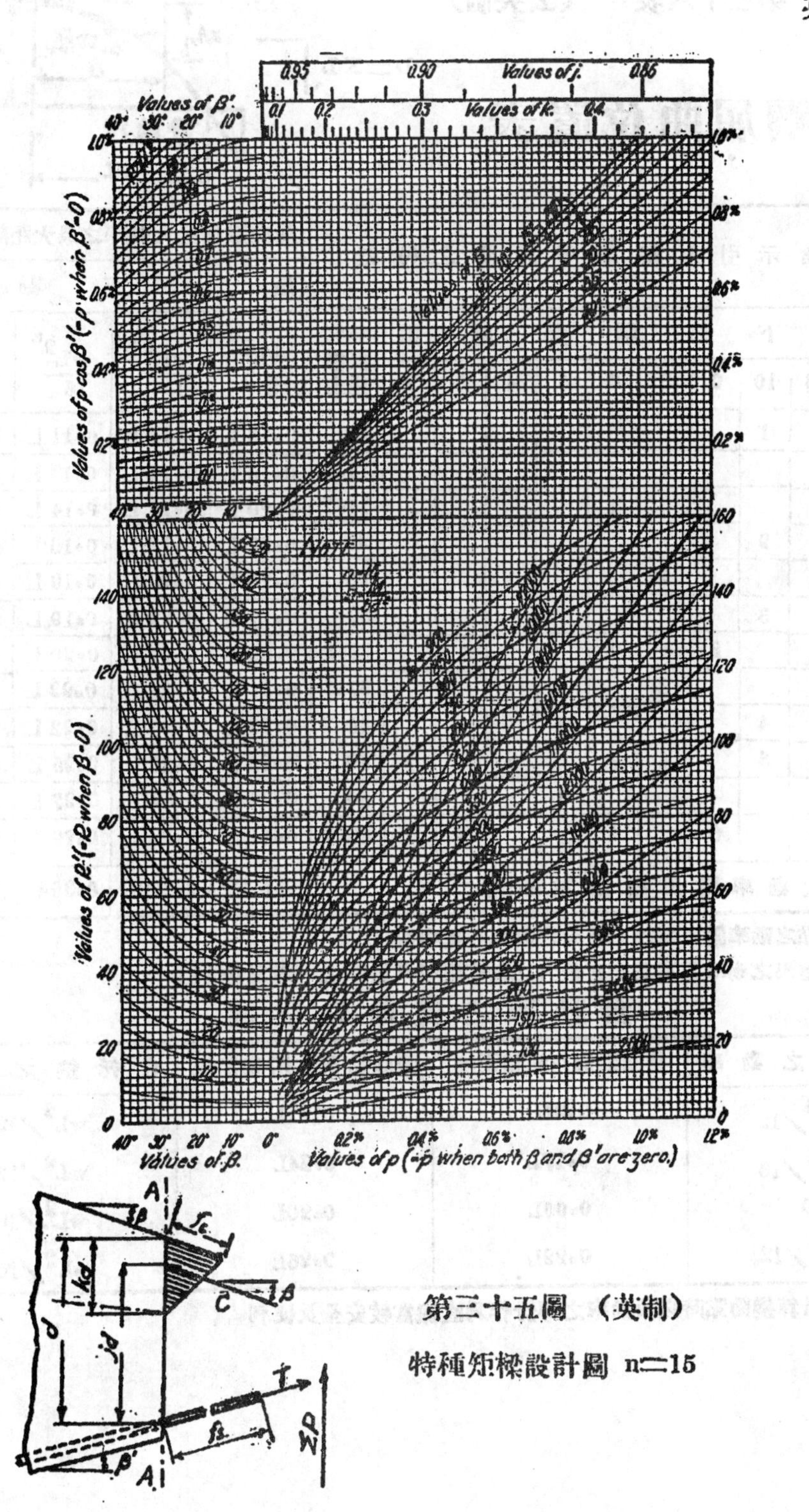

第三十五圖 （英制）

特種矩樑設計圖 n＝15

第三十六表　（公英制）

筋鋼屈曲位置表

$$D = XL\sqrt{x}$$

最大動率點

xA_s

屈曲點

A_s

D

變曲點

XL

此表指示引力鋼筋可屈曲之位置												屈曲點與最大動率之最大距離＝D			
												最大正動率			
樑桁下部鋼筋總數											x % 根數 屈上	$\frac{wL^2}{8}$ a	$\frac{wL^2}{12}$ b	$\frac{wL^2}{16}$ a	$\frac{wL^2}{10}$ b
16	15	12	10	9	8	6	5	4	3	2					
			1								10	0.16 L	0.13 L	0.11 L	0.14 L
2					1						12.5	0.18 L	0.14 L	0.13 L	0.16 L
		2				1					16.7	0.20 L	0.17 L	0.14 L	0.18 L
	3		2				1				20	0.22 L	0.18 L	0.16 L	0.20 L
4		3			2			1			25	0.25 L	0.20 L	0.18 L	0.22 L
			3								30	0.27 L	0.22 L	0.19 L	0.24 L
	5	4		3		2			1		33.3	0.29 L	0.24 L	0.20 L	0.26 L
6					3						37.5	0.31 L	0.25 L	0.22 L	0.27 L
	6		4				2				40	0.32 L	0.26 L	0.22 L	0.28 L
8		6	5		4	3		2		1	50	0.35 L	0.29 L	0.25 L	0.32 L
	9						3				60	0.39 L	0.32 L	0.27 L	0.35 L
	10	8		6		4			2		66.7	0.41 L	0.33 L	0.29 L	0.36 L
最大動率點與變曲點距離＝XL												0.5 L	0.408L	0.354L	0.447L

a 托墊點之動率假定相等。最大之動率在樑中點。

b 一托墊點之動率假定為0。最大之動率距離0動率之一端為0.44 L

負動率之變曲點位置

左托墊之動率	變曲點與左托墊距離	變曲點與右托墊距離	右托墊之動率
$wL^2/12$	0.21L	0.21L	$wL^2/12$
$wL^2/16$	0.17L	0.24L	$wL^2/10$
0	0.00L	0.20L	$wL^2/10$
$wL^2/12$	0.22L	0.26L	$wL^2/10$

附註：計算變曲點時以負動率之部份作為直線當較安全及便利

斜力鋼筋設計

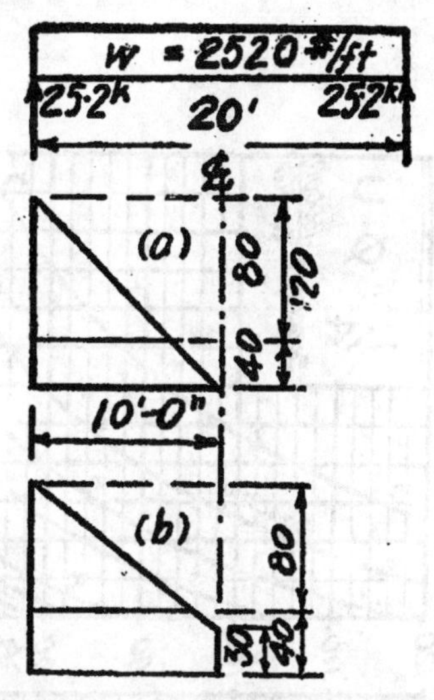

甲 1：均佈載重：樑中剪力爲零

設有樑如圖

fv＝18000

Vc＝40♯／sq. in.

b'＝12"

d ＝20"

鋼絡用3／8"U圓

在第三十八圖之左半用直線聯甲乙得A.線。甲點示樑端之單位剪力 ＝120，乙點示樑中點之單位剪力 ＝0。在甲點之水平線往右至樑寬 b'＝12吋 止。由右圖得鋼絡中距爲4吋。此4吋之中距乃在樑末之2.5呎處（甲乙段）用之。在乙丙段4--2•5＝1.5，可用6 吋之鋼絡距。在丙丁段 6.7—4＝2.7 呎，可用9吋中距。在丁戊段剪力少于40，不需鋼絡。所需鋼絡佈置如后：

2—4—4—4—4—4—4—4—6—6—6—9—9—9—9

甲 2：均佈載重：樑中剪力爲樑端之四份一。

設如甲1之樑，樑中剪力爲樑端之四份一 ＝30。 用3/8"圓U 鋼絡。如上法在第三十八圖之左部得甲己庚辛壬線。由甲己段（5.5呎）得鋼絡中距爲 4吋。在己庚段（5.3—3.5＝1.8呎）用6吋在庚辛段（9—5.3＝3.7呎）用 9吋中距。玆得下列鋼絡之佈置：

2—4—4—4—4—4—4—4—4—4—4—6—6—6—9—9—9—9—9—9

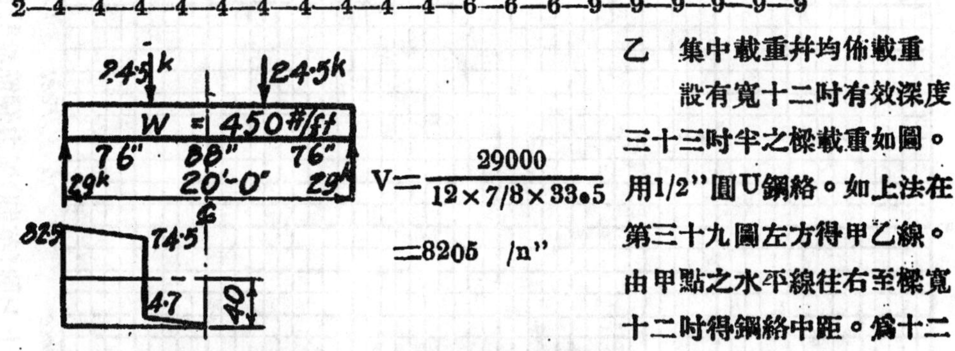

$$V＝\frac{29000}{12\times 7/8\times 33{\bullet}5}$$

＝8205 /n"

乙 集中載重幷均佈載重

設有寬十二吋有效深度三十三吋半之樑載重如圖。用1/2"圓U鋼絡。如上法在第三十九圖左方得甲乙線。由甲點之水平線往右至樑寬十二吋得鋼絡中距。爲十二吋。此十二吋中距之鋼絡，可用至離樑端76吋止。

注意如三合土單位准許剪力爲x磅／方吋。用圖時應將剪力圖下移x—40

設如(乙)樑，其三合土准許剪力爲60；用圖時樑端剪力應改作82•5—(60—40)＝62•5其他截面剪力，亦應減去60—40＝20

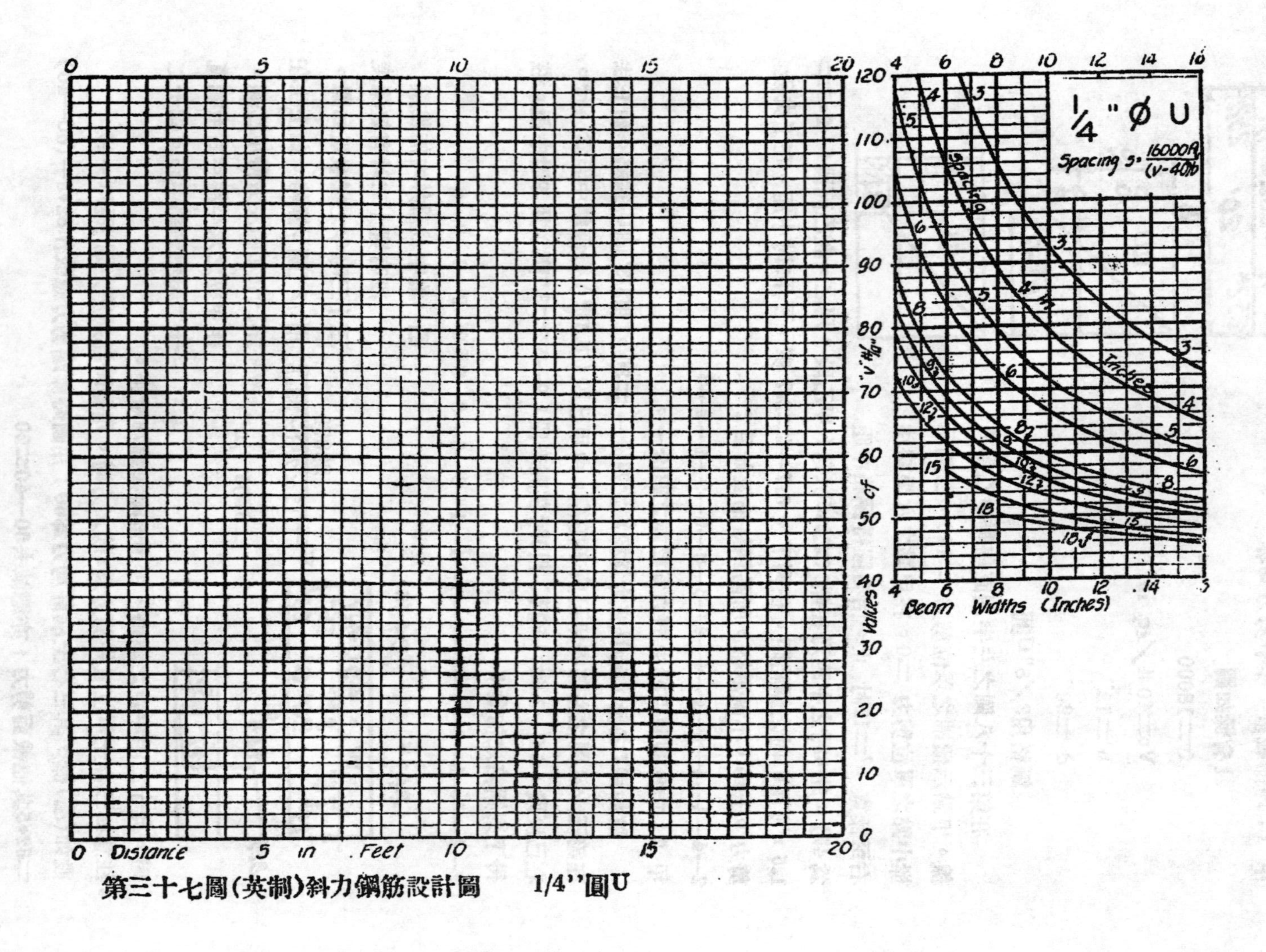

第三十七圖(英制)斜力鋼筋設計圖　1/4"圓U

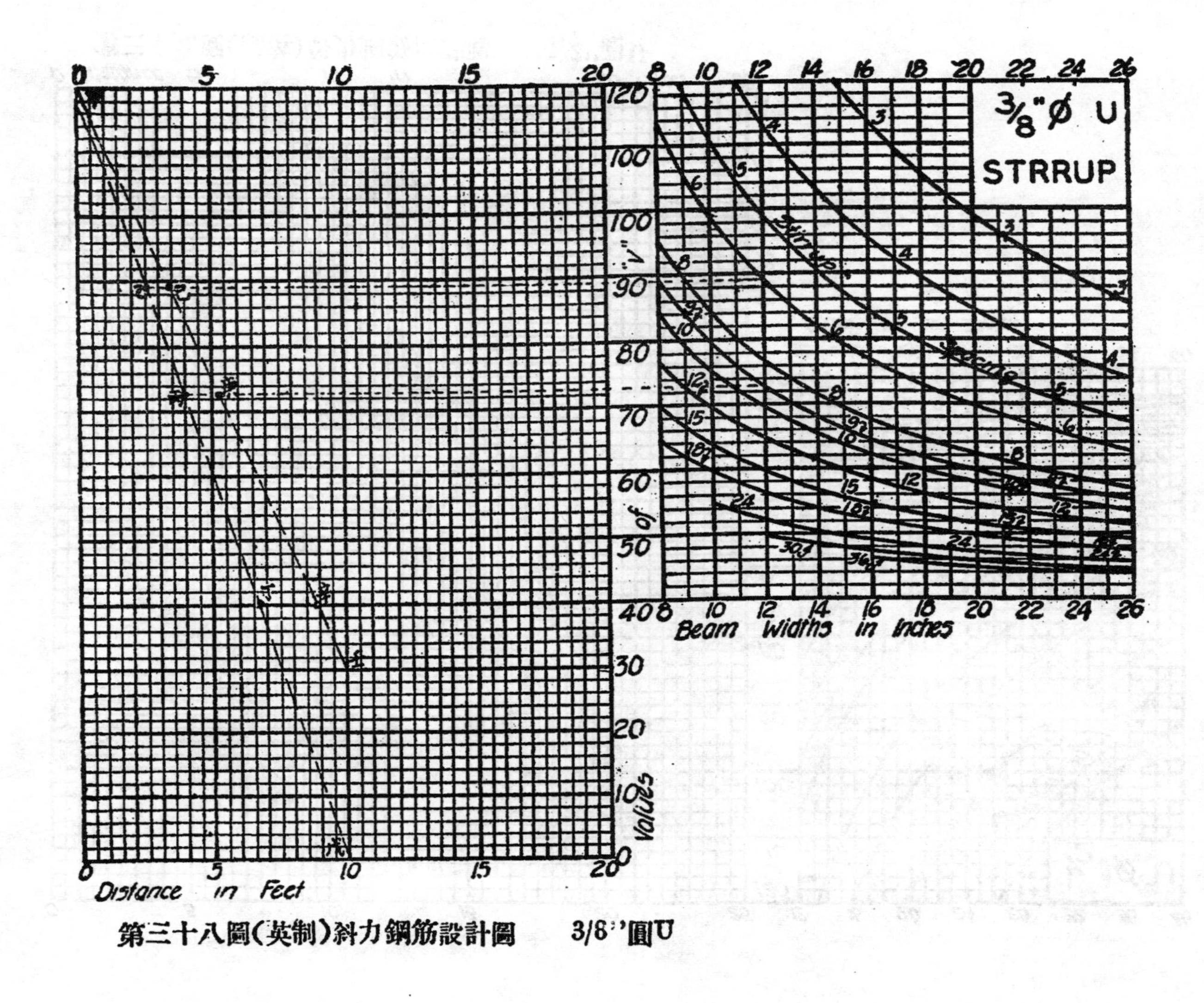

第三十八圖(英制)斜力鋼筋設計圖　3/8"圓U

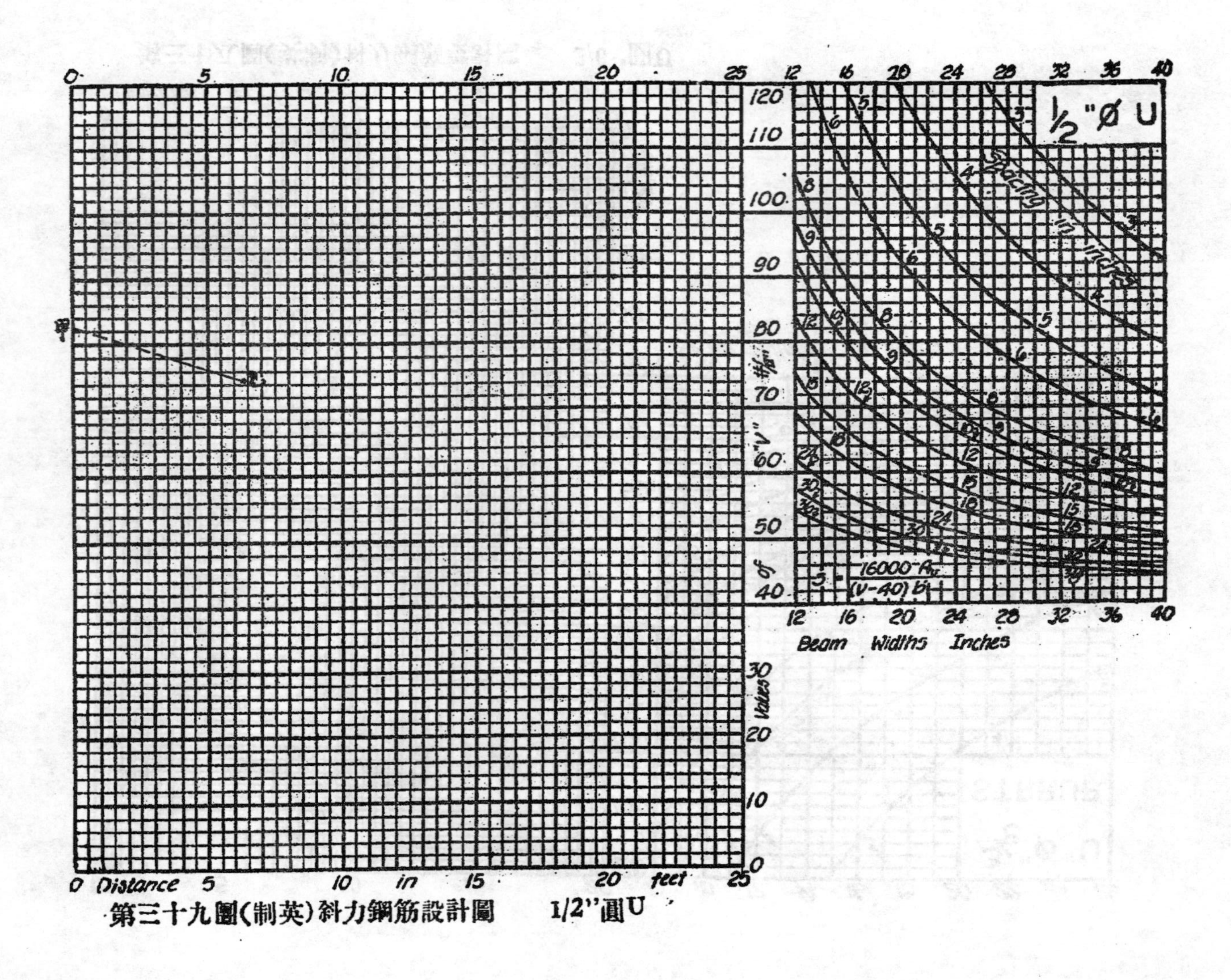

第三十九圖(制英)斜力鋼筋設計圖　1/2"圓U

第四十表(英制) 方柱設計表

安全載重以1000磅為單位

$\frac{柱長}{柱寬} \leq 11.5$

表中數目如在粗綫之右者其鋼筋面爲柱面積百分二至百分三之間

fc=400 #/口'' n=15 p=0.5—3%

如准許應力與此表不同，其安全載重可由比例求出

柱寬英吋	每呎重量	鋼筋根數	鋼筋橫斷面 1/2"○	1/2"□	5/8"○	3/4"○	7/8"○	1"○	1"□	1⅛"□	1¼"□
9	113	4	36.9	38.0	39.3	42.3	45.9				
10	125	4	44.5	45.6	46.9	49.9	53.5				
		4	52.9	54.0	55.3	57.3	6[illegible].9	66.0			
11	138	6	55.1	56.8	58.7	63.2	68.8				
		8	57.4	59.6	62.2	68.2					
		4	62.1	63.2	64.5	67.5	71.1	75.2	80.0		
12	150	6	64.3	66.0	67.9	72.4	77.8				
		8	66.6	68.8	71.4	77.4					
		4	——	73.2	74.5	77.5	81.1	85.2	90.0	96.0	
13	176	6	74.2	76.0	77.9	82.4	87.8	94.0			
		8	76.4	78.8	81.4	87.4	94.5				
		4	——	84.0	85.3	88.3	91.9	96.0	100.8	106.8	
14	204	6	85.0	86.8	88.7	93.2	98.6	104.8			
		8	87.2	89.6	92.2	98.2	105.3				
		4	——	——	96.9	99.9	103.5	107.2	112.4	118.4	125.0
15	234	6	96.6	98.4	100.3	104.8	110.2	116.4	123.6		
		8	98.8	101.2	103.8	109.8	116.[illegible]	125.2			
		4	——	——	——	112.3	115.9	120.0	124.8	139.8	137.4
16	267	6	——	110.8	112.7	117.2	122.6	128.8	136.0	144.9	
		8	111.2	113.6	116.2	112.2	129.1	137.6			
		4	——	——	——	125.5	129.1	133.2	138.0	144.0	150.6
17	301	6	——	124.0	125.9	130.4	135.8	142.0	14[illegible].2	158.1	
		8	124.4	126.8	129.4	135.4	142.5	150.8	160.4		
		4	——	——	——	139.5	143.1	149.2	152.0	158.0	164.6
18	337	6	——	——	139.9	144.4	149.8	156.0	163.2	172.1	182.1
		8	——	140.8	143.4	149.4	156.5	164.8	174.4		
		4	——	——	——	——	157.9	162.0	166.8	172.8	179.4
19	376	6	——	——	154.7	159.2	164.6	170.8	178.0	186.9	196.9
		8	——	155.6	158.2	164.2	171.3	179.6	189.2	201.1	
		4	——	——	——	——	173.5	177.6	182.4	188.4	195.0
20	416	6	——	——	——	174.8	180.2	186.4	193.6	202.5	212.5
		8	——	171.2	173.8	179.8	186.9	195.2	204.8	216.7	
		4	——	——	——	——	189.9	194.0	198.8	204.8	211.4
21	460	6	——	——	——	191.2	196.6	202.8	210.0	218.9	228.9
		8	——	——	190.2	196.2	203.3	211.6	221.2	233.1	246.4
		4	——	——	——	——	——	211.2	216.0	222.0	228.6
22	504	6	——	——	——	208.3	213.8	220.0	227.2	236.1	246.1
		8	——	——	207.4	213.4	220.5	228.8	238.4	250.3	263.6
		4	——	——	——	——	——	229.2	234.0	240.0	246.6
23	550	6	——	——	——	226.4	231.8	238.0	245.2	254.1	264.1
		8	——	——	——	231.4	238.5	246.8	256.4	268.3	281.6
		4	——	——		——	——	248.0	252.8	258.8	265.3
24	600	6	——	——	——	——	250.6	256.8	264.0	272.9	282.9
		8	——	——	——	250.2	257.3	265.6	275.2	287.1	300.4
		4	——	——	——	——	——	——	292.8	298.8	305.4
26	705	6	——	——	——	——	290.6	296.8	304.2	312.9	322.9
		8	——	——	——	290.2	297.3	305.6	315.2	327.1	340.4

方柱之旋幅 R=dc

鋼之百分率	c
0.5	0.307
1.0	0.323
2.0	0.347
3.0	0.365
4.0	0.379

長柱之載重遞減因數

h/R	P'/P
40	1.00
42	.983
44	.967
46	.950
48	.933
50	.917
52	.900
54	.883
56	.867
58	.850
60	.833
62	.817
64	.800
66	.783
68	.777
70	.750
72	.733
74	.717
76	.700
78	.683
80	.677
82	.650
84	.633
86	.617
88	.600
90	.583
92	.567
94	.550
96	.533
98	.517
100	.500
102	.483
104	.467
106	.450
108	.433
110	.417

第四十一表(英制) 方柱設計表

安全載重以1000磅爲單位　$p = C\cdot5 - 2\%$　$\frac{柱寬}{柱長} \leq 11\cdot5$　如准許應力與此表不同，其安全載重可由比例求出

邊長	每呎重量	鋼筋根數	n=12, fc=562·5磅/平方吋								n=10, fc=675磅/平方吋							
			1/2 圓	5/8 圓	3/4 圓	7/8 圓	1 圓	1 方	1 1/8 方	1 1/4 方	1/2 圓	5/8 圓	3/4 圓	7/8 圓	1 圓	1 方	1 1/8 方	1 1/4 方
12	150	4		88	92	96							105	108	112			
		6	88	92	97						104	108	113					
13	176	4		103	106	110	115					122	125	129	133			
		6	102	107	112						121	125	130					
14	204	4		118	121	125	130					140	143	147	151			
		6	118	122	127	133					139	143	148	154				
		8	120	126	132						142	147	154					
15	234	4		134	138	142	146	151				160	163	166	171	176		
		6	134	138	143	149					159	163	168	174				
		8	136	142	149						161	167	173					
16	267	4			155	159	164	169	175				184	187	192	197	204	
		6		156	161	166	173					184	189	194	201			
		8	154	159	166	174					182	188	194	202				
17	301	4			174	178	182	187	194				206	210	214	219	226	
		6		174	179	185	192					206	211	217	224			
		8	172	178	185	192					205	210	217	224				
18	337	4			193	197	202	207	214	221			229	233	238	243	249	257
		6		194	199	205	212	220				230	235	241	247	255		
		8		198	204	212	221					234	240	248	257			
19	376	4				218	223	228	235	242				258	263	268	274	282
		6		215	220	226	232	240				255	260	266	272	280		
		8		219	225	233	242					259	265	273	282			
20	416	4				240	245	250	257	264				285	289	294	301	308
		6			242	248	254	262	272				286	292	299	306	316	
		8		240	247	255	264	275				285	291	299	308	319		
		12	240	248	258	270					284	292	302	314				
21	460	4				263	268	273	280	287				312	317	322	328	336
		6			265	271	277	285	295				314	320	326	334	344	
		8		264	270	278	287	298				313	319	327	336	346		
		12	263	271	281	293					312	320	330	342				
22	504	4				287	292	297	304	311				341	346	351	358	365
		6			289	295	302	310	320	331			343	349	355	363	373	384
		8		288	294	302	311	322				342	348	356	365	375		
		12		295	305	317	331					349	359	371	384			
24	600	4					344	349	356	363					408	413	420	427
		6				347	353	358	371	382				411	417	425	435	446
		8			346	354	363	374	387				410	418	427	437	450	
		12		347	357	369	383					411	421	433	446			
26	705	4						405	412	419						481	487	494
		6				403	410	418	428	439				478	485	493	503	513
		8			402	410	419	430	443	458			478	486	494	505	518	532
		12		403	413	425	439	455				479	488	500	514	529		
28	818	4						466	473	480						554	560	567
		6					471	479	489	500					558	566	575	586
		8				471	480	491	504	519				558	567	578	591	605
		12			462	474	487	503					561	573	586	602		
30	937	4							538	545							638	645
		6					536	544	554	565					636	644	654	664
		8				536	546	556	569	584				637	646	656	669	683
		12			540	551	565	581	601				640	651	665	680	700	

第四十二表 方柱設計表（英制）

安全載重1000磅單位 $\frac{\text{柱長}}{\text{柱寬}}=11.5$ $f_c=426$ $n=15$ $p=0.5-3.0\%$

柱寬英寸	每呎重磅	鋼筋數	鋼筋尺度 1/2"○	1/2"□	5/8"○	5/8"□	3/4"○	3/4"□	7/8"○	1"○	1"□	1⅛"□	1¼"□
8	100	4	32.0	33.3	34.6	26.7	37.9						
9	113	4	39.2	40.6	41.8	43.9	45.2	43.0	49.0				
10	125	4	47.4	48.6	49.8	51.9	53.4	56.1	57.1				
		6	49.6	51.6	53.7	56.6	58.5						
		8	52.0	54.6	57.2								
11	138	4	56.3	57.5	59.0	60.8	62.4	65.1	66.0	70.4			
		6	58.5	60.5	62.5	65.6	67.5	71.9	73.4				
		8	61.0	63.5	66.2	70.2	72.9						
12	150	4	66.1	67.4	68.7	70.7	72.0	75.0	76.0	90.4	95.4		
		6	68.5	70.4	72.3	75.5	77.3	81.8	83.0				
		8	70.8	73.3	76.0	80.0	82.6						
13	176	4	—	78.0	79.5	81.5	82.7	85.6	86.5	91.6	96.0	102.1	
		6	79.1	81.0	83.2	86.0	88.0	92.3	93.3	100.0			
		8	81.5	84.1	86.7	90.0	93.2	99.1	101.0				
14	204	4	—	89.5	90.8	92.0	94.1	97.1	98.1	102.3	107.5	114.0	
		6	90.6	92.5	94.5	97.6	99.5	104.0	105.0	102.0			
		8	93.0	95.6	98.1	102.0	104.9	110.7	112.2				
15	234	4	—	—	103.0	105.1	106.7	109.7	110.3	114.8	120.0	126.5	133.2
		6	103.0	105.0	106.9	110.0	111.8	116.1	117.7	124.0	132.0		
		8	105.2	107.9	110.5	114.4	111.7	123.0	124.8	133.3			
16	267	4	—	—	—	118.4	120.0	122.9	123.8	128.0	133.0	139.8	146.5
		6	—	118.1	120.0	123.1	125.0	129.4	130.5	137.5	145.0	154.8	
		8	118.8	120.8	123.9	126.5	130.4	136.0	138.0	146.8			
17	301	4	—	—	—	132.2	134.0	137.0	137.9	142.0	147.1	153.2	160.5
		6	—	132.1	134.0	137.0	139.0	143.9	144.8	151.1	159.0	168.8	
		8	132.4	135.1	138.0	142.0	144.6	150.2	152.0	161.0	171.0		
18	337	4	—	—	—	—	149.0	152.0	152.8	157.0	162.0	168.6	175.8
		6	—	—	149.1	152.0	154.1	158.9	159.9	156.1	174.0	183.5	194.1
		8	—	150.1	153.0	157.0	159.2	165.3	167.1	176.0	186.0		
19	376	4	—	—	—	—	—	167.6	168.1	173.0	178.0	184.2	191.0
		6	—	—	165.0	168.0	170.0	174.1	176.0	182.0	190.0	198.1	210.0
		8	—	165.9	168.1	173.0	175.0	181.0	182.8	191.4	202.0	214.2	
20	416	4	—	—	—	—	—	184.3	185.0	189.1	294.5	200.2	207.8
		6	—	—	—	184.5	186.5	191.9	192.2	198.5	206.0	215.7	226.2
		8	—	—	—	189.1	192.0	198.0	199.5	208.0	218.3	230.3	
21	460	4	—	—	—	—	—	202.0	202.1	207.0	211.9	218.1	225.0
		6	—	—	—	202.0	204.0	208.0	209.9	216.5	224.0	234.0	244.0
		8	—	—	—	207.0	210.0	215.5	217.0	226.0	236.0	248.0	258.0
22	504	4	—	—	—	—	—	—	—	225.0	231.0	236.0	243.9
		6	—	—	—	—	222.5	227.0	228.0	234.3	242.0	255.2	262.1
		8	—	—	—	225.0	228.0	234.0	246.0	244.0	257.1	266.2	280.4
24	600	4	—	—	—	—	—	—	—	264.5	270.0	276.0	283.4
		6	—	—	—	—	—	266.0	267.5	274.0	282.0	291.0	302.0
		8	—	—	—	264.4	267.0	272.5	274.0	283.5	293.0	306.0	320.0
26	705	4	—	—	—	—	—	—	—	—	312.1	319.0	325.0
		6	—	—	—	—	—	308.5	310.0	316.5	324.0	334.0	344.0
		8	—	—	—	—	310.0	316.0	317.0	326.0	336.0	348.2	362.2
28	818	4	—	—	—	—	—	—	—	—	358.0	365.0	371.0
		6	—	—	—	—	—	—	—	362.5	370.5	380.0	390.0
		8	—	—	—	—	—	—	—	372.0	382.0	394.8	407.0

25487

第四十三表（英制）螺旋柱安全載重表

$P = Ac[1+(n-1)p]fc$ $fc = 300 + (0.10 + 4p)f'c$

有效直徑	每呎柱長之重量 圓	方	縱鋼筋百份數	安全載重單位以千磅計 f'c=2000	f'c=2500	f'c=3000	鋼筋平方面積吋	縱直鋼筋之數目 1/2圓	5/8圓	3/4圓	7/8圓	1圓	1方	1 1/8方	1 1/4方	螺旋絲 絲之直徑	節距
			1	75	82	89	1.13	6								1/4	2
			1.5	85	92	100	1.70	9	6							1/4	2
			2	96	101	112	2.26	12	8	6						1/4	2
			2.5	107	115	125	2.83	15	10	7						1/4	2
			3	119	128	138	3.39	17	11	8	6					1/4	2
12	209	267	3.5	131	141	152	3.96	20	13	9	7					1/4	1.75
			4	145	155	166	4.52		15	11	8	6				1/4	1.5
			4.5	159	169	181	5.09		17	12	9	7				5/16	2
			5	173	184	197	5.65		19	13	10	8	6			5/16	2
			5.5	188	200	213	6.22			14	11	8	7			5/16	1.75
			6	204	216	230	6.79			16	12	9	7			5/16	1.5
			1	88	96	104	1.33	7								1/4	2
			1.5	100	108	116	1.99	10	7							1/4	2
			2	112	121	132	2.65	14	9	6						1/4	2
			2.5	125	135	146	3.32	17	11	8	6					1/4	2
			3	140	150	162	3.98	20	13	9	7					1/4	2
13	237	301	3.5	154	166	178	4.65		15	11	8	6				1/4	1.75
			4	170	182	195	5.31		18	12	9	7				1/4	1.5
			4.5	186	198	213	5.97		20	14	10	8	6			5/16	2
			5	203	216	231	6.64			15	11	9	7			5/16	1.75
			5.5	221	234	250	7.30			17	13	10	8	6		5/16	1.5
			6	239	253	270	7.96			18	14	11	8	7		5/16	1.5
			1	102	111	121	1.54	8	6							1/4	2.25
			1.5	116	126	136	2.31	12	8	6						1/4	2.25
			2	130	141	153	3.08	16	10	7						1/4	2.25
			2.5	146	157	170	3.85	20	13	9	7					1/4	2.25
			3	162	174	188	4.62		15	11	8	6				1/4	1.75
14	265	337	3.5	179	192	207	5.39		18	13	9	7				1/4	1.5
			4	197	211	226	6.16		20	14	11	8	6			5/16	2
			4.5	216	230	247	6.93			16	12	9	7			5/16	1.75
			5	236	251	268	7.70			18	13	10	8	6		5/16	1.75
			5.5	256	272	290	8.47			19	14	11	9	7		3/8	2.25
			6	278	294	313	9.24				16	12	10	8	6	3/8	2
			1	117	128	139	1.77	9	6							1/4	2.5
			1.5	133	144	156	2.65	14	9	6						1/4	2.5
			2	149	162	175	3.53	18	12	8	6					1/4	2.5
			2.5	167	180	195	4.42		15	10	8	6				1/4	2
			3	186	200	216	5.30		18	12	9	7				1/4	1.75
15	296	376	3.5	205	220	237	6.19		20	14	11	8	6			5/16	2.25
			4	226	242	260	7.07			16	12	9	7	6		5/16	2
			4.5	248	264	283	7.95			18	13	10	8	7		5/16	1.75
			5	270	288	308	8.84			20	15	12	9	7	6	3/8	2.25
			5.5	294	312	333	9.72				17	13	10	8	7	3/8	2
			6	319	337	359	10.60				18	14	11	9	7	3/8	1.75
			1	133	145	158	2.01	11	7							1/4	2.5
			1.5	151	164	178	3.02	16	10	7						1/4	2.5
			2	170	184	199	4.02	20	14	9	7					1/4	2.5
			2.5	190	205	222	5.03		17	12	9	7				1/4	2
			3	211	227	245	6.03		20	14	10	8	6			5/16	2.5
16	327	416	3.5	234	251	270	7.04			16	12	9	7			5/16	2
			4	257	275	295	8.04			19	14	11	8	7		5/16	1.75
			4.5	282	301	322	8.95				15	12	9	8	6	3/8	2.25
			5	308	327	350	10.05				17	13	10	8	7	3/8	2
			5.5	335	355	379	11.06				19	14	11	9	7	3/8	2
			6	363	384	409	12.06				20	15	12	10	8	3/8	1.75

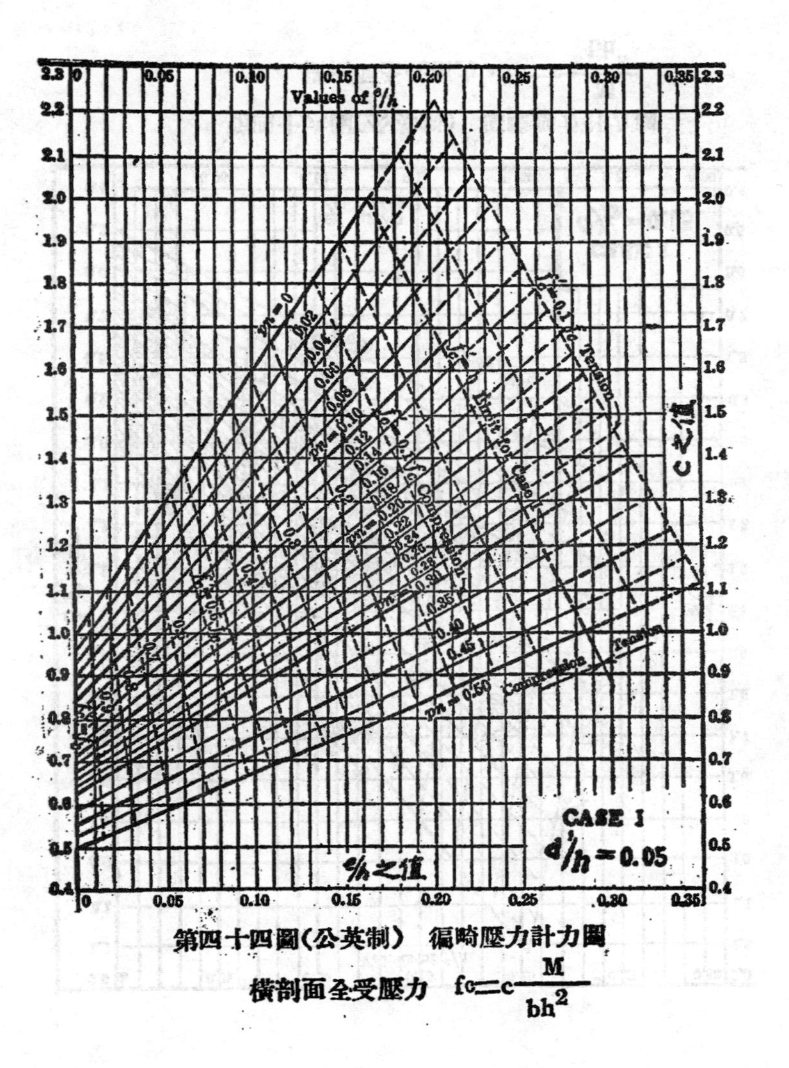

第四十四圖（公英制）　偏畸壓力計力圖

橫剖面全受壓力　$f_c = c\frac{M}{bh^2}$

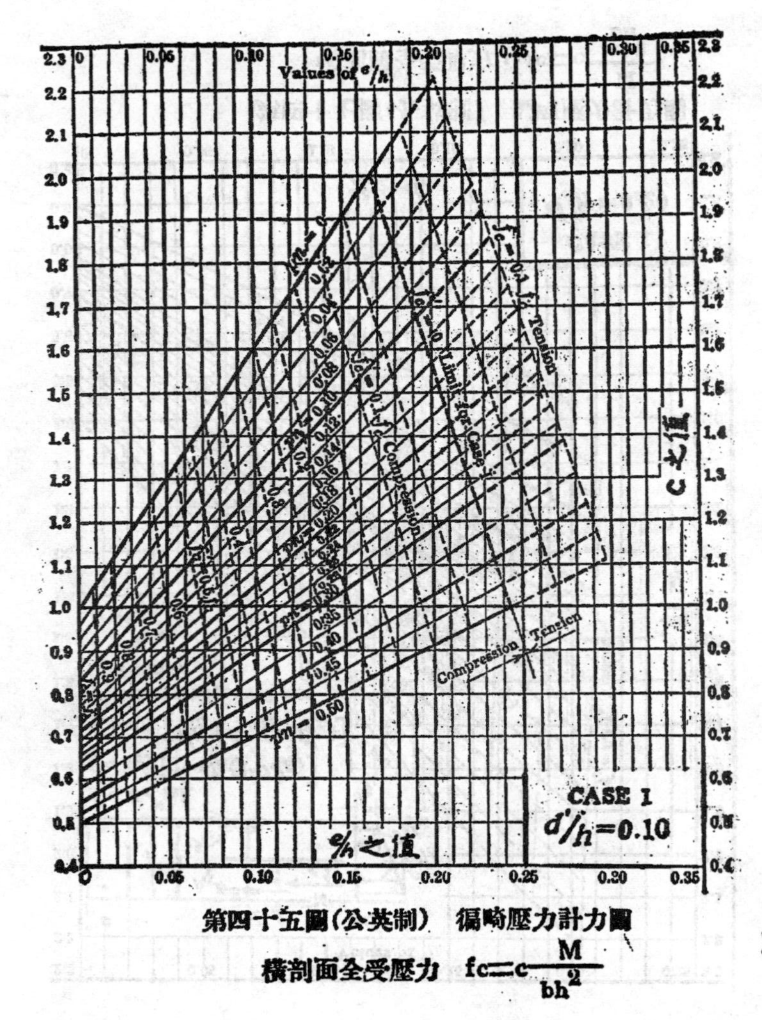

第四十五圖（公英制）　偏畸壓力計力圖

橫剖面全受壓力　$f_c = c\frac{M}{bh^2}$

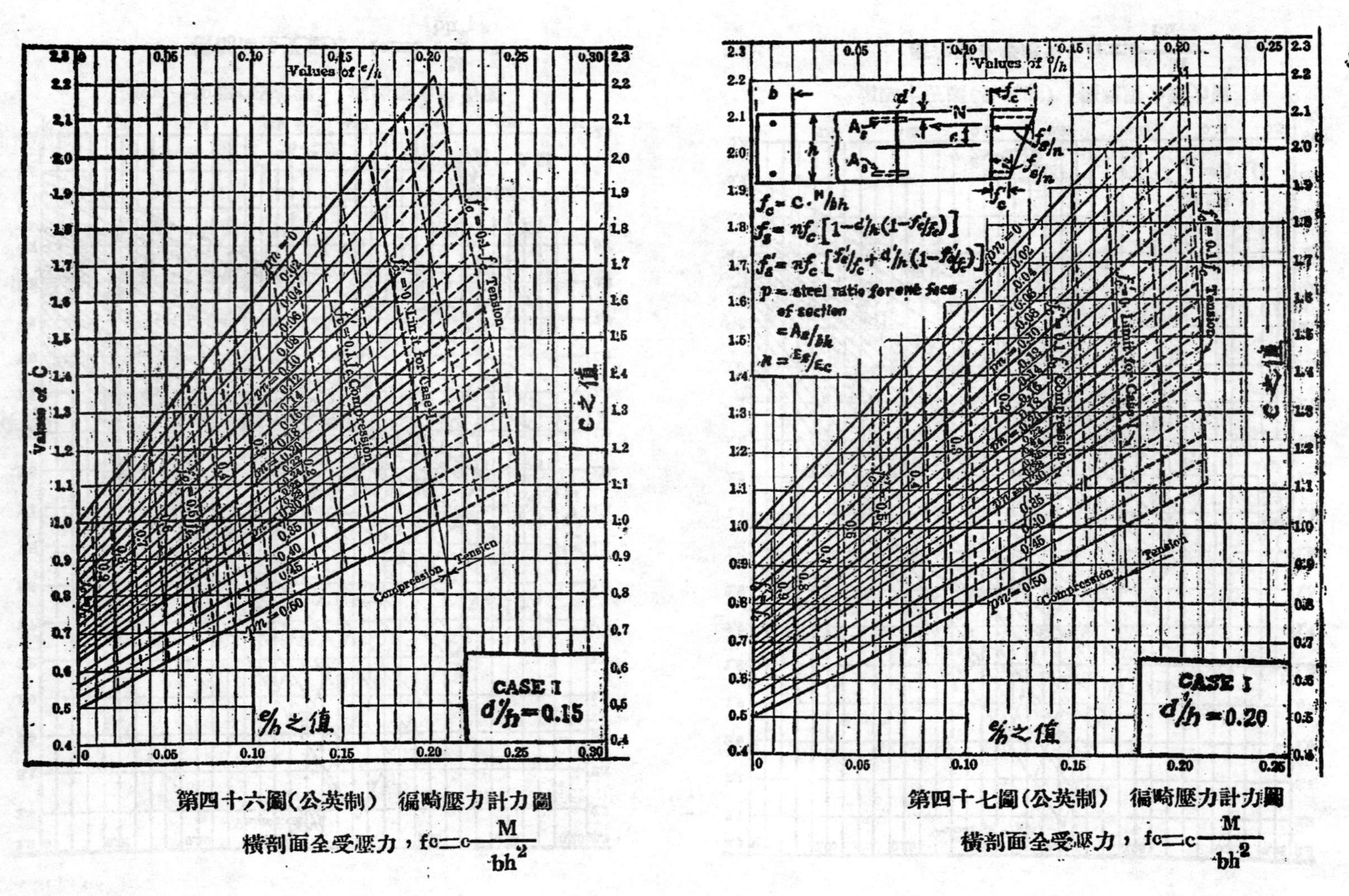

第四十六圖(公英制) 偏畸壓力計力圖

橫剖面全受壓力，$f_c = c\frac{M}{bh^2}$

第四十七圖(公英制) 偏畸壓力計力圖

橫剖面全受壓力，$f_c = c\frac{M}{bh^2}$

25491

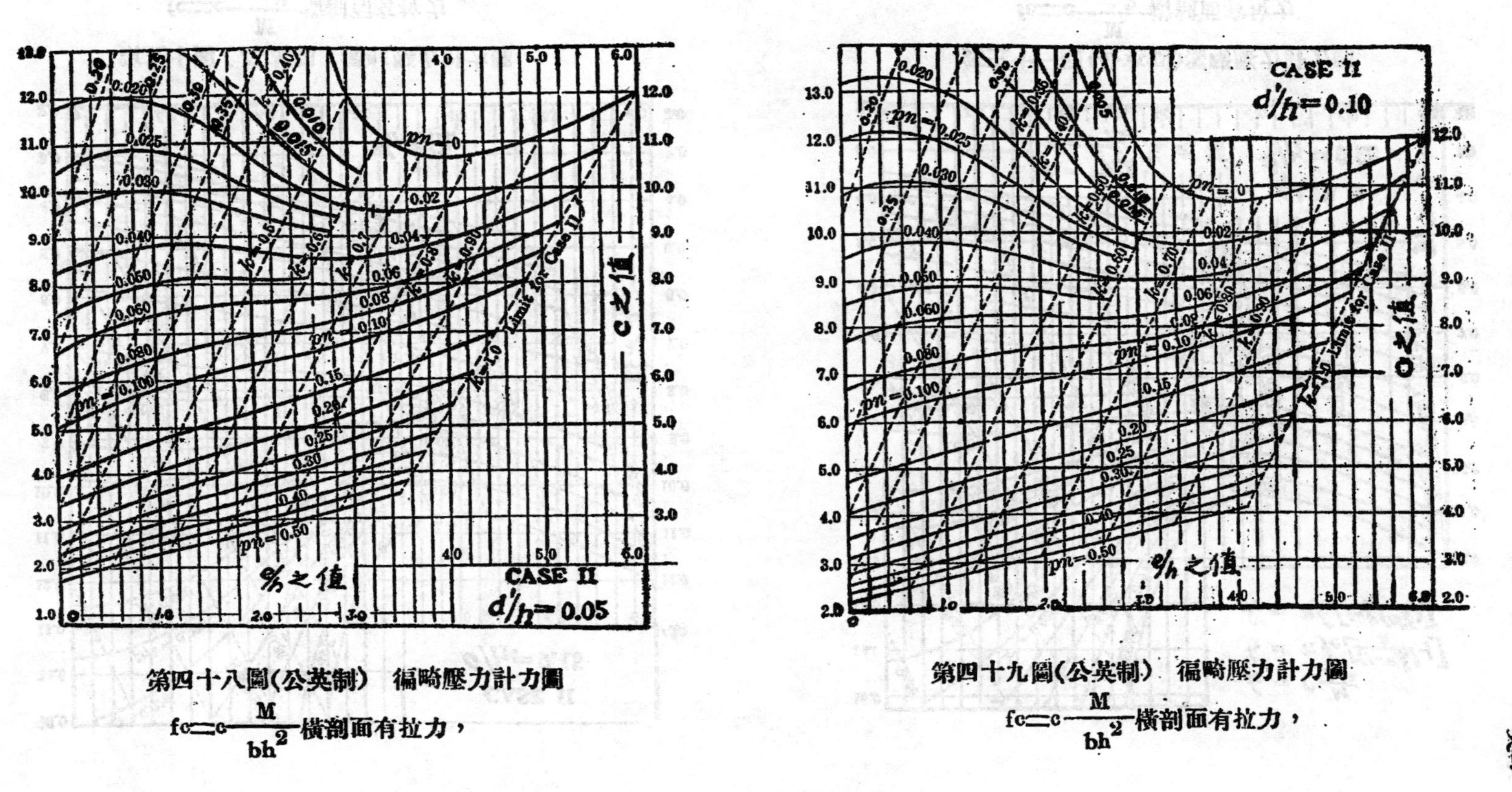

第四十八圖(公英制) 偏畸壓力計力圖

$fc = c\dfrac{M}{bh^2}$ 横剖面有拉力，

第四十九圖(公英制) 偏畸壓力計力圖

$fc = c\dfrac{M}{bh^2}$ 横剖面有拉力，

英47

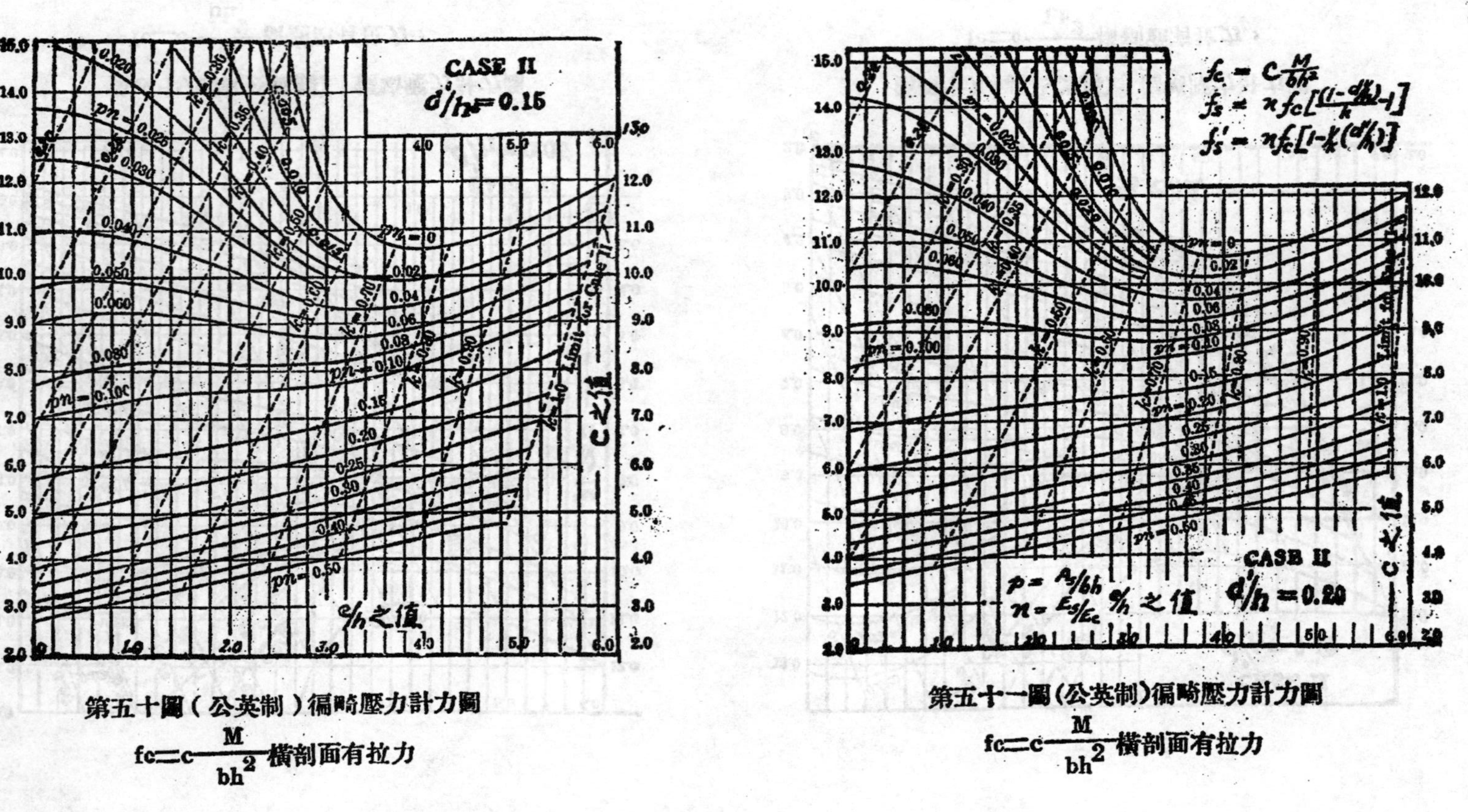

第五十圖（公英制）偏畸壓力計力圖

$f_c = c\frac{M}{bh^2}$ 横剖面有拉力

第五十一圖（公英制）偏畸壓力計力圖

$f_c = c\frac{M}{bh^2}$ 横剖面有拉力

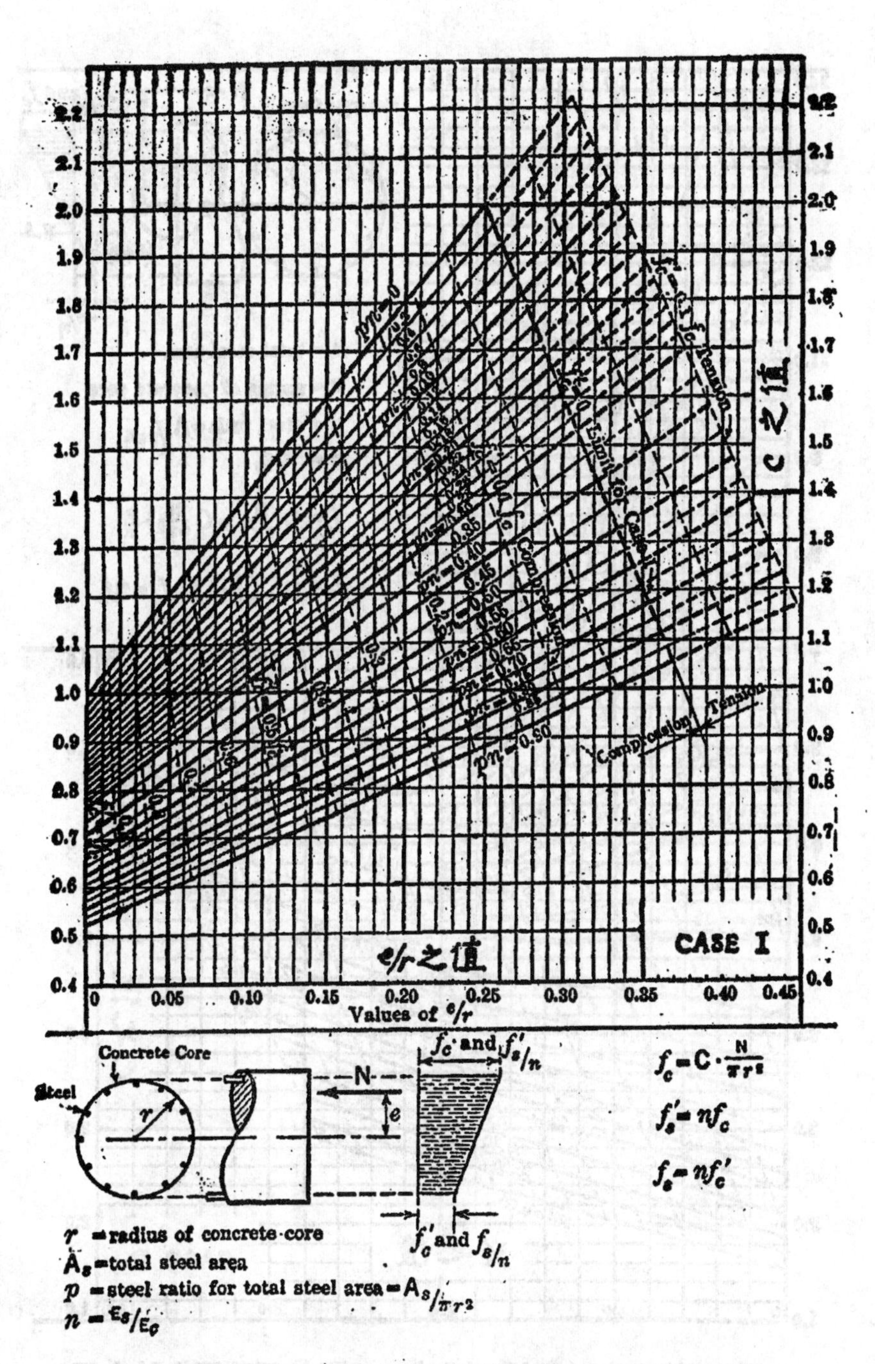

第五十二圖(公英制)　圓柱彎形壓力計力圖

橫剖面無拉力

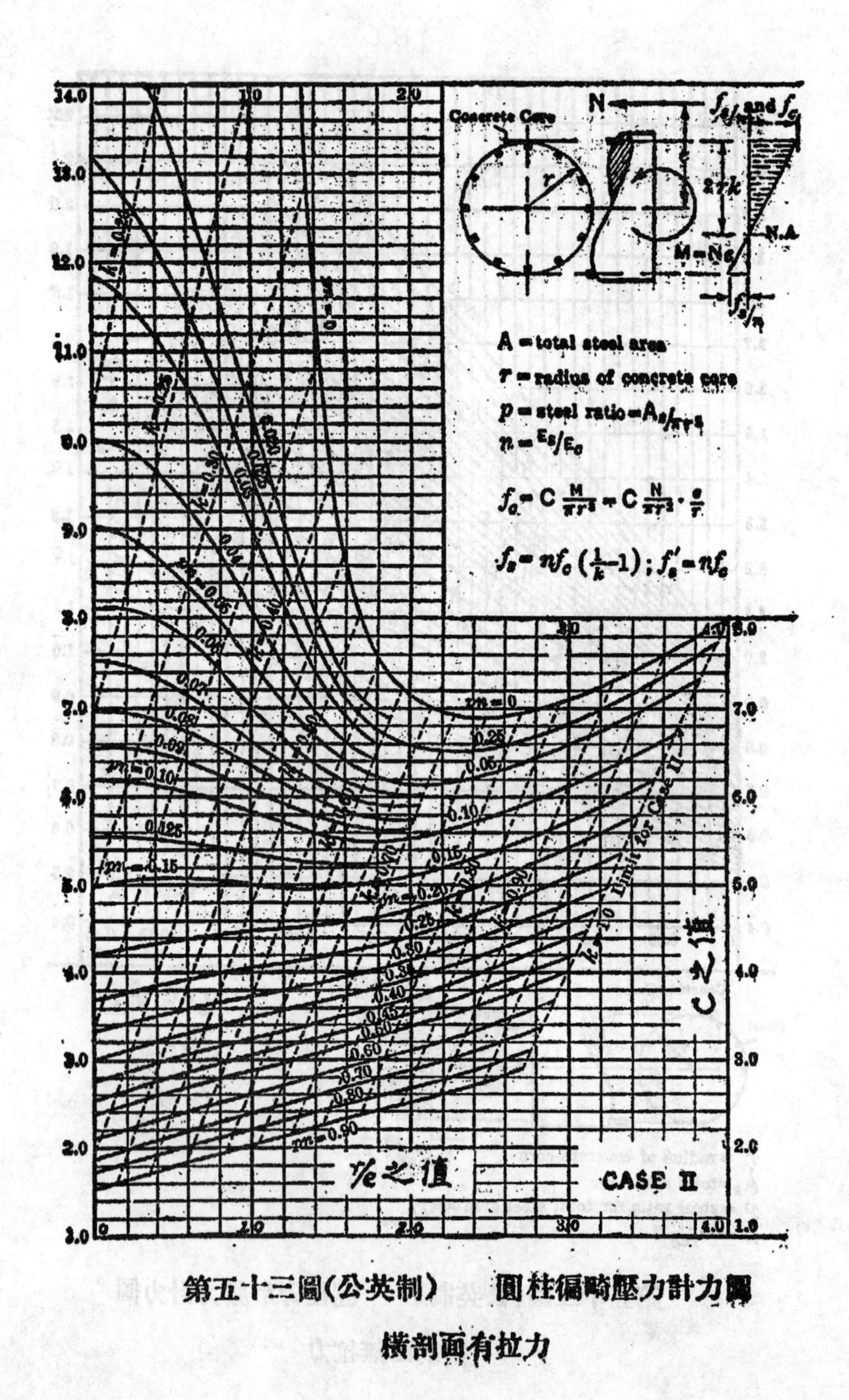

第五十三圖（公英制） 圓柱偏畸壓力計力圖

横剖面有拉力

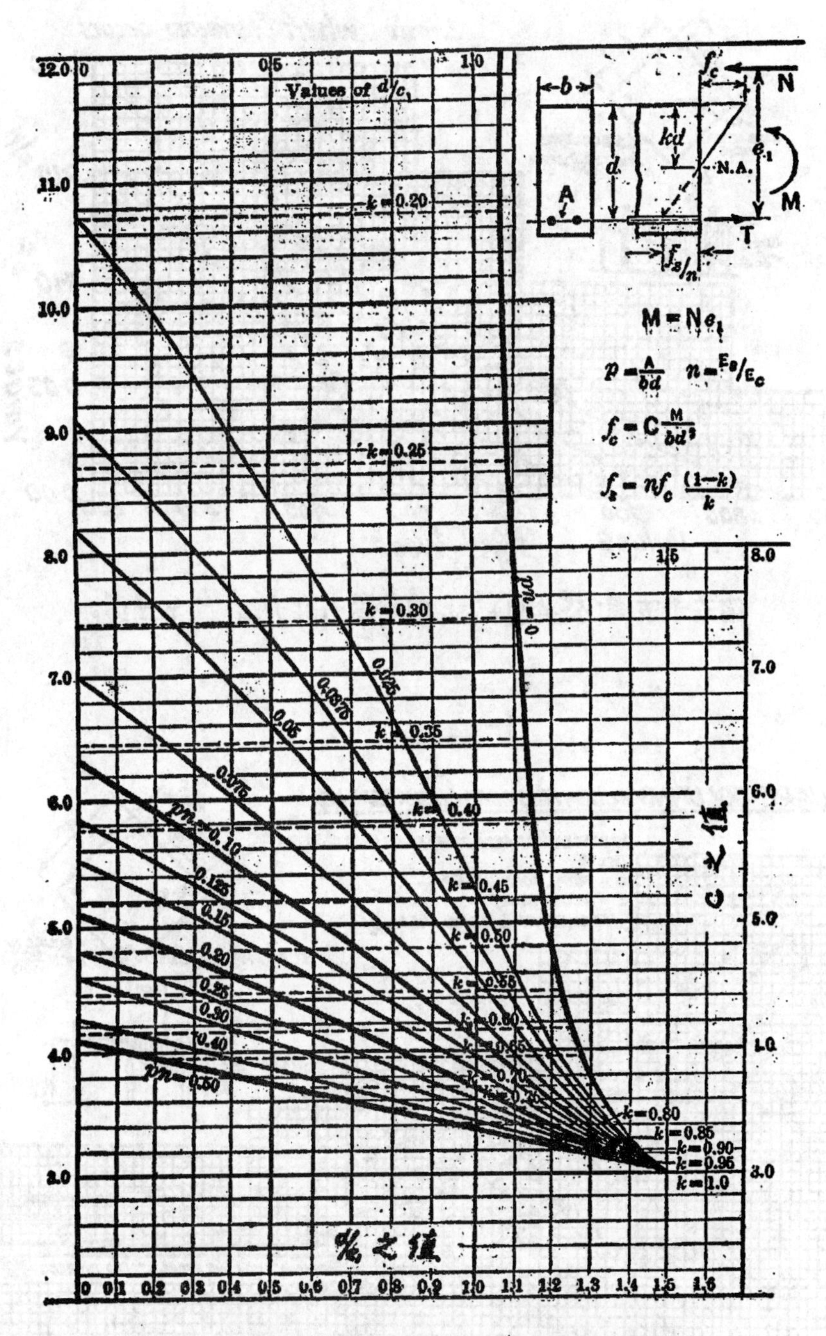

第五十四圖(公英制)　偏轴壓力計力圖

無壓力鋼筋

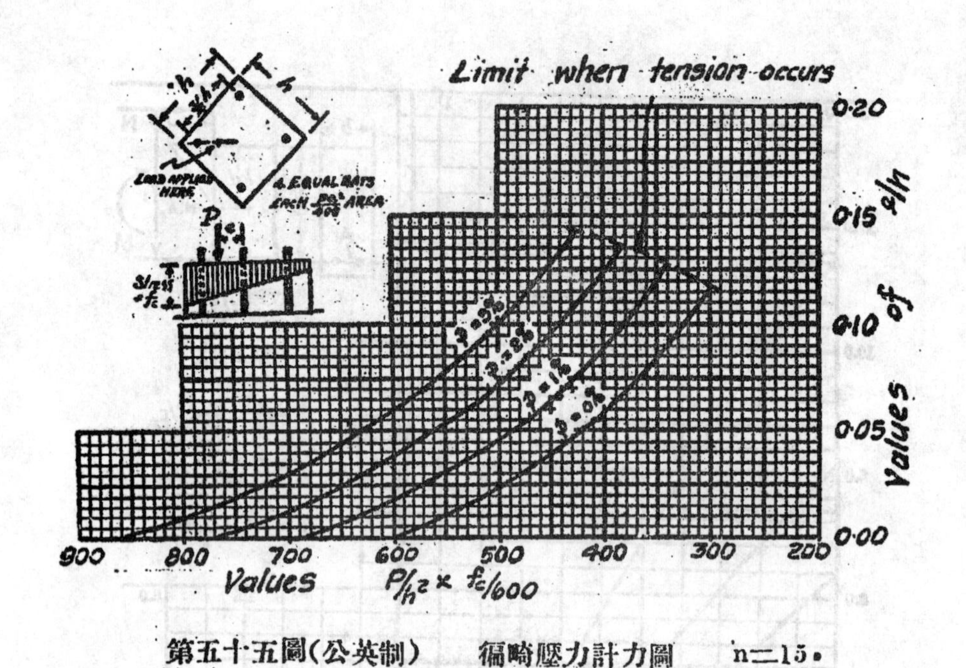

第五十五圖(公英制)　　偏畸壓力計力圖　　n=15.

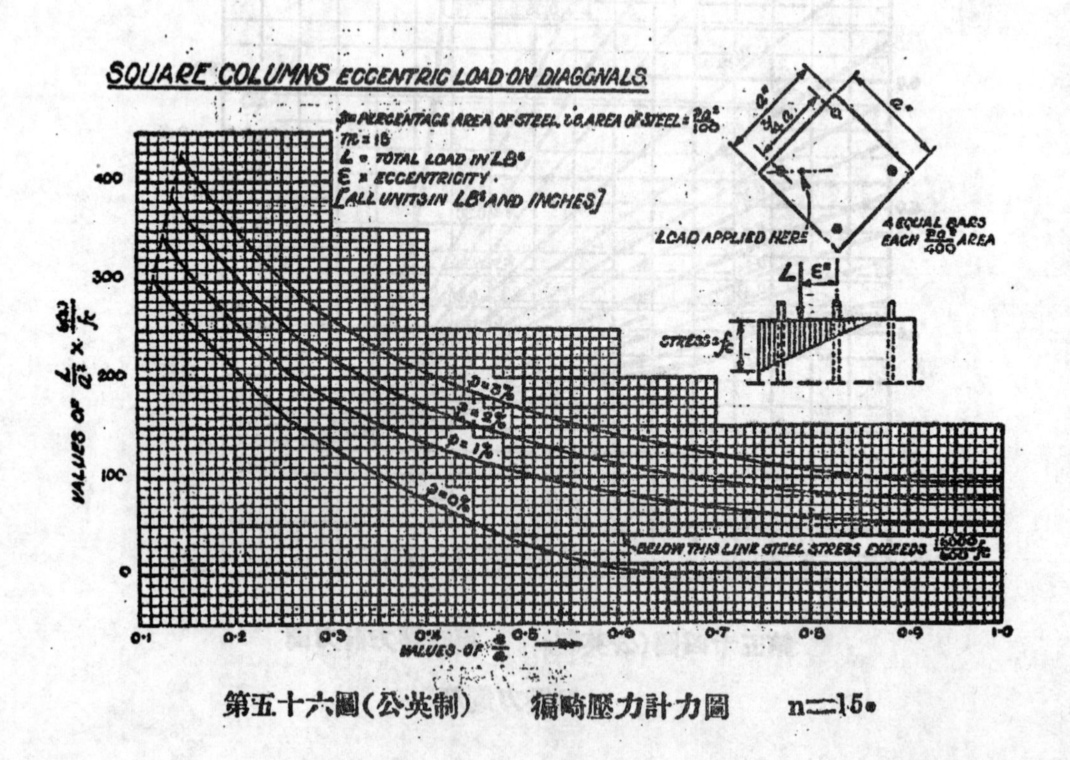

第五十六圖(公英制)　　偏畸壓力計力圖　　n=15.

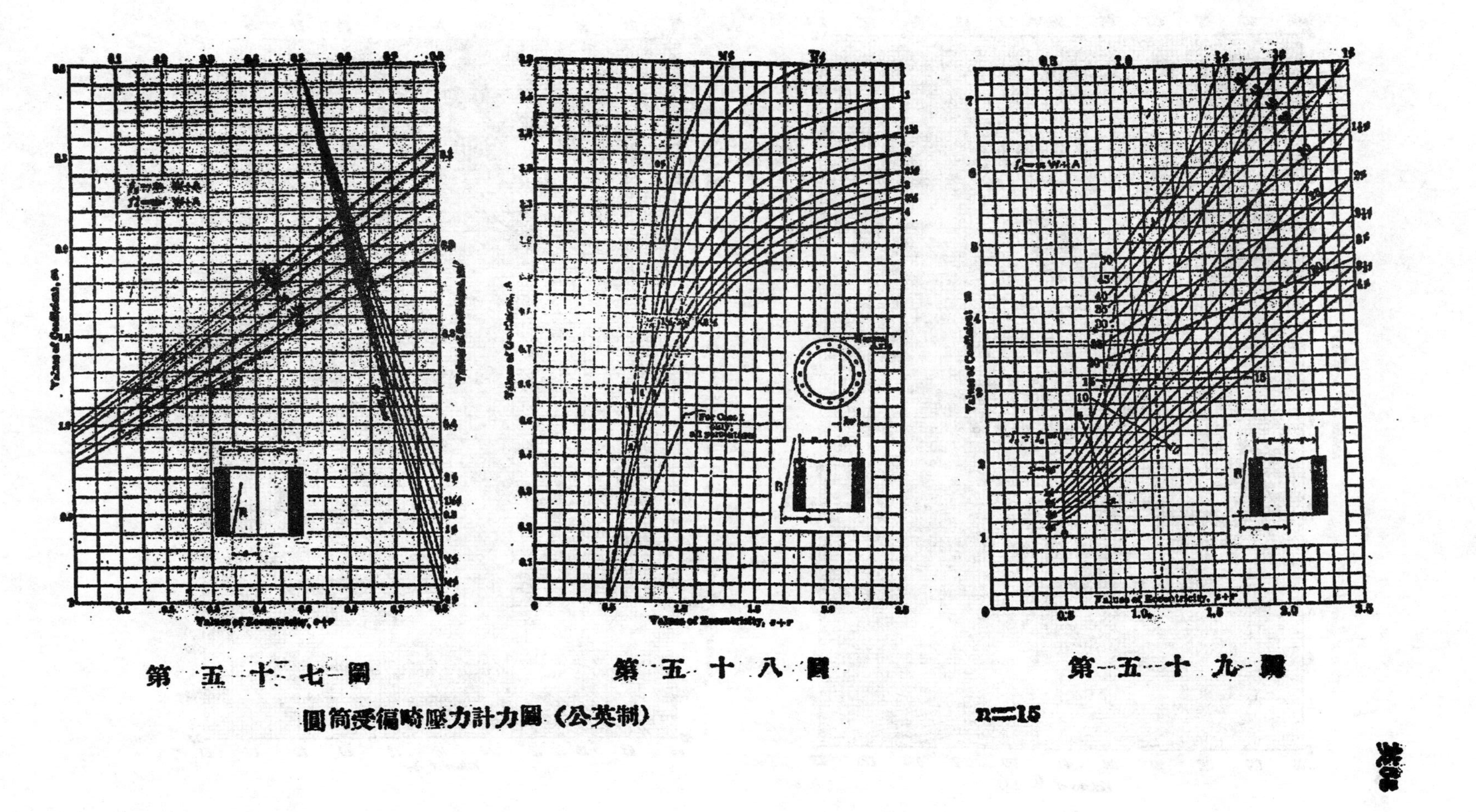

第五十七圖　第五十八圖　第五十九圖

圓筒受偏畸壓力計力圖（公英制）

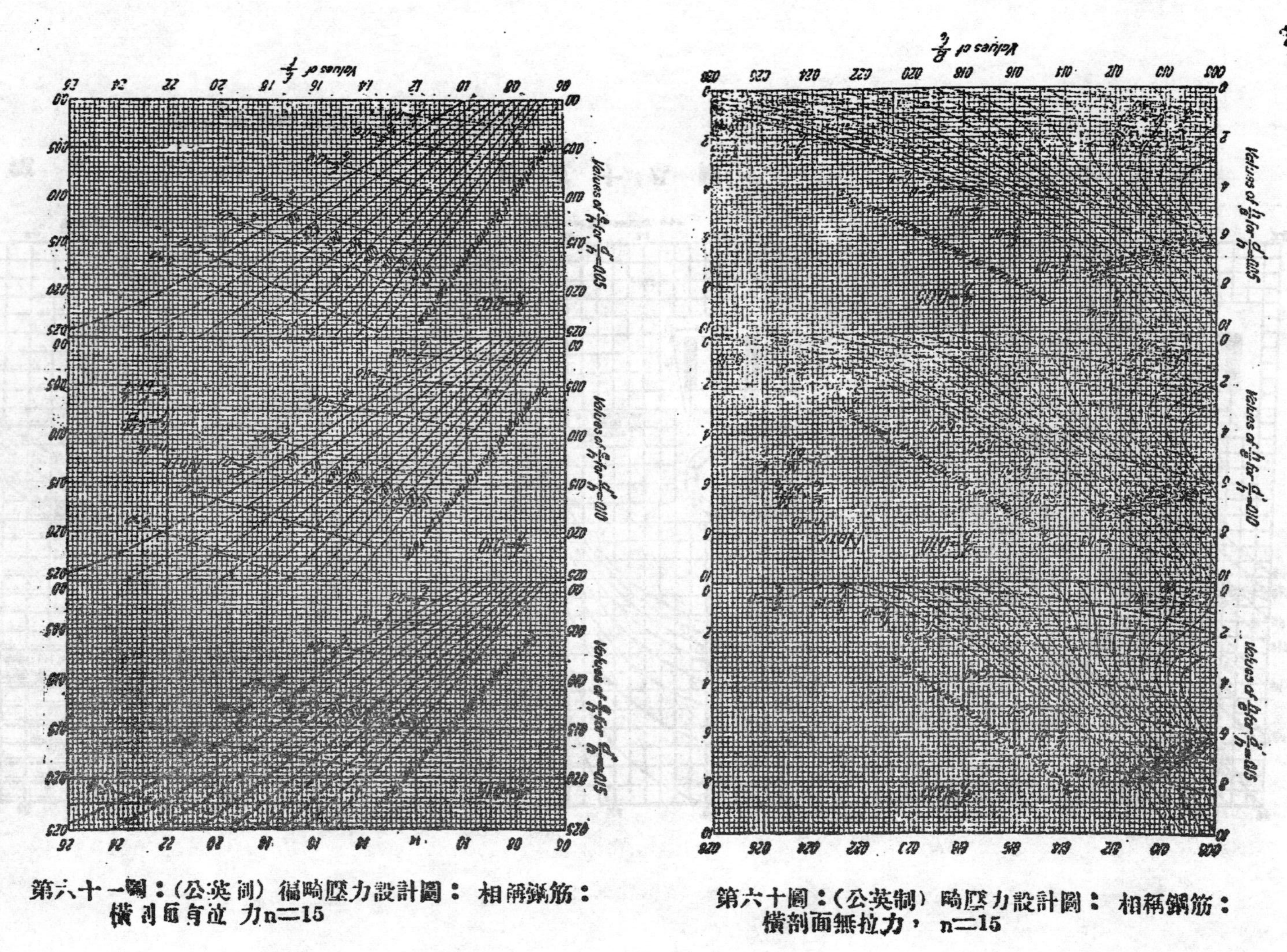

第六十圖：(公英制) 畸壓力設計圖：相稱鋼筋：橫剖面無拉力，n=15

第六十一圖：(公英制) 偏畸壓力設計圖：相稱鋼筋：橫剖面有拉力n=15

第六十二，六十三，六十四圖用法

設有受偏畸載重一萬磅及動率四萬尺磅之 10×20 矩形柱若 $d'=2\cdot5$吋 $f_s=18000$, $fc=600$ $n=15$ 求所需鋼筋

解法：$\frac{d'}{h}=\frac{2.5}{20}=0\cdot12, \frac{f_s}{fc}=\frac{18000}{600}=30, e=\frac{M}{N}=\frac{40000\times12}{10000}=48$吋

$$e_s=48+\frac{20}{2}-2\cdot5=55\cdot5\text{吋}$$

$$\frac{M_s}{fc\,bh^2}=\frac{Ne_s}{fc\,bh^2}=\frac{10000\times55\cdot5}{600\times10\times20^2}=0\cdot231$$

由第六十四圖之左部得 $p'=1\cdot73\%$ $A_s'=0\cdot0173\times10\times20=3\cdot46$方吋

$e'_s=48-\frac{20}{2}+2\cdot5=40\cdot5$吋 $\frac{M_s}{fc\,bh^2}=\frac{Ne_s'}{fc\,bh^2}=\frac{100000\times40\cdot5}{600\times10\times20^2}=0\cdot169$

由第六十四圖之右部得 $p=0\cdot73\%$ $A_s=0\cdot0073\times10\times20=1\cdot46$ 方吋

若欲顧及壓力鋼筋佔有之位置則 A_s 應改爲$\frac{n}{n-1}A_s=\frac{15}{14}\times3\cdot46$

所得壓力鋼筋應作 3•71 方吋

如 f_s 改爲 15000 其解法如下：

$\frac{f_s}{fc}=\frac{15000}{600}=25$ 其他數目如前由第六十四圖得

$p'=1\cdot43\%$ $p=0\cdot88\%$ 鋼筋之總面積爲$(0\cdot0143+0\cdot0088)10\times20=4\cdot62$方吋，比諸 $3\cdot46+1\cdot46=4\cdot92$ 方吋少0•3 方吋，注意此種設計，所用鋼筋應力以略小于准許應力爲經濟。

複筋矩樑設計：此項設計如准許應力與第十五，十六，十七，等圖相同時以用此數圖爲最簡便，如各項應力與該數圖不同，第六十二，六十三，六十四，等圖亦可作此種設計之用。設有矩樑闊十二吋深度二十吋 $d'=2\cdot5$ 吋動率爲四萬呎磅如 $f_s=18000$, $f_c=600$ $n=15$ 求所需鋼筋

$$\frac{d^1}{h}=\frac{2\cdot5}{20}=0\cdot12\ \frac{f_s}{f_c}=\frac{18000}{600}=30\ \frac{M}{f_c bh^2}=\frac{M's}{f_c bh^2}=\frac{M_s}{f_c bh^2}$$

$$=\frac{40000\times12}{600\times10\times20^2}=0\cdot2$$ 用第六十四圖可得 $p=0\cdot86\%$, $p'=0\cdot126\%$

$A_s=0\cdot0086\times10\times20=1\cdot32$ 方吋 $A_s'=\frac{15}{14}\times0\cdot0126\times10\times20=2\cdot7$ 方吋

注意：此數圖所用之 $n=\frac{E_s}{E_c}$爲 15，如 n 不等 15 時吋可用公制等圖

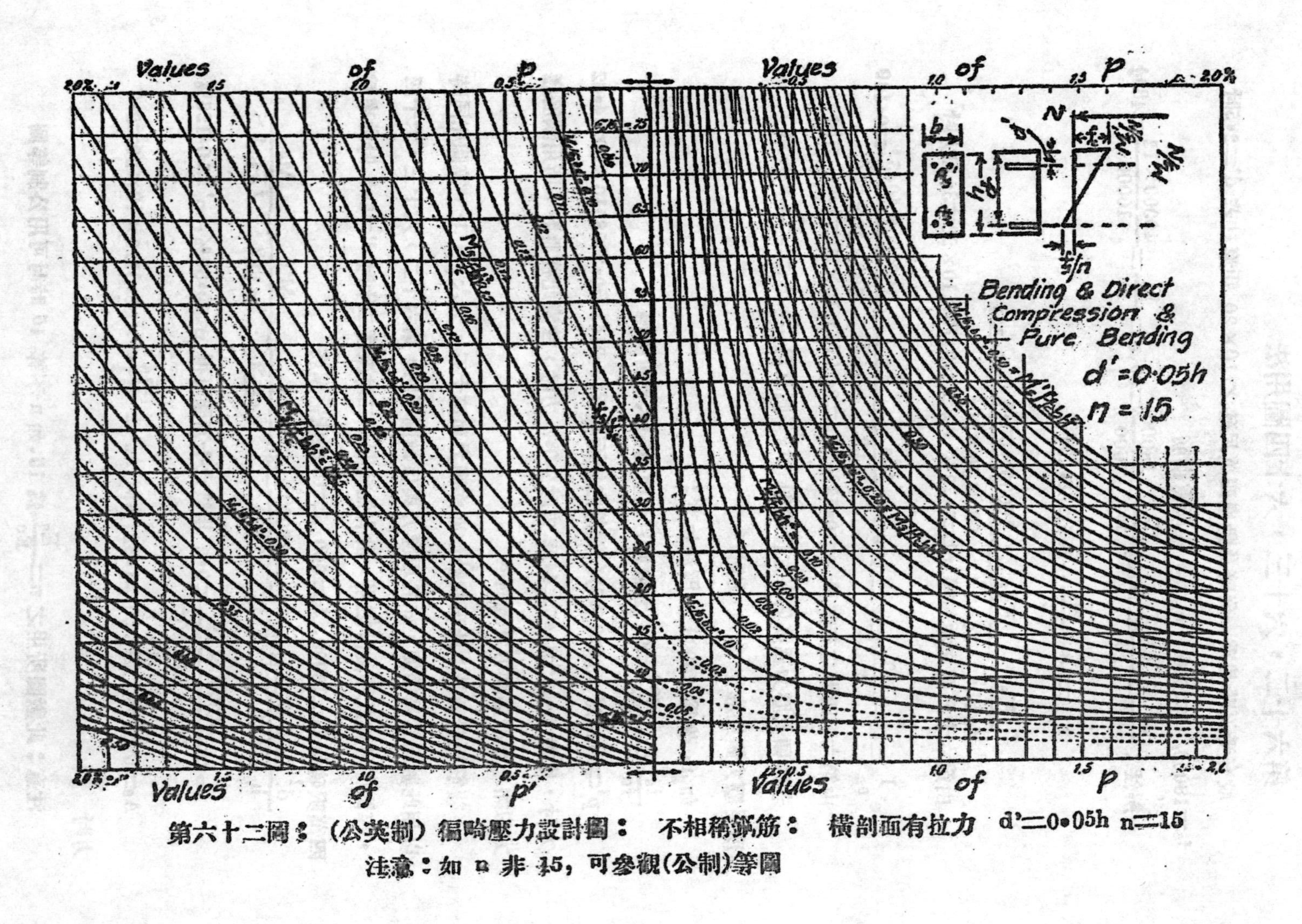

第六十三圖：（公英制）偏畸壓力設計圖：不相稱鋼筋：橫剖面有拉力 d'＝0•05h n＝15

注意：如 n 非 15，可參觀（公制）等圖

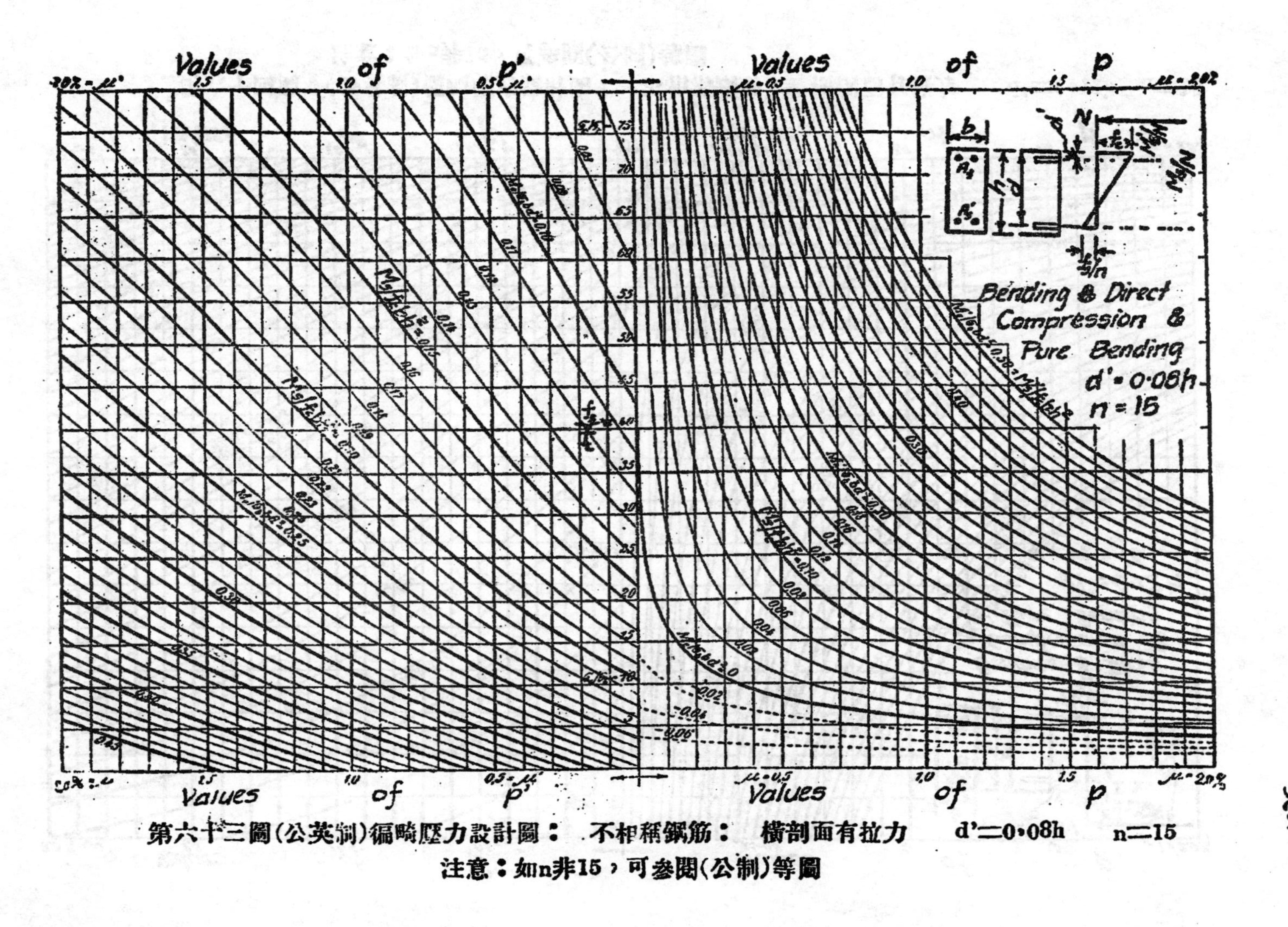

第六十三圖(公英制)偏畸壓力設計圖：　不相稱鋼筋：　橫剖面有拉力　　d'＝0·08h　　n＝15

注意：如n非15，可參閱(公制)等圖

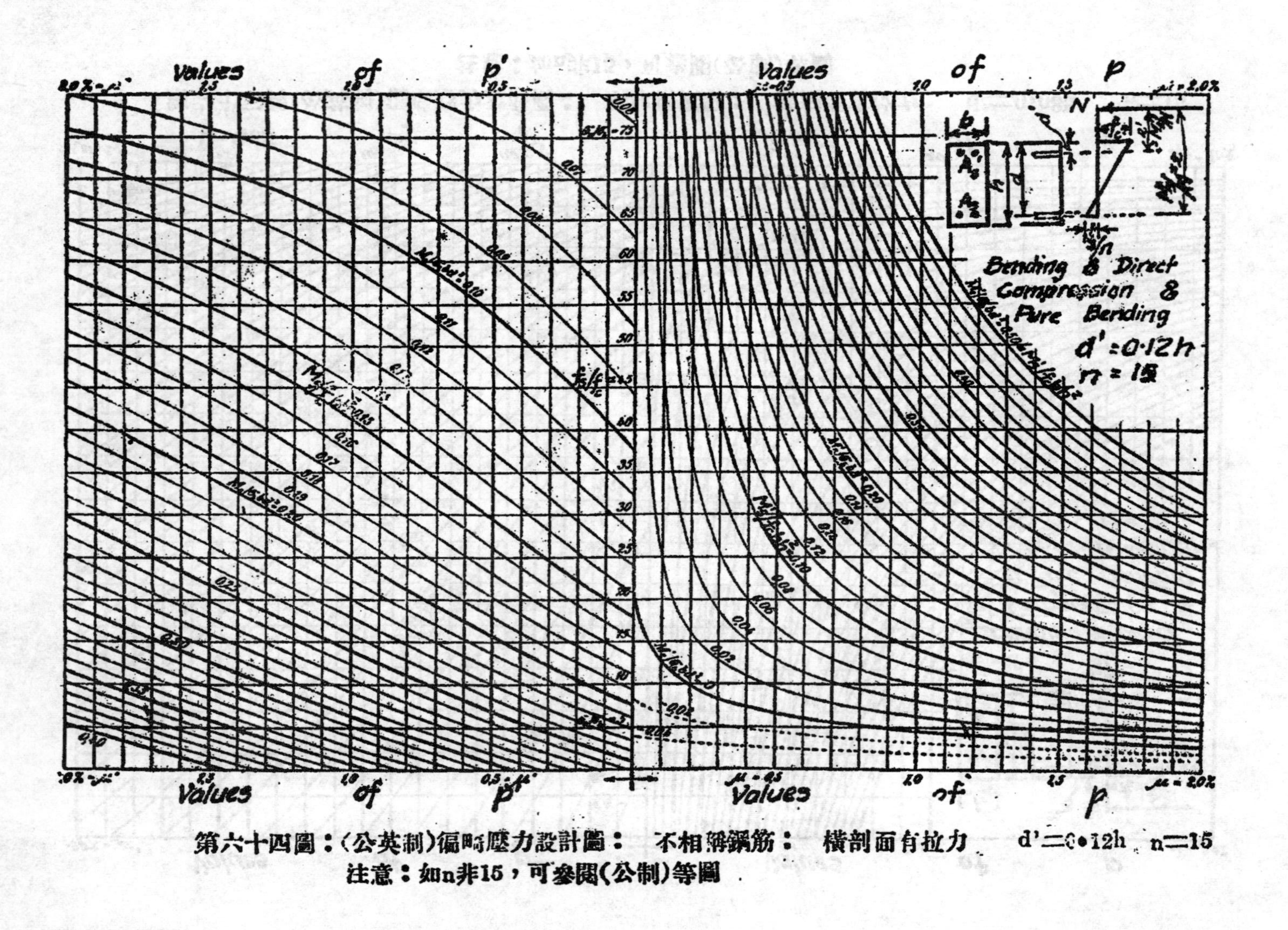

第六十四圖：（公英制）偏畸壓力設計圖：　不相稱鋼筋：　橫剖面有拉力　　d'＝0•12h　n＝15

注意：如n非15，可參閱（公制）等圖

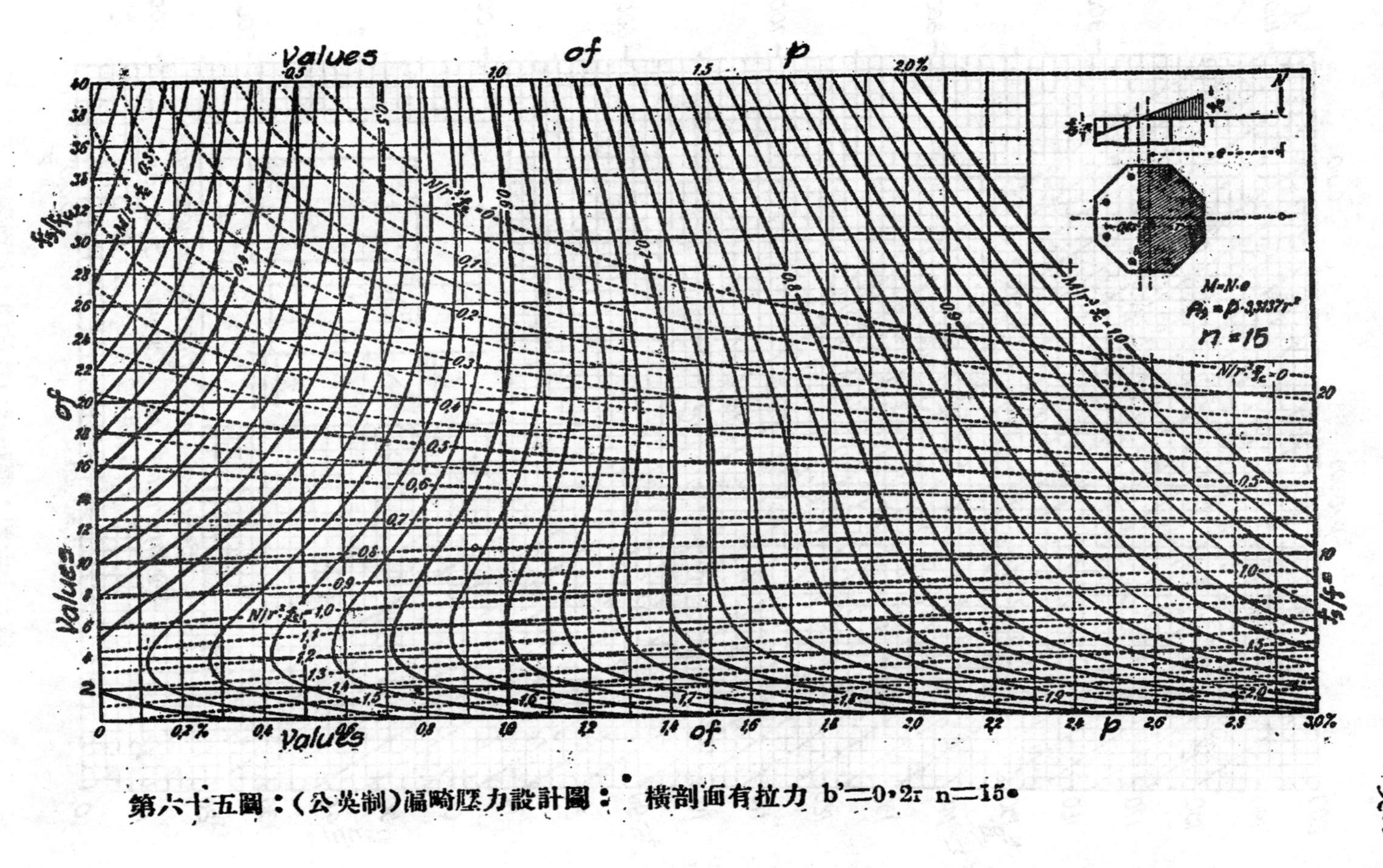

第六十五圖：（公英制）偏畸壓力設計圖： 橫剖面有拉力 b'＝0•2r n＝15•

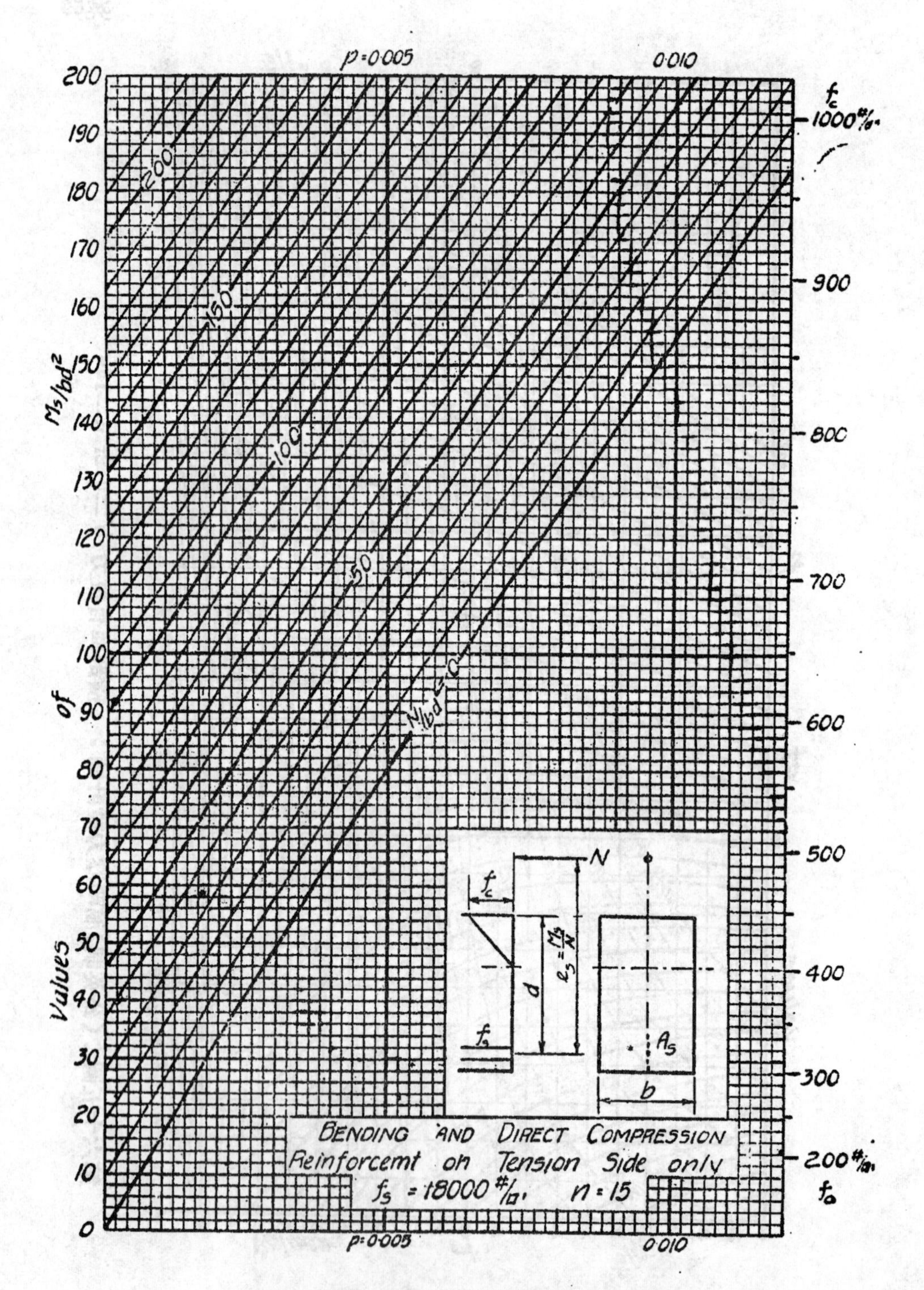

第六十六圖：偏畸壓力設計圖：單面鋼筋

橫剖面有拉力 fs＝18000，n＝15

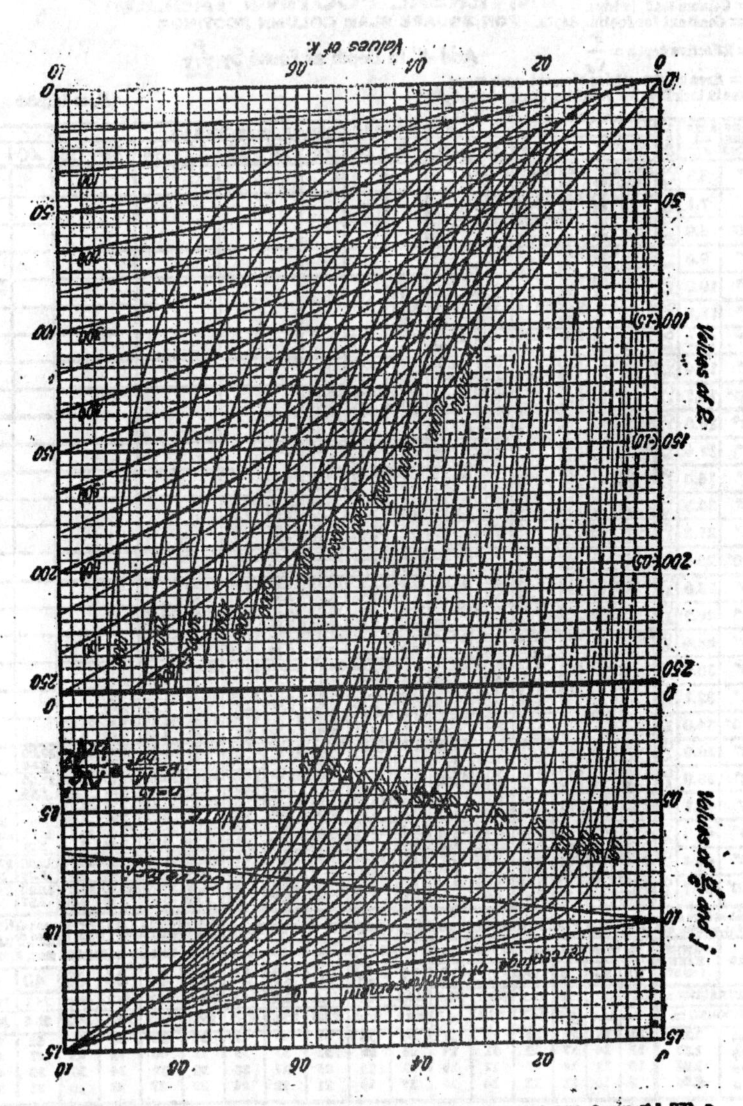

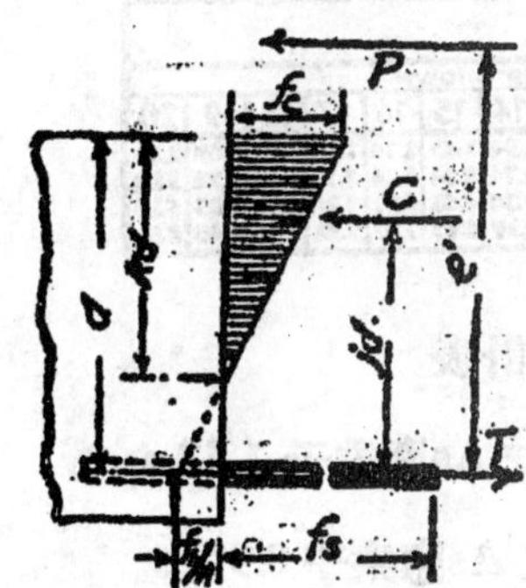

第六十七表(英制)：偏畸壓力設計圖：

單面鋼筋。 n=15

$$R=\frac{M}{bd^2}=\frac{Ne_s}{bd^2}$$

注意：偏畸壓力 P與鋼筋中心之距離爲 e_s。

UNIVERSAL FOOTING TABLES

FOR SQUARE SLAB COLUMN FOOTINGS

P = Column load, in kips.

C_d = Constant for footing depth.

d = Effective depth = $\frac{P}{C_d}$

Add 4" to Depth as Found by $\frac{P}{Cd}$

A_s = Area steel req'd for Moment, each way;
latter in inch lbs, for punching shear of 120 $/□"

F_s = 18,000

SIDE OF SQUARE FOOTING	AREA SQ. FT.	CONST	10	12	14	16	18	20	22	24	26	28	30	32	34	36	38	40	42
			SIZE SQUARE COLUMN, IN INCHES (C)																
2'-6"	6.3	Cd	5.40	6.85	8.60	10.70	13.50												
		As	.45	.47	.48	.47	.45												
2'-8"	7.1	Cd	5.34	6.70	8.33	10.20	12.60												
		As	.50	.53	.55	.55	.53												
2'-10"	8.0	Cd	5.26	6.57	8.10	9.85	12.00	14.70											
		As	.55	.59	.62	.62	.62	.58											
3'-0"	9.0	Cd	5.20	6.48	7.95	9.57	11.47	13.86											
		As	.60	.65	.68	.70	.70	.67											
3'-2"	10.0	Cd	5.15	6.40	7.80	9.35	11.15	13.25	15.90										
		As	.65	.71	.75	.78	.79	.77	.73										
3'-4"	11.1	Cd	5.12	6.33	7.63	9.15	10.84	12.75	15.15										
		As	.70	.77	.82	.86	.87	.86	.83										
3'-6"	12.3	Cd	5.09	6.25	7.56	9.00	10.56	12.40	14.60	17.00									
		As	.75	.83	.89	.93	.96	.96	.93	.90									
3'-8"	13.4	Cd	5.06	6.21	7.50	8.85	10.37	12.10	14.05	16.40									
		As	.80	.89	.96	1.01	1.04	1.05	1.04	1.01									
3'-10"	14.7	Cd	5.04	6.17	7.43	8.72	10.20	11.80	13.70	15.85	18.30								
		As	.85	.95	1.02	1.09	1.13	1.14	1.14	1.12	1.08								
4'-0"	16.0	Cd	5.02	6.14	7.35	8.62	10.02	11.60	13.40	15.30	17.70								
		As	.90	1.01	1.09	1.16	1.21	1.24	1.24	1.23	1.20								
4'-2"	17.4	Cd	5.00	6.10	7.28	8.55	9.90	11.42	13.05	15.00	17.15	19.50							
		As	.95	1.07	1.16	1.24	1.30	1.33	1.35	1.35	1.32	1.29							
4'-4"	18.8	Cd	4.98	6.07	7.25	8.48	9.80	11.28	12.80	14.67	16.60	18.90							
		As	1.00	1.13	1.23	1.32	1.39	1.43	1.49	1.46	1.44	1.41							
4'-6"	20.3	Cd	4.97	6.05	7.22	8.44	9.70	11.10	12.65	14.35	16.30	18.40	20.80						
		As	1.05	1.19	1.30	1.39	1.47	1.52	1.55	1.57	1.56	1.54	1.50						
4'-8"	21.8	Cd	4.96	6.03	7.18	8.36	9.65	11.00	12.47	14.10	15.90	17.85	20.20						
		As	1.10	1.25	1.36	1.47	1.56	1.61	1.65	1.68	1.69	1.67	1.64						
4'-10"	23.4	Cd	4.95	6.01	7.14	8.32	9.59	10.90	12.23	13.90	15.60	17.50	19.75	22.00					
		As	1.15	1.31	1.43	1.55	1.64	1.71	1.76	1.80	1.81	1.80	1.78	1.73					
5'-0"	25.0	Cd	4.94	6.00	7.10	8.28	9.53	10.80	12.20	13.70	15.35	17.20	19.20	21.40					
		As	1.20	1.37	1.50	1.63	1.73	1.80	1.86	1.91	1.93	1.93	1.92	1.88					
5'-2"	26.7	Cd	4.94	5.98	7.08	8.24	9.38	10.72	12.07	13.55	15.13	16.95	18.80	20.90	23.40				
		As	1.25	1.43	1.57	1.70	1.81	1.90	1.96	2.02	2.05	2.06	2.05	2.02	2.00				
5'-4"	28.4	Cd	4.93	5.97	7.06	8.20	9.34	10.65	11.95	13.40	14.92	16.60	18.40	20.40	22.60				
		As	1.30	1.49	1.64	1.78	1.90	1.99	2.07	2.13	2.17	2.19	2.19	2.17	2.15				
5'-6"	30.3	Cd	4.92	5.96	7.04	8.17	9.31	10.57	11.87	13.26	14.76	16.40	18.25	20.00	22.20	24.60			
		As	1.35	1.55	1.70	1.86	1.98	2.08	2.17	2.25	2.29	2.32	2.33	2.31	2.30	2.26			
5'-8"	32.1	Cd	4.92	5.95	7.02	8.14	9.28	10.50	11.80	13.12	14.60	16.20	17.90	19.70	21.70	24.00			
		As	1.40	1.61	1.77	1.93	2.07	2.18	2.27	2.36	2.41	2.45	2.47	2.46	2.45	2.42			
5'-10"	34.0	Cd	4.91	5.93	7.00	8.11	9.25	10.44	11.73	13.03	14.45	16.02	17.65	19.45	21.35	23.50	25.80		
		As	1.45	1.67	1.84	2.01	2.16	2.27	2.38	2.47	2.53	2.58	2.61	2.61	2.60	2.59	2.54		
6'-0"	36.0	Cd	4.90	5.92	6.98	8.08	9.22	10.38	11.66	12.94	14.30	15.85	17.40	19.20	21.00	23.00	25.40	27.80	
		As	1.50	1.73	1.91	2.09	2.24	2.37	2.48	2.58	2.65	2.70	2.75	2.76	2.76	2.75	2.71	2.68	
6'-2"	38.0	Cd	4.90	5.91	6.97	8.06	9.19	10.34	11.59	12.87	14.21	15.67	17.20	18.95	20.70	22.60	24.80	27.20	
		As	1.55	1.79	1.98	2.16	2.33	2.46	2.58	2.70	2.77	2.83	2.89	2.90	2.92	2.91	2.89	2.86	
6'-4"	40.1	Cd	4.90	5.90	6.96	8.04	9.16	10.30	11.52	12.80	14.12	15.50	17.05	18.70	20.40	22.20	24.25	26.60	29.00
		As	1.60	1.85	2.04	2.24	2.41	2.55	2.68	2.81	2.90	2.96	3.02	3.04	3.07	3.07	3.06	3.04	2.98
6'-6"	42.3	Cd	4.89	5.90	6.95	8.02	9.13	10.28	11.46	12.75	14.04	15.40	16.90	18.50	20.15	21.90	23.95	26.00	28.50
		As	1.65	1.91	2.11	2.32	2.50	2.65	2.79	2.92	3.02	3.09	3.16	3.19	3.22	3.24	3.23	3.22	3.16
6'-8"	44.4	Cd	4.89	5.89	6.94	8.00	9.10	10.24	11.40	12.70	13.96	15.30	16.75	18.30	19.90	21.65	23.60	25.50	27.80
		As	1.70	1.97	2.18	2.40	2.58	2.74	2.89	3.03	3.14	3.22	3.30	3.34	3.37	3.40	3.41	3.39	3.35
6'-10"	46.7	Cd	4.88	5.88	6.93	7.98	9.08	10.20	11.35	12.65	13.88	15.20	16.60	18.15	19.70	21.40	23.25	25.20	27.30
		As	1.75	2.03	2.25	2.47	2.67	2.84	2.99	3.15	3.26	3.35	3.44	3.48	3.53	3.56	3.59	3.57	3.55

If hooked bars are used, use half the number of bars shown. TOTAL PERIMETER BARS (Σ_0) REQ'D E. W. FOR BOND. Bond stress 75 $/□" for straight def. bar. Bond stress 150 $/□" for hooked def. bar.

SIZE	PERIMETER 1 BAR	10	12	14	16	18	20	22	24	26	28	30	32	34	36	38	40	42
		Σ_0 Given for Column Sizes Shown, for Both Straight and Hooked Bars. Number Given is for Straight Bars, E. W.																
Σ_0 STRAIGHT =		18.4	22.0	25.6	29.3	33.0	36.6	40.3	43.9	47.6	51.2	54.9	58.5	62.2	65.8	69.5	73.1	76.8
Σ_0 HOOKED =		9.2	11.0	12.8	14.7	16.5	18.3	20.2	22.0	23.8	25.6	27.5	29.3	31.1	32.9	34.8	36.6	38.4
⅜φ	1.18	16	19	22	25	28	31	34	38	41	44	47	50	53	56	59	62	65
½φ	1.57	12	14	17	19	21	24	26	28	31	33	35	38	40	42	45	47	49
⅝φ	1.96	10	12	14	15	17	19	21	23	25	27	28	30	32	34	36	38	40
¾φ	2.36	8	10	11	13	14	16	17	19	21	22	24	25	27	28	30	31	33

AREAS OF RENIFORCING BARS

SIZE	AREA 1 BAR	WT. 1 BAR	1	2	3	4	5	6	7	8	9	10	11	12	13	14	15	16	17	18	19	20
			AREAS OF VARIOUS NUMBERS OF BARS SHOWN																			
⅜φ	.11	.38	.11	.22	.33	.44	.55	.66	.77	.88	.99	1.10	1.21	1.32	1.43	1.54	1.65	1.76	1.87	1.98	2.09	2.20
½φ	.20	.68	.20	.39	.59	.78	.98	1.18	1.37	1.57	1.76	1.96	2.16	2.35	2.55	2.74	2.94	3.14	3.33	3.53	3.72	3.92
⅝φ	.31	1.06	.31	.61	.92	1.23	1.54	1.84	2.15	2.46	2.76	3.07	3.38	3.68	3.99	4.30	4.61	4.91	5.22	5.53	5.83	6.14
¾φ	.44	1.52	.44	.88	1.33	1.77	2.21	2.65	3.09	3.54	3.98	4.42	4.86	5.30	5.75	6.19	6.63	7.07	7.51	7.96	8.40	8.84

第六十八表(英制)　正方柱基設計表

fs＝18000　v_p＝120　如fs＝16,000　表中之As應乘于 1.12。

如fs＝20,000　表中之As應乘于 0.9

P = Column load, in kips.
C_d = Constant for footing depth.
d = Effective depth = $\frac{P}{Cd}$
A_s = Area steel req'd for Moment, each way; latter in inch lbs., for punching shear of 120 #/□".

UNIVERSAL FOOTING TABLES

FOR SQUARE SLAB COLUMN FOOTINGS

Add 4" to Depth as Found by $\frac{P}{Cd}$

F_s = 18,000

SIZE OF SQUARE FOOTING	AREA SQ. FT.	CONST.	SIZE SQUARE COLUMN IN INCHES. (C) 12	14	16	18	20	22	24	26	28	30	32	34	36	38	40	42	44	46	48
7'-0"	49.0	Cd	5.88	6.91	7.97	9.06	10.16	11.31	12.57	13.80	15.10	16.50	17.90	19.50	21.20	22.95	24.85	26.80	29.20	31.60	—
		As	2.06	2.31	2.53	2.75	2.93	3.10	3.26	3.38	3.48	3.58	3.63	3.68	3.72	3.75	3.75	3.71	3.70	3.63	
7'-2"	51.4	Cd	5.87	6.90	7.96	9.04	10.14	11.28	12.50	13.74	15.00	16.40	17.75	19.30	21.00	22.70	24.50	26.40	28.60	31.00	
		As	2.12	2.38	2.63	2.84	3.03	3.21	3.38	3.51	3.62	3.72	3.79	3.84	3.89	3.93	3.94	3.91	3.90	3.86	
7'-4"	53.6	Cd	5.87	6.89	7.95	9.02	10.12	11.25	12.45	13.66	14.92	16.30	17.63	19.15	20.80	22.45	24.20	26.10	28.10	30.40	32.90
		As	2.18	2.45	2.71	2.93	3.13	3.32	3.50	3.63	3.75	3.87	3.94	4.01	4.06	4.11	4.13	4.11	4.10	4.08	4.02
7'-6"	56.3	Cd	5.86	6.88	7.94	9.00	10.10	11.22	12.40	13.58	14.85	16.22	17.55	19.02	20.60	22.20	23.95	25.80	27.80	29.90	32.20
		As	2.24	2.52	2.79	3.02	3.23	3.43	3.62	3.76	3.89	4.01	4.10	4.17	4.23	4.29	4.31	4.30	4.30	4.29	4.24
7'-8"	58.8	Cd	5.86	6.88	7.93	8.98	10.08	11.18	12.35	13.52	14.78	16.13	17.48	18.90	20.45	22.00	23.63	25.50	27.40	29.30	31.80
		As	2.30	2.59	2.87	3.11	3.33	3.53	3.73	3.89	4.02	4.16	4.25	4.33	4.40	4.47	4.50	4.51	4.51	4.50	4.46
7'-10"	61.4	Cd	5.86	6.87	7.92	8.97	10.06	11.16	12.31	13.47	14.72	16.05	17.40	18.80	20.30	21.80	23.40	25.20	27.00	29.10	31.30
		As	2.36	2.66	2.93	3.20	3.43	3.64	3.85	4.01	4.16	4.30	4.41	4.50	4.58	4.65	4.69	4.70	4.72	4.72	4.68
8'-0"	64.0	Cd	5.85	6.87	7.91	8.96	10.04	11.14	12.28	13.43	14.66	15.98	17.31	18.70	20.15	21.65	23.20	24.90	26.70	28.70	30.60
		As	2.42	2.73	3.03	3.29	3.52	3.75	3.97	4.14	4.30	4.45	4.56	4.66	4.75	4.83	4.88	4.90	4.92	4.93	4.90
8'-2"	66.7	Cd	5.85	6.86	7.90	8.95	10.02	11.12	12.23	13.40	14.60	15.92	17.23	18.60	20.00	21.50	23.05	24.60	26.40	28.30	30.30
		As	2.48	2.80	3.11	3.38	3.62	3.86	4.09	4.27	4.43	4.59	4.72	4.82	4.92	5.01	5.07	5.10	5.12	5.14	5.12
8'-4"	69.4	Cd	5.85	6.86	7.89	8.94	10.00	11.10	12.20	13.37	14.53	15.85	17.15	18.50	19.85	21.38	22.90	24.40	26.15	28.00	30.00
		As	2.54	2.87	3.19	3.47	3.72	3.97	4.21	4.39	4.57	4.74	4.87	4.99	5.09	5.19	5.25	5.29	5.32	5.36	5.33
8'-6"	72.3	Cd	5.85	6.85	7.89	8.93	9.98	11.08	12.18	13.33	14.53	15.78	17.08	18.40	19.75	21.20	22.70	24.20	25.90	27.70	29.70
		As	2.60	2.94	3.27	3.56	3.82	4.08	4.33	4.58	4.70	4.88	5.03	5.15	5.26	5.37	5.44	5.49	5.53	5.57	5.55
8'-8"	75.1	Cd	5.84	6.85	7.88	8.92	9.96	11.06	12.16	13.30	14.50	15.70	17.00	18.30	19.64	21.05	22.55	24.05	25.70	27.40	29.40
		As	2.66	3.01	3.35	3.65	3.92	4.19	4.45	4.63	4.84	5.03	5.18	5.32	5.43	5.55	5.65	5.69	5.73	5.79	5.77
8'-10"	78.0	Cd	5.84	6.84	7.87	8.91	9.95	11.04	12.14	13.26	14.44	15.62	16.92	18.20	19.52	20.90	22.40	23.90	25.50	27.15	29.05
		As	2.72	3.08	3.43	3.74	4.02	4.29	4.56	4.77	4.96	5.17	5.34	5.48	5.60	5.73	5.82	5.89	5.93	6.00	5.99
9'-0"	81.0	Cd	5.83	6.84	7.86	8.90	9.94	11.02	12.12	13.24	14.40	15.56	16.82	18.10	19.40	20.80	22.22	23.70	25.30	26.90	28.75
		As	2.78	3.15	3.51	3.83	4.12	4.40	4.68	4.90	5.11	5.31	5.49	5.64	5.77	5.91	6.01	6.09	6.14	6.21	6.21
9'-2"	84.0	Cd	5.83	6.84	7.86	8.89	9.93	11.00	12.10	13.21	14.36	15.51	16.75	18.00	19.34	20.71	22.05	23.55	25.10	26.70	28.50
		As	2.84	3.22	3.59	3.92	4.22	4.51	4.80	5.02	5.23	5.46	5.65	5.81	5.94	6.09	6.19	6.28	6.34	6.43	6.43
9'-4"	87.1	Cd	5.82	6.83	7.85	8.88	9.92	10.98	12.08	13.18	14.33	15.47	16.70	17.94	19.27	20.62	21.90	23.42	24.95	26.50	28.25
		As	2.90	3.29	3.67	4.01	4.32	4.62	4.92	5.15	5.38	5.60	5.80	5.97	6.11	6.27	6.38	6.48	6.54	6.64	6.65
9'-6"	90.3	Cd	5.82	6.83	7.85	8.87	9.91	10.96	12.06	13.15	14.30	15.43	16.65	17.88	19.20	20.54	21.80	23.30	24.80	26.35	28.05
		As	2.96	3.36	3.75	4.10	4.42	4.73	5.04	5.28	5.52	5.75	5.96	6.13	6.29	6.45	6.57	6.68	6.75	6.85	6.87
9'-8"	93.4	Cd	5.81	6.82	7.85	8.86	9.90	10.94	12.04	13.14	14.27	15.39	16.60	17.82	19.14	20.45	21.70	23.20	24.68	26.20	27.83
		As	3.02	3.43	3.83	4.19	4.51	4.84	5.16	5.40	5.66	5.89	6.11	6.30	6.46	6.63	6.76	6.88	6.95	7.07	7.09
9'-10"	96.7	Cd	5.81	6.82	7.84	8.85	9.89	10.92	12.02	13.12	14.24	15.35	16.56	17.77	19.07	20.37	21.62	23.10	24.53	26.05	27.63
		As	3.08	3.50	3.91	4.28	4.61	4.94	5.27	5.53	5.79	6.04	6.27	6.46	6.63	6.81	6.95	7.06	7.15	7.28	7.31
10'-0"	100.0	Cd	5.80	6.81	7.84	8.84	9.88	10.90	12.00	13.10	14.21	15.31	16.52	17.72	19.00	20.30	21.53	23.00	24.40	25.85	27.45
		As	3.14	3.57	3.99	4.37	4.71	5.05	5.39	5.66	5.93	6.18	6.42	6.62	6.80	6.99	7.13	7.27	7.35	7.49	7.52
10'-2"	103.4	Cd	5.80	6.81	7.83	8.84	9.87	10.88	11.99	13.08	14.18	15.27	16.48	17.68	18.94	20.22	21.46	22.90	24.30	25.70	27.25
		As	3.20	3.64	4.07	4.46	4.81	5.16	5.51	5.78	6.06	6.33	6.58	6.79	6.97	7.17	7.32	7.47	7.56	7.71	7.74
10'-4"	106.8	Cd	5.80	6.81	7.83	8.83	9.86	10.87	11.98	13.06	14.16	15.24	16.44	17.64	18.87	20.15	21.40	22.80	24.20	25.55	27.05
		As	3.26	3.71	4.15	4.55	4.91	5.27	5.63	5.91	6.20	6.47	6.73	6.95	7.14	7.35	7.51	7.67	7.76	7.92	7.96
10'-6"	110.3	Cd	5.80	6.80	7.82	8.83	9.85	10.86	11.97	13.04	14.14	15.22	16.40	17.60	18.80	20.10	21.32	22.70	24.10	25.42	26.90
		As	3.34	3.78	4.23	4.64	5.01	5.38	5.75	6.04	6.34	6.61	6.89	7.11	7.31	7.53	7.70	7.87	7.96	8.13	8.18
10'-8"	113.8	Cd	5.80	6.80	7.82	8.82	9.85	10.86	11.96	13.02	14.12	15.20	16.36	17.55	18.75	20.05	21.25	22.63	24.00	25.30	26.75
		As	3.40	3.85	4.31	4.73	5.11	5.49	5.87	6.16	6.47	6.76	7.04	7.28	7.48	7.71	7.89	8.07	8.17	8.35	8.40
10'-10"	117.4	Cd	5.80	6.80	7.81	8.82	9.84	10.85	11.95	13.00	14.10	15.18	16.33	17.50	18.70	20.00	21.20	22.55	23.90	25.20	26.63
		As	3.46	3.92	4.39	4.82	5.21	5.60	5.98	6.29	6.61	6.90	7.19	7.44	7.65	7.90	8.07	8.26	8.37	8.56	8.62

If hooked bars are used, use half the number of bars shown.

TOTAL PERIMETER BARS (Σ_0) REQ'D E. W. FOR BOND

Bond stress 75 #/□" for straight bars
Bond stress 150 #/□" for hooked bars

SIZE	PERIMETER 1 BAR	Σ_0 Given for Column Sizes Shown, Both Straight and Hooked Bars. Number Given is for Straight Bars, E. W. 12	14	16	18	20	22	24	26	28	30	32	34	36	38	40	42	44	46	48
Σ_0 STRAIGHT =		22.0	25.6	29.3	33.0	36.6	40.3	43.9	47.6	51.2	54.9	58.5	62.2	65.8	69.5	73.1	76.8	80.4	84.9	87.7
Σ_0 HOOKED =		11.0	12.8	14.7	16.5	18.3	20.2	22.0	23.8	25.6	27.5	29.3	31.1	32.9	34.8	36.6	38.4	40.2	42.0	43.9
3/8φ	1.18	19	22	25	28	31	34	38	41	44	47	50	53	56	59	62	65	69	72	75
1/2φ	1.57	14	17	19	21	24	26	28	31	33	35	38	40	42	45	47	49	52	54	56
5/8φ	1.96	12	14	15	17	19	21	23	25	27	28	30	32	34	36	38	40	41	43	45
3/4φ	2.36	10	11	13	14	16	17	19	21	22	24	25	27	28	30	31	33	34	36	38

AREAS OF REINFORCING BARS.

SIZE	AREA 1 BAR	WT. 1 BAR	AREAS OF VARIOUS NUMBERS OF BARS SHOWN 10	11	12	13	14	15	16	17	18	19	20	21	22	23	24	25	26
3/8φ	.11	.38	1.10	1.21	1.32	1.43	1.54	1.65	1.76	1.87	1.98	2.09	2.20	2.31	2.42	2.53	2.64	2.75	2.86
1/2φ	.20	.68	1.96	2.16	2.35	2.55	2.74	2.94	3.14	3.33	3.53	3.72	3.92	4.12	4.31	4.51	4.70	4.90	5.10
5/8φ	.31	1.06	3.07	3.38	3.68	3.99	4.30	4.61	4.91	5.22	5.53	5.83	6.14	6.45	6.75	7.06	7.37	7.68	7.98
3/4φ	.44	1.52	4.42	4.86	5.30	5.75	6.19	6.63	7.07	7.51	7.96	8.40	8.84	9.28	9.72	10.17	10.61	11.05	11.49

第六十九表(英制)　正方柱基設計表

fs＝18000　vp＝120　如fs＝16000　表中之 As 應乘于1.120

如fs＝20,000表中之 As應乘于0.90

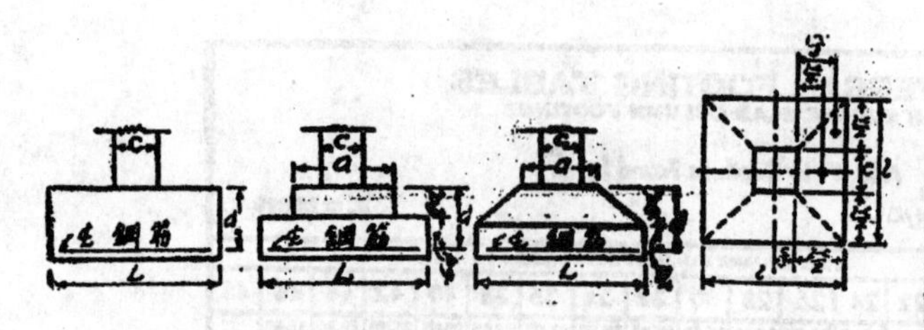

第七十表(公英制)

K₁ 與 K'₁ 之值							
c/L	K_1	c/L	K_1	c/L	K_1	c/L	K_1
•06	•249	•20	•240	•34	•221	•48	•192
•07	•249	•21	•239	•35	•220	•49	•190
•08	•248	•22	•238	•36	•218	•50	•188
•09	•248	•23	•237	•37	•216	•52	•182
•10	•248	•24	•236	•38	•214	•54	•177
•11	•247	•25	•234	•39	•212	•56	•172
•12	•246	•26	•233	•40	•210	•58	•166
•13	•246	•27	•232	•41	•208	•60	•160
•14	•245	•28	231	•42	•206	•62	•154
•15	•244	•29	•229	•43	•204	•64	•148
•16	•244	•30	•227	•44	•202	•66	•141
•17	•243	31	•226	•45	•199	•68	•134
•18	•242	•32	•224	•46	•197	•70	•128
•19	•241	•33	•223	•47	•195	•72	•120

第七十一表(公英制)

$\frac{c}{L}$	級分柱基 "a"	斜面柱基 "a"	斜面柱基 "d_2"
0•10	0•36L	0•30L	0•2d
0•15	0•38L	0•31L	0•2d
0•20	0•40L	0•32L	0•2d
0•25	0•45L	0•34L	0•2d
0•30	0•50L	0•35L	0•2d
0•35	0•52L	0•40L	0•2d
0•40	0•55L	0•45L	0•2d

〔平面地基〕 設有方柱其邊長， C＝28''×28'' 荷重 P＝350,000 磅

地質安承力 ＝4000 磅一方呎 fs＝18000 v＝40

設地基重量為0.1×350,000＝35000 (地基重百份數約由.08至.12)

承托面積約$=\frac{350,000+35,000}{4000}=$96.2方呎

用邊長10 呎 方基

由第六十九表得$C_d=14.21$

有效深度，$d=\frac{P}{C_d}=\frac{350,000}{14.21}=24.6''$ 總深＝24.6＋4＝29''

$A_s=5.93$方吋

(用20根1/2吋圓鋼筋)

倘用竹節直鋼筋准許粘合力為75磅一方吋，則鋼筋圓周之總數由表中 28 吋方柱之欄得51.2方吋共需33根1/2吋圓鋼筋。此數比之用動率求得者多 4根。

若用屈曲鋼筋，單位粘合力可用至 150 磅一方吋 。 圓周 ＝ 25.6 吋，共需 17 根1/2 吋圓鋼筋。有時地基可以加深，以減小斜引力，或減少粘合力，或減少鋼筋面積。今設用有效深度30吋，則所需鋼筋面積為$As=\frac{24•6}{30}\times5•93=4•87$吋方需10根1/2吋圓。

粘力所需之鋼筋為：$\left(\frac{24•6}{30}\times51•2\right)\div1•57=$27根1/2吋圓

25508

〔分級柱基〕

設柱形及荷重等如上述，而柱基在柱邊截面處之有效深度為24.6吋 $\frac{c}{L}=\frac{28}{120}=.238$.

由第七十一表得柱基上層有效寬度＝•45L ＝54吋

柱基下層之有效厚度＝0.6d ＝0.6×24.6吋 ＝14.75吋

在柱邊截面處所有動率鋼筋面積及粘合力均與平面柱基相催

在上層基邊之粘合力必須顧及。如上層用寬度54吋

從第七十二表得圓周＝98•6(直鋼筋)或＝49•3(曲鋼筋)。

如欲將圓周減少，上層之寬度必須增加。

在下層c＋2d或77吋處之斜引力必須計算之。有效深度＝24•6吋，

下層之深度＝14•75吋 $\frac{c+2d}{L}=\frac{77}{120}=\cdot 64$，由第七十一表得 K_1 ＝0•143

$$u=\frac{\cdot 143\times 350,000}{77\times \cdot 875\times 14\cdot 75}=50\cdot 3\text{磅一方吋}$$

如用曲鋼筋，則此數已適合，苟用直鋼筋，則下層之厚度必須增加。

〔斜面柱基〕

斜面柱基之設計約與其他相同，但 $\frac{c}{L}=\frac{28}{120}=\cdot 234$ 基頂寬度＝•34L＝40吋。

在c＋2d＝77吋之處，其斜引力必需顧及。$u=\frac{K_1P}{(c+2d)jd}$，$d_1=\frac{K_1P}{(c+2d)jv}$

如用曲鋼筋則 $d_1=\frac{\cdot 143\times 350000}{77\times \cdot 875\times 60}=11\cdot 94$吋

第七二表(英制)粘合力所需之直鋼筋個數如用曲鋼筋表中個數可折半

鋼		筋	柱之邊長	
尺寸		每根鋼筋之圓周	54	60
圓周直筋＝			98•6	109•6
圓周曲筋＝			49•3	54•8
1/2	圓	1•57	63	70
5/8	圓	1•96	51	56
3/4	圓	2•36	42	47
7/8	圓	2•75	36	40

矩形柱基

$$K3=\frac{1}{2}\left(1-\frac{e}{L}\cdot\frac{c}{s}\right)$$

$$K4=\frac{1}{4}\left(1-\frac{e}{L}\right)\left(1-\frac{c}{s}\right)$$

$$K5=\frac{1}{24}\left(1-\frac{e}{L}\right)^2\left(2+\frac{c}{s}\right)$$

有效深度 $=d=K_3\dfrac{P}{(c+e)v}$

$$As=\frac{K_5\,PL}{jdts}\qquad u=\frac{K_4\,P}{\Sigma ojd}$$

$$v=\frac{K_4\,P}{(c+2d)jd}$$

第七十三表(公英制)　　K_3 之值

e/L	c/s 之值												
	•10	•15	•20	•25	•30	•35	•40	•45	•50	•55	•60	•65	•70
•10	•49	•49	•49	•49	•48	•48	•48	•48	•47	•47	•47	•47	•46
•15	•49	•49	•48	•48	•48	•47	•47	•47	•46	•46	•45	•45	•45
•20	•49	•48	•48	•47	•47	•46	•46	•45	•45	•44	•44	•43	•43
•25	•49	•48	•47	•47	•46	•46	•45	•44	•44	•43	•42	•42	•41
•30	•48	•48	•47	•46	•45	•45	•44	•43	•42	•42	•41	•40	•39
•35	•48	•47	•46	•46	•45	•44	•43	•42	•41	•40	•39	•39	•38
•40	•48	•47	•46	•45	•44	•43	•42	•41	•40	•39	•38	•37	•36
•45	•48	•47	•46	•44	•43	•42	•41	•40	•39	•38	•36	•35	•34
•50	•47	•46	•45	•44	•42	•41	•40	•39	•37	•36	•35	•34	•32
•55	•47	•46	•44	•43	•42	•40	•39	•38	•36	•35	•33	•32	•31
•60	•47	•45	•44	•42	•41	•39	•38	•36	•35	•33	•32	•30	•29
•65	•47	•45	•43	•42	•40	•39	•37	•35	•34	•32	•30	•29	•27
•70	•46	•45	•43	•41	•39	•38	•36	•34	•32	•31	•29	•27	•25

第七十四表(公英制)　　K_4 之值

e/L	c/s 之值												
	•10	•15	•20	•25	•30	•35	•40	•45	•50	•55	•60	•65	•70
•10	•25	•23	•22	•20	•19	•18	•16	•15	•14	•12	•11	•09	•98
•15	•26	•24	•23	•21	•20	•19	•17	•16	•14	•13	•11	•10	•08
•20	•27	•25	•24	•22	•21	•19	•18	•16	•15	•13	•12	•10	•09
•25	•28	•26	•25	•23	•22	•20	•19	•17	•15	•14	•12	•11	•09
•30	•29	•27	•26	•24	•23	•21	•19	•18	•16	•14	•13	•11	•10
•35	•30	•29	•27	•25	•24	•22	•20	•19	•17	•15	•13	•12	•10
•40	•31	•30	•28	•26	•24	•23	•21	•20	•17	•16	•14	•12	•10
•45	•33	•31	•29	•27	•25	•23	•22	•20	•18	•16	•14	•13	•11
•50	•34	•32	•30	•28	•26	•24	•22	•21	•19	•17	•15	•13	•11
•55	•35	•33	•31	•29	•27	•25	•23	•22	•19	•17	•15	•13	•11
•60	•36	•34	•32	•30	•28	•26	•24	•22	•20	•18	•16	•14	•12
•65	•37	•35	•33	•31	•29	•27	•25	•23	•20	•18	•16	•14	•12
•70	•38	•36	•34	•32	•30	•27	•25	•23	•21	•19	•17	•15	•13

第七十五表（公英制）　　K_5 之值

$\frac{e}{L}$	c/s 之值 •10	•15	•20	•25	•30	•35	•40	•45	•50
0•10	•071	•073	•074	•076	•078	•079	•081	•083	•085
0•15	•063	•065	•066	•068	•069	•071	•072	•074	•075
0•20	•056	•057	•059	•060	•061	•063	•064	•065	•067
0•25	•049	•050	•052	•053	•054	•055	•056	•057	•059
0•30	•043	•044	•045	•046	•047	•048	•049	•050	•051
0•35	•037	•038	•039	•040	•041	•041	•042	•043	•044
0•40	•032	•032	•033	•034	•035	•035	•036	•037	•038
0•45	•027	•027	•028	•028	•029	•030	•030	•031	•032
0•50	•022	•022	•023	•024	•024	•024	•025	•026	•026
0•55	•018	•018	•019	•019	•019	•020	•020	•021	•021
0•60	•014	•014	•015	•015	•015	•016	•016	•016	•017
0•65	•011	•011	•011	•012	•012	•012	•012	•013	•013
0•70	•008	•008	•008	•008	•009	•009	•009	•009	•009

矩形柱基說明

設有22×36"吋之柱荷重爲P＝350000磅。泥土安承力＝4000磅一方呎

承托面積如上法求得約＝100方呎。用9'—6"×10'—6"矩形基。

$$\frac{e}{L}=\frac{36}{126}=•286\qquad \frac{c}{s}=\frac{22}{114}=•193$$

受穿鑽力限制之$d=K_3\dfrac{P}{(c+e)v}=.47\times\dfrac{350,000}{(22+36)120}=23•7$" 用24吋　由第75表得 $K_5=•0466$

縱置鋼筋$As=\dfrac{K_5\ PL}{jdfs}=\dfrac{0•0466\times350•000\times10•5\times12}{.875\times18000\times24}=5•45$方吋

用28根1/2"Q 10'-0，長圓周＝44.0吋

將$\frac{e}{L}=0.286$，$\frac{c}{s}=0.193$互換，即$\frac{e}{L}=0.193$，$\frac{c}{s}=0.286$　　$K_5=.0621$

橫置鋼筋$As=\dfrac{.0621\times350,000\times9.5\times12}{.875\times18000\times24}=6.55$方吋

用34根1/2圓9'-0"，長圓周＝53.4方吋

粘合力在柱之長邊處$\frac{e}{L}=.286$，$\frac{c}{s}=0.193$，$K_4=.26$　圓周＝53.4吋

$$u=\frac{0.26\times 350,000}{53.4\times .875\times 24}=81\text{磅一方吋}$$

在短邊處 $\frac{e}{L}=.193,\ \frac{c}{s}=\cdot 236\qquad K_3=.213$

$$u=\frac{.213\times 350,000}{44.0\times .875\times 24}=81\text{磅一方吋}$$

若用直鋼筋，其准許應力=75磅一方吋，其根數應乘于$\frac{81}{75}$

斜引力：計算此力時仍用 K_4 表但將 $\frac{e+2d}{L}$ 代$\frac{e}{L}$，$\frac{c+2d}{s}$ 代 $\frac{c}{L}$

在柱之長邊 $\frac{c+2d}{L}=\frac{84}{126}=.67\quad K_4=.17$

$$v=\frac{.17\times 350000}{(36\times 48)\times .875\times 24}=33.8\text{磅一方吋}$$

在柱之短邊 $\frac{c+2d}{s}=\frac{70}{114.}=.61\quad K_4=.14$

$$v=\frac{\cdot 14\times 350000}{(22\times 48)\times .875\times 24}=33\cdot 4\text{磅一方吋}$$

就斜引力言，可用直鋼筋。

角柱之重心點

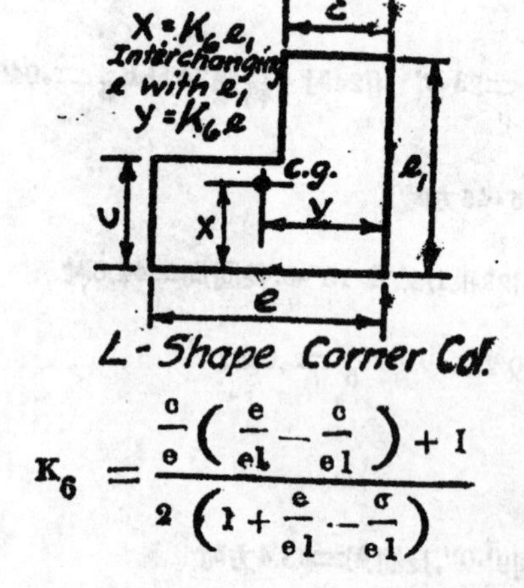

$$K_6=\frac{\frac{c}{e}\left(\frac{e}{e1}-\frac{c}{e1}\right)+1}{2\left(1+\frac{e}{e1}-\frac{c}{e1}\right)}$$

第七十六表(公英制) K6之值

$\frac{c}{e_1}$	c/e_1 之值								
	•20	•25	•30	•35	•40	•45	•50	•55	•60
0•2	•433	•451	•468	•484					
0•4	•385	•403	•419	•434	•450	•464	•477	•489	
0•6	•350	•367	•383	•399	•414	•428	•442	•454	•466
0•8	•322	•339	•356	•372	•387	•402	•416	•430	•443
1•0	•300	•317	•334	•350	•366	•382	•397	•411	•425
1•2	•282	•299	•316	•333	•350	•366	•381	•396	•411
1•4	•266	•284	•302	•319	•336	•352	•369	•384	•400
1•6	•253	•272	•290	•307	•325	•342	•358	•375	•390
1•8	•242	•261	•279	•297	•315	•332	•350	•366	•383
2•0	•233	•252	270	•289	•307	•325	•342	•359	•377
2•4	•225	•244	•263	•281	•300	•318	•336	•353	•371
2•6	•217	•237	•256	•275	•293	•312	•330	•348	•366
2•8	•211	•230	•250	•269	•288	•307	•325	•344	•362

第七十七表(英制) 樓梯現面尺度表

各種梯級高度及踏脚

梯級數目	梯級高度(呎)						
	梯級高						
	6 1/2	6 3/4	7	7 1/8	7 1/4	7 3/8	7 1/2
2	1—1	1—1 1/2	1—2	1—2 1/4	1—2 1/2	1—2 3/4	1—3
3	1—7 1/2	1—8 1/4	1—9	1—9 3/8	1—9 3/4	1—10 1/8	1—10 1/2
4	2—2	2—3	2—4	2—4 1/2	2—5	2—5 1/2	2—6
5	2—8 —/2	2—9 3/4	2—11	2—11 5/8	3—0 1/4	3—0 7/8	3—1 1/2
6	3—3	3—4 1/2	3—6	3—6 3/4	3—7 1/2	3—8 1/4	3—9
7	3—9 1/2	3—11 1/4	4—1	4—1 7/8	4—2 3/4	4—3 5/8	4—4 1/2
8	4—4	4—6	4—8	4—9	4—10	4—11	5—0
9	4—10 1/2	5—0 3/4	5—3	5—4 1/8	5—5 1/4	5—6 3/8	5—7 1/2
10	5—5	5—7 1/2	5—10	5—11 1/4	6—0 1/2	6—1 3/4	6—3
11	5—11 1/2	6—2 1/4	6—5	6—6 3/8	6—7 3/4	6—9 1/8	6—10 1/2
12	6—6	6—9	7—0	7—1 1/2	7—3	7—4 1/2	7—6
13	7—0 1/2	7—3 3/4	7—7	7—8 5/8	7—10 1/4	7—11 7/8	8—1 1/2
14	7—7	7—10 1/2	8—2	8—3 3/4	8—5 1/2	8—7 1/4	8—9
15	8—1 1/2	8—5 1/4	8—9	8—10 7/8	9—0 3/4	9—2 5/8	9—4 1/2
16	8—8	9—0	9—4	9—6	9—8	9—10	10—0
17	9—2 1/2	9—6 3/4	9—11	10—1 1/8	10—3 1/4	10—5 3/8	10—7 1/2
18	9—9	10—1 1/2	10—6	10—8 1/4	10—10 1/2	11—0 3/4	11—3
19	10—3 1/2	10—8 1/4	11—1	11—3 3/8	11—5 3/4	11—8 1/8	11—10 1/2
20	10—10	11—3	11—8	11—10 1/2	12—1	12—3 1/2	12—6

踏脚數目	踏脚之寬度(呎)						
	梯級踏脚寬度						
	9 1/4	9 1/2	9 3/4	10	10 1/4	10 1/2	10 3/4
2	1—6 1/2	1—7	1—7 1/2	1—8	1—8 1/2	1—9	1—9 1/2
3	2—3 3/4	2—4 1/2	2—5 1/4	2—6	2—6 3/4	2—7 1/2	2—8 1/4
4	3—1	3—2	3—3	3—4	3—5	3—6	3—7
5	3—10 1/4	3—11 1/2	4—0 3/4	4—2	4—3 1/4	4—4 1/2	4—5 3/4
6	4—7 1/2	4—9	4—10 1/2	5—0	5—1 1/2	5—3	5—4 1/2
7	5—4 3/4	5—6 1/2	5—8 1/4	5—10	5—11 3/4	6—1 1/2	6—3 1/4
8	6—2	6—4	6—6	6—8	6—10	7—0	7—2
9	6—11 1/4	7—1 1/2	7—3 3/4	7—6	7—8 1/4	7—10 1/2	8—0 3/4
10	7—8 1/2	7—11	8—1 1/2	8—4	8—6 1/2	8—9	8—11 1/2
11	8—5 3/4	8—8 1/2	8—11 1/4	9—2	9—4 3/4	9—7 1/2	9—10 1/4
12	9—3	9—6	9—9	10—0	10—3	10—6	10—9
13	10—0 1/4	10—3 1/2	10—6 3/4	10—10	11—1 1/4	11—4 1/2	11—7 3/4
14	10—9 1/2	11—1	11—4 1/2	11—8	11—11 1/2	12—3	12—6 1/2
15	11—6 3/4	11—10 1/2	12—2 1/4	12—6	12—9 3/4	13—1 1/2	13—5 1/4
16	12—4	12—8	13—0	13—4	13—8	14—0	14—4
17	13—1 1/4	13—5 1/2	13—9 3/4	14—2	14—6 1/4	14—10 1/2	15—2 3/4
18	13—10 1/2	14—3	14—7 1/2	15—0	15—4 1/2	15—9	16—1 1/2
19	14—7 3/4	15—0 1/2	15—5 1/4	15—10	16—2 3/4	16—7 1/2	17—0 1/4
20	15—5	15—10	16—3	16—8	17—1	17—6	17—11

第七十八表 樓梯塊面重度表 （英制）

三合土樓梯塊面每平面方呎重量　　（三合土及批盪重度，活重不計在內）

塊面 t"	無批盪 樓級高度乘于高度 6x12	6¼x11¾	6½x11¼	6¾x10¾	7x10½	7¼x10	7½x9¾	7¾x9½	8x9
3	90	92	96	98	99	104	107	110	114
3.5	97	99	102	106	109	113	116	119	122
4	104	106	110	114	116	121	123	126	132
4.5	111	114	117	121	123	128	130	134	140
5	118	121	125	129	131	136	139	143	148
5.5	125	128	131	135	138	140	146	151	156
6	132	135	139	143	146	151	155	159	165
6.5	139	142	146	150	153	158	163	167	173
7	146	149	154	158	161	167	170	175	181
7.5	153	157	160	166	168	174	178	183	191
8	160	164	168	172	176	182	186	191	199
8.5	167	171	173	180	183	190	194	198	207
9	174	178	182	188	191	198	202	207	215
9.5	181	185	189	197	198	205	210	215	224
10	188	192	197	202	206	212	218	224	232
10.5	195	198	204	210	214	221	225	232	240
11	202	206	212	217	220	228	234	239	248
11.5	209	213	218	225	228	236	241	248	258
12	216	220	225	231	235	244	250	256	266
12.5	223	227	232	238	242	252	258	264	274
13	230	234	239	246	250	259	266	272	283
13.5	237	241	247	253	257	267	274	280	291
14	244	248	254	260	265	275	282	288	300
14.5	251	256	261	268	272	283	289	296	308
15	258	263	268	275	280	290	297	304	317
15.5	265	270	275	282	287	298	305	313	325
16	272	277	283	290	295	306	313	321	333

3/4吋 批 盪

塊面 t"	6x12	6¼x11¾	6½x11¼	6¾x10¾	7x10½	7¼x10	7½x9¾	7¾x9½	8x9
3	96	98	102	105	108	112	114	118	122
3.5	103	105	109	113	115	120	123	126	133
4	110	113	116	121	123	128	130	134	140
4.5	117	120	124	128	130	136	138	142	148
5	124	127	131	135	138	143	147	151	156
5.5	131	134	138	142	145	151	154	158	164
6	138	141	145	150	153	158	162	167	173
6.5	145	148	153	157	160	166	170	174	182
7	152	156	160	164	168	174	177	182	189
7.5	159	163	166	172	175	181	186	191	199
8	166	170	174	179	183	190	192	199	206
8.5	173	177	181	187	189	197	202	206	215
9	180	184	189	195	198	205	209	215	223
9.5	187	192	194	204	205	213	218	223	232
10	194	199	203	209	213	219	225	232	240
10.5	201	205	210	217	221	228	233	239	248
11	208	212	218	221	225	235	241	246	256
11.5	215	219	224	231	235	240	249	256	266
12	222	226	232	238	242	251	256	264	273
12.5	229	233	239	245	249	259	264	272	281
13	234	240	247	253	257	266	272	280	290
13.5	243	247	254	260	264	274	280	288	298
14	250	254	261	267	272	282	288	296	307
14.5	257	262	268	275	279	290	295	304	315
15	264	269	275	282	287	297	303	312	324
15.5	271	276	282	289	294	305	311	321	332
16	278	283	290	297	302	313	319	329	340

第七十九表　鋼合土樓梯設計表　（英制）

T＝塊面厚度　　俱用圓鋼筋

fc＝650　　由鋼筋中心至三合土外面之距離爲1吋

f_s＝18000　　"鋼筋"欄指示鋼筋直徑及其中距

n＝15　　梯級面無批盪

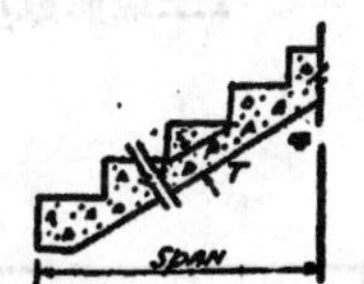

跨度（呎）	活重60磅						活重80磅					
	梯級高度與踏脚之比率						梯級高度與踏脚之比率					
	6 ×12 6 1/4×11 3/4 6 1/2×11 1/4 6 3/4×10 3/4		7 ×10 1/2 7 1/4×10 7 1/2× 9 3/4 7 3/4× 9 1/2		8 × 9		6 ×12 6 1/4×11 3/4 6 1/2×11 1/4 6 3/4×10 3/4		7 ×10 1/2 7 1/4×10 7 1/2× 9 3/4 7 3/4× 9 1/2		8 × 9	
	T	鋼筋	T	鋼筋	T	鋼筋	T	鋼筋	T	鋼筋	T	鋼筋
4	3	3/8—7	3	3/8—7	3	3/8—7	3	3/8—7	3 1/2	3/8—6	3 1/2	3/8—6
5	3 1/2	3/8—6	3 1/2	3/8—6	3 1/2	3/8—6	3 1/2	3/8—6	4	3/8—5	4	3/8—5
6	4	3/8—5	4	3/8—5	4 1/2	1/2—8	4	3/8—5	4 1/2	1/2—8	4 1/2	1/2—8
7	4 1/2	1/2—8	5	1/2—7	5	1/2—7	5	1/2—7	5	1/2—7	5	1/2—7
8	5	1/2—7	5 1/2	1/2—6	5 1/2	1/2—6	5 1/2	1/2—6	5 1/2	1/2—6	5 1/2	1/2—6
9	5 1/2	1/2—6	6	1/2—5 1/2	6	1/2—5 1/2	5 1/2	1/2—6	6	1/2—5 1/2	6 1/2	1/2—5
10	6 1/2	1/2—5	6 1/2	1/2—5	6 1/2	1/2—5	6 1/2	1/2—5	7	5/8—7	7	5/8—7
11	7	5/8—7	7	5/8—7	7 1/2	5/8—6 1/2	7	5/8—7	7 1/2	5/8—6 1/2	7 1/2	5/8—6 1/2
12	7 1/2	5/8—6 1/2	8	5/8—6	8	5/8—6	8	5/8—6	8	5/8—6	8 1/2	5/8—5 1/2
13	8	5/8—6	8 1/2	5/8—5 1/2	9	5/8—5	8 1/2	5/8—5 1/2	9	5/8—5	9	5/8—5
14	9	5/8—5	9 1/2	5/8—5	9 1/2	5/8—5	9	5/8—5	9 1/2	5/8—5	10	3/4—7
15	9 1/2	5/8—5	10	3/4—7	10 1/2	3/4—6 1/2	10	3/4—7	10 1/2	3/4—6 1/2	10 1/2	3/4—6 1/2
16	10 1/2	3/4—6 1/2	11	3/4—6	11	3/4—6	10 1/2	3/4—6 1/2	11	3/4—6	11 1/2	3/4—5 1/2
17	11	3/4—6	11 1/2	3/4—5 1/2	12	3/4—5 1/2	11 1/2	3/4—5 1/2	12	3/4—5 1/2	12 1/2	3/4—5
18	12	3/4—5 1/2	12 1/2	3/4—5	12 1/2	3/4—5	12 1/2	3/4—5	13	3/4—5	13	3/4—5
19	12 1/2	3/4—5	13	3/4—5	13 1/2	7/8—6 1/2	13	3/4—5	13 1/2	7/8—6 1/2	14	7/8—6 1/2
20	13 1/2	7/8—6 1/2	14	7/8—6 1/2	14 1/2	7/8—6 1/2	14	7/8—6 1/2	14 1/2	7/8—6 1/2	15	7/8—6

跨度（呎）	活重100磅						活重120磅					
4	3 1/2	3/8—6	3 1/2	3/8—6	3 1/2	3/8—6	3 1/2	3/8—6	3 1/2	3/8—6	3 1/2	3/8—6
5	4	3/8—5	4	3/8—5	4	3/8—5	4	3/8—5	4	3/8—5	4	3/8—5
6	4 1/2	1/2—8	4 1/2	1/2—8	4 1/2	1/2—8	4 1/2	1/2—8	4 1/2	1/2—8	4 1/2	1/2—8
7	5	1/2—7	5	1/2—7	5	1/2—7	5	1/2—7	5	1/2—7	5 1/2	1/2—6
8	5 1/2	1/2—6	5 1/2	1/2—6	6	1/2—5 1/2	5 1/2	1/2—5 1/2	6	1/2—5 1/2	6	1/2—5 1/2
9	6	1/2—5 1/2	6 1/2	1/2—5	6 1/2	1/2—5	6 1/2	1/2—5	6 1/2	1/2—5	6 1/2	1/2—5
10	7	3/8—7	7	5/8—7	7	5/8—7	7	5/8—7	7 1/2	5/8—7	7 1/2	5/8—6 1/2
11	7 1/2	5/8—6 1/2	7 1/2	5/8—6 1/2	8	5/8—6	7 1/2	5/8—6 1/2	8	5/8—6	8	5/8—6
12	8	5/8—6	8 1/2	5/8—5 1/2	8 1/2	5/8—5 1/2	8 1/2	5/8—5 1/2	8 1/2	5/8—5 1/2	9	5/8—5
13	9	5/8—5	9	5/8—5	9 1/2	5/8—5	9	5/8—5	9 1/2	5/8—5	9 1/2	5/8—5
14	9 1/2	5/8—5	10	3/4—7	10	3/4—7	10	3/4—7	10 1/2	3/4—6 1/2	10 1/2	3/4—6 1/2
15	10 1/2	3/4—6 1/2	10 1/2	3/4—6 1/2	11	3/4—6	10 1/2	3/4—6 1/2	11	3/4—6	11	3/4—6
16	11	3/4—6	11 1/2	3/4—5 1/2	12	3/4—5 1/2	11 1/2	3/4—5 1/2	12	3/4—5 1/2	12	3/4—5 1/2
17	12	3/4—5 1/2	12 1/2	3/4—5	12 1/2	3/4—5	12 1/2	3/4—5	13	3/4—5	13	3/4—5
18	12 1/2	3/4—5	13	3/4—5	13 1/2	7/8—6 1/2	13	7/8—5	13 1/2	3/4—6 1/2	14	7/8—6 1/2
19	13 1/2	7/8—6 1/2	14	7/8—6 1/2	14 1/2	7/8—6 1/2	14	7/8—6 1/2	14	7/8—6 1/2	15	7/8—6
20	14 1/2	7/8—6 1/2	15	7/8—6	15 1/2	7/8—5 1/2	15	7/8—5 1/2	15 1/2	7/8—5 1/2	16	7/8—5 1/2

第八十表(英制) 鋼合土樓梯設計表

T＝塊面厚度　　fc＝700　　fs＝18000　　n＝15

由鋼筋中心至三合外面之距離爲1吋

"鋼筋"項指示鋼筋直徑及其中距　　梯級面3/4吋批盪

跨度呎	活重60磅						活重80磅					
	梯級高度與踏脚之比率						梯級高度與踏脚之比率					
	6 x12 6 1/4x11 3/4 6 1/2x11 1/4 6 3/4x10 3/4		7 x10 1/2 7 1/4x10 7 1/2x 9 3/4 7 3/4x 9 1/2		8 x 9		6 x12 6 1/4x11 3/4 6 1/2x11 1/4 6 3/4x10 3/4		7 x10 1/2 7 1/4x10 7 1/2x 9 3/4 7 3/4x 9 1/2		8 x 9	
	T	鋼筋	T	鋼筋	T	鋼筋	T	鋼筋	T	鋼筋	T	鋼筋
4	3	3/8_7	3	3/8_7	3	3/8_7	3	3/8_7	3	3/8_7	3	3/8_7
5	3.5	3/8_6	3.5	3/8_6	3.5	3/8_6	3.5	3/8_6	3.5	3/8_6	3.5	3/8_6
6	4	3/8_5	4	3/8_5	4	3/8_5	4	3/8_5	4	3/8_5	4	3/8_5
7	4.5	1/2_8	4.5	1/2_8	4.5	1/2_8	4.5	1/2_8	4.5	1/2_8	5〔	1/2_7
8	5	1/2_7	5	1/2_7	5	1/2_7	5	1/2_7	5.5	1/2_6	5.5	1/2_6
9	5.5	1/2_6	5.5	1/2_6	5.5	1/2_6	5.5	1/2_6	6	1/2_5.5	6	1/2_5.5
10	6〔	1/2_5.5	6.5	1/2_5	〔6.5	1/2_5	〔6.5	1/2_5	6.5	1/2_5	6.5	1/2_5
11	6.5	1/2_5	7	5/8_7	7	5/8_7	7	5/8_7	7	5/8_7	7	5/8_7
12	7	5/8_7	7.5	5/8_6.5	7.5	5/8_6.5	7.5	5/8_6.5	〔8	5/8_6	8	5/8_6
13	8	5/8_6	8	5/8_6	〔8.5	5/8_5.5	8	5/8_6	8.5	5/8_5.5	8.5	5/8_5.5
14	8.5	5/8_5.5	9〔	5/8_5	9〔	5/8_5	8.5	5/8_5.5	9	5/8_5	9.5	5/8_5
15	9	5/8_5	9.5	5/8_5	10〔	5/8_7	9.5	5/8_5	10	3/4_7	10	3/4_7
16	10	3/4_7	10	3/4_7	10.5	3/4_6.5	〔10.5	3/4_6.5	10.5	3/4_6.5	11〔	3/4_6
17	10.5	3/4_6.5	11	3/4_6	11	3/4_6	〔11.5	3/4_5.5	11.5	3/4_6.5	12〔	3/4_5.5
18	11.5	3/4_5.5	12〔	3/4_5.5	12	3/4_5.5	12〔	3/4_5.5	12	3/4_5.5	12.5	3/4_5
19	12	3/4_5.5	13〔	3/4_5	13	3/4_5	12.5	3/4_5	13	3/4_5	13.5	7/8_6.5
20	13〔	3/4_5	13.5	7/8_6.5	14〔	7/8_6.5	〔13.5	7/8_6.5	14	7/8_6.5	〔14.5	7/8_6.5

跨度呎	活重100磅						活重120磅					
4	3	3/8_7	3	3/8_7	3	3/8_7	3.5	3/8—6	3.5	3/8—6	3.5	3/8—6
5	3.5	3/8_6	〔4	3/8_5	〔4	3/8_5	4	3/8—5	4	3/8—5	4	3/8—5
6	4	3/8_5	〔4.5	1/2_8	4.5	1/2_8	4.5	1/2—8	4.5	1/2—8	4.5	1/2—8
7	5〔	1/2_7	5	1/2_7	5	1/2_7	5	1/2—7	5	1/2—7	5	1/2—7
8	5.5	1/2_6	5.5	1/2_6	5.5	1/2_6	5.5	1/2—6	5.5	1/2—6	5.5	1/2—6
9	6	1/2_5.5	6	1/2_5.5	6	1/2_5.5	6	1/2—5.5	6.5	1/2—5	6.5	1/2—5
10	6.5	1/2_5	6.5	1/2_5	〔7	5/8_7	7〔	5/8—7	7	5/8—7	7	5/8—7
11	7	5/8_7	7.5	5/8_6.5	7.5	5/8_6.5	7.5	5/8—6.5	7.5	5/8—6.5	8〔	5/8—6
12	8	5/8_6	8	5/8—6	8	5/8_6	8	5/8—6	8.5	5/8—5.5	8.5	5/8—5.5
13	8.5	5/8_5.5	〔9	5/8_5	9	5/8_5	9〔	5/8—5	9	5/8—5	〔9.5	5/8—5
14	9	5/8_5	9.5	5/8_5	9.5	5/8_5	9.5	5/8—5	10〔	3/4—7	10	3/4—7
15	10	3/4_7	〔10.5	3/4_6.5	10.5	3/4_6.5	〔10.5	3/4—6.5	10.5	3/4—6.5	11〔	3/4—6
16	10.5	3/4_6.5	11	3/4_6	〔11.5	3/4_5.5	11	3/4—6	11.5	3/4—5.5	11.5	3/4—5 1/2
17	11.5	3/4_5.5	12	3/4_5.5	12	3/4_5.5	12〔	3/4—5.5	〔12.5	3/4—5	12.5	3/4—5
18	〔12.5	3/4_5	13〔	3/4_5	13	3/4_5	12.5	3/4—5	13	3/4—5	〔13.5	7/8_6.5
19	13	3/4_5	13.5	7/8_6.5	13〔	7/8_6.5	13.5	7/8—6.5	14	7/8—6.5	〔14.5	7/8_6.5
20	14〔	7/8_6.5	14.5	7/8_6.5	14.5	7/8_6.5	〔14.5	7/8—6.5	15	7/8—6	15	7/8_6

若梯級面無批盪者，此表仍可用，但遇有〔之T，可減半吋，鋼筋則仍舊。

源記成建築公司
承造大小工程
廣州白雲路三十五號
電話一四四三八

卷 後 語

本期因鋼筋三合土設計手册占篇幅極多，所以在收到之著作中，尙有三數篇如棐福安先生之 Structural Effect of Cracks in Concrete Pavements''，本校材料試驗所之『木材及磚試驗結果』，及梁鼎燊先生之『俄國建築學』等，未能刊出。龍寶鎏先生之』窗牖光綫問題及其計算』，亦祗能刋出一部份。以上各篇，下期定當繼續刊出。此實限於篇幅之故，不得已也。編者謹向讀者及著者道歉。

蒙李院長權亨，羅教授石麟，梁教授綽餘任本期顧問，指導有方；楊錫宗先生，李寶榮博士及全體會員鼎力勷助，致有成功。同人等銘感之餘，特此致謝。

本期內容，社會人士自有批評，編者毋須一一介紹耳。苟有謬誤，尙祈指正。

奠 原

南大工學會出版部

THE JOURNAL
OF
THE LINGNAN ENGINEERING ASSOCIATION

VOL. 4, NO.1

ADVISERS

K. H. LEI　SHIH LIN LO　C. Y. LIANG

EDITOR-IN-CHIEF　PANG TIEN YUEN

ASSOCIATE EDITORS　ADVERTISING MANAGER

LAM KA CHAU　CHIU CHONG YOU

CHAN SOU KAN　CIRCULATING MANAGER

LEONG TENG SUN　YANG TIN YUNG

中華民國二十五年六月一日出版

南大工程

第四卷　第一期

每冊二元

郵費：本市二分：國內五分：國外三角

編輯者……………………………………彭　眞　原

出版者……………………………………嶺南大學工程學會

發行者……………………………………嶺南大學工程學會

印刷者……………………………………維新路超摩印務館

廣告價目表 ADVERTISING RATES PER ISSUE

地　位 Position	全頁Full Page	半頁 Half Page	四分一頁 Quarter Page
封底外面 Outside backcover	一百六十元 $160.00	不登	不登
封面之裏面 Inside front cover	九十元$90.00	不登	不登
普通地位 Ordinary position	八十元$ 80.00	四十元$40.00	廿元$ 20.00

廣告概用白紙黑字，電版自備，告白稿最遲於本刊通知之截稿期三日前寄到：如有稽延，本刊祇代登出公司名稱，費用照規收取。各告白費均於本刊出版之日收足。

嶺南大學工程學會啓事

啓事一

本會離校會員請按期將會費交到本會（離校會員會費經第五次會務會議議決每年二元）。如路途跋涉，可由郵寄下。倘會員通訊地址有更改，或本會刊物寄失時，祈即通知，以便補寄。

啓事二

本期南大工程，因鋼筋三合土設計手册所用印刷費浩繁，經本會議決將本期定價略爲增加，每册二元，以資彌補。但長期定閱者，仍不加價，以示優待。望讀者原諒。